Lecture Notes in Computer Science 14579

Advanced Research in Computing and Software Science
Subline of Lecture Notes in Computer Science

More information about this series at https://link.springer.com/bookseries/558

José A. Soto · Andreas Wiese
Editors

LATIN 2024:
Theoretical Informatics

16th Latin American Symposium
Puerto Varas, Chile, March 18–22, 2024
Proceedings, Part II

Editors
José A. Soto

Andreas Wiese

 Springer

Editors
José A. Soto (iD)
DIM-CMM, Universidad de Chile
Santiago, Chile

Andreas Wiese (iD)
Technical University of Munich
Munich, Germany

ISSN 0302-9743   ISSN 1611-3349   (electronic)
Lecture Notes in Computer Science
ISBN 978-3-031-55600-5   ISBN 978-3-031-55601-2   (eBook)
https://doi.org/10.1007/978-3-031-55601-2

This Springer imprint is published by the registered company Springer Nature Switzerland AG
The registered company address is: Gewerbestrasse 11, 6330 Cham, Switzerland

Paper in this product is recyclable.

Preface

This volume contains the papers presented at the 16th Latin American Theoretical Informatics Symposium (LATIN 2024), held during March 18–22, 2024, in Puerto Varas, Chile. Previous editions of LATIN took place in São Paulo, Brazil (1992), Valparaíso, Chile (1995), Campinas, Brazil (1998), Punta del Este, Uruguay (2000), Cancún, Mexico (2002), Buenos Aires, Argentina (2004), Valdivia, Chile (2006), Búzios, Brazil (2008), Oaxaca, Mexico (2010), Arequipa, Peru (2012), Montevideo, Uruguay (2014), Ensenada, Mexico (2016), Buenos Aires, Argentina (2018), São Paulo, Brazil (2021) and Guanajuato, Mexico (2022). The symposium received 93 submissions from around the world. Each submission was double-blind reviewed by three or four program committee members, and carefully evaluated on quality, originality, and relevance to the conference. Committee members often reviewed the submissions with the help of additional external referees. Based on an extensive electronic discussion, the committee selected 44 papers. In addition to the accepted contributions, the symposium featured keynote talks by Pablo Barceló (Universidad Católica de Chile, Chile), Pierre Fraigniaud (Université Paris Cité and CNRS, France), Penny Haxell (University of Waterloo, Canada), Eunjung Kim (Korea Advanced Institute of Science and Technology, South Korea) and Jon Kleinberg (Cornell University, USA).

Also, LATIN 2024 featured two awards: the Imre Simon Test-of-Time Award and the Alejandro López-Ortiz Best Paper Award. In this edition, the Imre Simon Test-of-Time Award winners were Pierre Fraigniaud, Leszek Gąsieniec, Dariusz R. Kowalski and Andrzej Pelc for their paper "Collective Tree Exploration" which appeared in LATIN 2004. For the Alejandro López-Ortiz Best Paper Award, the program committee selected the paper "Faster Combinatorial k-Clique Algorithms" by Yarin Shechter, Amir Abboud and Nick Fischer. We thank our sponsor Springer for supporting both awards.

Our heartfelt thanks go to the authors for their excellent papers and cooperation, to the program committee members for their insightful discussions, to the subreferees for their careful reports, and to the steering committee for their valuable advice and feedback.

We would also like to recognize Nikhil Bansal, Conrado Martínez and Yoshiko Wakabayashi for their work on the 2024 Imre Simon Test-of-Time Award Selection Committee. Finally, the conference would not have been possible without our generous sponsors: ANID-Chile through grants BASAL FB210005 and ANILLO ACT210005, the Center for Mathematical Modelling (CMM), the research group Information and Computation in Market Design (ICMD) and Springer. We are also grateful for the facilities provided by EasyChair for the evaluation of the submitted papers and the discussions of the program committee.

January 2024

José A. Soto
Andreas Wiese

The Imre Simon Test-of-Time Award

The winner of the 2024 Imre Simon Test-of-Time Award, considering papers up to the 2014 edition of the Latin American Theoretical INformatics Symposium (LATIN), is

> Collective Tree Exploration by Pierre Fraigniaud, Leszek Gąsieniec, Dariusz R. Kowalski and Andrzej Pelc, LATIN 2004, LNCS 2976, 141–151, 2004,

which later appeared as part of the journal article *Collective Tree Exploration*, by the same authors, and published in *Networks* 48(3): 166–177, 2006.

Collective Tree Exploration is an important milestone in the more general context of exploring an unknown environment. The problem arises in many applications, and over the years several researchers have studied efficient ways to explore different environments consisting of closed regions of the plane or graphs (typically used to model complex physical environments involving obstacles) under several computational/communication models.

For connected undirected graphs, it is well known that *depth-first search* (DFS) examines every single vertex and edge in an optimal way, provided that we can mark vertices and edges already explored. However, despite much interest, the amount of memory needed and the role of communication were much less understood. *Collective Tree Exploration* was one of the first contributions which addressed the issue of how much improvement can be obtained using $k \geq 2$ agents or *robots* instead of just $k = 1$, and which communication capabilities made a difference (and by how much).

In their work, Fraigniaud, Gąsieniec, Kowalski and Pelc consider an n-node tree that has to be explored by a team of k robots, starting from the root. All edges (and nodes) have to be visited by at least one robot, and this must be completed as quickly as possible. In each *synchronous* round, each of the k robots can stay where it is, or traverse one incident edge (either leading towards the root or towards some leaf). Two extreme communication scenarios are studied in the paper: in one, robots can share instantaneously all the information they have gathered until that moment (exploration with complete communication), in the other the robots share no information. A third scenario which is of interest is *exploration with write-read communication*, where robots can leave all the information that they have gathered for others to read. In the paper, an efficient algorithm is developed for the stronger scenario (complete communication), then the authors show that the same time complexity can be achieved in the less demanding and more practical model of write-read communication.

The first important contribution of the paper was to show that computing an optimal schedule for the exploration is NP-hard even if the full tree is known in advance; the proof was omitted in the conference extended abstract, but given in full detail in the journal version. The optimal algorithm with full knowledge of the tree is however an important piece of the investigation, as it sets the minimum exploration cost (= number of synchronous rounds) with which we can compare online exploration algorithms, that is, those that do not know the tree in advance and explore it in rounds.

The second fundamental result is the simple and elegant algorithm called "Collective Exploration" (CE) in the paper, which solves the problem on a tree of diameter D in $O(D + n/\log k)$ rounds. The *overhead* of CE is $O(k/\log k)$; it is the maximum competitive ratio between the exploration cost of the algorithm and that of the optimum, which is $\Theta(\max\{2n/k, D\})$, and taking the maximum over all possible trees of size n and all possible roots (starting points). This result was first established for the scenario with complete communication, then the authors showed how to simulate CE in the write-read communication scenario, while the time complexity remains the same.

The third main result of the work was to show that the overhead of any exploration algorithm (with no knowledge of the tree) is $\Omega(k)$ if there is no communication among the robots. Thus, combining these results, the authors give an interesting separation: without any form of communication k robots are essentially not better than one robot, on the other hand even a limited (and reasonable) amount of communication improves the exploration process, and allows us to take (some) advantage of having $k \geq 2$ robots.

Collaborative exploration of trees has been an important problem since its introduction in the ISTT 2024 awarded paper, cited many times by authors coming from different areas of Computer Science. *Collective Tree Exploration* has become a well-known and recognized reference by theoretical computer scientists working on exploration algorithms, but also for researchers in more practical areas like distributed robotics. Since the publication of the LATIN paper in 2004, and two years later of the journal version, many authors have studied variants, proposed new exploration algorithms for trees and other graphs, and analyzed the consequences of different communication capabilities, asynchronous settings and different ways to compare the collaborative online algorithms against the optimal algorithm that has complete knowledge of the tree. The area has been quite active during these years since the publication of *Collective Tree Exploration*, with the first improvement of the original overhead $O(k/\log k)$ obtained as recently as 2023.

The relevance of the problem addressed, the originality of the techniques used to solve it, the clarity of presentation and the continued and widespread recognition of this contribution throughout the years since its publication weighed heavily in the committee's choice.

The committee for the 2024 Imre Simon Test-of-Time Award,

Nikhil Bansal
Conrado Martínez
Yoshiko Wakabayashi

Organization

Program Committee Chairs

José A. Soto   Universidad de Chile, Chile
Andreas Wiese   Technical University of Munich, Germany

Steering Committee

Jacques Sakarovitch   CNRS and Télécom Paris, France
Armando Castañeda   Universidad Nacional Autónoma de México (UNAM), Mexico
Conrado Martínez   Universitat Politècnica de Catalunya, Spain
Flávio Keidi Miyazawa   Universidade Estadual de Campinas, Brazil
Cristina G. Fernandes   Universidade de São Paulo, Brazil
Michael A. Bender   Stony Brook University, USA

Program Committee

Shaull Almagor   Technion, Israel
Gabriela Araujo   Universidad Nacional Autónoma de México, Mexico
Flavia Bonomo   Universidad de Buenos Aires, Argentina
Fabio Botler   Universidade Federal do Rio de Janeiro, Brazil
Mario Bravo   Universidad Adolfo Ibáñez, Chile
Igor Carboni Oliveira   University of Warwick, UK
Timothy Chan   University of Illinois Urbana-Champaign, USA
Mark de Berg   TU Eindhoven, The Netherlands
Franziska Eberle   London School of Economics and Political Science, UK
Celina Figueiredo   Universidade Federal do Rio de Janeiro, Brazil
Johannes Fischer   TU Dortmund University, Germany
Emily Fox   University of Texas at Dallas, USA
Paweł Gawrychowski   University of Wrocław, Poland
Cristóbal Guzmán   Pontificia Universidad Católica de Chile, Chile
Christoph Haase   University of Oxford, UK
Adriana Hansberg   Universidad Nacional Autónoma de México, Mexico
Tobias Harks   University of Passau, Germany
Christoph Hertrich   London School of Economics and Political Science, UK
Martin Hoefer   Goethe University Frankfurt, Germany
Bart Jansen   TU Eindhoven, The Netherlands
Artur Jeż   University of Wrocław, Poland
Andrea Jiménez   Universidad de Valparaíso, Chile

Michael Kerber	Graz University of Technology, Austria
Thomas Kesselheim	University of Bonn, Germany
Arindam Khan	Indian Institute of Science, India
Stefan Kratsch	Humboldt University of Berlin, Germany
Jan Kretinsky	Technical University of Munich, Germany, and Masaryk University, Czech Republic
Ian Mertz	University of Warwick, UK
Pedro Montealegre	Universidad Adolfo Ibáñez, Chile
Ryuhei Mori	Nagoya University, Japan
Gonzalo Navarro	Universidad de Chile, Chile
Alantha Newman	Université Grenoble Alpes, France
Harumichi Nishimura	Nagoya University, Japan
André Nusser	University of Copenhagen, Denmark
Joël Ouaknine	Max Planck Institute for Software Systems, Germany
Dana Pizarro	Universidad de O'Higgins, Chile
Sergio Rajsbaum	Universidad Nacional Autónoma de México, Mexico
Andrea Richa	Arizona State University, USA
Saket Saurabh	Institute of Mathematical Sciences, India, and University of Bergen, Norway
Kevin Schewior	University of Southern Denmark, Denmark
Ildikó Schlotter	Centre for Economic and Regional Studies, Hungary
Sebastian Siebertz	University of Bremen, Germany
Jose A. Soto (Co-chair)	Universidad de Chile, Chile
Maya Stein	Universidad de Chile, Chile
Kavitha Telikepalli	Tata Institute of Fundamental Research, India
Roei Tell	Institute for Advanced Study, Princeton, USA, and Rutgers University, USA
Erik Jan van Leeuwen	Utrecht University, The Netherlands
Rob van Stee	University of Siegen, Germany
Jose Verschae	Pontificia Universidad Católica de Chile, Chile
Seeun William Umboh	University of Melbourne, Australia
Andreas Wiese (Co-chair)	Technical University of Munich, Germany

Organization Committee

Waldo Gálvez	Universidad de O'Higgins, Chile
José A. Soto	Universidad de Chile, Chile
Victor Verdugo	Universidad de O'Higgins, Chile
Andreas Wiese	Technical University of Munich, Germany

Additional Reviewers

Maximilian J. Stahlberg
Nicole Megow
Martín Ríos-Wilson
Lydia Mirabel Mendoza Cadena
Armando Castaneda
Kaustav Bose
Marta Grobelna
Pierre Vandenhove
Antonio Casares
Youssouf Oualhadj
Simon Weber
Sudebkumar Prasant Pal
Matt Gibson
Andrew Ryzhikov
Maël Le Treust
Stavros Kolliopoulos
Carolina Gonzalez
Luis Cunha
Lehilton L. C. Pedrosa
Abhinav Chakraborty
Lasse Wulf
André van Renssen
Leonidas Theocharous
Sanjana Dey
Tatsuya Gima
Bartlomiej Dudek
Bruno Netto
Yasuaki Kobayashi
Lucas De Meyer
Akira Suzuki
Alexandre Vigny
Torsten Mütze
Wanderson Lomenha
Jan Petr
Julien Portier
Sariel Har-Peled
Saladi Rahul
João Pedro de Souza Gomes da Costa
Aritra Banik
Anja Schedel
Raul Lopes
Tesshu Hanaka
François Dross

Hans Bodlaender
Augusto Modanese
Victor Larsen
Jens Schlöter
Matheus Pedrosa
Madhusudhan Reddy Pittu
Karol Pokorski
Sharma V. Thankachan
Julian Mestre
Alexander Braun
Maximilian Fichtl
Sugata Gangopadhyay
Eric Pérez
Ran Duan
Arturo Merino
K. Somasundaram
Asaf Yeshurun
Florian Dorfhuber
Vincent Froese
Andrei Draghici
Marc Vinyals
Torsten Ueckerdt
Elmar Langetepe
Martín Ríos-Wilson
Aditya Subramanian
Tobias Hofmann
Óscar C. Vásquez
Abdolhamid Ghodselahi
Jacob Calvert
Stefano Gogioso
Martin Koutecky
Syamantak Das
Sarita de Berg
Yuan Sha
Ge Xia
Shaily Verma
Andrea Marino
Neta Dafni
Venkatesh Raman
Benjamin Jauregui
Juan L. Reutter
Patrick Dinklage
Claudson Bornstein

Moses Ganardi
Jonas Ellert
Travis Gagie
Tomasz Kociumaka
Ernesto Araya Valdivia
Fahad Panolan
Valmir Barbosa
Stefan Schirra
Tassio Naia
Manuel Cáceres
Hadas Shachnai
Markus Bläser
Akanksha Agrawal
Mohammad Sadegh Mohagheghi
Eduardo Moreno
Giovanna Varricchio
Jamison Weber
Dolores Lara
César Hernández-Cruz
Vikash Tripathi
Ivan Bliznets
Pranabendu Misra
Ioan Todinca
Roohani Sharma
Patrick Eades
Alexandra Weinberger
Shaohua Li
Adam Kasperski
Nadia Brauner
Bertrand Simon
Łukasz Jeż
Ivan Rapaport
Daniel Rehfeldt
Andre Schidler
Anahi Gajardo
Pacôme Perrotin

Alberto Dennunzio
Mingyu Xiao
Benjamin Raichel
Britta Peis
Juan Pablo Contreras
Bart de Keijzer
Andrés Cristi
Michael Kaufmann
Philipp Kindermann
Juan Gutiérrez
Felix Schröder
Nicola Prezza
Jonas Ellert
Robert Bredereck
Claudio Telha Cornejo
Sung-Hwan Kim
Tomasz Kociumaka
Nikhil Balaji
Gerth Stølting Brodal
Jakub Łacki
Dominik Kempa
Adam Karczmarz
Nidhi Purohit
Kirill Simonov
Maximilian Prokop
Sven Jäger
Cristian Urbina
Bartlomiej Dudek
David Eppstein
Jean Cardinal
Mikkel Abrahamsen
Felipe A. Louza
Sabine Rieder
Zhouningxin Wang
Santiago Guzman Pro
Kathryn Nurse

Sponsors

ANID-Chile through grants BASAL FB210005 and ANILLO ACT210005
Center for Mathematical Modelling (CMM)
Information and Computation in Market Design (ICMD)
Springer

Contents – Part II

Automata Theory and Formal Languages

Game Theory and Fairness

Contents – Part I

Approximation and Online Algorithms

Computational Geometry

Complexity Theory

Combinatorics and Graph Theory

Self-complementary (Pseudo-)Split Graphs

Yixin Cao[✉][iD], Haowei Chen, and Shenghua Wang

Department of Computing, Hong Kong Polytechnic University, Hong Kong, China
{yixin.cao,haowei.chen}@polyu.edu.hk, shenghua.wang@connect.polyu.hk

Abstract. We study split graphs and pseudo-split graphs that are isomorphic to their complements. These special subclasses of self-complementary graphs are actually the core of self-complementary graphs. Indeed, we show that all realizations of forcibly self-complementary degree sequences are pseudo-split graphs. We also give formulas to calculate the number of self-complementary (pseudo-)split graphs of a given order, and show that Trotignon's conjecture holds for all self-complementary split graphs.

Keywords: self-complementary graph · split graph · pseudo-split graph · degree sequence

1 Introduction

The *complement* of a graph G is a graph defined on the same vertex set of G, where a pair of distinct vertices are adjacent if and only if they are not adjacent in G. In this paper, we study the graph that is isomorphic to its complement, hence called *self-complementary*. A graph is a *split graph* if its vertex set can be partitioned into a clique and an independent set.

These two families of graphs are connected by the following observation. An elementary counting argument convinces us that the order of a nontrivial self-complementary graph is either $4k$ or $4k+1$ for some positive integer k. Consider a self-complementary graph G of order $4k$, where L (resp., H) represents the set of $2k$ vertices with smaller (resp., higher) degrees. Note that $d(x) \leq 2k-1 < 2k \leq d(y)$ for every pair of vertices $x \in L$ and $y \in H$. Xu and Wong [17] observed that the subgraphs of G induced by L and H are complementary to each other. More importantly, the bipartite graph spanned by the edges between L and H is closed under *bipartite complementation*, i.e., reversing edges in between but keeping both L and H independent. When studying the connection between L and H, it is more convenient to add all the missing edges among H and remove all the edges among L, thereby turning G into a self-complementary split graph. In this sense, every self-complementary graph of order $4k$ can be constructed from a self-complementary split graph of the same order and a graph of order $2k$. For a self-complementary graph of an odd order, the self-complementary

Supported by RGC grant 15221420 and NSFC grants 61972330 and 62372394.

J. A. Soto and A. Wiese (Eds.): LATIN 2024, LNCS 14579, pp. 3–18, 2024.
https://doi.org/10.1007/978-3-031-55601-2_1

split graph is replaced by a self-complementary pseudo-split graph. A pseudo-split graph is either a split graph or a split graph plus a five-cycle such that every vertex on the cycle is adjacent to every vertex in the clique of the split graph and is nonadjacent to any vertex in the independent set of the split graph.

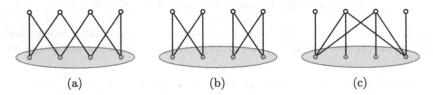

(a)   (b)   (c)

Fig. 1. Self-complementary split graphs with eight vertices. Vertices in I are represented by empty nodes on the top, while vertices in K are represented by filled nodes on the bottom. For clarity, edges among vertices in K are omitted. Their degree sequences are (a) $(5^4, 2^4)$, (b) $(5^4, 2^4)$, and (c) $(6^2, 4^2, 3^2, 1^2)$.

The decomposition theorem of Xu and Wong [17] was for the construction of self-complementary graphs, another ingredient of which is the degree sequences of these graphs (the non-increasing sequence of its vertex degrees). Clapham and Kleitman [3,5] present a necessary condition for a degree sequence to be that of a self-complementary graph. However, a realization of such a degree sequence may or may not be self-complementary. A natural question is to ask about the degree sequences all of whose realizations are necessarily self-complementary, called *forcibly self-complementary.* All the degree sequences for self-complementary graphs up to order five, (0^1), $(2^2, 1^2)$, (2^5), and $(3^2, 2^1, 1^2)$, are forcibly self-complementary. Of the four degree sequences for the self-complementary graphs of order eight, only $(5^4, 2^4)$ and $(6^2, 4^2, 3^2, 1^2)$ are focibly self-complementary; see Fig. 1. All the realizations of these forcibly self-complementary degree sequences turn out to be pseudo-split graphs. As we will see, this is not incidental.

We take p graphs $S_1, S_2, \ldots, S_p$, each being either a four-path or one of the first two graphs in Fig. 1. Note that the each of them admits a unique decomposition into a clique K_i and an independent set I_i. For any pair of i, j with $1 \le i < j \le p$, we add all possible edges between K_i and $K_j \cup I_j$. It is easy to verify that the resulting graph is self-complementary, and can be partitioned into a clique $\bigcup_{i=1}^p K_i$ and an independent set $\bigcup_{i=1}^p I_i$. By an *elementary self-complementary pseudo-split graph* we mean such a graph, or one obtained from it by adding a single vertex or a five-cycle and make them complete to $\bigcup_{i=1}^p K_i$. For example, we end with the graph in Fig. 1(c) with $p = 2$ and both S_1 and S_2 being four-paths. It is a routine exercise to verify that the degree sequence of an elementary self-complementary pseudo-split graph is forcibly self-complementary. We show that the other direction holds as well, thereby fully characterizing forcibly self-complementary degree sequences.

Theorem 1. *A degree sequence is forcibly self-complementary if and only if every realization of it is an elementary self-complementary pseudo-split graph.*

Our result also bridges a longstanding gap in the literature on self-complementary graphs. Rao [10] has proposed another characterization for forcibly self-complementary degree sequences (we leave the statement, which is too technical, to Sect. 3). As far as we can check, he never published a proof of his characterization. It follows immediately from Theorem 1.

All self-complementary graphs up to order five are pseudo-split graphs, while only three out of the ten self-complementary graphs of order eight are. By examining the list of small self-complementary graphs, Ali [1] counted self-complementary split graphs up to 17 vertices. Whether a graph is a split graph can be determined solely by its degree sequence. However, this approach needs the list of all self-complementary graphs, and hence cannot be generalized to large graphs. Answering a question of Harary [8], Read [11] presented a formula for the number of self-complementary graphs with a specific number of vertices. Clapham [4] simplified Read's formula by studying the isomorphisms between a self-complementary graph and its complement. We take an approach similar to Clapham's for self-complementary split graphs with an even order, which leads to a formula for the number of such graphs. For other self-complementary pseudo-split graphs, we establish a one-to-one correspondence between self-complementary split graphs on $4k$ vertices and those on $4k + 1$ vertices, and a one-to-one correspondence between self-complementary pseudo-split graphs of order $4k + 1$ that are not split graphs and self-complementary split graphs on $4k - 4$ vertices.

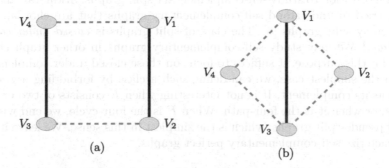

Fig. 2. The (a) rectangle and (b) diamond partitions. Each node represents one part of the partition. A solid line indicates that all the edges between the two parts are present, a missing line indicates that there is no edge between the two parts, while a dashed line imposes no restrictions on the two parts.

We also study a conjecture of Trotignon [16], which asserts that if a self-complementary graph G does not contain a five-cycle, then its vertex set can be partitioned into four nonempty sets with the adjacency patterns of a rectangle or a diamond, as described in Fig. 2. He managed to prove that certain special graphs satisfy this conjecture. The study of rectangle partitions in self-complementary graphs enabled Trotignon to present a new proof of Gibbs' theorem [7, Theorem 4]. We prove Trotignon's conjecture on self-complementary split graphs, with a stronger statement. We say that a partition of $V(G)$ is *self-complementary* if it forms the same partition in the complement of G, illustrated in Fig. 3. Every self-complementary split graph of an even order admits a diamond partition that is self-complementary. Moreover, for each positive integer k, there is a single graph of order $4k$ that admits a rectangle partition. Note that under the setting as Trotignon's, the graph always admits a partition that is self-complementary, while in general, there are graphs that admit a partition, but do not admit any partition that is self-complementary [2].

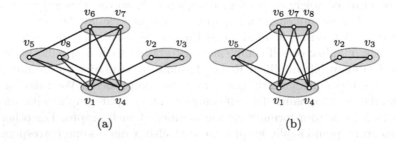

(a)   (b)

Fig. 3. Two diamond partitions, of which only the first is self-complementary.

Before closing this section, let us mention related work. There is another natural motivation to study self-complementary split graphs. Sridharan and Balaji [15] tried to understand self-complementary graphs that are chordal. They are precisely split graphs [6]. The class of split graphs is *closed under complementation.*[1] We may study self-complementary graphs in other graph classes. Again, for this purpose, it suffices to focus on those closed under complementation. In the simplest case, we can define such a class by forbidding a graph F as well as its complement. It is not interesting when F consists of two or three vertices, or when it is the four-path. When F is the four-cycle, we end with the class of pseudo-split graphs, which is the simplest in this sense. We leave it open to characterize self-complementary perfect graphs.

2 Preliminaries

All the graphs discussed in this paper are finite and simple. The vertex set and edge set of a graph G are denoted by, respectively, $V(G)$ and $E(G)$. The two ends of an edge are *neighbors* of each other, and the number of neighbors of a vertex v, denoted by $d_G(v)$, is its *degree*. We may drop the subscript G if the graph is clear from the context. For a subset $U \subseteq V(G)$, let $G[U]$ denote the subgraph of G induced by U, whose vertex set is U and whose edge set comprises all the edges with both ends in U, and let $G - U = G[V(G) \setminus U]$, which is simplified to $G - u$ if U comprises a single vertex u. A *clique* is a set of pairwise adjacent

---

[1] Some authors call such graph classes "self-complementary," e.g., graphclasses.org.

vertices, and an *independent set* is a set of vertices that are pairwise nonadjacent. For $\ell \geq 1$, we use P_ℓ and K_ℓ to denote the path graph and the complete graph, respectively, on ℓ vertices. For $\ell \geq 3$, we use C_ℓ to denote the ℓ-cycle. We say that two sets of vertices are *complete* or *nonadjacent* to each other if there are all possible edges or if there is no edge between them, respectively.

An *isomorphism* between two graphs G_1 and G_2 is a bijection between their vertex sets, i.e., $\sigma \colon V(G_1) \to V(G_2)$, such that two vertices u and v are adjacent in G_1 if and only if $\sigma(u)$ and $\sigma(v)$ are adjacent in G_2. Two graphs with an isomorphism are *isomorphic*. A graph is *self-complementary* if it is isomorphic to its *complement* $\overline{G}$, the graph defined on the same vertex set of G, where a pair of distinct vertices are adjacent in $\overline{G}$ if and only if they are not adjacent in G. An isomorphism between G and $\overline{G}$ is a permutation of $V(G)$, called an *antimorphism*.

We represent an antimorphism as the product of disjoint cycles $\sigma = \sigma_1 \sigma_2 \cdots \sigma_p$, where $\sigma_i = (v_{i1} v_{i2} \cdots)$ for all i. Sachs and Ringel [12,14] independently showed that there can be at most one vertex v fixed by an antimorphism σ, i.e., $\sigma(v) = v$. For any other vertex u, the smallest number k satisfying $\sigma^k(u) = u$ has to be a multiplier of four. Gibbs [7] observed that if a vertex v has d neighbors in G, then the degree of $\sigma(v)$ in G is $n - 1 - d$ where n is the order of G. It implies that if v is fixed by σ, then its degree in G is $(n-1)/2$. The vertices in every cycle of σ with a length of more than one alternate in degrees d and $n - 1 - d$.

Lemma 1 ([12,14]). *In an antimorphism of a self-complementary graph, the length of each cycle is either one or a multiplier of four. Moreover, there is a unique cycle of length one if and only if the order of the graph is odd.*

For any subset of cycles in σ, the vertices within those cycles induce a subgraph that is self-complementary. Indeed, the selected cycles themselves act as an antimorphism for the subgraph.

Proposition 1 ([7]). *Let G be a self-complementary graph and σ an antimorphism of G. For any subset of cycles in σ, the vertices within those cycles induce a self-complementary graph.*

A graph is a *split graph* if its vertex set can be partitioned into a clique and an independent set. We use $K \uplus I$, where K being a clique and I an independent set, to denote a *split partition*. A split graph may have more than one split partition.

Lemma 2. *A self-complementary split graph on $4k$ vertices has a unique split partition and it is $\{v \mid d(v) \geq 2k\} \uplus \{v \mid d(v) < 2k\}$.*

Proof. Let G be a self-complementary split graph with $4k$ vertices, and σ an antimorphism of G. By definition, for any vertex $v \in V(G)$, we have $d(v) + d(\sigma(v)) = 4k - 1$. Thus,

$$\min(d(v), d(\sigma(v))) \leq 2k - 1 < 2k \leq \max(d(v), d(\sigma(v))).$$

As a result, G does not contain any clique or independent set of order $2k + 1$. Suppose for contradiction that there exists a split partition $K \uplus I$ of G different from the given. There must be a vertex $x \in I$ with $d(x) \geq 2k$. We must have $d(x) = 2k$ and $N(x) \subseteq K$. But then there are at least $|N[x]| = 2k + 1$ vertices having degree at least $2k$, a contradiction. $\qquad\square$

We correlate self-complementary split graphs having even and odd orders.

Proposition 2 ($\star$). *Let G be a split graph on $4k + 1$ vertices. If G is self-complementary, then G has exactly one vertex v of degree $2k$, and $G - v$ is also self-complementary.*[2]

It is obvious that every self-complementary split graph admits a diamond partition. We prove a stronger statement.

Lemma 3. *Every self-complementary split graph G admits a diamond partition. If G has an even order, then it admits a diamond partition that is self-complementary.*

Proof. Let $K \uplus I$ be a split partition of G. For any proper and nonempty subset $K' \subseteq K$ and proper and nonempty subset $I' \subseteq I$, the partition $\{K', I', K \setminus K', I \setminus I'\}$ is a diamond partition. Now suppose that the order of G is $4k$. We fix an arbitrary antimorphism $\sigma = \sigma_1 \sigma_2 \cdots \sigma_p$ of G. We may assume without loss of generality that for all $i = 1, \ldots, p$, the first vertex in σ_i is in K. For $j = 1, \ldots, |\sigma_i|$, we assign the jth vertex of σ_i to V_j (mod 4). For $j = 1, \ldots, 4$, we have $\sigma(V_j) = V_{j+1}$ (mod 4). Moreover, $V_1 \cup V_3 = K$ and $V_2 \cup V_4 = I$. Thus, $\{V_1, V_2, V_3, V_4\}$ is a self-complementary diamond partition of G. $\qquad\square$

For a positive integer k, let Z_k denote the graph obtained from a P_4 as follows. We substitute each degree-one vertex with an independent set of k vertices, and each degree-two vertex with a clique of k vertices.

Lemma 4. *A self-complementary split graph admits a rectangle partition if and only if it is isomorphic to Z_k.*

Proof. The sufficiency is trivial, and we consider the necessity. Suppose that G is a self-complementary split graph and it has a rectangle partition $\{V_1, V_2, V_3, V_4\}$. Let $K \uplus I$ be a split partition of G. There are at least one edge and at least one missing edge between any three parts. Thus, vertices in K are assigned to precisely two parts in the partition. By the definition of rectangle partition, K is either $V_2 \cup V_3$ or $V_1 \cup V_4$. Assume without loss of generality that $K = V_2 \cup V_3$. Since V_2 is complete to V_1 and nonadjacent to V_4, any antimorphism of G maps V_2 to either V_1 or V_4. If $|V_2| \neq |V_3|$, then the numbers of edges between K and I in G and $\overline{G}$ are different. This is impossible. $\qquad\square$

---

[2] Proofs of statements marked with $\star$ can be found in the full version (arXiv:2312.10413).

A *pseudo-split graph* is either a split graph, or a graph whose vertex set can be partitioned into a clique K, an independent set I, and a set C that (1) induces a C_5; (2) is complete to K; and (3) is nonadjacent to I. We say that $K \uplus I \uplus C$ is a *pseudo-split partition* of the graph, where C may or may not be empty. If C is empty, then $K \uplus I$ is a split partition of the graph. Otherwise, the graph has a unique pseudo-split partition. Similar to split graphs, the complement of a pseudo-split graph remains a pseudo-split graph.

Proposition 3. *Let G be a self-complementary pseudo-split graph with a pseudo-split partition $K \uplus I \uplus C$. If $C \neq \emptyset$, then $G - C$ is a self-complementary split graph of an even order.*

Proof. Let σ be an antimorphism of G. In both G and its complement, the only C_5 is induced by C. Thus, $\sigma(C) = C$. Since C is complete to K and nonadjacent to I, it follows that $\sigma(K) = I$ and $\sigma(I) = K$. Thus, $G - C$ is a self-complementary graph. It is clearly a split graph and has an even order. $\qquad\square$

3 Forcibly Self-complementary Degree Sequences

The *degree sequence* of a graph G is the sequence of degrees of all vertices, listed in non-increasing order, and G is a *realization* of this degree sequence. For our purpose, it is more convenient to use a compact form of degree sequences where the same degrees are grouped:

$$(d_i^{n_i})_{i=1}^{\ell} = (d_1^{n_1}, \ldots, d_\ell^{n_\ell}) = \Big(\underbrace{d_1, \ldots, d_1}_{n_1}, \underbrace{d_2, \ldots, d_2}_{n_2}, \ldots, \underbrace{d_\ell, \ldots, d_\ell}_{n_\ell} \Big).$$

Note that we always have $d_1 > d_2 > \cdots > d_\ell$. For example, the degree sequences of the first two graphs in Fig. 1 are both $(5^4, 2^4) = (5, 5, 5, 5, 2, 2, 2, 2)$.

For four vertices v_1, v_2, v_3, and v_4 such that v_1 is adjacent to v_2 but not to v_3 while v_4 is adjacent to v_3 but not to v_2, the operation of replacing v_1v_2 and v_3v_4 with v_1v_3 and v_2v_4 is a *2-switch*, denoted as $(v_1v_2, v_3v_4) \rightarrow (v_1v_3, v_2v_4)$. It is easy to check that this operation does not change the degree of any vertex.

Lemma 5 ([13])**.** *Two graphs have the same degree sequence if and only if they can be transformed into each other by a series of 2-switches.*

The subgraph induced by the four vertices involved in a 2-switch operation must be a $2K_2$, P_4, or C_4. Moreover, after the operation, the four vertices induce an isomorphic subgraph. Since a split graph G cannot contain any $2K_2$ or C_4 [6], a 2-switch must be done on a P_4. In any split partition $K \uplus I$ of G, the two degree-one vertices of P_4 are from I, while the others from K. The graph remains a split graph after this operation. Thus, if a degree sequence has a realization that is a split graph, then all its realizations are split graphs [6]. A similar statement holds for pseudo-split graphs [9].

We do not have a similar claim on degree sequences of self-complementary graphs. Clapham and Kleitman [5] have fully characterized all such degree sequences, called *potentially self-complementary degree sequences*. A degree sequence is *forcibly self-complementary* if all of its realizations are self-complementary.

Proposition 4. *The following degree sequences are all forcibly self-complementary:* (0^1), $(2^2, 1^2)$, (2^5), *and* $(5^4, 2^4)$.

Proof. It is trivial for (0^1). Applying a 2-switch operation to a realization of $(2^2, 1^2)$ or (2^5) leads to an isomorphic graph. A 2-switch operation transforms the graph in Fig. 1(a) into Fig. 1(b), and vice versa. Thus, the statement follows from Lemma 5.   □

We take p vertex-disjoint graphs S_1, S_2, …, S_p, each of which is isomorphic to P_4, or one of the graphs in Fig. 1(a, b). For $i = 1, \ldots, p$, let $K_i \uplus I_i$ denote the unique split partition of S_i. Let C be another set of 0, 1, or 5 vertices. We add all possible edges among $\bigcup_{i=1}^{p} K_i$ to make it a clique, and for each $i = 1, \ldots, p$, add all possible edges between K_i and $\bigcup_{j=i+1}^{p} I_j$.[3] Finally, we add all possible edges between C and $\bigcup_{i=1}^{p} K_i$, and add edges to make C a cycle if $|C| = 5$. Let $\mathcal{E}$ denote the set of graphs that can be constructed as above.

Lemma 6. *All graphs in $\mathcal{E}$ are self-complementary pseudo-split graphs, and their degree sequences are forcibly self-complementary.*

Proof. Let G be any graph in $\mathcal{E}$. It has a split partition $(\bigcup_{i=1}^{p} K_i \cup C) \uplus \bigcup_{i=1}^{p} I_i$ when $|C| \leq 1$, and a pseudo-split partition $\bigcup_{i=1}^{p} K_i \uplus \bigcup_{i=1}^{p} I_i \uplus C$ otherwise. To show that it is self-complementary, we construct an antimorphism σ for it. For each $i = 1, \ldots, p$, we take an antimorphism σ_i of S_i, and set $\sigma(x) = \sigma_i(x)$ for all $x \in V(S_i)$. If C consists of a single vertex v, we set $\sigma(v) = v$. If $|C| = 5$, we take an antimorphism σ_{p+1} of C_5 and set $\sigma(x) = \sigma_{p+1}(x)$ for all $x \in C$. It is easy to verify that a pair of vertices u, v are adjacent in G if and only if $\sigma(u)$ and $\sigma(v)$ are adjacent in $\overline{G}$.

For the second assertion, we show that applying a 2-switch to a graph G in $\mathcal{E}$ leads to another graph in $\mathcal{E}$. Since G is a split graph, a 2-switch can only be applied to a P_4. For two vertices $v_1 \in K_i$ and $v_2 \in K_j$ with $i < j$, we have $N[v_2] \subseteq N[v_1]$. Thus, there cannot be any P_4 involving both v_1 and v_2. A similar argument applies to two vertices in I_i and I_j with $i \neq j$. Therefore, a 2-switch can be applied either *inside C* or *inside S_i* for some $i \in \{1, \ldots, p\}$. By Proposition 4, the resulting graph is in $\mathcal{E}$.   □

---

[3] The reader familiar with threshold graphs may note its use here. If we contract K_i and I_i into two vertices, the graph we constructed is a threshold graph. Threshold graphs have a stronger characterization by degree sequences. Since a threshold graph free of $2K_2$, P_4, and C_4, no 2-switch is possible on it. Thus, the degree sequence of a threshold graph has a unique realization.

We refer to graphs in $\mathcal{E}$ as *elementary self-complementary pseudo-split graphs*. The rest of this section is devoted to showing that all realizations of forcibly self-complementary degree sequences are elementary self-complementary pseudo-split graphs. We start with a simple observation on potentially self-complementary degree sequences with two different degrees. It can be derived from Clapham and Kleitman [5]. We provide a direct and simple proof here.

Proposition 5 ($\star$). *There is a self-complementary graph of the degree sequence $(d^{2k}, (4k - 1 - d)^{2k})$ if and only if $2k \leq d \leq 3k - 1$. Moreover, there exists a self-complementary graph with a one-cycle antimorphism.*

The next proposition considers the parity of the number of vertices with a specific degree. It directly follows from [5] and [17, Theorem 4.4].

Proposition 6 ([5,17]). *Let G be a graph of order $4k$ and v an arbitrary vertex of G. Let H and L be the $2k$ vertices of the largest and smallest degrees, respectively in G. If G is self-complementary, then all the following are even: the number of vertices with degree $d_G(v)$ in G, the number of vertices with degree $d_{G[H]}(v)$ in $G[H]$, and the number of vertices with degree $d_{G[L]}(v)$ in $G[L]$.*

In general, it is quite challenging to verify that a degree sequence is indeed forcibly self-complementary. On the other hand, to show that a degree sequence is not forcibly self-complementary, it suffices to construct a realization that is not self-complementary. We have seen that degree sequences (0^1), (2^5) and $(2^2, 1^2)$, $(5^4, 2^4)$ are forcibly self-complementary. They are the only ones of these forms.

Proposition 7 ($\star$). *The following degree sequences are not forcibly self-complementary.*

i) $((2k)^{4k+1})$, *where* $k \geq 2$.
ii) $(d^{2k}, (n - 1 - d)^{2k})$, *where* $k \geq 2$ *and* $d \neq 5$.
iii) $(d^{2k_1}, (d - 1)^{2k_2}, (n - d)^{2k_2}, (n - 1 - d)^{2k_1})$, *where* $k_1, k_2 > 0$.

Let G be a self-complementary graph with ℓ different degrees $d_1, \ldots, d_\ell$. For each $i = 1, \ldots, \ell$, let $V_i(G) = \{v \in V(G) \mid d(v) = d_i\}$, and we define the *$i$th slice* of G as the induced subgraph $S_i(G) = G[V_i \cup V_{\ell+1-i}]$. We may drop (G) when the graph is clear from the context. Note that $V_i = V_{\ell+1-i}$ and $S_i = G[V_i]$ when ℓ is odd and $i = (\ell + 1)/2$. Each slice must be self-complementary, and more importantly, its degree sequence is forcibly self-complementary.

Lemma 7. *Let τ be a forcibly self-complementary degree sequence, G a realization of τ, and σ an antimorphism of G. The degree sequence of every slice of G is forcibly self-complementary.*

Proof. Let $\tau = (d_i^{n_i})_{i=1}^{\ell}$. Since $d_1 > d_2 > \cdots > d_\ell$, the antimorphism σ maps the vertices from V_i to $V_{\ell+1-i}$, and vice versa. Therefore, $n_i = n_{\ell+1-i}$, and the cycles of σ consisting of vertices from $V_i \cup V_{\ell+1-i}$ is an antimorphism of S_i. Therefore, S_i is self-complementary.

We now verify any graph S with the same degree sequence as S_i is self-complementary. By Lemma 5, we can transform S_i to S by a sequence of 2-switches applied on vertices in $V_i \cup V_{\ell+1-i}$. We can apply the same sequence of 2-switches to G, which lead to a graph G' with degree sequence τ. Note that S is the ith slice of G', hence self-complementary. □

The following result follows from Lemma 7.

Corollary 1. *Let G be a graph with ℓ different degrees. If the degree sequence of G is forcibly self-complementary, then there cannot be a 2-switch that changes the number of edges in S_i or between V_i and $V_{\ell+1-i}$ for every $i \in \{1, 2, \ldots, \lfloor \ell/2 \rfloor\}$.*

Proof. Since $S_i(G)$ is a self-complementary graph, the number of edges is fixed. Since there exists an antimorphism of $S_i(G)$ that maps $V_i(G)$ to $V_{\ell+1-i}(G)$, the number of edges between them is fixed. Suppose there exists a 2-switch that changes the number of edges in $S_i(G)$ or between $V_i(G)$ and $V_{\ell+1-i}(G)$ for some $i \in \{1, 2, \ldots, \lfloor \ell/2 \rfloor\}$. Let G' be the resulting graph. Consequently, $S_i(G')$ is not self-complementary. Since the 2-switch operation does not change the degree of any vertex, the degree sequence of $S_i(G')$ should be forcibly self-complementary by Lemma 7. We encounter a contradiction. □

Two vertices in V_i share the same degree *in the ith slice.*

Lemma 8. *Let G be a graph with ℓ different degrees. If the degree sequence of G is forcibly self-complementary, then for each $i \in \{1, \ldots, \ell\}$, the vertices in V_i share the same degree in S_i.*

Proof. Suppose for contradiction that vertices in V_i have different degrees in $S_i(G)$. By Lemma 7, the degree sequence of $S_i(G)$ is forcibly self-complementary. It cannot be of the form $(d^{2k_1}, (d-1)^{2k_2}, (n-d)^{2k_2}, (n-1-d)^{2k_1})$ by Proposition 7(iii). Thus, there must be two vertices v_1 and v_2 in V_i such that

$$d = d_{S_i(G)}(v_1) > d_{S_i(G)}(v_2) + 1.$$

There exists a vertex

$$x_1 \in V(S_i(G)) \cap N(v_1) \setminus N(v_2).$$

On the other hand, since $d_G(v_1) = d_G(v_2)$, there must be a vertex

$$x_2 \in N(v_2) \setminus (N(v_1) \cup V(S_i(G))).$$

We apply the 2-switch $(x_1 v_1, x_2 v_2) \to (x_1 v_2, x_2 v_1)$ to G and denote by G' the resulting graph. By assumption, G' is also self-complementary. By Lemma 7, $S_i(G)$ is self-complementary, and hence the number of vertices with degree d in $S_i(G)$ is even by Proposition 6. The degree of a vertex x in $S_i(G')$ is

$$\begin{cases} d_{S_i(G)}(x) - 1 & x = v_1, \\ d_{S_i(G)}(x) + 1 & x = v_2, \\ d_{S_i(G)}(x) & \text{otherwise.} \end{cases}$$

Thus, the number of vertices with degree d in $S_i(G')$ is odd. Hence, $S_i(G')$ is not self-complementary by Proposition 6, which contradicts Lemma 7. □

We next show all possible configurations for the slices of G.

Lemma 9. *Let G be a graph with ℓ different degrees. If the degree sequence of G is forcibly self-complementary, then*

i) S_i is a P_4 or one of Fig. 1(a, b) for every $i \in \{1, 2, \ldots, \lfloor \ell/2 \rfloor\}$, and
ii) $S_{(\ell+1)/2}$ is either a C_5 or contains exactly one vertex if ℓ is odd.

Proof. For all $i = 1, \ldots, \ell$, the induced subgraph S_i of G is self-complementary by Lemma 7. Furthermore, S_i is either a regular graph or has two different degrees (Lemma 8). By considering the number of edges in S_i, we can deduce that S_i is a regular graph if and only if its order is odd. From the proof of Lemma 7, we know that the order of S_i is odd if and only if ℓ is odd and $i = (\ell + 1)/2$.

If ℓ is odd, then $S_{(\ell+1)/2}$ is a regular graph. Let $|V(S_i)| = 4k + 1$ for some integer $k \geq 0$. It can be derived that the degree sequence of $S_{(\ell+1)/2}$ is $((2k)^{4k+1})$. By Lemma 7 and Proposition 7(i), we can obtain that $k \leq 1$ and the degree sequence of $S_{(\ell+1)/2}$ can either be (2^5) or (0^1). Therefore, $S_{(\ell+1)/2}$ is either a C_5 or contains exactly one vertex.

For every $i \in \{1, 2, \ldots, \lfloor \ell/2 \rfloor\}$, we may assume that S_i has $4k$ vertices for some positive integer k and the degree sequence of S_i is $(d^{2k}, (4k - 1 - d)^{2k})$ for some positive integer d. By Lemma 7 and Proposition 7(ii), we can obtain that $k = 1$ or $d = 5$. Since S_i is a self-complementary graph, the degree sequence of S_i can either be $(5^4, 2^4)$ or $(2^2, 1^2)$ by Proposition 5. Consequently, S_i is either a P_4 or one of the graphs in Fig. 1(a, b). $\square$

By Lemma 9, the induced subgraph S_i is a self-complementary split graph for every $i \in \{1, 2, \ldots, \lfloor \ell/2 \rfloor\}$, We use $K_i \uplus I_i$ to denote the unique split partition of S_i (Lemma 2). Moreover, no vertex in I_i is adjacent to all the vertices in K_i.

Lemma 10. *Let G be a graph with ℓ different degrees. If the degree sequence of G is forcibly self-complementary, then for every $i \in \{1, 2, \ldots, \lfloor \ell/2 \rfloor\}$,*

i) V_i is a clique, and $V_{\ell+1-i}$ an independent set; and
ii) if a vertex in $V(G) \setminus V_i$ has a neighbor in $V_{\ell+1-i}$, then it is adjacent to all the vertices in $V_i \cup V_{\ell+1-i}$.

Proof. (i) Suppose for contradiction that there is a vertex $v_1 \in V_i \cap I_i$. By Lemma 9(i), S_i is either a P_4 or one of the graphs in Fig. 1(a, b). We can find a vertex $v_2 \in K_i \setminus N(v_1)$ and a vertex $x_2 \in N(v_2) \cap I_i$. Note that x_2 is not adjacent to v_1. Since $d_G(v_1) > d_G(v_2)$ while $d_{S_i}(v_1) < d_{S_i}(v_2)$, we can find a vertex x_1 in $V(G) \setminus V(S_i)$ that is adjacent to v_1 but not v_2. The applicability of 2-switch $(x_1 v_1, x_2 v_2) \to (x_1 v_2, x_2 v_1)$ violates Corollary 1.

(ii) Let $x_1 \in V(G) \setminus V_i$ be adjacent to $v_1 \in V_{\ell+1-i}$. Since $V_{\ell+1-i}$ is an independent set, it does not contain x_1. Suppose that there exists $v_2 \in V_{\ell+1-i} \setminus N(x_1)$. Every vertex $x_2 \in N(v_2) \cap V_i$ is adjacent to v_1. Otherwise, we may conduct the 2-switch $(x_1 v_1, x_2 v_2) \to (x_1 v_2, x_2 v_1)$, and denote by G' the resulting graph. It can be seen that $S_i(G')$ is neither a P_4 nor one of the graphs in Fig. 1(a, b), contradicting Lemma 9(i). Therefore, $S_i(G)$

can only be the graph in Fig. 1(b). Let x_3 be a non-neighbor of v_1 in V_i and v_3 a neighbor of x_3 in $V_{\ell+1-i}$. Note that neither x_2v_3 nor x_3v_1 is an edge. We may either conduct the 2-switch $(x_1v_3, x_2v_2) \rightarrow (x_1v_2, x_2v_3)$ or $(x_1v_1, x_3v_3) \rightarrow (x_1v_3, x_3v_1)$ to G, depending on whether x_1 is adjacent to v_3, and denote by G' the resulting graph. The ith slice of G' contradicts Lemma 9(i).

Suppose that there exists a vertex $v_2 \in V_i \setminus N(x_1)$. The vertex v_2 is not adjacent to v_1; otherwise, the applicability of the 2-switch $(x_1x_2, v_1v_2) \rightarrow (x_1v_2, x_2v_1)$ where x_2 is a non-neighbor of v_2 in $V_{\ell+1-i}$ violates Corollary 1. Let x_3 be a neighbor of v_2 in $V_{\ell+1-i}$. Note that v_1 is not adjacent to x_3. The applicability of the 2-switch $(x_1v_1, x_3v_2) \rightarrow (x_1v_2, x_3v_1)$ violates Corollary 1. □

We are now ready to prove the main lemma.

Lemma 11. *Any realization of a forcibly self-complementary degree sequence is an elementary self-complementary pseudo-split graph.*

Proof. Let G be an arbitrary realization of a forcibly self-complementary degree sequence and σ an antimorphism of G. Lemmas 9(i) and 10(i) imply that $V_i = K_i$ and $V_{\ell+1-i} = I_i$ for every $i \in \{1, 2, \ldots, \lfloor \ell/2 \rfloor\}$. Let i, j be two distinct indices in $\{1, 2, \ldots, \lfloor \ell/2 \rfloor\}$. We argue that there cannot be any edge between K_i and I_j if $i > j$. Suppose for contradiction that there exists $x \in K_i$ that is adjacent to $y \in I_j$ for some $i > j$. By Lemma 10(ii), x is adjacent to all the vertices in S_j. Consequently, $\sigma(x)$ is in I_i and has no neighbor in S_j. Let v_1 be a vertex in K_j. Since v_1 is not adjacent to $\sigma(x)$, it has no neighbor in I_i by Lemma 10(ii). Note that S_i is either a P_4 or one of the graphs in Fig. 1(a, b) and so does S_j. If we focus on the graph induced by $V(S_i) \cup V(S_j)$, we can observe that

$$d_{G[V(S_i) \cup V(S_j)]}(v_1) < d_{G[V(S_i) \cup V(S_j)]}(x).$$

Since $d_G(v_1) > d_G(x)$, we can find a vertex x_1 in $V(G) \setminus (V(S_i) \cup V(S_j))$ that is adjacent to v_1 but not x. Let v_2 be a neighbor of x in I_i. Note that v_2 is not adjacent to v_1. We can conduct the 2-switch $(x_1v_1, xv_2) \rightarrow (x_1x, v_1v_2)$, violating Corollary 1. Therefore, I_i is nonadjacent to $\bigcup_{p=i+1}^{\lfloor \ell/2 \rfloor} K_p$ for all $i = 1, \ldots, \lfloor \ell/2 \rfloor$. Since $\sigma(I_i) = K_i$ and $\sigma(\bigcup_{p=i+1}^{\lfloor \ell/2 \rfloor} K_p) = \bigcup_{p=i+1}^{\lfloor \ell/2 \rfloor} I_p$, we can obtain that K_i is complete to $\bigcup_{p=i+1}^{\lfloor \ell/2 \rfloor} I_p$. Moreover, K_i is complete to $\bigcup_{p=i+1}^{\lfloor \ell/2 \rfloor} K_p$ by Lemma 10(ii), and hence I_i is nonadjacent to $\bigcup_{p=i+1}^{\lfloor \ell/2 \rfloor} I_p$.

We are done if ℓ is even. In the rest, we assume that ℓ is odd. By Lemma 9(ii), the induced subgraph $S_{(\ell+1)/2}$ is either a C_5 or contains exactly one vertex. It suffices to show that $V_{(\ell+1)/2}$ is complete to K_i and nonadjacent to I_i for every $i \in \{1, 2, \ldots, \lfloor \ell/2 \rfloor\}$. Suppose $\sigma(v) = v$. When $V_{(\ell+1)/2} = \{v\}$, the claim follows from Lemma 10 and that $\sigma(v) = v$ and $\sigma(V_i) = V_{\ell+1-i}$. Now $|V_{(\ell+1)/2}| = 5$. Suppose for contradiction that there is a pair of adjacent vertices $v_1 \in V_{(\ell+1)/2}$ and $x \in I_i$. Let $v_2 = \sigma(v_1)$. By Lemma 10(ii), v_1 is adjacent to all the vertices in S_i. Accordingly, v_2 has no neighbor in S_i. Since $S_{(\ell+1)/2}$ is a C_5, we can find $v_3 \in V_{(\ell+1)/2}$ that is adjacent to v_2 but not v_1. We can conduct the 2-switch

$(xv_1, v_2v_3) \rightarrow (xv_2, v_1v_3)$ and denote by G' as the resulting graph. It can be seen that $S_{(\ell+1)/2}(G')$ is not a C_5, contradicting Lemma 9(ii). □

Lemmas 6 and 11 imply Theorem 1 and Rao's characterization of forcibly self-complementary degree sequences [10].

Theorem 2 ([10]). *A degree sequence* $(d_i^{n_i})_{i=1}^{\ell}$ *is forcibly self-complementary if and only if for all* $i = 1, \ldots, \lfloor \ell/2 \rfloor$,

$$n_{\ell+1-i} = n_i \qquad \in \{2,4\}, \tag{1}$$

$$d_{\ell+1-i} = n - 1 - d_i = \sum_{j=1}^{i} n_j - \frac{1}{2}n_i, \tag{2}$$

and $n_{(\ell+1)/2} \in \{1,5\}$ *and* $d_{(\ell+1)/2} = \frac{1}{2}(n-1)$ *when* ℓ *is odd.*

Proof The sufficiency follows from Lemma 6: note that an elementary self-complementary pseudo-split graph in which S_i has $2n_i$ vertices satisfies the conditions. The necessity follows from Lemma 11. □

4   Enumeration

In this section, we consider the enumeration of self-complementary pseudo-split graphs and self-complementary split graphs. The following corollary of Propositions 2 and 3 focuses us on self-complementary split graphs of even orders. Let λ_n and λ'_n denote the number of split graphs and pseudo-split graphs, respectively, of order n that are self-complementary. For convenience, we set $\lambda_0 = 1$.

Corollary 2. *For each* $k \geq 1$, *it holds* $\lambda_{4k+1} = \lambda_{4k}$. *For each* $n > 0$,

$$\lambda'_n = \begin{cases} \lambda_n & n \equiv 0 \pmod 4, \\ \lambda_{n-1} + \lambda_{n-5} & n \equiv 1 \pmod 4. \end{cases}$$

Let $\sigma = \sigma_1 \ldots \sigma_p$ be an antimorphism of a self-complementary graph of $4k$ vertices. We find the number of ways in which edges can be introduced so that the result is a self-complementary split graph with σ as an antimorphism. We need to consider adjacencies among vertices in the same cycle and the adjacencies between vertices from different cycles of σ. For the second part, we further separate into two cases depending on whether the cycles have the same length. We use G to denote a resulting graph and denote by G_i the graph induced by the vertices in the ith cycle, for $i = 1, \ldots, p$. By Lemma 2, G has a unique split partition and we refer to it as $K \uplus I$.

(i) The subgraph G_i is determined if it has been decided whether v_{i1} is to be adjacent or not adjacent to each of the following $\frac{|\sigma_i|}{2}$ vertices in σ_i. Among those $\frac{|\sigma_i|}{2}$ vertices, half of them are odd-numbered in σ_i. Therefore, v_{i1} is either adjacent to all of them or adjacent to none of them by Lemma 2. The number of adjacencies to be decided is $\frac{|\sigma_i|}{4} + 1$.

(ii) The adjacencies between two subgraphs G_i and G_j of the same order are determined if it has been decided whether v_{i1} is to be adjacent or not adjacent to each of the vertices in G_j. By Lemma 2, the vertex v_{i1} and half of vertices of G_j are decided in K or in I after (i). The number of adjacencies to be decided is $\frac{|\sigma_j|}{2}$.

(iii) We now consider the adjacencies between two subgraphs G_i and G_j of different orders. We use $\gcd(x, y)$ to denote the greatest common factor of two integers x and y. The adjacencies between G_i and G_j are determined if it has been decided whether v_{i1} is to be adjacent or not adjacent to each of the first $\gcd(|\sigma_i|, |\sigma_j|)$ vertices of G_j. Among those $\gcd(|\sigma_i|, |\sigma_j|)$ vertices of G_j, half of them are decided in the same part of $K \uplus I$ as v_{i1} after (i). The number of adjacencies to be decided is $\frac{1}{2}\gcd(|\sigma_i|, |\sigma_j|)$.

By Lemma 1, $|\sigma_i| \equiv 0 \pmod 4$ for every $i = 1, \ldots, p$. Let c be the cycle structure of σ. We use c_q to denote the number of cycles in c with length $4q$ for every $q = 1, 2, \ldots, k$. The total number of adjacencies to be determined is

$$P = \sum_{q=1}^{k}(c_q(q+1) + \frac{1}{2}c_q(c_q - 1) \cdot 2q) + \sum_{1 \le r < s \le k} c_r c_s \cdot \frac{1}{2}\gcd(4r, 4s)$$

$$= \sum_{q=1}^{k}(qc_q^2 + c_q) + 2\sum_{1 \le r < s \le k} c_r c_s \gcd(r, s).$$

For each adjacency, there are two choices. Therefore, the number of labeled self-complementary split graphs with this σ as an antimorphism is 2^P.

The number of distinct permutations of the cycle structure c consisting of c_q cycles of length $4q$ for every $q = 1, 2, \ldots, k$ is

$$\frac{(4k)!}{\prod_{q=1}^{k}(4q)^{c_q} \cdot c_q!},$$

and it is the number of possible choices for σ [4]. Let C_{4k} be the set that contains all cycle structures c that satisfy $\sum_{q=1}^{k} c_q \cdot 4q = 4k$. Then the number of antimorphisms with all possible labeled self-complementary split graphs with $4k$ vertices corresponding to each is

$$\sum_{c \in C_{4k}} \frac{(4k)!}{\prod_{q=1}^{k}(4q)^{c_q} \cdot c_q!} 2^P. \tag{3}$$

For a graph G with $4k$ vertices, let A_G be the set of automorphisms of G. Then, the number of different labelings of G is $(4k)!/|A_G|$. If G is self-complementary, then the number of antimorphisms of G is equal to the number of automorphisms of G. Let S be the set of all non-isomorphic self-complementary split graphs with $4k$ vertices and let $\lambda_{4k} = |S|$. The number of labeled self-complementary split graphs with all possible antimorphisms corresponding to each is equal to

$$\sum_{G \in S} |A_G| \frac{(4k)!}{|A_G|} = \lambda_{4k} (4k)!.$$ (4)

Let Eq. (3) equals to Eq. (4) and we solve for λ_{4k}:

$$\lambda_{4k} = \sum_{c \in C_{4k}} \frac{2^P}{\prod_{q=1}^{k}(4q)^{c_q} \cdot c_q!}.$$

We list below the number of self-complementary (pseudo-)split graphs on up to 21 vertices.

n	4	5	8	9	12	13	16	17	20	21
split	1	1	3	3	16	16	218	218	9608	9608
pseudo-split	1	2	3	4	16	19	218	234	9608	9826
all	1	2	10	36	720	5600	703760	11220000	9168331776	293293716992

References

1. Ali, P.: Study of Chordal graphs. Ph.D. thesis, Aligarh Muslim University, India (2008)
2. Cao, Y., Chen, H., Wang, S.: On Trotignon's conjecture on self-complementary graphs. Manuscript (2023)
3. Clapham, C.R.J.: Potentially self-complementary degree sequences. J. Comb. Theor. Ser. B **20**(1), 75–79 (1976)
4. Clapham, C.R.J.: An easier enumeration of self-complementary graphs. Proc. Edinburgh Math. Soc. **27**(2), 181–183 (1984)
5. Clapham, C.R.J., Kleitman, D.J.: The degree sequences of self-complementary graphs. J. Comb. Theor. Ser. B **20**(1), 67–74 (1976)
6. Foldes, S., Hammer, P.L.: Split graphs. In: Proceedings of the Eighth Southeastern Conference on Combinatorics, Graph Theory and Computing, pp. 311–315 (1977)
7. Gibbs, R.A.: Self-complementary graphs. J. Comb. Theor. Ser. B **16**, 106–123 (1974)
8. Harary, F.: Unsolved problems in the enumeration of graphs. Magyar Tud. Akad. Mat. Kutató Int. Közl. **5**, 63–95 (1960)
9. Maffray, F., Preissmann, M.: Linear recognition of pseudo-split graphs. Discret. Appl. Math. **52**(3), 307–312 (1994)
10. Rao, S.B.: A survey of the theory of potentially P-graphic and forcibly P-graphic degree sequences. In: Rao, S.B. (ed.) Combinatorics and Graph Theory. LNM, vol. 885, pp. 417–440. Springer, Heidelberg (1981). https://doi.org/10.1007/BFb0092288
11. Read, R.C.: On the number of self-complementary graphs and digraphs. J. Lond. Math. Soc. **38**, 99–104 (1963)
12. Ringel, G.: Selbstkomplementäre Graphen. Arch. Math. (Basel) **14**, 354–358 (1963)

13. Ryser, H.J.: Combinatorial properties of matrices of zeros and ones. Can. J. Math. **9**, 371–377 (1957)
14. Sachs, H.: Über selbstkomplementäre Graphen. Publ. Math. Debrecen **9**, 270–288 (1962)
15. Sridharan, M.R., Balaji, K.: Characterisation of self-complementary chordal graphs. Discret. Math. **188**(1–3), 279–283 (1998)
16. Trotignon, N.: On the structure of self-complementary graphs. Electron. Notes Discret. Math. **22**, 79–82 (2005)
17. Xu, J., Wong, C.K.: Self-complementary graphs and Ramsey numbers. I. The decomposition and construction of self-complementary graphs. Discret. Math. **223**(1–3), 309–326 (2000)

Schnyder Woods and Long Induced Paths in 3-Connected Planar Graphs

Institute of Computer Science, University of Rostock, Rostock, Germany

Abstract. In the recent 30 years, Schnyder woods have become an invaluable asset in the study of planar graphs. We contribute to this research with a brief and comprehensible proof of a new structural feature: every Schnyder wood of a 3-connected planar graph on n vertices has a tree of depth at least $\lfloor 1/6 \log_2 n \rfloor$. As a simple implication, our result improves the previous hard-won lower bound on the length of an induced path in such a graph to $1/6 \log_2 n$.

Keywords: Induced paths · 3-connected planar graphs · Schnyder woods · depth

1 Introduction

The problem of finding a long induced path has first been investigated by Erdős et al. [5] already in 1986. Let $p(G)$ be the size, i.e. the number of vertices, of a longest induced path of G. Erdős et al. [5] gave the lower bound $p(G) \geq 2r(G) - 1$ for a connected graph G in terms of the radius $r(G)$ of G. In 2000, Arocha and Valencia [3] found a lower bound of $\log_\Delta(n)$ on the diameter (and hence on $p(G)$) of a 3-connected planar graph G with bounded maximum degree Δ, and the number of vertices n. If Δ is not bounded, they show that there is an induced path of size $\sqrt{\log_3(\Delta)}$. In 2016, Di Giacomo et al. [4] gave the lower bound $p(G) \geq \frac{\log_2 n}{12 \log_2 \log_2 n}$ for 3-connected planar graphs G. They also showed that there exist 3-connected planar graphs G with $p(G) \leq 1.3 \log_2(n) + 5$. Also in 2016, Esperet et al. [6] could improve the lower bound of Di Giacomo et al. [4] to $1/6(\log_2(n) - 3 \log_2 \log_2(n))$ choosing a similar approach.

Here, we show the slightly better lower bound $p(G) \geq 1/6 \log_2(n)$. But, our proof is by far simpler than the previous ones of Di Giacomo et al. [4] and Esperet et al. [6] and also uses a completely different approach which yields new structural insights. These new insights are discussed at the end of the introduction. The next paragraph gives the reader an intuition about Schnyder woods.

The concept of Schnyder woods is widely used in graph drawing and related areas [1,2,7,8,11]. A Schnyder wood of a 3-connected planar graph is a triple of directed spanning trees such that every edge is in at least one and at most two

---

This research is supported by the grant SCHM 3186/2-1 (401348462) from the Deutsche Forschungsgemeinschaft (DFG, German Research Foundation).

J. A. Soto and A. Wiese (Eds.): LATIN 2024, LNCS 14579, pp. 19–30, 2024.
https://doi.org/10.1007/978-3-031-55601-2_2

trees. The three trees are colored red, blue and green. In every tree, the edges are oriented towards the root. If an edge is in two different trees it has opposite orientations in the two trees. Consider for example the red edges in the graph in Fig. 1a. They form a spanning tree rooted at r_1. Additionally, around each vertex the outgoing and ingoing edges need to obey a certain pattern. In the clockwise sector between the outgoing red and the outgoing green edge only ingoing blue edges occur. For the ingoing red and green edges symmetric properties hold. See Fig. 1b for illustration. A formal definition is given in Sect. 2. In this paper we exploit the fact that a path from a leaf to the root in a tree of a Schnyder wood needs to be an induced path. See Corollary 1 for this argument. So in order to find a long induced path, we investigate the depth, i.e. the length of a longest path from a leaf to the root, of all three trees in a Schnyder wood of a given 3-connected planar graph.

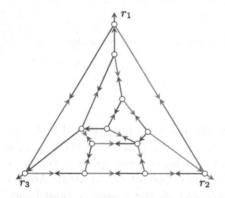

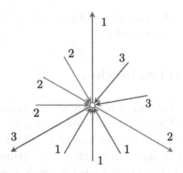

(a) A Schnyder wood of the suspension of a 3-connected planar graph.

(b) Example for Condition 3 at a vertex in a Schnyder wood. The ingoing edges in color i are in the clockwise sector between the outgoing edge in color $i + 1$ and the outgoing edge in color $i - 1$.

Fig. 1. Illustrations for the definition of Schnyder woods.

Given a planar embedding of a 3-connected planar graph and a Schnyder wood on this embedding, we show that at least one of the three trees has depth at least $1/6 \log_2(n)$, which, as mentioned above, directly implies that $p(G) \geq 1/6 \log_2(n)$. To the best of our knowledge, we are the first to investigate the depth of the trees of Schnyder woods. This new structural property of Schnyder woods is not only of theoretical interest, but also yields the following extra information on the problem of finding long induced paths.

We easily obtain a linear time approach computing an induced path of the desired size. In fact, given a 3-connected, planar graph, we compute a Schnyder wood in linear time [7,10,11]. Then, we compute a longest leaf-to-root path for each of the three trees of the Schnyder wood in linear time by traversing the three trees e.g. with a breadth-first search. The longest such path has size at least $1/6 \log_2(n)$.

Furthermore, we know that for every choice of three vertices which are on the boundary of a common face, we can find an induced path of size at least $1/6 \log_2(n)$ that ends in one of the three vertices. It is easy to derive that there are at least $f/(2\Delta)$ different such paths, where f is the number of faces and Δ the maximum degree.

Finally, for each such path there exists a grid drawing such that the path is monotone in both coordinates. Such a drawing can be found using for example the algorithm of Felsner [7] which is based on Schnyder woods and yields a drawing on the $(f-1) \times (f-1)$ grid.

To conclude the paper, we show that, for every $n \geq 3$, there exists a 3-connected planar graph with a Schnyder wood such that the three trees have depth at most $\log_3(2n-5)+1 \approx 0.63 \cdot \log_2(n) + 2$ using the approach of Di Giacomo et al. [4].

2   Preliminaries

We use standard graph notation. The graphs G we consider in this paper are simple, planar, 3-connected and come with a fixed embedding into the plane, that is, G is plane.

We use the definition of Schnyder woods as given by Felsner [8]. The *suspension* G^σ of G is obtained by choosing three different vertices r_1, r_2 and r_3 which appear in clockwise order on the outer face and by adding adjacent to each of those vertices a half-edge which reaches into the outer face. With a little abuse of notation, we define a *half-edge* as an arc starting at a vertex but with no defined end vertex. The special vertices r_1, r_2 and r_3 are called *roots*.

Given a suspension G^σ, a *Schnyder wood* rooted at r_1, r_2 and r_3 is an orientation and coloring of the edges with colors 1, 2 and 3 satisfying the following conditions. Indices indicating colors are modulo 3. This means that e.g. $3+1 \equiv 1$.

1. Every edge is either oriented in one direction (unidirected edge) or in both directions (bidirected edge). Every edge receives a distinct color for every direction. So unidirected edges receive one color and bidirected edges two different colors.
2. For every $i \in \{1, 2, 3\}$ the half-edge at r_i is directed away from r_i and colored i.
3. For every vertex v and every color $i \in \{1, 2, 3\}$ there is exactly one incident outgoing (half-)edge of color i. The outgoing edges e_1, e_2 and e_3 of v in colors 1, 2 and 3, respectively, occur clockwise around v. The ingoing edges of v in color i are in the clockwise sector from e_{i+1} to e_{i-1}.
4. No interior face has a boundary which is a directed cycle in one color.

See Fig. 1b and a for illustration. For ease of notation we define a Schnyder wood of G to be a Schnyder wood of a suspension of G. Throughout the paper we use red, green and blue synonymously for color 1, 2 and 3, respectively. Denote by T_i the directed graph induced by the (uni- and bidirected) edges that have color i. T_1, T_2 and T_3 are called the *trees* of the Schnyder wood.

The following properties of Schnyder woods are used in this paper.

Lemma 1 (Felsner [7]). *Every 3-connected plane graph has a Schnyder wood.*

Lemma 2 (Felsner [7]). *T_i is a directed tree rooted at r_i for every $i \in \{1, 2, 3\}$.*

For a directed graph H, let H^{-1} be the graph obtained from H by reversing the orientation of all edges.

Lemma 3 (Felsner [7]). *For all $i \in \{1, 2, 3\}$ $T_i \cup T_{i-1}^{-1} \cup T_{i+1}^{-1}$ does not have an oriented cycle.*

For every vertex $v \in V(G)$ denote by $P_i(v)$ the path from v to the root in the tree T_i. The *depth* of T_i is the length (number of edges) of a longest path from a vertex to the root in T_i. For ease of notation, $P_i(v)$ also denotes the vertex set of the path $P_i(v)$. Hence, $|P_i(v)|$ is the number of vertices of $P_i(v)$ and $P_j(w) \cap P_i(v)$ is the set of vertices $P_j(w)$ and $P_i(v)$ have in common.

Denote by $R_i(v)$ the region bounded by and including $P_{i-1}(v)$, $P_{i+1}(v)$ and the clockwise path from r_{i+1} to r_{i-1} on the outer face. See Fig. 2 for an illustration. For example, vertices on the path $P_{i-1}(v)$ belong to both $R_i(v)$ and $R_{i+1}(v)$.

Lemma 4 (Felsner [8]). *Let $i \in \{1, 2, 3\}$. If $u \in R_i(v)$ then $R_i(u) \subseteq R_i(v)$.*

Similar observations as in Felsner's proof of Lemma 4 directly yield the following lemma.

Lemma 5. *Let $i \in \{1, 2, 3\}$. If $u \in R_i(v) \setminus P_{i+1}(v)$, then $R_i(u) \setminus P_{i+1}(u) \subsetneq R_i(v) \setminus P_{i+1}(v)$ and $v \notin R_i(u) \setminus P_{i+1}(u)$.*

Proof. We prove the claim for $i = 1$. For $i \in \{2, 3\}$, the proof is symmetric. Let $u \in R_1(v) \setminus P_2(v)$. By Lemma 4, $R_1(u) \subseteq R_1(v)$. We now need to prove that $P_2(v)$ intersects $R_1(u)$ only in $P_2(u)$. By definition, $R_1(u)$ is bounded by $P_2(u)$, $P_3(u)$ and the clockwise path P from r_2 to r_3 on the outer face. Since $P_2(v)$ is on the boundary of $R_1(v)$ and $R_1(u) \subseteq R_1(v)$, we know that $P_2(v)$ can intersect $R_1(u)$ only on its boundary, meaning $P_2(v) \cap R_1(u) \subseteq P_3(u) \cup P_2(u) \cup P$.

In the following, we show that $P_2(v) \cap (P \setminus P_2(u)) = \emptyset$. Let x_v be the first vertex of $P_2(v)$, starting at v, which is also on P and let x_u be the first vertex of $P_2(u)$, starting at u, which is also on P. Assume, for the sake of contradiction, that x_v comes after x_u on P starting at r_2. Since $u \in R_1(v)$, there needs to be a vertex y at which $P_2(u)$ leaves $R_1(v)$. But $P_2(u) \subseteq R_1(u) \subseteq R_1(v)$, so we arrive at a contradiction. So x_u comes after x_v on P starting at r_2 and $P_2(v)$ does not intersect $P \setminus P_2(u)$. Thus $P_2(v) \cap R_1(u) \subseteq P_3(u) \cup P_2(u)$.

For the sake of contradiction, assume that $P_2(v) \cap P_3(u)$ is non-empty. Let x be the first vertex on the path $P_3(u)$ starting at u which is also in $P_2(v)$. If

$x = u$, then $u \in P_2(v)$, contradicting the assumption of the lemma. So $x \neq u$. Since $u \in R_1(v)$, there is either an outgoing 2-colored, an ingoing 3-colored and an outgoing 3-colored edge around x in that clockwise order or there is an outgoing 3-colored, an ingoing 3-colored and an ingoing 2-colored edge around x in that clockwise order. This both contradicts Condition 3 at x.

Remember that $P_2(v) \cap R_1(u) \subseteq P_3(u) \cup P_2(u)$. Additionally, we now have that $P_2(v) \cap P_3(u) = \emptyset$, which then yields $P_2(v) \cap R_1(u) \subseteq P_2(u)$. Together with the above mentioned fact that $R_1(u) \subseteq R_1(v)$, this yields that $R_1(u) \setminus P_2(u) \subseteq R_1(v) \setminus P_2(v)$. Since $u \in R_1(v) \setminus P_2(v)$ but $u \notin R_1(u) \setminus P_2(u)$, we have $R_1(u) \setminus P_2(u) \subsetneq R_1(v) \setminus P_2(v)$. As $v \in P_2(v)$ and $P_2(v) \cap R_1(u) \subseteq P_2(u)$, we obtain $v \notin R_1(u) \setminus P_2(u)$. □

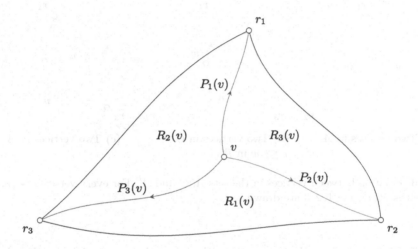

Fig. 2. Illustration of the definition of regions and paths for a vertex v.

3 Lower Bound on the Maximum Depth of a Tree

In this section we bound the maximum depth of a tree in a Schnyder wood of a 3-connected planar graph G on n vertices from below by $\lfloor 1/6 \log_2(n) \rfloor$. As a corollary we derive that G has an induced path of size at least $\lfloor 1/6 \log_2(n) \rfloor + 1$.

Theorem 1. *Let G be a 3-connected plane graph and let T_1, T_2 and T_3 be the trees of a Schnyder wood of G. Then at least one, T_1, T_2 or T_3, has depth at least $\lfloor 1/6 \log_2(n) \rfloor$.*

The general idea of the proof is the following. For any set $C \subseteq V(G)$ and $v \in C$, we consider the set of vertices which are on $P_1(v)$ and also on a path $P_3(w)$ for some $w \in C$. Naturally, this set $\bigcup_{w \in C}(P_1(v) \cap P_3(w))$ is a subset of

$P_1(v)$. The size of this set gives a natural lower bound on the depth on T_1 and is later denoted as $l_1^C(v)$.

Now, we need to find a set of vertices C and a vertex $x \in C$ such that $\bigcup_{w \in C}(P_1(x) \cap P_3(w))$ (or $\bigcup_{w \in C}(P_3(x) \cap P_1(w))$) is large enough, i.e. has size at least $\lfloor 1/6 \log_2(n) \rfloor + 1$. Let C be a set such that for two vertices $v, w \in C$, $v \neq w$, we have that either $v \in R_1(w) \setminus P_2(w)$ and $w \in R_3(v) \setminus P_1(v)$ or vice versa $v \in R_3(w) \setminus P_1(w)$ and $w \in R_1(v) \setminus P_2(v)$. See Fig. 3b for illustration. We show that such a set C of size at least $n^{1/3}$ exists, and that there is a vertex $x \in C$ such that either $\bigcup_{w \in C}(P_1(x) \cap P_3(w))$ or $\bigcup_{w \in C}(P_3(x) \cap P_1(w))$ has size at least $\lfloor 1/2 \log_2(|C|) \rfloor + 1$, as required. Together this then yields the desired lower bound of $\lfloor 1/6 \log_2(n) \rfloor$ on the depth of either T_1 or T_3, respectively.

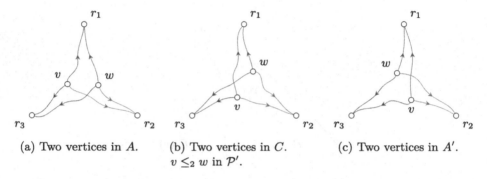

(a) Two vertices in A.   (b) Two vertices in C.   (c) Two vertices in A'.
                            $v \leq_2 w$ in $\mathcal{P}'$.

Fig. 3. Relation between vertices in the sets A, C and A'. For every vertex $x \in \{v, w\}$ the paths $P_i(x)$, $i = 1, 2, 3$ are drawn.

Proof. First, we show that a set C with properties symmetric to the above properties exists. Those properties are: $|C| \geq n^{1/3}$ and for some $i \in \{1, 2, 3\}$ we have for every $v, w \in C$, $v \neq w$, that either $v \in R_i(w) \setminus P_{i+1}(w)$ and $w \in R_{i+2}(v) \setminus P_i(v)$ or vice versa $v \in R_{i+2}(w) \setminus P_i(w)$ and $w \in R_i(v) \setminus P_{i+1}(v)$.

We define the following relation $\leq_1$ on $V(G)$: For $v, w \in V(G)$ let $v \leq_1 w$ if $v \in R_1(w) \setminus P_2(w)$ or $v = w$. The relation $\leq_1$ is obviously reflexive. By Lemma 5, $\leq_1$ is antisymmetric, since $w \notin R_1(v) \setminus P_2(v)$ for $v \leq_1 w$ and $v \neq w$. Also by Lemma 5, for $u \leq_1 v$, $v \leq_1 w$ with $v \neq w$, we have either $u \in R_1(v) \setminus P_2(v) \subseteq R_1(w) \setminus P_2(w)$ or $u = v$. Therefore $u \leq_1 w$. Hence $\leq_1$ is transitive and, thus, $\mathcal{P} := (V(G), \leq_1)$ is a poset.

By Mirsky's theorem [9], we either find a chain L of size at least $n^{2/3}$ or we can decompose $\mathcal{P}$ into at most $n^{2/3}$ antichains. In the latter case, we find, by the pigeonhole principle, an antichain A of size $n/n^{2/3} = n^{1/3}$. Consider the set A. For any two vertices $v, w \in A$, $v \neq w$, we have that $v \notin R_1(w) \setminus P_2(w)$ and $w \notin R_1(v) \setminus P_2(v)$. Thus v is either in $R_3(w) \setminus P_1(w)$ or $R_2(w) \setminus P_3(w)$. Assume w.l.o.g. that the latter applies. Then $w \notin R_2(v) \setminus P_3(v)$, by Lemma 5. And thus w needs to be in $R_3(v) \setminus P_1(v)$. This is symmetric to the property which we are aiming for and thus A is already fine for us. See Fig. 3a for illustration.

Otherwise, if the chain L of $\mathcal{P}$ exists, we define the relation $\leq_2$ on L: For $v, w \in L$, let $v \leq_2 w$ if $w \in R_3(v) \setminus P_1(v)$ or $v = w$. As above, $\mathcal{P}' := (L, \leq_2)$ is a poset and we either find a chain C of size $n^{1/3}$ or an antichain A' of size $n^{2/3}/n^{1/3} = n^{1/3}$. For two vertices $v, w \in C$, $v \neq w$, we have w.l.o.g. $v \leq_2 w$ and hence $w \in R_3(v) \setminus P_1(v)$. Since v and w are in L, a chain in $\mathcal{P}$, we either have $w \leq_1 v$ or $v \leq_1 w$. If $w \leq_1 v$ then $w \in R_1(v) \setminus P_2(v)$, contradicting $w \in R_3(v) \setminus P_1(v)$. So $v \leq_1 w$ and hence $v \in R_1(w) \setminus P_2(w)$.

Similarly we obtain for two vertices $v, w \in A'$, $v \neq w$, that $v \in R_1(w) \setminus P_2(w)$ and $w \in R_2(v) \setminus P_3(v)$ or vice versa. Again, see Fig. 3. Since the relations on A, A' and C are symmetric, we assume w.l.o.g. that C exists.

Let $S \subseteq C$. For all $v \in S$ and all $(i, j) \in \{(1, 3), (3, 1)\}$ define

$$l_i^S(v) := \left| \bigcup_{x \in S} (P_i(v) \cap P_j(x)) \right|,$$

that is, the number of vertices on $P_i(v)$ that are also on a path $P_j(x)$ for some $x \in S$. For example, $l_1^S(v)$ is the number of vertices on $P_1(v)$ that are also on a path $P_3(x)$ for any $x \in S$.

Let, furthermore,

$$l_i^S := \max_{v \in S} l_i^S(v).$$

and let ω_i^S be a vertex of S, that realizes this maximum value, i.e. $l_i^S(\omega_i^S) = l_i^S$.

In the following, we prove by induction that $l_3^C + l_1^C \geq \lfloor \log_2(|C|) \rfloor + 2$. Our bound then follows by pigeonhole principle. Clearly, if $|C| = 1$, we have $l_3^C + l_1^C = 2 = \log_2(1) + 2$ and the claim holds.

Let $|C| \in \{2, 3\}$. Then we have two vertices $v, w \in C$, $v \neq w$, such that $v \leq_2 w$ in $\mathcal{P}'$. So $v \in R_1(w) \setminus P_2(w)$ and $w \in R_3(v) \setminus P_1(v)$. Especially $w \notin P_1(v)$ and hence $P_1(v)$ intersects $P_3(w)$ in a vertex different from w. See Fig. 3b for illustration. This yields $l_3^C(w) \geq 2$. Thus

$$l_3^C + l_1^C \geq l_3^C(w) + l_1^C(v) \geq 2 + 1 = 3 = \lfloor \log_2(3) \rfloor + 2 \geq \lfloor \log_2(|C|) \rfloor + 2.$$

So assume that $|C| \geq 4$. Partition $C = C_1 \cup C_2 \cup X$ such that $|C_1| = |C_2| = 2^z$, z is maximal, and $C_1 = \{v_1, \ldots, v_s\}$, $C_2 = \{v_{s+1}, \ldots, v_{2s}\}$ with $v_1 \leq_2 \ldots \leq_2 v_{2s}$ in $\mathcal{P}'$. By induction, $l_3^{C_k} + l_1^{C_k} \geq \lfloor \log_2(|C_k|) \rfloor + 2 = \log_2(|C_k|) + 2$ for $k = 1, 2$.

So $v_l \in R_1(\omega_3^{C_2}) \setminus P_2(\omega_3^{C_2})$ and $\omega_3^{C_2} \in R_3(v_l) \setminus P_1(v_l)$ for $l = 1, \ldots, s$. Hence $P_1(v_l)$ intersects $P_3(\omega_3^{C_2})$ for $l = 1, \ldots, s$. Observe that, by Condition 3, this intersection is a set of vertices that appears consecutively on both $P_1(v_l)$ and $P_3(\omega_3^{C_2})$. And, by Condition 1 and 3, the first vertex of $P_1(v_l)$, starting at v_l, that is also on $P_3(\omega_3^{C_2})$ is the vertex where $P_3(\omega_3^{C_2})$ leaves $R_3(v_l)$.

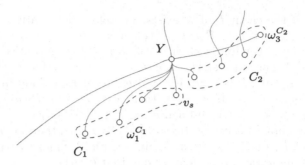

Fig. 4. Situation as in Case 1. The blue path from $\omega_3^{C_2}$ to the root intersects the red path from $\omega_1^{C_1}$ to the root. (Color figure online)

Define Y to be the intersection of the 1-colored path starting at v_s and the 3-colored path starting at $\omega_3^{C_2}$, i.e. $Y := P_1(v_s) \cap P_3(\omega_3^{C_2})$. For $l = 1, \ldots, s-1$ we have that $v_s \in R_3(v_l) \setminus P_1(v_l) \subseteq R_3(v_l)$. Hence, by Lemma 4, $P_1(v_s) \subseteq R_3(v_s) \subseteq R_3(v_l)$. Let $p \in P_1(v_l)$ and $q \in P_1(v_s)$ be the first vertex, starting at v_l and v_s, of $P_1(v_l)$ and $P_1(v_s)$, respectively, that is also on $P_3(\omega_3^{C_2})$. As observed above, $P_3(\omega_3^{C_2})$ leaves $R_3(v_s)$ at q and $R_3(v_l)$ at p. And since $R_3(v_s) \subseteq R_3(v_l)$ the vertex q occurs before p on $P_3(\omega_3^{C_2})$, starting at $\omega_3^{C_2}$. Since T_1 is a tree, we know that if $P_1(v_l) \cap P_1(v_s) \cap P_3(\omega_3^{C_2}) \neq \emptyset$, then the first vertex of $P_3(\omega_3^{C_2})$, starting at $\omega_3^{C_2}$, that is also on $P_1(v_s)$ coincides with the first vertex of $P_3(\omega_3^{C_2})$, starting at $\omega_3^{C_2}$, that is also on $P_1(v_l)$. So if $P_1(v_l) \cap P_3(\omega_3^{C_2}) \setminus Y = \emptyset$, then $p = q$ and $P_1(v_l) \cap P_3(\omega_3^{C_2}) = Y$. Thus the following case distinction is exhaustive. Either all 1-colored paths starting at vertices of C_1 intersect $P_3(\omega_3^{C_2})$ in the set Y or there is a vertex in C_1 such that its 1-colored path intersects $P_3(\omega_3^{C_2})$ in a vertex not in Y. We distinguish those two cases.

Case 1: Assume that $P_1(v_l) \cap P_3(\omega_3^{C_2}) = Y$ for $l = 1, \ldots, s-1$, see Fig. 4. So especially $P_1(\omega_1^{C_1}) \cap P_3(\omega_3^{C_2}) = Y$. We now show that no 3-colored path starting at a vertex in C_1 intersects Y. So assume for the sake of contradiction that there is an $l \in \{1, \ldots, s\}$ such that $P_3(v_l) \cap Y \neq \emptyset$. If $v_l \notin Y$ then $P_3(v_l)$ intersects $P_1(v_l)$ in a vertex different from v_l. This directly yields a directed cycle in $T_1 \cup T_2^{-1} \cup T_3^{-1}$, contradicting Lemma 3. So, assume $v_l \in Y$. Then $v_l \in P_1(v_t)$ for some $t \in \{1, \ldots, s\} \setminus \{l\}$, contradicting the definition of C.

So $P_3(v_l) \cap Y = \emptyset$ for $l = 1, \ldots, s$. Remember that Y equals the intersection of $P_1(\omega_1^{C_1})$ and $P_3(\omega_3^{C_2})$. So we have that

$$l_1^C(\omega_1^{C_1}) = \left| \bigcup_{x \in C} (P_1(\omega_1^{C_1}) \cap P_3(x)) \right|$$

$$\geq \left| \bigcup_{x \in C_1} (P_1(\omega_1^{C_1}) \cap P_3(x)) \right| + |Y| = l_1^{C_1}(\omega_1^{C_1}) + |Y|$$

$$\geq l_1^{C_1}(\omega_1^{C_1}) + 1 = l_1^{C_1} + 1,$$

and we obtain

$$l_3^C + l_1^C \geq l_3^{C_1} + l_1^C(\omega_1^{C_1}) \geq l_3^{C_1} + l_1^{C_1} + 1$$
$$\geq \log_2(|C_1|) + 2 + 1 = \log_2(2|C_1|) + 2$$
$$= \lfloor \log_2(|C|) \rfloor + 2.$$

Case 2: Assume that there is a vertex v_l, $l \in \{1, \ldots, s-1\}$ such that $Z := (P_1(v_l) \cap P_3(\omega_3^{C_2})) \setminus Y \neq \emptyset$, see Fig. 5. We now show that no 1-colored path starting at a vertex in C_2 intersects Z. Assume for the sake of contradiction that there is a vertex v_t, $t \in \{s+1, \ldots, 2s\}$ such that $P_1(v_t) \cap Z \neq \emptyset$.

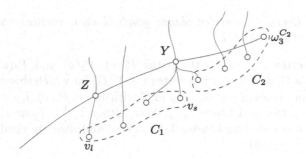

Fig. 5. Situation as in Case 2. The red path from v_l to the root intersects the blue path from $\omega_3^{C_2}$ to the root. (Color figure online)

In the following we show that $Z \cap R_3(v_s) = \emptyset$. Then, since $v_t \in R_3(v_s) \setminus P_1(v_s)$, the 1-colored path $P_1(v_t)$ starting at v_t starts in $R_3(v_s)$ and leaves it at some vertex. This will lead to a contradiction.

We have that $v_l \leq_2 v_s$ in $\mathcal{P}'$ and $v_l \neq v_s$ so $v_s \in R_3(v_l) \setminus P_1(v_l)$. By Lemma 5, $R_3(v_s) \setminus P_1(v_s) \subsetneq R_3(v_l) \setminus P_1(v_l)$. Assume there exists a vertex $z \in Z \cap R_3(v_s)$. Either $z \in P_1(v_s)$, contradicting the definition of Z, or $z \in R_3(v_s) \setminus P_1(v_s)$. Then we have $z \notin R_3(v_l) \setminus P_1(v_l) \supsetneq R_3(v_s) \setminus P_1(v_s) \ni z$, a contradiction. So $Z \cap R_3(v_s) = \emptyset$. As mentioned above $v_t \in R_3(v_s) \setminus P_1(v_s)$ since $v_s \leq_2 v_t$ and $v_s \neq v_t$. This situation requires the 1-colored path $P_1(v_t)$ starting at v_t to leave $R_3(v_s)$ at some vertex $w \in R_3(v_s)$.

If this vertex $w \in P_1(v_s)$, it would have two outgoing 1-colored edges. And if $w \in P_2(v_s) \setminus \{v_s\}$, it would have an outgoing 2-colored edge, an outgoing 1-colored edge and an ingoing 2-colored edge in that clockwise order. In both cases Condition 3 would be violated at w.

So, $P_1(v_t) \cap Z = \emptyset$ for $t = s+1, \ldots, 2s$. And as in Case 1 we have $l_3^C(\omega_3^{C_2}) \geq l_3^{C_2}(\omega_3^{C_2}) + |Z| \geq l_3^{C_2}(\omega_3^{C_2}) + 1 = l_3^{C_2} + 1$. And we obtain that

$$l_3^C + l_1^C \geq l_3^C(\omega_3^{C_2}) + l_1^{C_2} \geq l_3^{C_2} + 1 + l_1^{C_2}$$
$$\geq \log_2(|C_2|) + 2 + 1 = \log_2(2|C_2|) + 2$$
$$= \lfloor \log_2(|C|) \rfloor + 2.$$

By pigeonhole principle, there is an $i \in \{1,3\}$ such that $l_i^C \geq \lceil 1/2\lfloor \log_2(|C|) \rfloor + 1 \rceil \geq \lfloor 1/2 \log_2(n^{1/3}) \rfloor + 1 = \lfloor 1/6 \log_2(n) \rfloor + 1$. As $l_i^C - 1$ is a lower bound on the depth of T_i the claim follows. □

In 2016, Esperet et al. [6] showed that in a 3-connected planar graph there exists an induced path with at least $1/2(1/3\log_2(n) - \log_2\log_2(n))$ vertices. Using Theorem 1, we can do slightly better.

Corollary 1. *Every 3-connected planar graph G on n vertices has an induced path of size $\lfloor 1/6 \log_2(n) \rfloor + 1$.*

Proof. For a vertex $v \in V(G)$ the paths $P_1(v)$, $P_2(v)$ and $P_3(v)$ are always induced. Assume that there exists a vertex $v \in V(G)$ for which this does not hold. Then there is an i-colored edge $e = xy$ in G with $x, y \in P_j(v)$, $i, j \in \{1, 2, 3\}$. By Lemma 2, T_i is a tree and hence $i \neq j$. Now, either $T_i^{-1} \cup T_j$ or $T_i \cup T_j$ has an oriented cycle, contradicting Lemma 3. So, Theorem 1 directly yields an induced path of size $\lfloor 1/6 \log_2(n) \rfloor + 1$. □

4   On the Tightness of the Lower Bound

The natural question is, if there are graphs G_n of size n with Schnyder woods such that the maximum depth of a tree in the Schnyder wood is bounded from above by a function of n.

Di Giacomo et al. [4] considered this kind of question for induced paths. They showed that for every $n \geq 3$ there exists a 3-connected planar graph G on n vertices such that the longest induced path in G has size at most $2\log_3(2n - 5) + 3 \approx 1.3 \cdot \log_2(n) + 5$ using planar 3-trees. We use the same graphs to show that there are 3-connected planar graphs G_n of size n with Schnyder woods such that the maximum depth of any of its trees is at most $\log_3(2n - 5) + 1 \approx 0.63 \cdot \log_2(n) + 2$.

We use the definition of *(almost) complete planar 3-trees* as given by Di Giacomo et al. [4]: G_0 is a triangle and is defined to be a complete planar 3-tree. We obtain G_{i+1} from G_i by placing a vertex into each internal face and connecting this vertex to the vertices on the face boundary. Concerning the Schnyder wood of those graphs, we observe: G_0 has a unique Schnyder wood S_0. We obtain S_{i+1} from S_i by the following operation: The edges already in G_i retain their orientation and coloring. We assign the only possible orientation and coloring to the new edges which does not violate Condition 3. See Fig. 6 for illustration. An *almost complete planar 3-tree* $\hat{G}_i$ is a graph which is constructed as follows. We take a complete planar 3-tree G_{i-1} and only add in a subset of the internal faces new vertices. They are connected as above and also the Schnyder wood of $\hat{G}_i$ is obtained as above. Complete planar 3-trees are also almost complete planar 3-trees. Now, take an almost complete planar 3-tree $\hat{G}_i$ and, for every vertex $v \notin V(G_0)$, define the *level* of v to be the smallest integer i such that v is in $\hat{G}_i$ but not in G_{i-1}. For vertices of G_0 define the level to be 0.

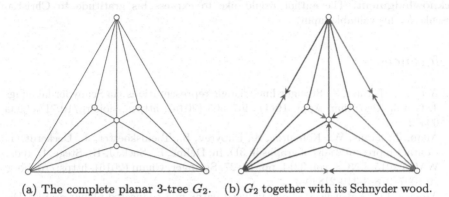

(a) The complete planar 3-tree G_2.   (b) G_2 together with its Schnyder wood.

Fig. 6. Illustration for the definition of complete planar 3-trees.

Lemma 6. *For every $n \geq 3$ there exists a 3-connected planar graph with a Schnyder wood such that the trees T_1, T_2 and T_3 all have depth at most $\log_3(2n - 5) + 1$.*

Proof. Let $\hat{G}_i$ be an almost complete planar 3-tree with n vertices. G_{i-1} has $\frac{3^{i-1}+5}{2}$ vertices and hence $n \geq \frac{3^{i-1}+5}{2}$. Every leaf of a tree of the Schnyder wood of $\hat{G}_i$ has level at most i and the path from a vertex to the root in a tree of the Schnyder wood is strictly decreasing in level. So the maximum depth of a tree in $\hat{G}_i$ is at most $i \leq \log_3(2n - 5) + 1$. $\qquad\square$

5 Conclusion

We showed that every Schnyder wood of a 3-connected planar graph G on n vertices has a tree of depth at least $1/6 \log_2(n)$. As a leaf-to-root path in a tree of a Schnyder wood is an induced path in G, this yields an induced path of size at least $1/6 \log_2(n)$. Schnyder woods are well investigated objects. So without any additional effort we obtain a linear time algorithm which finds such an induced path of size at least $1/6 \log_2(n)$. Also, there is a grid drawing such that the long induced path is monotone in both coordinates and we know that there are at least $f/(2\Delta)$ different such paths. Here f is the number of faces of G and Δ its maximum degree. Furthermore, every improvement of the bound on the depth of trees in Schnyder woods directly leads to the improvement of the bound on the length of induced paths. We are confident, that via this approach further improvements are possible.

We also showed that there exists a graph with a Schnyder wood such that each of the three trees has depth at most $\log_3(2n - 5) + 1 \approx 0.63 \cdot \log_2(n) + 2$. This gives an idea where the limits of this new method might be.

Acknowledgment. The author would like to express his gratitude to Christian Rosenke for his valuable input.

References

1. Aerts, N., Felsner, S.: Straight-line triangle representations via Schnyder labelings. J. Graph Algorithm. Appl. **19**(1), 467–505 (2015). https://doi.org/10.7155/jgaa. 00372
2. Alam, J., Evans, W., Kobourov, S., Pupyrev, S., Toeniskoetter, J., Ueckerdt, T.: Contact representations of graphs in 3D. In: Dehne, F., Sack, J.-R., Stege, U. (eds.) WADS 2015. LNCS, vol. 9214, pp. 14–27. Springer, Cham (2015). https://doi.org/ 10.1007/978-3-319-21840-3_2
3. Arocha, J.L., Valencia, P.: Long induced paths in 3-connected planar graphs. Discuss. Math. Graph Theor. **20**(1), 105–107 (2000). https://doi.org/10.7151/dmgt. 1110
4. Di Giacomo, E., Liotta, G., Mchedlidze, T.: Lower and upper bounds for long induced paths in 3-connected planar graphs. Theoret. Comput. Sci. **636**, 47–55 (2016). https://doi.org/10.1016/j.tcs.2016.04.034
5. Erdős, P., Saks, M., Sós, V.T.: Maximum induced trees in graphs. J. Combin. Theor. Ser. B **41**(1), 61–79 (1986). https://doi.org/10.1016/0097-3165(86)90115-9
6. Esperet, L., Lemoine, L., Maffray, F.: Long induced paths in graphs. Eur. J. Combin. **62**, 1–14 (2017). https://doi.org/10.1016/j.ejc.2016.11.011
7. Felsner, S.: Convex drawings of planar graphs and the order dimension of 3-polytopes. Order **18**(1), 19–37 (2001). https://doi.org/10.1023/A:1010604726900
8. Felsner, S.: Geometric Graphs and Arrangements. Advanced Lectures in Mathematics. Friedr. Vieweg & Sohn, Wiesbaden (2004). https://doi.org/10.1007/978-3-322-80303-0
9. Mirsky, L.: A dual of Dilworth's decomposition theorem. Am. Math. Mon. **78**, 876–877 (1971). https://doi.org/10.2307/2316481
10. Miura, K., Azuma, M., Nishizeki, T.: Canonical decomposition, realizer, Schnyder labeling and orderly spanning trees of plane graphs. Int. J. Found. Comput. Sci. **16**(1), 117–141 (2005). https://doi.org/10.1142/S0129054105002905
11. Schnyder, W.: Embedding planar graphs on the grid. In: Proceedings of the First Annual ACM-SIAM Symposium on Discrete Algorithms, pp. 138–148 (1990)

Bi-arc Digraphs: Recognition Algorithm and Applications

Pavol Hell[1], Akbar Rafiey[2], and Arash Rafiey[1,3(✉)]

[1] Simon Fraser University, Burnaby, BC, Canada
{arashr,pavol}@sfu.ca
[2] University of California San Diego, San Diego, CA, USA
arafiey@ucsd.edu
[3] Indiana State University, Terre Haute, IN, USA
arash.rafiey@indstate.edu

Abstract. We study the class of *bi-arc digraphs*, important from two seemingly unrelated perspectives. On the one hand, they are a broad generalization of interval graphs that include other popular generalizations of interval graphs, such as co-threshold tolerance graphs and adjusted interval digraphs. On the other hand, they are precisely the digraphs that admit the so-called conservative semilattice polymorphisms, also known as min orderings or X-underbar enumerations. These digraphs are generally interesting in studying graph homomorphisms and constraint satisfaction problems.

Our main result is a forbidden obstruction characterization of the class of bi-arc digraphs and a polynomial-time recognition algorithm. In addition, we show that they are precisely the digraphs that admit certain other kinds of conservative polymorphisms, thereby collapsing these polymorphism types in the class of digraphs.

We complement our result by providing a complete dichotomy classification of which general relational structures have polynomial or NP-complete recognition problems for the existence of conservative semilattice polymorphisms.

Keywords: Min ordering · Polymorphisms · Graph Homomorphism · Interval digraphs

1 Background and Motivation

1.1 Graph Theoretic Motivation

Digraph Generalization of Interval Graphs: Part of our motivation stems from a wish to generalize interval graphs. A graph H is an *interval graph* if there is a family of intervals $I_v, v \in V(H)$, such that $uv \in E(H)$ if and only if $I_u \cap I_v \neq \emptyset$. Interval graphs constitute one of the most important graph classes; they admit efficient recognition algorithms, and elegant obstruction characterizations and frequently occur in applications [2,8,14,15,24]. The classical digraph

---

Full version on arXiv [19].

© The Author(s), under exclusive license to Springer Nature Switzerland AG 2024
J. A. Soto and A. Wiese (Eds.): LATIN 2024, LNCS 14579, pp. 31–45, 2024.
https://doi.org/10.1007/978-3-031-55601-2_3

version of interval graphs [9] lacks many of these desirable attributes. A more successful generalization is given in [11]: we say that H is an *adjusted interval digraph* if there are two families of real intervals $I_v, J_v, v \in V(H)$, where for each $v \in V(H)$ the intervals I_v, J_v have the same left endpoint, such that $uv \in A(H)$ if and only if $I_u \cap J_v \neq \emptyset$. Adjusted interval digraphs have many of the desirable algorithmic attributes of interval graphs, including efficient recognition algorithms and forbidden structure characterizations [11].

It is useful to view both interval graphs and adjusted interval digraphs as being *reflexive*, i.e., each vertex having a loop. (This is consistent with their definition as each I_v intersects itself, or the corresponding J_v.) The adjusted interval digraphs appear to be the right generalization of interval graphs for reflexive digraphs. For general (not necessarily reflexive) digraphs, the right analog was less clear. Another special class of digraphs is *bipartite* digraphs, which are just bipartite graphs with all edges oriented from one part of the bipartition to the other part. It turns out there is a natural generalization of interval graphs amongst bipartite digraphs, namely the two-directional orthogonal ray digraphs [27], which have many equivalent definitions [16,18], and also share several of the desirable properties of interval graphs.

One particular property that has been noticed in studying these classes of graphs and digraphs is the notion of *min ordering*. An ordering $<$ of the vertices of digraph H is min ordering if whenever uv and $u'v'$ with $u < u'$ and $v' < v$ are arcs of H then uv' is also an arc of H. For graph H, ordering $<$, is min ordering, if whenever $u < v < w$ and uw is an edge, then uv is also an edge of H. A reflexive graph has a *min ordering* if and only if it is an interval graph; a reflexive digraph has a min ordering if and only if it is an adjusted interval digraph, and a bipartite digraph has a min ordering if and only if it is a two-directional orthogonal ray graph [11,16,18,27]. Thus it was long believed that *min-orderable* digraphs are the right overall generalization of interval graphs. However, it was not known whether this class of digraphs could be recognized in polynomial time, whether it has an obstruction characterization, and whether it has any geometric meaning. Recently, two geometric representations of the class of digraphs with a min ordering have been given in [17]. Min-orderable digraphs are shown there to be exactly the same as signed-interval digraphs, which arise as a natural extension of another well-studied graphs class, the complements of so-called threshold tolerance graphs. They are also shown to be exactly the same digraphs as *bi-arc digraphs*, which are defined as a digraph analogue of the previously studied class of bi-arc graphs [10]. Both these classes are defined by the intersection or inclusion of intervals or circular arcs. Thus it remained to find a forbidden structure characterization for, and a polynomial time recognition algorithm of, min-orderable digraphs. This is what we accomplish in this paper, thus contributing to the argument that min-orderable digraphs are the right general digraph analog of interval graphs.

1.2   CSPs, a Meta-question, and an Algebraic Motivation

Another part of our motivation stems from the study of *Constraint Satisfaction Problems (CSPs)* and the so-called algebraic approach to them. A CSP involves deciding, given a set of variables and a set of constraints on the variables, whether or not there is an assignment to the variables satisfying all of the constraints.

A *relational structure* is a tuple $\mathbb{H} = \langle V, R_1, \ldots, R_s \rangle$ where V is a non-empty finite set, called the universe, and each R_i is a relation of arity r_i on V. For instance, a digraph H with vertex set $V(H)$ and arc set $A(H)$ is a relational structure with universe $V(H)$ and a single binary relation $A(H)$ i.e., $H = \langle V(H), A(H) \rangle$. A *homomorphism* from a relational structure $\mathbb{G}$ to relational structure $\mathbb{H}$ is a mapping from the universe of $\mathbb{G}$ to the universe of $\mathbb{H}$ so that the image of every r-tuples in $\mathbb{G}$ is an r-tuple in $\mathbb{H}$.

The CSP can be formulated in terms of homomorphisms as follows. Given a pair $(\mathbb{G}, \mathbb{H})$ of (similar) relational structures, decide whether there is a homomorphism from the first structure to the second structure. A common way to restrict this problem is to fix the second structure $\mathbb{H}$ so that each structure $\mathbb{H}$ gives rise to a problem $\mathrm{CSP}(\mathbb{H})$. The most effective approach to the study of the $\mathrm{CSP}(\mathbb{H})$ is the so-called algebraic approach that associates every $\mathbb{H}$ with its *polymorphisms*.

A polymorphism of a structure $\mathbb{H}$ is defined as a finite operation $f : V^k \to V$ that is a homomorphism from $\mathbb{H}^k$ to $\mathbb{H}$. That is for every k tuples $\tau_1, \ldots, \tau_k$ from relation R_i (of arity r_i), we have $(x_1, x_2, \ldots, x_{r_i}) \in R_i$ such that $x_j, 1 \le j \le r_i$ is of form $x_j = f(\tau_1[j], \tau_2[j], \ldots, \tau_k[j])$ where $\tau_t[j], 1 \le t \le k$ is the j-element of τ_t. A polymorphism f is *conservative* if each value $f(x_1, x_2, \ldots, x_k)$ is one of the arguments $x_1, x_2, \ldots, x_k$. A binary (arity two) polymorphism $f : V^2 \to V$ that is conservative and *commutative* ($f(x, y) = f(y, x)$ for all vertices x, y) is called a CC polymorphism. Notice that by definition any binary CC polymorphism is *idempotent* i.e., $f(x, x) = x$. If f is additionally *associative* then it is called a *conservative semilattice* or a CSL polymorphism. That is, it satisfies the following *identities*, $f(f(x, y), z) = f(x, f(y, z))$, and $f(x, y) = f(y, x) \in \{x, y\}$ for all $x, y, z \in V$.

Roughly speaking, the presence of nice enough polymorphisms leads directly to the polynomial time tractability of $\mathrm{CSP}(\mathbb{H})$, while their absence leads to hardness [4, 28]. Besides decision CSPs, polymorphisms have been used extensively for approximating CSPs, robust satisfiability of CSPs, testing solutions, and the study of the Ideal Membership Problems [5, 23, 26].

An interesting question arising from these studies, is known as the *meta-question*. Given a relational structure $\mathbb{H}$, decide whether or not $\mathbb{H}$ admits a polymorphism from a class–for various classes of polymorphisms. In many cases, hardness results are known [7]. One particular case, that is, the study of this paper, is deciding whether or not $\mathbb{H}$ admits a CSL polymorphism. The presence of semilattice polymorphisms leads to many positive results. As an example, it is now a classic theorem in the area that for any structure $\mathbb{H}$ having a semilattice polymorphism, the problem $\mathrm{CSP}(\mathbb{H})$ is polynomial time decidable [22]. In terms of approximation algorithms, the Minimum Cost Homomorphism problem to $\mathbb{H}$ (when $\mathbb{H}$ is a digraph) is approximable within a constant factor if $\mathbb{H}$ admits a CSL polymorphism [18, 26]. In terms of robust satisfiability, given a $(1 - \varepsilon)$-satisfiable instance of $\mathrm{CSP}(\mathbb{H})$, it is easy to find a $(1 - O(1/\log(1/\varepsilon)))$-satisfying assignment if $\mathbb{H}$ admits a semilattice polymorphism (in fact, the result holds for width-1 CSPs). However, on the negative side, there are instances where $\mathbb{H}$ admits a semilattice polymorphism and it is hard to find a $(1 - o(1/\log(1/\varepsilon)))$-satisfying assignment [23].

For a single binary relation, i.e., a digraph, the meta-question often turns out to be better behaved. For instance, there are forbidden induced structure characterizations for the existence of conservative *majority* [20] and conservative *Maltsev* [6,20] polymorphisms in digraphs. The question of whether the existence of conservative semilattice polymorphism is polynomial was explicitly raised in [1,21]. This problem is polynomial for reflexive digraphs [11] and bipartite digraphs [18]. In this paper, we give a forbidden obstruction characterization for digraphs admitting a conservative semilattice polymorphism. Observe that if a digraph H admits a CSL polymorphism then the CSL polymorphism naturally defines an ordering on the vertices of H. It turns out that a digraph admits a CSL polymorphism if and only if it has a min ordering. Other questions about the existence of polymorphisms of various kinds have also turned out to be interesting [3,12,20,25]. In particular, the existence of conservative polymorphisms is a hereditary property (if H has a particular kind of conservative polymorphism, then so does any induced subgraph of H). Thus, these questions present interesting problems in graph theory.

1.3   Our Contribution

In this paper, we study the problem of deciding if a relational structure $\mathbb{H}$ admits a conservative semilattice (CSL) polymorphism. That is, we study for which relational structures PROBLEM 1 is polynomial-time decidable and for which ones it is NP-complete.

> *Problem 1.*
> *Input:* A relational structure $\mathbb{H} = \langle V, R_1, \ldots, R_s \rangle$,
> *Goal:* Decide if $\mathbb{H}$ admits a conservative semilattice (CSL) polymorphism.

Note that any unary relation R admits a CSL polymorphism. This is because if $a, b \in R$, then applying CSL polymorphism f on a, b, would give either a or b, and hence, R is closed under f. So the interesting cases are when the arity of R is at least two. On the positive side, we present a polynomial time algorithm that, given a relational structure with a single binary relation $\mathbb{H} = \langle V, A(V) \rangle$ i.e., digraph, decides if $\mathbb{H}$ admits a CSL polymorphism.

Theorem 1 (Main Theorem). *There exists a polynomial time algorithm that, given a digraph H, decides if H admits a CSL polymorphism or not.*

We also have a structural characterization of digraphs with a CSL polymorphism, in terms of a forbidden structure we call a *strong circuit*. Recall that the class of digraphs that admit a CSL polymorphism is exactly the class of digraphs admitting a min ordering (also called bi-arc digraphs).

The class of digraphs admitting a min ordering coincides with the class of signed-interval digraphs. We therefore have the following corollary.

Corollary 1. *The class of min-orderable digraphs, bi-arc digraphs and signed-interval digraphs can be recognized in polynomial time.*

Furthermore, we show that there is quite a bit of collapse for digraph classes in the conservative case. We will point out that the class of digraphs with a min ordering is included in the class of digraphs with a *conservative set* polymorphism, which is included in the class of digraphs with a *conservative and commutative* polymorphism (called CC polymorphism).

Formally, we prove the following:

Theorem 2. *Let $\mathbb{H}$ be a digraph, then $\mathbb{H}$ admits a CSL polymorphism if and only if $\mathbb{H}$ admits a conservative set polymorphism if and only if $\mathbb{H}$ admits conservative cyclic polymorphisms of all arities.*

On the negative side, we prove that it is NP-complete to decide if a relational structure $\mathbb{H} = \langle V, R \rangle$ where R is a ternary relation (arity of R is three) admits a CSL polymorphism.

Theorem 3. *Deciding if a relational structure with a single ternary relation admits a CSL polymorphism is NP-complete.*

Moreover, we prove PROBLEM 1 remains NP-complete even for two binary relations i.e., two digraphs. This leads us to the following dichotomy classification of the complexity of PROBLEM 1.

Theorem 4 (Dichotomy Theorem). *Deciding if a relational structure $\mathbb{H} = \langle V, R_1, \ldots, R_k \rangle$ admits a CSL polymorphism is polynomial-time solvable if all relations R_i are unary, except possibly one binary relation. In all other cases, the problem is NP-complete.*

2 Bi-arc Digraphs and Min-orderable Digraphs

A *digraph* H consists of a finite vertex set $V(H)$ and an arc set $A(H)$, each arc being an ordered pair of vertices. We say that $uv \in A(H)$ is an arc from u to v. Sometimes we emphasize this by saying that uv is a *forward* arc of H, and also say vu is a *backward* arc of H. We say that u, v are *adjacent* in H if uv is a forward or a backward arc of H (either $uv \in A(H)$ or $vu \in A(H)$). A *symmetric* arc is an arc $uv \in A(H)$ such that $vu \in A(H)$; thus, a symmetric arc is both a forward arc and a backward arc.

A *graph* H is a symmetric digraph (the binary relation $A(H)$ is symmetric), where we identify each pair of opposite arcs ab, ba into one edge $ab = ba$.

Let C be a circle with two distinguished points N and S. A *bi-arc* is a pair of arcs I, J on C such that I contains N but not S and J contains S but not N. The following definition unites many disparate geometric representations, although we know little about the corresponding class of digraphs.

A *weak bi-arc representation* of a digraph H is a family of bi-arcs $I_v, J_v, v \in V(H)$, such that $ab \in A(H)$ if and only if I_a and J_b are disjoint. A digraph H is a *weak bi-arc digraph* if it admits a weak bi-arc representation.

As mentioned above, we do not know which digraphs admit a weak bi-arc representation and believe they may be interesting. However, several well-studied graph and digraph classes are characterized by the existence of special kinds of weak bi-arc representations. A weak bi-arc representation is *consistent* if the clockwise end of I_a precedes, in the clockwise order on C, the clockwise end of I_b if and only if the clockwise end of J_a precedes (in the clockwise order) the clockwise end of J_b. A consistent weak bi-arc representation will be called simply a *bi-arc representation*, and a digraph admitting a bi-arc representation will be called a *bi-arc digraph*.

It turns out bi-arc digraphs are precisely the digraphs that admit a min ordering [10,17] (see Fig. 1, for min ordering definition see page 2). We add further statements, namely, we prove the following theorem.

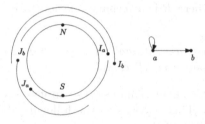

Fig. 1. A min ordering $a < b$ for digraph H and its bi-arc representation

Theorem 5. *Let H be a bi-arc digraph. Then H admits a conservative semi-lattice polymorphism, admits cyclic polymorphisms of all arities, and admits a conservative set polymorphism.*

A *set polymorphism* of H is a mapping f of the non-empty subsets of $V(H)$ to $V(H)$, such that $f(S)f(T) \in A(H)$ whenever S, T are non-empty subsets of $V(H)$ with the property that for each $s \in S$ there is a $t \in T$ with $st \in A(H)$ and also for every $t \in T$ there is an $s \in S$ with $st \in A(H)$. It is easy to see, cf. [13], that H has a conservative set polymorphism if and only if it has conservative totally symmetric (CTS) polymorphisms of all arities k. A polymorphism f of arity k on digraph H is called *cyclic* if $f(x_1, x_2, \ldots, x_k) = f(x_2, x_3, \ldots, x_k, x_1)$ for all $x_1, x_2, \ldots, x_k \in V(H)$.

3   Obstructions to Min Ordering

A *walk* in H is a sequence $P = x_0, x_1, \ldots, x_n$ of consecutively adjacent vertices of H; note that a walk has a designated first and last vertex. A *path* $P = x_0, x_1, \ldots,$

x_n is a walk in which all x_i are distinct. A walk $P = x_0, x_1, \ldots, x_n$ is *closed* if $x_0 = x_n$ and a *cycle* if all other x_i are distinct. A walk is *directed* if all its arcs are forward. We define two walks $P = x_0, x_1, \ldots, x_n$ and $Q = y_0, y_1, \ldots, y_n$ in H to be *congruent*, if they follow the same pattern of forward and backward arcs, i.e., $x_i x_{i+1}$ is a forward (backward) arc if and only if $y_i y_{i+1}$ is a forward (backward) arc (respectively). Suppose the walks P and Q as above are congruent. We say an arc $x_i y_{i+1}$ is *a faithful arc from P to Q*, if it is a forward (backward) arc when $x_i x_{i+1}$ is a forward (backward) arc (respectively), and we say an arc $y_i x_{i+1}$ is *a faithful arc from Q to P*, if it is a forward (backward) arc when $x_i x_{i+1}$ is a forward (backward) arc (respectively). We say that P *avoids* Q if there is no faithful arc from P to Q at all. We now introduce a basic tool for this paper.

Definition 1 (The pair digraph H^+). *The vertices of H^+ are all ordered pairs (x, y) of distinct vertices of H. There is an arc from pair (x, y) to pair (x', y') if and only if*

1. $xx', yy' \in A(H)$ but $xy' \notin A(H)$, or
2. $x'x, y'y \in A(H)$ but $y'x \notin A(H)$.

To avoid confusion with the vertices of H, we will refer to the vertices of H^+ as *pairs*. Arcs $(x, y)(x', y') \in A(H^+)$ arising from case (1) are called *positive arcs*, and those arising from case (2) are called *negative arcs*.

Note that in H^+ we have an arc from (x, y) to (x', y') if and only if there is an arc from (y', x') to (y, x). We call this the *skew property* of H^+, and call the pair (y, x) the *dual* of the pair (x, y). From the skew property, $(x, y)(x', y')$ is a positive arc in H^+ if and only if $(y', x')(y, x)$ is a negative arc. Note that when $(x, y)(x', y')$ is an arc of H^+ then in any min ordering $<$ of H, if $x < y$ then $x' < y'$. More generally, if there is a directed path from (x, y) to (x', y') in H^+, then in any min ordering $<$ (of H) having $x < y$ implies that $x' < y'$.

Definition 2 (Circuit, Strong Circuit). *Let D be a subset of $V(H^+)$. A circuit in D is a set of pairs $(x_0, x_1), (x_1, x_2), \ldots, (x_{n-1}, x_n), (x_n, x_0)$ in D. A strong circuit of H^+ is a circuit in C, where C is a strongly connected component (in short, strong component) of H^+. When $n = 1$, and $(x_0, x_1), (x_1, x_0)$ form a strong circuit, then (x_0, x_1) is called an* invertible *pair.*

Thus, in a strong circuit, there are directed paths (in H^+) from (x_{i-1}, x_i) to (x_i, x_{i+1}) for all $i = 1, 2, \ldots, n + 1$, modulo $n + 1$. If H^+ contains a strong circuit, then H cannot have a min ordering, since $x_0 < x_1$ implies $x_0 < x_1 < x_2 < \cdots < x_n < x_0$ (and similarly for $x_0 > x_1$) contradicting the transitivity of $<$. We have proved that if a digraph H admits a min ordering, then H^+ does not contain a strong circuit. It turns out that the converse also holds.

Theorem 6. *A digraph H admits a min ordering if and only if H^+ does not contain a strong circuit.*

This is our main result, giving a polynomially testable characterization of min-orderable digraphs. We provide an algorithm that outputs a min ordering when H^+ does not have a strong circuit to prove the theorem. We can also use the algorithm to find a min ordering if one exists by pre-processing the input digraph to check for strong circuits. The time complexity of our algorithm is $O(|A(H)|^2)$. Detecting a strong circuit in H^+ amounts to testing, for each strong connected component C of H^+, the acyclicity of a digraph on $V(H)$ whose arcs are the pairs in C.

From now on, we write strong component for strongly connected component.

4   The Algorithm

In this section, we introduce an algorithm to construct a min ordering $<$ of H, provided H^+ contains no strong circuit. We first give the necessary definitions and terminology in the following subsection and provide the algorithm's descriptions in the subsequent subsection.

4.1   Necessary Definitions

Paths and walks. A vertex u' is said to be *reachable* from a vertex u in H if there is a directed path from u to u' in H; a set U' is *reachable* from a set U if every vertex of U' is reachable from some vertex of U. Note that every vertex is reachable from itself by a directed path of length zero. A path can also be a graph on its own, consisting of all the vertices and arcs needed for the definition. We note that our terms *path* and *walk* correspond to what is sometimes called *oriented path* and *oriented walk*.

Net Length and (Un)balanced Digraphs. The *net length* of a walk is the number of forward arcs minus the number of backward arcs. A closed walk is *balanced* if it has net length zero; otherwise, it is *unbalanced*. Note that in an unbalanced closed walk, we may always choose a direction in which the net length is positive (or negative). A digraph is *unbalanced* if it contains an unbalanced closed walk (or, equivalently, an unbalanced cycle); otherwise, it is *balanced*. It is easy to see that a digraph is balanced if and only if it admits a *labeling* of vertices by non-negative integers so that each arc goes from a vertex with a label i to a vertex with a label $i + 1$.

We now focus on properties of H^+. Reachability in H^+ is defined in the usual way by the existence of directed paths in H^+. We use the following notation.

Definition 3 (Reachability Notation). *We write* $(u, v) \rightsquigarrow (u', v')$ *in* H^+ *if* (u', v') *is reachable from* (u, v) *in* H^+, *and, otherwise, we write* $(u, v) \not\rightsquigarrow (u', v')$ *in* H^+.

Definition 4 (Closure of S). *Suppose* $S \subseteq V(H^+)$. *The* closure *of* S, *denoted by* $\widehat{S}$, *is the set of all pairs in* H^+ *that are reachable from* S *in* H^+.

Note that $\widehat{S}$ contains S. We say S is *closed under reachability* if $\widehat{S} = S$.

Net value of a path in H^+. In H^+ when we mention a path, we mean a directed path. A (directed) path W in H^+ corresponds precisely to a pair of congruent walks P, Q in H such that P avoids Q. We occasionally write $W = (P, Q)$ and also denote the path W from (x, y) to (u, v) by $W : (x, y) \rightsquigarrow (u, v)$. The *net value* of the path W is defined to be the net length of the walk P (or equivalently the net length of Q). It is the difference between the number of positive and negative arcs of W. Walk W is called symmetric if P and Q avoid each other.

(Un)balanced components in H^+. A closed walk of H^+ is *balanced* if has net value zero, and *unbalanced* otherwise. A strong component of H^+ is *balanced* if it does not contain an unbalanced closed walk, and *unbalanced* otherwise. A strong component S of H^+, is balanced if every directed cycle of S has net value zero. Finally, a pair is called *balanced* if it is in a balanced strong component otherwise, it is called *unbalanced*.

4.2   Description of the Algorithm

We will be choosing pairs of H^+ to decide the ordering. Specifically, if a pair (x, y) of H^+ is chosen, we will set $x < y$. Note that choosing a pair requires choosing all pairs reachable from it. The process of choosing is different for pairs with balanced and unbalanced strong components. However, the chosen pairs will be closed under reachability in each case.

Then all the duals of the chosen pairs will be discarded. At any stage of the algorithm, we will have a set V_c of *chosen pairs*, and a set V_d of *discarded pairs*; the pairs in the set $\mathcal{R} = V(H^+) \setminus (V_c \cup V_d)$ will be called the *remaining pairs*. Initially, we will have $V_c = V_d = \emptyset$, and throughout the algorithm, we will maintain the following properties:

1. $(a, b) \in V_c$ if and only if $(b, a) \in V_d$;
2. if $(a, b) \in V_c$ and $(a, b)(a', b') \in A(H^+)$ then $(a', b') \in V_c$;
3. V_c does not contain a circuit.

Note that we will always have $V_c \cap V_d = \emptyset$, and each strong component of H^+ lies entirely in one of the three sets $V_c, V_d, \mathcal{R}$. Moreover, at the end of the algorithm, the set $\mathcal{R}$ will be empty; this ensures that $<$ is a total ordering. Therefore, property (3) will then imply the following transitivity on the chosen pairs:

– if $(a, b) \in V_c$ and $(b, c) \in V_c$ then $(a, c) \in V_c$.

This fact, together with property (2), ensures that the chosen pairs do define a min ordering, by setting $x < y$ for all chosen pairs (x, y).

Algorithm 1 has two phases.

PHASE ONE: In the first phase we reduce the problem to a balanced subdigraph $H^\#$ of H^+. We accomplish this by dealing with all the unbalanced strong components of H^+ first.

At each step, we consider an unbalanced strong component $C \not\subseteq (V_c \cup V_d)$ and its dual component C'. In Theorem 8, we prove that if $\widehat{C} \cup V_c$ contains a circuit, then $\widehat{C'} \cup V_c$ does not contain a circuit. Therefore, if $(\widehat{C} \cup V_c)$ does not contain a circuit, then we add $\widehat{C}$ into V_c and add the dual pair of $\widehat{C}$ into V_d, update $\mathcal{R}$, and proceed to the next unbalanced strong component. Otherwise, we remove C from further consideration and add $\widehat{C'}$ into V_c and update V_d and $\mathcal{R}$ accordingly.

PHASE TWO: For the balanced strong components we need a different strategy because of the different structural properties of balanced and unbalanced strong components. In particular, unbalanced strong components have walks of *unbounded* net value.

Now consider the induced sub-digraph $H^\#$ of H^+ consisting of all pairs in the balanced strong components of H^+. Thus, $H^\#$ is itself balanced. (This is true, since each closed walk lies in a strong component of $H^\#$; recall that in H^+ balance refers to the equality of the number of positive and negative arcs in each closed walk.)

We partition the vertices of $H^\#$ into *layers* as follows. Consider an auxiliary digraph D with $V(D) = V(H^\#)$ and $(a,b)(c,d) \in A(D)$ if and only if (c,d) is reachable from (a,b) by a path in $H^\#$ with negative net value. Since all directed cycles in $H^\#$ are balanced, D is acyclic. Layer 0 of $H^\#$, denoted by $\mathcal{L}_0$, consists of all vertices that have out-degree zero in D. Having defined layers $\mathcal{L}_1, \mathcal{L}_2, \ldots, \mathcal{L}_j$, layer $\mathcal{L}_{j+1}$ of $H^\#$ consists of all vertices of out-degree zero in the digraph obtained from D by removing all the vertices in layers $\mathcal{L}_1, \mathcal{L}_2, \ldots, \mathcal{L}_j$.

We handle the pairs in $\mathcal{L}_0, \mathcal{L}_1, \ldots$, consecutively, one at a time.

To proceed with the current layer $\mathcal{L}_k, k \geq 0$, we seek a vertex $p \in V(H)$ such that there exists **no** $q' \in V(H)$ so that $(q', p) \in V_c \cap \mathcal{L}_k$, and

– there exists a q such that $(p, q) \in \mathcal{R} \cap \mathcal{L}_k$ and $(p, q) \not\rightsquigarrow (q, p)$,

The existence of such p is justified in Lemma 1. For each choice of p, as long as there exists some pair $(p, r) \in \mathcal{R} \cap \mathcal{L}_k$ so that $(p, r) \not\rightsquigarrow (r, p)$ we add (p, r) into V_c. (This process can start with r being the vertex q from above and then continue as long as further r can be found.) We now define the transitivity-reachability (TR) closure of V_c as follows.

Definition 5 (Transitivity+Reachability (TR) Closure). *The transitivity+reachability closure of V_c, $Tr(V_c)$, is the smallest set of pairs containing V_c that is closed under reachability and transitivity. In other words, if $(x, y) \in Tr(V_c)$, and $(x, y) \rightsquigarrow (x', y')$ then $(x', y') \in Tr(V_c)$. Moreover, if $(x, y), (y, z) \in Tr(V_c)$ then $(x, z) \in Tr(V_c)$.*

Note that $V_c \subseteq Tr(V_c)$. Then we update the set V_c to be $Tr(V_c)$. Of course, we also update $\mathcal{R}$ by removing all the dual pairs of V_c from $\mathcal{R}$, and all the pairs of V_c from $\mathcal{R}$. Note that during the computation of $Tr(V_c)$ we may add $(q', p) \in \mathcal{L}_k$ into $Tr(V_c)$ and p no longer satisfies the condition that there exists no $(q', p) \in V_c \cap \mathcal{L}_k$. Lemma 2 of Sect. 5 shows $Tr(V_c)$ doesn't contain a circuit.

Once we are done with p, we look for another vertex p_1 on layer $\mathcal{L}_k$ satisfying the aforementioned conditions and repeat. Once we finish processing all the pairs in $\mathcal{L}_k \cap \mathcal{R}$, we go on to the next layer and consider the remaining pairs in $\mathcal{L}_{k+1}$. The details are provided in Algorithm 1.

Algorithm 1. Algorithm to find a min ordering of input digraph H

1: **function** MINORDERING(H)   ▷ PHASE ONE: Handling unbalanced components
2:     Construct H^+ and compute its strong components
3:     **if** H^+ contains a strong circuit **then return** False
4:     Set $V_c = V_d = \emptyset$ and let $\mathcal{R} = V(H^+)$
5:     **while** $\mathcal{R}$ contains an unbalanced strong component C **do**
6:         **if** $\widehat{C} \cup V_c$ has no circuit **then**
7:             Add $\widehat{C}$ into V_c, and add all the dual pairs of $\widehat{C}$ into V_d.
8:             Remove from $\mathcal{R}$ all the pairs in $\widehat{C}$ and their dual pairs.
9:         **else** ($\widehat{C'} \cup V_c$ has no circuit)
10:             Add $\widehat{C'}$ into V_c, and add all the dual pairs of $\widehat{C'}$ into V_d.
11:             Remove from $\mathcal{R}$ all the pairs in $\widehat{C'}$ and their dual pairs.
                    ▷ PHASE TWO: Handling the remaining balanced components
12:     Let $H^\#$ be the set of all balanced pairs, and let $\mathcal{R} = V(H^\#) \setminus V_c$
13:     Compute the layers of $H^\#$; $\mathcal{L}_0, \mathcal{L}_1, \ldots$, and set $k = 0$
14:     **while** $\mathcal{R} \neq \emptyset$ **do**
15:         **while** $\mathcal{R} \cap \mathcal{L}_k \neq \emptyset$ **do**
16:             Find $p \in V(H)$ s.t. no $(q', p) \in V_c \cap \mathcal{L}_k$ and $\exists (p, q) \in \mathcal{R} \cap \mathcal{L}_k$ with
                  $(p, q) \not\rightarrow (q, p)$
17:             **while** $\exists (p, r) \in \mathcal{R} \cap \mathcal{L}_k$ s.t. $(p, r) \not\rightarrow (r, p)$ **do**
                  ▷ at least one (p, r) exists, i.e. $r = q$ in line 16, and empty while loop avoided
18:                 Add (p, r) into V_c and set $V_c = Tr(V_c)$
19:                 Remove all the dual pairs of V_c from $\mathcal{R}$, and add them into V_d.
20:                 Set $\mathcal{R} = \mathcal{R} \setminus V_c$.
21:             Increase k by one
            return V_c

---

5   Correctness

To justify the correctness of PHASE ONE we first define the concept of a dual-free set and a minimal circuit. A subset of H^+ is called *dual-free* if it does not contain a pair and its dual.

Definition 6 (Minimal Circuit). *Suppose $S_0, S_1, \ldots, S_n$ (not necessarily distinct) are strong components in $T \subseteq V(H^+)$ where $\widehat{T}$ is dual-free. Let C : $(a_0, a_1), (a_1, a_2), \ldots, (a_n, a_0)$ be a circuit where $(a_i, a_{i+1}) \in \widehat{S_i}, 0 \leq i \leq n$. We say C is minimal if there is no other circuit $(a'_0, a'_1), (a'_1, a'_2), \ldots, (a'_m, a'_0), m < n$, where each (a'_j, a'_{j+1}) belongs to some $\widehat{S_i}, 0 \leq i \leq n$.*

We need some technical definition to state and prove Theorem 8. For walks P from a to b, and Q from b to c, we denote by $P + Q$ the walk from a to c which is the concatenation of P and Q. We denote by P^{-1} the walk P traversed in the opposite direction, from b to a; we call P^{-1} the *reverse* of P. Notice that if walk P avoids walk Q then Q^{-1} avoids P^{-1}.

For a closed walk C, we denote by C^a the concatenation of C with itself a times. The *height* of H is the maximum net length of a walk in H. Note that an unbalanced digraph has infinite height, and the height of a balanced digraph is the greatest label in non-negative labeling in which some vertex has label zero.

For a walk $P = x_0, x_1, \ldots, x_n$ and any $i \leq j$, we denote by $P[x_i, x_j]$ the walk $x_i, x_{i+1}, \ldots, x_j$. We call $P[x_i, x_j]$ a *prefix* of P if $i = 0$. Suppose $P = x_0, x_1, \ldots, x_n$ is a walk in H of net length $k \geq 0$. We say that P is *constricted from below* if the net length of any prefix $P[x_0, x_j]$ is non-negative and is *constricted from above* if the net length of any prefix is at most k. We also say that P is *constricted* if it is constricted both from below and from above. Moreover, we say that P is *strongly* constricted from below or above if the corresponding net lengths are strictly positive or smaller than k. For a walk P of net length $k < 0$, we say that P is (strongly or not) constricted from below, or above, or both if the above definitions apply to the reverse walk P^{-1}.

Definition 7 (Extremal Vertex). *Consider a cycle C in H of positive net length k. A vertex v is* extremal *in C if traversing C, in the positive direction, from v yields a walk constricted from below.*

A cycle of H is *induced* if H contains no other arcs on the vertices of the cycle. An induced cycle with more than one vertex does not contain a loop.

Let $W = (P, Q)$ be a path in H^+. We say W is *constricted* if the walk P (or Q) is constricted, i.e., if each prefix of W has a net value between zero and the net value of W. Paths (in H^+) constricted from below or above are defined similarly. Other notions for H^+ are also defined in the manner corresponding to the notions in H. Consider, for instance, the above notion of an extremal vertex. We define *extremal pair* of a cycle C in H^+ as a pair $\bar{v}$ such that traversing C from $\bar{v}$ in the positive direction yields a walk with values constricted from below.

The correctness of the first phase relies on the following technical theorem.

Theorem 7. *Let T be the vertices of a set of unbalanced strong components (in H^+) where $\widehat{T}$ is dual-free, and assume that $\widehat{T}$ contains a minimal circuit $C : (a_0, a_1), (a_1, a_2), \ldots, (a_n, a_0)$. Then $n > 1$ and the following statements hold.*

1. *There exists some minimal circuit (Definition 6) with extremal pairs $(b_0, b_1), (b_1, b_2), \ldots, (b_{n-1}, b_n), (b_n, b_0)$ in T such that $(a_i, a_{i+1}), (b_i, b_{i+1}), 0 \leq i \leq n$, are in the same strong component, and (a_i, a_{i+1}) is reachable from (b_i, b_{i+1}) by a symmetric walk of non-negative net value, and constricted from below.*
2. *For each i, $0 \leq i \leq n$, there exists an infinite walk P_i that starts from b_i and has unbounded positive net length. P_i is obtained by winding around a cycle in H containing b_i. Furthermore, for every i, j, $0 \leq i < j \leq n$, P_i and P_j avoid each other[1].*
3. *In statement 1, for a given $0 \leq i \leq n$, we can choose (b_i, b_{i+1}) to be any given extremal pair from its corresponding strong component.*
4. *There is no path in H^+ from (b_i, b_{i+1}) to any of (b_j, b_{j+1}) $i \neq j$, and to any of (b_{j+1}, b_j).*
5. *There is no path in H^+ from any of (b_{i+1}, b_i), $0 \leq i \leq n$ to (b_i, b_{i+1}).*

---

[1] When we say two infinite walks P, Q avoid each other it means for every prefix of P there exists a prefix of Q that avoid each other.

Theorem 8. *Suppose* $C \not\subseteq (V_c \cup V_d)$ *is an unbalanced strong component and* V_c *does not contain a circuit. If* $\widehat{C} \cup V_c$ *contains a circuit, then* $\widehat{C'} \cup V_c$ *does not contain a circuit.*

Proof. Since $C \not\subseteq (V_c \cup V_d)$, skew property implies $C' \not\subseteq (V_c \cup V_d)$. Suppose for contradiction that $\widehat{C} \cup V_c$ contains a circuit $(b_0, b_1), (b_1, b_2), \ldots, (b_n, b_0)$

and $\widehat{C'} \cup V_c$ contains a circuit $(d_0, d_1), (d_1, d_2), \ldots, (d_m, d_0)$. We may assume that both are minimal circuits. Notice that Algorithm 1 selects unbalanced strong components one at a time and adds their closure into V_c. Thus, if $\widehat{C} \cup V_c$ contains a circuit, then that circuit would be at $\widehat{T}$ where T is a set of unbalanced strong components in $C \cup V_c$. A similar statement is true for $\widehat{C'} \cup V_c$. Observe that since V_c does not contain a circuit, at least one of the (b_i, b_{i+1}) pairs should be in $\widehat{C}$. The same holds for $\widehat{C'}$, and at least one of the (d_j, d_{j+1}) pairs is in $\widehat{C'}$. Hence, without loss of generality, we assume that $(b_n, b_0) \in \widehat{C}$, and $(d_m, d_0) \in \widehat{C'}$.

We first assume that both $m, n > 1$. Thus, there is no $(p, q) \in C \cup V_c$ so that $(p, q) \rightsquigarrow (q, p)$, as otherwise, we have $(p, q), (q, p) \in \widehat{C} \cup V_c$ which contradicts the minimality assumption and the assumption that $n > 1$. Similarly, there is no $(p', q') \in C' \cup V_c$ so that $(p', q') \rightsquigarrow (q', p') \in \widehat{C'} \cup V_c$. Therefore, $\widehat{C} \cup V_c$, and $\widehat{C'} \cup V_c$ are dual-free. Thus, according to the statement (1) of Theorem 7, we may also assume that all the pairs on these two circuits are extremal pairs in H^+. Moreover, by statement (3) of Theorem 7, we assume that $(b_n, b_0) \in C$ and $(d_m, d_0) \in C'$, i.e., $(d_0, d_m) \in C$, and that $(b_n, b_0) = (d_0, d_m)$.

Moreover, according to statement (4) of Theorem 7, we may assume that (b_n, b_0) is the only pair of the first circuit in C and (d_m, d_0) is the only pair of the second circuit in C'. Now, consider the following circuit (where $(b_{n-1}, b_n) = (b_{n-1}, d_0)$, $(d_{m-1}, d_m) = (d_{m-1}, b_0)$)

$$(b_0, b_1), (b_1, b_2), \ldots, (b_{n-1}, d_0), (d_0, d_1), (d_1, d_2), \ldots, (d_{m-1}, b_0)$$

all pairs of which are in V_c. This contradicts the assumption that V_c has no circuit. In what follows we consider separately the cases when n or m is 1.

Observation 9. *If* $\widehat{C} \cup V_c$ *contains a circuit* $(b_0, b_1), (b_1, b_0)$ *(i.e.,* $n = 1$*) then by definition we have* $C \cup V_c \rightsquigarrow (b_0, b_1)$, *and* $C \cup V_c \rightsquigarrow (b_1, b_0)$. *Now by skew property, we have* $(b_1, b_0) \rightsquigarrow C' \cup V_d$ *and* $(b_0, b_1) \rightsquigarrow C' \cup V_d$. *Therefore,* $C \rightsquigarrow C'$, *and hence, there is also a circuit* $(p, q), (q, p)$ *where* $(p, q) \in C$, *and* $(p, q) \rightsquigarrow (q, p)$ *(it is not possible that,* (p, q) *or* (q, p) *in* V_c *because* $C \cup C' \not\subseteq V_c \cup V_d$*).*

If both circuits have $n = m = 1$ then by the above observation we have $C \rightsquigarrow C'$ and also $C' \rightsquigarrow C$, implying a strong circuit in H^+, a contradiction. Finally, if $n = 1$, but $m > 1$, then the first circuit is $(b_0, b_1), (b_1, b_0)$ and by Observation 9 and skew property we have $(b_0, b_1) \rightsquigarrow C'$ and $C' \rightsquigarrow (b_0, b_1)$. Now again since $m > 1$, by statement (3) of Theorem 7 we may assume that C' contains (b_1, b_0). This means $(b_0, b_1) \rightsquigarrow (b_1, b_0)$ which is in contradiction to statement (5) of Theorem 7 (i.e., reverse of a pair on the circuit does not reach that pair).

The following two lemmas justify the computation in PHASE TWO. (Lemma 1 justifies Line 16, and Lemma 2 justifies Line 17.)

Lemma 1. *Suppose V_c does not contain a circuit, and furthermore, $\mathcal{R} \cap \mathcal{L}_k \neq \emptyset$. Then there exists a vertex $p \in V(H)$ such that there exists **no** $(q', p) \in V_c \cap \mathcal{L}_k$, and the following condition is satisfied:*

– *there exists a q such that $(p, q) \in \mathcal{R} \cap \mathcal{L}_k$ and $(p, q) \not\rightarrow (q, p)$.*

Lemma 2. *Suppose V_c does not contain a circuit, and furthermore, $\mathcal{R} \cap \mathcal{L}_k \neq \emptyset$. Then after executing the entire while loop at line 17, V_c does not contain a circuit. In other words, after adding all the (p, r) pairs on line 17 and computing $Tr(V_c)$ and setting $V_c = Tr(V_c)$, there will not be a circuit in V_c.*

Theorem 10. *Algorithm 1 correctly decides if a digraph H admits a min ordering or not and it correctly outputs a min ordering for H if one exists.*

Proof. The proof follows from Theorem 8, Lemma 1, and Lemma 2.

6   Conclusions

We have provided a polynomial-time algorithm, and an obstruction characterization of digraphs that admit a min ordering, i.e., a CSL polymorphism. We believe they are a useful generalization of interval graphs, encompassing adjusted interval digraphs, monotone proper interval digraphs, complements of circular arc graphs of clique covering number two, two-directional orthogonal ray graphs, and other well-known classes. We also study this problem beyond digraphs, and consider the general case of relational structures. We fully classify the polynomial-time cases (see Theorem 4). Due to space limit, this part is presented in the full version [19].

References

1. Bagan, G., Durand, A., Filiot, E., Gauwin, O.: Efficient enumeration for conjunctive queries over x-underbar structures. In: Dawar, A., Veith, H. (eds.) CSL 2010. LNCS, vol. 6247, pp. 80–94. Springer, Heidelberg (2010). https://doi.org/10.1007/978-3-642-15205-4_10
2. Booth, K.S., Lueker, G.S.: Testing for the consecutive ones property, interval graphs, and graph planarity using PQ-tree algorithms. J. Comput. Syst. Sci. **13**(3), 335–379 (1976)
3. Brewster, R.C., Feder, T., Hell, P., Huang, J., MacGillivray, G.: Near-unanimity functions and varieties of reflexive graphs. SIAM J. Discret. Math. **22**(3), 938–960 (2008)
4. Bulatov, A.A.: A dichotomy theorem for nonuniform CSPs. In: FOCS, pp. 319–330. IEEE (2017)
5. Bulatov, A.A., Rafiey, A.: On the complexity of csp-based ideal membership problems. In: STOC, pp. 436–449. ACM (2022)
6. Carvalho, C., Egri, L., Jackson, M., Niven, T.: On Maltsev digraphs. Electron. J. Comb. **16**, 1–21 (2015)
7. Chen, H., Larose, B.: Asking the metaquestions in constraint tractability. ACM Trans. Comput. Theor. (TOCT) **9**(3), 11 (2017)

8. Corneil, D.G., Olariu, S., Stewart, L.: The LBFS structure and recognition of interval graphs. SIAM J. Discret. Math. **23**(4), 1905–1953 (2009)
9. Das, S., Sen, M., Roy, A.B., West, D.B.: Interval digraphs: An analogue of interval graphs. J. Graph Theor. **13**(2), 189–202 (1989)
10. Feder, T., Hell, P., Huang, J.: Bi-arc graphs and the complexity of list homomorphisms. J. Graph Theor. **42**(1), 61–80 (2003)
11. Feder, T., Hell, P., Huang, J., Rafiey, A.: Interval graphs, adjusted interval digraphs, and reflexive list homomorphisms. Discret. Appl. Math. **160**(6), 697–707 (2012)
12. Feder, T., Hell, P., Loten, C., Siggers, M., Tardif, C.: Graphs admitting k-nu operations. Part 1: the reflexive case. SIAM J. Discret. Math. **27**(4), 1940–1963 (2013)
13. Feder, T., Vardi, M.Y.: Monotone monadic SNP and constraint satisfaction. In: STOC, pp. 612–622 (1993)
14. Fulkerson, D., Gross, O.: Incidence matrices and interval graphs. Pac. J. Math. **15**(3), 835–855 (1965)
15. Golumbic, M.C.: Algorithmic Graph Theory and Perfect Graphs, vol. 57. Elsevier (2004)
16. Hell, P., Huang, J.: Interval bigraphs and circular arc graphs. J. Graph Theor. **46**(4), 313–327 (2004)
17. Hell, P., Huang, J., McConnell, R.M., Rafiey, A.: Min-orderable digraphs. SIAM J. Discret. Math. **34**(3), 1710–1724 (2020)
18. Hell, P., Mastrolilli, M., Nevisi, M.M., Rafiey, A.: Approximation of minimum cost homomorphisms. In: Epstein, L., Ferragina, P. (eds.) ESA 2012. LNCS, vol. 7501, pp. 587–598. Springer, Heidelberg (2012). https://doi.org/10.1007/978-3-642-33090-2_51
19. Hell, P., Rafiey, A., Rafiey, A.: Bi-arc digraphs and conservative polymorphisms. arXiv preprint arXiv:1608.03368 (2016)
20. Hell, P., Rafiey, A.: The dichotomy of list homomorphisms for digraphs. In: SODA, pp. 1703–1713 (2011)
21. Hell, P., Rafiey, A.: Monotone proper interval digraphs and min-max orderings. SIAM J. Discret. Math. **26**(4), 1576–1596 (2012)
22. Jeavons, P., Cohen, D.A., Gyssens, M.: Closure properties of constraints. J. ACM **44**(4), 527–548 (1997)
23. Kun, G., O'Donnell, R., Tamaki, S., Yoshida, Y., Zhou, Y.: Linear programming, width-1 CSPs, and robust satisfaction. In: ITCS, pp. 484–495 (2012)
24. Lekkeikerker, C., Boland, J.: Representation of a finite graph by a set of intervals on the real line. Fundam. Math. **51**(1), 45–64 (1962)
25. Maróti, M., McKenzie, R.: Existence theorems for weakly symmetric operations. Algebra Universalis **59**(3), 463–489 (2008)
26. Rafiey, A., Rafiey, A., Santos, T.: Toward a dichotomy for approximation of h-coloring. In: ICALP, pp. 91:1–91:16 (2019)
27. Shrestha, A.M.S., Tayu, S., Ueno, S.: On orthogonal ray graphs. Discret. Appl. Math. **158**(15), 1650–1659 (2010)
28. Zhuk, D.: A proof of CSP dichotomy conjecture. In: FOCS, pp. 331–342. IEEE (2017)

Pebbling in Kneser Graphs

Matheus Adauto[1,3](✉)[ID], Viktoriya Bardenova[3][ID], Mariana da Cruz[1][ID], Celina de Figueiredo[1][ID], Glenn Hurlbert[3][ID], and Diana Sasaki[2][ID]

[1] Programa de Engenharia de Sistemas e Computação, Universidade Federal do Rio de Janeiro, Rio de Janeiro, Brazil
{adauto,celina,mmartins}@cos.ufrj.br
[2] Instituto de Matemática e Estatística, Universidade do Estado do Rio de Janeiro, Rio de Janeiro, Brazil
diana.sasaki@ime.uerj.br
[3] Department of Mathematics and Applied Mathematics, Virginia Commonwealth University, Richmond, VA, USA
{bardenovav,ghurlbert}@vcu.edu

Abstract. Graph pebbling is a game played on graphs with pebbles on their vertices. A pebbling move removes two pebbles from one vertex and places one pebble on an adjacent vertex. The pebbling number $\pi(G)$ is the smallest t so that from any initial configuration of t pebbles it is possible, after a sequence of pebbling moves, to place a pebble on any given target vertex. We consider the pebbling number of Kneser graphs, and give positive evidence for the conjecture that every Kneser graph has pebbling number equal to its number of vertices.

Keywords: graph pebbling · Kneser graphs · odd graphs · weight function method

1 Introduction

Graph pebbling is a network model for studying whether or not a given supply of discrete pebbles can satisfy a given demand via pebbling moves. A pebbling move across an edge of a graph takes two pebbles from one endpoint and places one pebble at the other endpoint; the other pebble is lost in transit as a toll. The pebbling number of a graph is the smallest t such that every supply of t pebbles can satisfy every demand of one pebble by a vertex. The number of vertices is a sharp lower bound, and graphs where the pebbling number equals the number of vertices is a topic of much interest [7,9].

Pebbling numbers of many graphs are known: cliques, trees, cycles, cubes, diameter 2 graphs, graphs of connectivity exponential in its diameter, and others [11]. The pebbling number has also been determined for subclasses of chordal graphs: split graphs [2], semi-2-trees [3], and powers of paths [4], among others. Other well-known families of graphs (e.g. flower snarks [1]) have been investigated; here we continue the study on Kneser graphs. In order to state our main results in Sect. 2, we first introduce graph theoretic definitions, followed by graph pebbling terminology, and then present some context for these results.

J. A. Soto and A. Wiese (Eds.): LATIN 2024, LNCS 14579, pp. 46–60, 2024.
https://doi.org/10.1007/978-3-031-55601-2_4

1.1 General Definitions

In this paper, $G = (V, E)$ is always a simple connected graph. The numbers of vertices and edges of G as well as its diameter, are denoted by $n(G)$, $e(G)$, and $D(G)$, respectively, or simply n, e, and D, when it is clear from the context. For a vertex w and positive integer d, denote by $N_d(w)$ the set of all vertices that are at distance exactly d from w, with $N_d[w] = \cup_{i=0}^{d} N_i(w)$ being the set of all vertices that are at distance at most d from w.

Given two positive integers m and t, the *Kneser graph* $K(m, t)$ is the graph whose vertices represent the t-subsets of $\{1, \ldots, m\}$, with two vertices being adjacent if, and only if, they correspond to disjoint subsets. Thus, $K(m, t)$ has $\binom{m}{t}$ vertices and is regular, with degree $\deg(K(m, t)) = \binom{m-t}{t}$. When $m = 2t$, each vertex is adjacent to just one other vertex and the Kneser graph $K(2t, t)$ is a perfect matching. Therefore we assume that $m \geq 2t + 1$ so that $K(m, t)$ is connected. For $m \geq 1$, $K(m, 1)$ is the complete graph on m vertices, so we assume that $t > 2$. The special case $K(2t + 1, t)$ is known as the *odd graph* O_t; in particular, $O_2 = K(5, 2)$ is the Petersen graph. The odd graphs constitute the sparsest case of connected Kneser graphs. A graph G is k-connected if it contains at least $k + 1$ vertices but does not contain a set of $k - 1$ vertices whose removal disconnects the graph, and the connectivity $\kappa(G)$ is the largest k such that G is k-connected.

Since Kneser graphs are regular and edge-transitive, their vertex connectivity equals their degree $\binom{m-t}{t} \geq t + 1 \geq 3$ (see [13]). The diameter of $K(m, t)$ is given in [15] to be $\lceil \frac{t-1}{m-2t} \rceil + 1$. Notice that this value equals t for $m = 2t + 1$ and equals 2 for $m \geq 3t - 1$.

1.2 Graph Pebbling Definitions

A *configuration* C on a graph G is a function $C : V(G) \to \mathbb{N}$. The value $C(v)$ represents the number of pebbles at vertex v. The size $|C|$ of a configuration C is the total number of pebbles on G. A *pebbling move* consists of removing two pebbles from a vertex and placing one pebble on an adjacent vertex. For a target vertex r, a configuration C is r-*solvable* if one can place a pebble on r after a sequence of pebbling moves, and is r-*unsolvable* otherwise. Also, C is *solvable* if it is r-solvable for all r. The *pebbling number* $\pi(G, r)$ is the minimum number t such that every configuration of size t is r-solvable. The pebbling number of G equals $\pi(G) = \max_r \pi(G, r)$.

The basic lower and upper bounds for every graph are as follows.

Fact 1 ([5, 10]). *For every graph G we have* $\max\{n(G), 2^{D(G)}\} \leq \pi(G) \leq (n(G) - D(G))(2^{D(G)} - 1) + 1$.

A graph is called *Class 0* if $\pi(G) = n(G)$. For example, complete graphs, hypercubes, and the Petersen graph are known to be Class 0 [10].

1.3 Context

The upper bound in Fact 1 is due to the pigeonhole principle. The simplest pigeonhole argument yields an upper bound of $(n(G) - 1)(2^{D(G)} - 1) + 1$: a configuration of this size guarantees that either the target vertex r has a pebble on it or some other vertex has at least $2^{D(G)}$ pebbles on it, which can then move a pebble to r without assistance from pebbles on other vertices. The improvement of Chan and Godbole [5] combines the vertices on a maximum length induced path from r into one "pigeon hole", recognizing that $2^{D(G)}$ pebbles on that path is enough to move one of them to r. Generalizing further, one can take any spanning tree T of G and realize that the same pigeonhole argument yields the upper bound $|L(T)|(2^{D(G)} - 1) + 1$, where $L(T)$ is the set of leaves of T. Then Chung [6] found that the paths from the leaves to r, which typically overlap, could instead be shortened in a special way so as to partition the edges of T, thereby decreasing the exponent of 2 for most of the leaves. (The proof of her result needed double induction, however, rather than the pigeonhole principle.) In short, she defined the *maximum path partition* of T and used it to derive the exact formula for $\pi(T, r)$. We will not need to use this formula here, but we will record the resulting upper bound.

Fact 2. *If T is a spanning tree of G and r is a vertex of G, then $\pi(G, r) \leq \pi(T, r)$.*

Moreover, Fact 2 holds if T is any spanning subgraph of G. However, it is mostly used when T is a tree because we have Chung's formula for trees. In Sect. 3.1 we describe a powerful generalization from [12] that uses many (not necessarily spanning) trees instead of just one, and utilizes linear optimization as well.

2 Results

Here we briefly present known results on the pebbling numbers of Kneser graphs, followed by our new theorems, which we will prove in Sect. 4 after describing the tools used for them in Section 3.

2.1 Historical Contributions

It was proved in [14] that every diameter two graph G has pebbling number at most $n(G) + 1$, and in [7] the authors characterize which diameter two graphs are Class 0. As a corollary they derive the following result.

Theorem 3 ([7]). *If $D(G) = 2$ and $\kappa(G) \geq 3$, then G is Class 0.*

As those authors pointed out, since almost every graph is 3-connected with diameter 2, it follows that almost all graphs are Class 0. Additionally, since $K(m, t)$ is 3-connected with diameter two for every $m \geq 3t - 1$ and $t \geq 2$, one obtains the following corollary.

Corollary 4. *If $t \geq 2$ and $m \geq 3t - 1$, then $K(m, t)$ is Class 0.*

A much better asymptotic result was obtained in [9].

Theorem 5 ([9]). *For any constant c there is a t_0 such that, for all $t \geq t_0$ and $s \geq c(t/\lg t)^{1/2}$ and $m = 2t + s$, we have that $K(m, t)$ is Class 0.*

Based on this evidence, the following was raised as a question in [7], which has since been conjectured in numerous talks on the subject by Hurlbert.

Conjecture 6. *If $m \geq 2t + 1$, then $K(m, t)$ is Class 0.*

2.2 Our Contributions

From Corollary 4 we see that the smallest three open cases for Kneser graphs are $K(7, 3)$, $K(9, 4)$, and $K(10, 4)$. In every case, the lower bound of $K(m, t) \geq \binom{m}{t}$ comes from Fact 1. Conjecture 4 posits that these graphs have pebbling numbers equal to their number of vertices, namely $\binom{7}{3} = 35$, $\binom{9}{4} = 126$, and $\binom{10}{4} = 210$, respectively. Our main results in this paper address the upper bounds for these cases.

Fact 1 delivers upper bounds of 224, 1830, and 1449, respectively. By using breadth-first-search spanning trees, Fact 2, and Chung's tree formula, it is not difficult to derive the improved upper bounds of 54, 225, and 247, respectively. However, our Theorems 7, 8, and 9, below, are significantly stronger. Besides the infinite family $K(m, 2)$ and the Kneser graphs satisfying Theorem 5, Theorem 9 gives further positive evidence to Conjecture 6.

Theorem 7. *For $K(7, 3)$ we have $35 \leq \pi(K(7, 3)) \leq 36$.*

Theorem 8. *For $K(9, 4)$ we have $126 \leq \pi(K(9, 4)) \leq 141$.*

Theorem 9. *For $K(10, 4)$ we have $\pi(K(10, 4)) = 210$; i.e., $K(10, 4)$ is Class 0.*

Additionally, the most obvious infinite family of open cases for Kneser graphs are the odd graphs $K(2t + 1, t)$ for $t \geq 3$. We note that the number of vertices of $G = K(2t + 1, t)$ is $n = \binom{2t+1}{t}$, which Stirling's formula implies is asymptotic to $4^{t+1}/\sqrt{\pi t}$, so that t is roughly (in fact greater than) $(\lg n)/2$. Observe also that for odd graphs, we have $D(G) = t$. Thus Fact 1 yields an upper bound on $\pi(G)$ on the order of $n^{1.5}$. Here we improve this exponent significantly.

Theorem 10. *For any $t \geq 3$, let $n = n(K(2t + 1, t))$ and $\alpha = \log_4((5e)^{2/3}) \approx 1.25$. Then we have $n \leq \pi(K(2t + 1, t)) \leq .045n^\alpha (\lg n)^{\alpha/2} < .045n^{1.26} (\lg n)^{0.63}$.*

We will also prove in Theorem 19 below that a well-known lower bound technique (Lemma 13) will not produce a lower bound for odd graphs that is higher than that of Fact 1.

3 Techniques

3.1 Upper Bound

Here we describe a linear optimization technique invented in [12] to derive upper bounds on the pebbling numbers of graphs.

Let T be a subtree of a graph G rooted at the vertex r, with at least two vertices. For a vertex $v \in V(T)$, a *parent* of v, denoted by v^+, is the unique neighbor of v in T whose distance to r is one less than that of v. Moreover v is called a *child* of v^+). We say that T is an r-*strategy* if we assign to it a non-negative *weight function* w having the properties that $w(r) = 0$ and $w(v^+) \geq 2w(v)$ for every vertex $v \in V(T)$ that is not a neighbor of r. In addition, $w(v) = 0$ for vertices not in T.

Now set $\boldsymbol{T}$ to be the configuration defined by $\boldsymbol{T}(r) = 0$, $\boldsymbol{T}(v) = 1$ for all $v \in V(T) - \{r\}$, and $\boldsymbol{T}(v) = 0$ for all $v \in V(G) - V(T)$. Then the *weight* of any configuration C, including $\boldsymbol{T}$, is defined to be $w(C) = \sum_{v \in V} w(v)C(v)$. The following Lemma 11 provides an upper bound on $\pi(G)$.

Lemma 11 (Weight Function Lemma, [12]). *Let T be an r-strategy of G with associated weight function w. Suppose that C is an r-unsolvable configuration of pebbles on $V(G)$. Then $w(C) \leq w(\boldsymbol{T})$.*

The main use of Lemma 11 is as follows. Given a collection of r-strategies, the Weight Function Lemma delivers a corresponding set of linear equations. From these, one can use linear optimization to maximize the size of a configuration, subject to those constraints. If α is the result of that optimization, then the size of every r-unsolvable configuration is at most $\lfloor \alpha \rfloor$ and so $\pi(G, r) \leq \lfloor \alpha \rfloor + 1$.

A special instance of Lemma 11 yields the following result.

Lemma 12 (Uniform Covering Lemma, [12]). *Let $\mathcal{T}$ be a set of strategies for a root r of a graph G. If there is some q such that, for each vertex $v \neq r$, we have $\sum_{T \in \mathcal{T}} T(v) = q$, then $\pi(G, r) = n(G)$.*

3.2   Lower Bound

Now we turn to a technique introduced in [8] to derive lower bounds for the pebbling numbers of graphs.

Lemma 13 (Small Neighborhood Lemma [8]). *Let G be a graph and $u, v \in V(G)$. If $N_a[u] \cap N_b[v] = \emptyset$ and $|N_a[u] \cup N_b[v]| < 2^{a+b+1}$, then G is not Class 0.*

The idea behind Lemma 13 is that one considers the configuration that places $2^{a+b+1} - 1$ pebbles on u, 1 pebble on each vertex of $V(G) - (N_a[u] \cup N_b[v])$, and no pebbles elsewhere. It is not difficult to argue that this configuration is v-unsolvable and, under the hypotheses of Lemma 13, has size at least $n(G)$. Thus, what the idea behind the Small Neighborhood Lemma delivers is slightly stronger: if $N_a[u] \cap N_b[v] = \emptyset$ then $\pi(G, u) \geq n(G) + 2^{a+b+1} - |N_a[u] \cup N_b[v]|$.

With this in mind, when attempting to prove that a graph is not Class 0, one always checks if the Small Neighborhood Lemma applies. We show in Theorem 19 below that this lemma cannot apply to odd graphs. Thus, if one attempts to prove that some odd graph is not Class 0, a different method would be required.

4   Proofs

We begin with an important result that describes the distance structure of Kneser graphs. This result and its consequent corollary will be used in both the upper and lower bound arguments that follow.

Lemma 14 ([15]). *Let A and B be two different vertices of $K(m,t)$, where $t \geq 2$ and $m \geq 2t+1$. If $|A \cap B| = s$, then $\text{dist}(A,B) = \min\{2\lceil \frac{t-s}{m-2t} \rceil, 2\lceil \frac{s}{m-2t} \rceil + 1\}$. In particular, $D(K(m,t)) = \lceil \frac{t-1}{m-2t} \rceil + 1$.*

For odd graphs, this yields the following characterization of vertices at a fixed distance from any given vertex, a corollary that is easily proved by induction.

Corollary 15. *Let $A \in V(K(2t+1,t))$. For each $0 \leq d \leq t$ we have $B \in N_d(A)$ if and only if $|B \cap A| = t - d/2$ for even d and $|B \cap A| = \lfloor d/2 \rfloor$ for odd d. Consequently $|N_d(A)| = \binom{t}{\lfloor d/2 \rfloor}\binom{t+1}{\lceil d/2 \rceil}$ for all d.*

4.1   Upper Bounds

Because Kneser graphs are vertex-transitive, we know that for every vertex r we have $\pi(K(m,t)) = \pi(K(m,t), r)$. Thus we may set $r = \{1, \ldots, t\}$ in each case.

Proof of Theorem 7

Proof. Let $G = K(7,3)$. We describe a particular r-strategy T (see Fig. 1, with weights in red). From this, we set T to be the set of all r-strategies determined by the set of automorphisms of G that fix r. The result of summing together all the corresponding inequalities given by Lemma 11 is that every pair of vertices having the same distance from r will have the same coefficient.

Thus, note that T is a set of $3!4! = 144$ r-strategies, one for each permutation of $\{1, \ldots, 7\}$ that fixes r. As $D(G) = 3$, and considering the structure of G from Corollary 15, we see that $|N_1(r)| = 4$, $|N_2(r)| = 12$, and $|N_3(r)| = 18$. For each d define c_d to be the average of the coefficients in $N_d(r)$: $c_1 = 16/4 = 4$, $c_2 = [3(8) + 6(2)]/12 = 3$, and $c_3 = [9(4) + 9(2)]/18 = 3$. We now consider the sum of all these inequalities and then re-scale by dividing the result by 144. The result is that if $v \in N_d(r)$ then the coefficient of $C(v)$ in the re-scaled inequality equals c_d. Thus we derive

$$3|C| = \sum_{v \neq r} 3C(v)$$

$$\leq \sum_{v \in N_1(r)} 4C(v) + \sum_{v \in N_2(r)} 3C(v) + \sum_{v \in N_3(r)} 3C(v)$$

$$\leq \sum_{v \in N_1(r)} 4 + \sum_{v \in N_2(r)} 3 + \sum_{v \in N_3(r)} 3 \qquad \text{(by Lemma 11)}$$

$$= 3(n(K(7,3)) - 1) + |N_1(r)|$$

$$= 3(35) + 1.$$

Hence $|C| \leq 35$ and so $\pi(K(7,3)) \leq 36$. $\qquad \square$

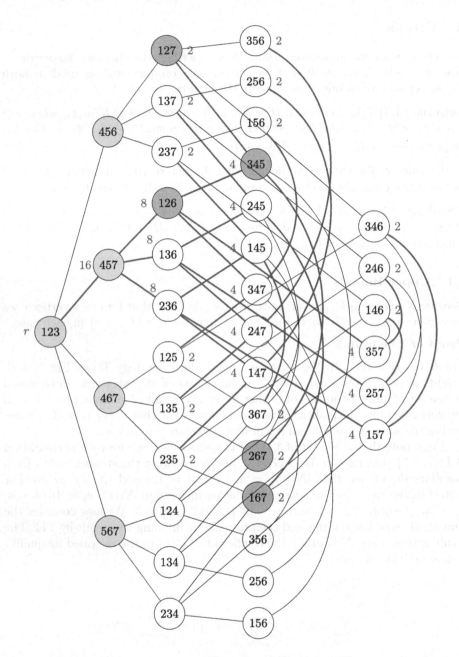

Fig. 1. The Kneser graph $K(7,3)$, with red edges showing the strategy $\boldsymbol{T}$ defined in the proof of Theorem 7, and yellow and green vertices illustrating $N_1[123] \cup N_1[345]$ in the proof of Theorem 19. It is easy to see that $N_2[123] \cup N_0[345]$ is a much larger set, containing 18 vertices instead of 10. Note that vertices 356, 256, and 156 have been drawn twice (near the top and the bottom) for ease in drawing their edges and that $N_3(123)$ has been drawn in the rightmost two columns for similar reasons.

Proof of Theorem 8

Proof. Let $G = K(9,4)$. We describe a particular r-strategy $\boldsymbol{T}$ (see Fig. 2), using the tree T defined as follows: choose a vertex $v \in N_1(r)$ and set $T_0 = \{r\}$ and $T_1 = \{v\}$; for each $d \in \{2,3,4\}$ set $T_d = \{u \in N_d(r) \cap N_1(w) \mid w \in T_{d-1}\}$; then set $T_5 = \{u \in (N_5(r) - T_4) \cap N_1(w) \mid w \in T_4\}$. Note that $|T_1| = 1$, $|T_2| = \binom{4}{3}\binom{1}{1} = 4$, $|T_3| = \binom{4}{1}\binom{4}{3} = 16$, $|T_4| = \binom{4}{2}\binom{4}{1} = 24$, and $|T_5| = \binom{4}{2}\binom{4}{2} = 36$. Indeed, these calculations are derived from observing that the distance from a vertex $u \in T_d$ to r is d, while its distance to v is $d-1$, and using Lemma 14 for both instances. Now define $\boldsymbol{T}$ by giving weight $160/2^d$ to each vertex in T_d.

From this, we set $\mathcal{T}$ to be the set of all r-strategies determined by the set of automorphisms of $K(9,4)$ that fix r. The result of summing together all the corresponding inequalities given by Lemma 11 is that every pair of vertices having the same distance from r will have the same coefficient.

Thus, note that $\mathcal{T}$ is a set of $4!5! = 2880$ r-strategies, one for each permutation of $\{1,\ldots,9\}$ that fixes r. As $D(G) = 4$, and considering the structure of G from Corollary 15, we see that $|N_1(r)| = 5$, $|N_2(r)| = 20$, $|N_3(r)| = 40$, and $|N_4(r)| = 60$ (see Table 1). For each d define c_d to be the average of the coefficients in $N_d(r)$: $c_1 = 80/5 = 16$, $c_2 = [4(40)]/20 = 8$, $c_3 = [16(20)]/40 = 8$, and $c_4 = [24(10) + 36(5)]/60 = 7$. We now consider the sum of all these inequalities and then re-scale by dividing the result by 2880. The result is that if $v \in N_d(r)$ then the coefficient of $C(v)$ in the re-scaled inequality equals c_d. Thus we derive

$$
\begin{aligned}
7|C| = \sum_{v \neq r} 7C(v) & \\
\leq \sum_{v \in N_1(r)} 16C(v) + \sum_{v \in N_2(r)} 8C(v) &+ \sum_{v \in N_3(r)} 8C(v) + \sum_{v \in N_4(r)} 7C(v) \\
\leq \sum_{v \in N_1(r)} 16 + \sum_{v \in N_2(r)} 8 + \sum_{v \in N_3(r)} 8 &+ \sum_{v \in N_3(r)} 7 \qquad \text{(by Lemma 11)} \\
= (5)(16) + (20)(8) + (40)(8) &+ (60)(7) \\
= 980.&
\end{aligned}
$$

Hence $|C| \leq 140$ and so $\pi(K(9,4)) \leq 141$. $\qquad\qquad\qquad\qquad\qquad\square$

Proof of Theorem 9

Proof. Let $G = K(10,4)$. We describe a particular r-strategy $\boldsymbol{T}$ (see Fig. 3), using the tree T defined as follows: choose vertex $v = \{5,6,7,8\} \in N_1(r)$ and define the set $Z = \{9,0\}$. We assign the label (x,y,z) to a vertex u if u shares x, y, and z elements with r, v, and Z, respectively; $V(x,y,z)$ will denote the set of vertices with such a label. We add edges in T from v to all its neighbors in $N_2(r)$; i.e. to $V(3,0,1) \cup V(2,0,2)$. Because $V(2,2,0) \subset N_2(r) \cap N_1(u)$ for some $u \in V(2,0,2)$, we extend T with edges from $V(2,0,2)$ to $V(2,2,0)$. Finally, we add edges in T from $V(3,0,1)$ to $V(1,2,1) \cup V(1,3,0)$. Note that $|V(3,0,1)| = 8$,

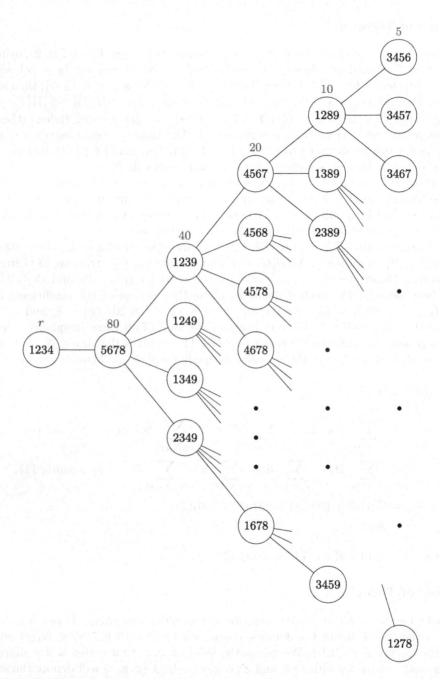

Fig. 2. A schematic diagram of the strategy T in $K(9,4)$ defined in the proof of Theorem 8, with weights in red. As in Fig. 1, the vertices in both of the rightmost two columns have maximum distance (4) from r (1234); the two columns differ, however, in their distances (3 and 4, respectively) from 5678.

Table 1. Number and structure of $K(9,4)$ vertices per distance to the root $\{1,2,3,4\}$, according to Corollary 15. Columns 3 and 4 show the numbers of elements chosen from the sets $\{1,2,3,4\}$ and $\{5,6,7,8,9\}$, respectively, for vertices at each distance; e.g., $A \in N_2(\{1,2,3,4\})$ if and only if $|A \cap \{1,2,3,4\}| = 3$ and $|A \cap \{5,6,7,8,9\}| = 1$.

Distance i	$|N_i(\{1,2,3,4\})|$	$\{1,2,3,4\}$	$\{5,6,7,8,9\}$
0	$\binom{4}{4}\cdot\binom{5}{0} = 1$	4	0
1	$\binom{4}{0}\cdot\binom{5}{4} = 5$	0	4
2	$\binom{4}{3}\cdot\binom{5}{1} = 20$	3	1
3	$\binom{4}{1}\cdot\binom{5}{3} = 40$	1	3
4	$\binom{4}{2}\cdot\binom{5}{2} = 60$	2	2

$|V(2,0,2)| = 6$, $|V(2,2,0)| = 36$, $|V(1,2,1)| = 48$, and $|V(1,3,0)| = 16$, while $|N_1(r)| = 15$, $|N_2(r)| = 114$, and $|N_3(r)| = 80$.

Now we define T by giving weight 60 to v, weight 30 to each vertex of $V(3,0,1) \cup V(2,0,2)$, weight 5 to each vertex of $V(1,2,1) \cup V(1,3,0)$, and weight 1 to each vertex of $V(2,2,0)$. From this, we set $\mathcal{T}$ to be the set of all r-strategies determined by the set of 4!6! automorphisms of $K(10,4)$ that fix r. The result of summing together all the corresponding inequalities given by Lemma 11 is that every pair of vertices having the same distance from r will have the same coefficient.

For each d define c_d to be the average of the coefficients in $N_d(r)$: $c_1 = [60(1)]/15 = 4$, $c_2 = [30(8)+30(6)+1(36)]/114 = 4$, and $c_3 = [5(48)+5(16)]/80 = 4$. We now consider the sum of all these inequalities and then re-scale by dividing the result by 4!6!. The result is that if $v \in N_d(r)$ then the coefficient of $C(v)$ in the re-scaled inequality equals c_d. By Lemma 12 we have that $\pi(G,r) = n(G) = 210$; i.e. $K(10,4)$ is Class 0.   □

Proof of Theorem 10

Proof. Let $G = K(2t+1,t)$. Theorems 7 and 8 already have better bounds, so we will assume that $t \geq 5$. As in the proof of Theorem 8, we set $r = \{1,\ldots,t\}$, choose some $v \in N_1(r)$, and define the tree T by $T_0 = \{r\}$, $T_1 = \{v\}$, and for each $d \in \{2,\ldots,t\}$ set $T_d = \{u \in N_d(r) \cap N_1(w) \mid w \in T_{d-1}\}$, with $T_{t+1} = \{u \in (N_t(r)-T_t) \cap N_1(w) \mid w \in T_t\}$. We note that $|N_d(r)| = \binom{t}{\lfloor d/2 \rfloor}\binom{t+1}{\lceil d/2 \rceil}$ for $1 \leq d \leq t$ and that $|T_d| = \binom{t}{\lfloor d/2 \rfloor}\binom{t}{\lceil d/2 \rceil - 1}$ for $1 \leq d \leq t-1$, with $|T_t| = \binom{t}{\lfloor t/2 \rfloor}\binom{t}{\lceil t/2 \rceil - 1}$ and $|T_{t+1}| = \binom{t}{\lfloor t/2 \rfloor}\binom{t}{\lceil t/2 \rceil}$ when t is even and $|T_t| = \binom{t}{\lfloor t/2 \rfloor}\binom{t}{\lceil t/2 \rceil}$ and $|T_{t+1}| = \binom{t}{\lfloor t/2 \rfloor}\binom{t}{\lceil t/2 \rceil - 1}$ when t is odd. Now define T by giving weight $w_d = (t+1)2^{t+1-d}$ to each vertex in T_d for all $d > 0$.

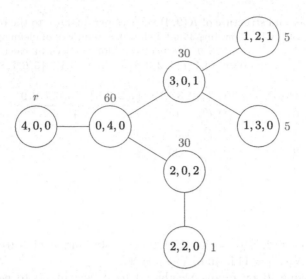

Fig. 3. A schematic diagram of the strategy T in $K(10, 4)$ defined in the proof of Theorem 9, with weights in red. Each vertex of the tree represents a set of vertices of the form (x, y, z), where x, y, and z are the numbers of digits chosen from $\{1, 2, 3, 4\}$, $\{5, 6, 7, 8\}$, and $\{9, 0\}$, respectively.

From this, we set $\mathcal{T}$ to be the set of all r-strategies determined by the set of $t!(t + 1)!$ automorphisms of G that fix r. The result of summing together all the corresponding inequalities given by Lemma 11 is that every pair of vertices having the same distance from r will have the same coefficient. For each d define c_d to be the average of the coefficients in $N_d(r)$:

$$c_d = \begin{cases} |T_d| w_d / |N_d(r)| = \lceil d/2 \rceil 2^{t+1-d} \text{ for } d < t \text{ and} \\ (|T_t| w_t + |T_{t+1}| w_{t+1}) / |N_t(r)| = \begin{cases} 3t/2 + 1 \text{ for even } t \text{ and} \\ 3\lceil t/2 \rceil \text{ for odd } t. \end{cases} \end{cases}$$

We now consider the sum of all these inequalities and then re-scale by dividing the result by $t!(t+1)!$. The result is that if $v \in N_d(r)$ then the coefficient of $C(v)$ in the re-scaled inequality equals c_d. Because c_t is the smallest coefficient when $t \geq 6$ (it is c_4 when $t = 5$ but we can add some edges from T_5 into $N_4(r) - T_4$, as in the proof of Theorem 7, with sufficiently chosen weights to remedy this without effecting the calculations below), and using Lemma 11, we derive

$$c_t |C| = \sum_{v \neq r} c_t C(v) \leq \sum_{d=1}^{t} \sum_{v \in N_d(r)} c_d C(v)$$

$$\leq \sum_{d=1}^{t} \sum_{v \in N_d(r)} c_d = \sum_{d=1}^{t} |N_d(r)| c_d. \tag{1}$$

By computing the ratios $r_d = |N_d(r)|c_d/|N_{d-1}(r)|c_{d-1}$ we find that $r_d = (t - a + 1)/2a$ for $d \in \{2a, 2a+1\}$, showing that the sequence $|N_1(r)|c_1, \ldots, |N_t(r)|c_t$ is unimodal with its maximum occurring when $d = j := 2\lceil (t+1)/3 \rceil - 1$. Hence from Inequality 1 we obtain the upper bound

$$|C| < t|N_j(r)|c_j/c_t, \tag{2}$$

which yields that

$$\pi(G) \leq t|N_j(r)|c_j/c_t = t\binom{t}{\lfloor j/2 \rfloor}\binom{t+1}{\lceil j/2 \rceil}\lceil j/2 \rceil 2^{t+1-j}\Big/(3t/2). \tag{3}$$

Because the bound in Inequality 2 is so generous, we can dispense with floors and ceilings and addition/subtraction by one, approximating $j/2$ by $t/3$ and considering it to be an integer, thereby replacing the right side of Inequality 3 by the following. We use the notation $x^{\underline{h}} = x(x-1)\cdots(x-h+1)$, observe that $x^{\underline{h}} \leq (x - h/2)^h$, and make use of the lower bound $x! \geq (x/e)^x\sqrt{2ex}$, which works for all x, rather than using Stirling's asymptotic formula.

$$t\binom{t}{t/3}^2\frac{(t/3)2^{2t/3}}{(3t/2)} = \left(\frac{t^{\underline{t/3}}}{(t/3)!}\right)^2\frac{(2t)2^{2t/3}}{9} \leq \left(\frac{5t}{6}\right)^{2t/3}\left(\frac{3e}{t}\right)^{2t/3}\frac{(2t)2^{2t/3}}{9(2et/3)}$$

$$= \frac{(5e)^{2t/3}}{3e} < 5.7^t/8 < n\left(\frac{\pi^{\alpha/2}}{2^{3+2\alpha}}\right)t^{\alpha/2} < .045n^\alpha(\lg n)^{\alpha/2},$$

where $\alpha = \log_4((5e)^{2/3})$. This completes the proof. □

4.2   Lower Bound Attempt

The following two claims will be useful in proving Theorem 19.

Claim 16. *The following inequalities hold for all $k \geq 1$.*

1. $\binom{4k+1}{k}/\binom{4k-3}{k-1} > 4$.
2. $\binom{4k+2}{k}\binom{4k+4}{k+2}/\binom{4k-2}{k-1}\binom{4k}{k+1} > 16$.
3. $\binom{4k+1}{k}\binom{4k+2}{k}/\binom{4k-3}{k-1}\binom{4k-2}{k-1} > 16$.
4. $\binom{4k+3}{k}\binom{4k+4}{k+1}/\binom{4k-1}{k-1}\binom{4k}{k} > 16$.

Proof. We only display the proof for the first inequality, as the others use identical techniques. Indeed, we calculate

$$\frac{\binom{4k+1}{k}}{\binom{4k-3}{k-1}} = \frac{(4k+1)\cdots(4k-k+2)(k-1)!}{(4k-3)\cdots(4k-k-1)k!}$$

$$= \frac{(4k+1)\cdots(4k-2)}{(4k-k+1)\cdots(4k-k-1)k}$$

$$= \left(\frac{4k+1}{k}\right)\left(\frac{4k}{3k+1}\right)\left(\frac{4k-1}{3k}\right)\left(\frac{4k-2}{3k-1}\right) > \frac{4k}{k} = 4$$

since $k \geq 1$. □

The next corollary follows from Claim 16 by induction.

Corollary 17. *The following inequalities hold for all $k \geq 1$.*

1. $\binom{4k+1}{k} > 4^k$.
2. $\binom{4k+2}{k}\binom{4k+4}{k+2} > 16^k$.
3. $\binom{4k+1}{k}\binom{4k+2}{k} > 16^k$.
4. $\binom{4k+3}{k}\binom{4k+4}{k+1} > 16^k$.

For any vertex r in $K(2t+1, t)$ and any $0 \leq d \leq t$ define $g_d(t) = |N_d(r)|$.

Claim 18. *The following inequalities hold for every $0 \leq d \leq t$.*

1. $g_d(2d+2) + g_{d+1}(2d+2) \geq 2^{2d+2}$.
2. $2g_d(2d+1) \geq 2^{2d+1}$.

Proof. Recall the formulas from Corollary 15: $g_d(t) = \binom{t}{d/2}\binom{t+1}{d/2}$ for even d and $g_d(t) = \binom{t}{\lfloor d/2 \rfloor}\binom{t+1}{\lceil d/2 \rceil}$ for odd d. We will only display the proof for the first inequality, as the second uses identical techniques.

When $d = 2s - 1$ is odd we have by Corollary 17 that

$$g_{2s-1}(4s) + g_{2s}(4s) = \binom{4s}{s-1}\binom{4s+1}{s} + \binom{4s}{s}\binom{4s+1}{s}$$

$$= \binom{4s+1}{s}^2 > 4^{2s} = 2^{2d+2}.$$

The case when $d = 2s$ is even is proven similarly. $\qquad\square$

Theorem 19. *The hypotheses of Lemma 13 are not satisfied for any odd graph.*

Proof. Recall from above that $D(K(2t+1, t)) = t$ and define $f_a(t) = |N_d[A]| = \sum_{i=0}^{d} g_d(t)$, where A is any vertex. Then we must show that $f_a(t) + f_b(t) \geq 2^{a+b+1}$ for every $a \geq b$ and $t \geq a + b + 1$.

We first reduce to the "balanced" case, in which $a \leq b + 1$; that is, we prove that $f_a(t) + f_b(t) \geq f_x(t) + f_y(t)$, where $x = \lceil \frac{a+b}{2} \rceil$ and $y = \lfloor \frac{a+b}{2} \rfloor$. It is sufficient to show that $f_a(t) + f_b(t) \geq f_{a-1}(t) + f_{b+1}(t)$ whenever $a > b+1$. This inequality is equivalent to

$$g_a(t) = f_a(t) - f_{a-1}(t) \geq f_{b+1}(t) - f_b(t) = g_{b+1},$$

which is trivial since Corollary 15 states that $g_d(t) = \binom{t}{\lfloor d/2 \rfloor}\binom{t+1}{\lceil d/2 \rceil}$ and binomial coefficients increase up to $\lfloor t/2 \rfloor$.

Second, we reduce to the case in which $t = a + b + 1$; that is, we prove that $f_x(t) + f_y(t) \geq f_x(a+b+1) + f_y(a+b+1)$. This inequality follows simply from the property that $\binom{t}{d}$ is an increasing function in t when d is fixed.

Third, we note the obvious relation that $f_x(a+b+1) + f_y(a+b+1) \geq g_x(a+b+1) + g_y(a+b+1)$, since each $g_d(t)$ is merely the final term of the summation $f_d(t)$.

Thus it suffices to show that

1. $g_y(2y + 2) + g_{y+1}(2y + 2) \geq 2^{2y+2}$ and
2. $2g_y(2y + 1) \geq 2^{2y+1}$,

which follows from Claim 18. The above arguments yield

$$f_a(t) + f_b(t) \geq f_x(t) + f_y(t)$$
$$\geq f_x(a + b + 1) + f_y(a + b + 1)$$
$$\geq f_x(x + y + 1) + f_y(x + y + 1)$$
$$\geq g_x(x + y + 1) + g_y(x + y + 1) \geq 2^{a+b+1}.$$

The final inequality follows from the identity $x + y = a + b$ and the two cases that $x = y$ or $x = y + 1$.   □

5   Concluding Remarks

As discussed above, all diameter two Kneser graphs are Class 0, and we verified in Theorem 9 that the diameter three Kneser graph $K(10, 4)$ is also Class 0, while the diameter three Kneser graph $K(7, 3)$ remains undecided. By Lemma 14 we see that $D(K(m, t)) \leq 3$ for all $m \geq (5t - 1)/2$. The following theorem shows that graphs with high enough connectivity are Class 0 (the value $2^{2D(G)+3}$ is not thought to be best possible, but cannot be smaller than $2^{D(G)}/D(G)$). It is this theorem that was used to prove Theorem 5.

Theorem 20 ([9]). *If G is $2^{2D(G)+3}$-connected, then G is Class 0.*

Accordingly, Theorem 20 implies that diameter three Kneser graphs with connectivity at least 2^9 are Class 0, which occurs when $\binom{m-t}{t} \geq 512$ because connectivity equals degree for Kneser graphs. This begs the following subproblem of Conjecture 6.

Problem 21. *For all $t \geq 5$ and $m \geq (5t - 1)/2$, if $\binom{m-t}{t} \leq 511$ then is $K(m, t)$ Class 0? In particular, can the Weight Function Lemma be used to prove so?*

For example, if $t = 5$, then the interval of interest in Problem 21 is $12 \leq m \leq 16$.

We also see from our work both the power and the limitations of the usage of the Weight Function Lemma. For example, it did not produce a very close bound for $K(9, 4)$, which has diameter 4, but did produce the actual pebbling number for $K(10, 4)$, which has diameter 3. Indeed the power of strategies weakens as the diameter grows. Curiously, though, it did not yield a Class 0 result for $K(7, 3)$, which also has diameter 3. (Conceivably, it did give the right answer, but we do not believe this.) The trees we used in the proof for this case were very simple and structured and were all isomorphic. In trying to improve the result, we had a computer generate hundreds of thousands of tree strategies and fed them into linear programming software and even used integer programming. No results were better than the bound we presented in Theorem 7.

For fixed t the Kneser graphs $K(m, t)$ with the largest diameter (t) have $m = 2t + 1$; the odd graphs. We see that weight functions produce a fairly large upper bound in this case, with the multiplicative factor of $n^{.26}(\lg n)^{.63}$ attached, where $n = n(K(2t + 1, t))$. Nonetheless, this is the best known bound. Along these lines we offer the following additional subproblem of Conjecture 6.

Problem 22. *Find a constant c such that $\pi(K(2t + 1, t)) \leq cn$, where $n = n(K(2t + 1, t))$.*

References

1. Adauto, M., Cruz, M., Figueiredo, C., Hurlbert, G., Sasaki, D.: On the pebbling numbers of some snarks. arXiv:2303.13292 [math.CO] (2023)
2. Alcón, L., Gutierrez, M., Hurlbert, G.: Pebbling in split graphs. SIAM J. Discret. Math. **28**(3), 1449–1466 (2014)
3. Alcón, L., Gutierrez, M., Hurlbert, G.: Pebbling in semi-2-trees. Discret. Math. **340**(7), 1467–1480 (2017)
4. Alcón, L., Hurlbert, G.: Pebbling in powers of paths. Discret. Math. **346**(5), 113315 (20 p.) (2023)
5. Chan, M., Godbole, A.P.: Improved pebbling bounds. Discret. Math. **308**(11), 2301–2306 (2008)
6. Chung, F.R.K.: Pebbling in hypercubes. SIAM J. Discret. Math. **2**(4), 467–472 (1989)
7. Clarke, T., Hochberg, R., Hurlbert, G.: Pebbling in diameter two graphs and products of paths. J. Graph Theor. **25**(2), 119–128 (1997)
8. Cranston, D.W., Postle, L., Xue, C., Yerger, C.: Modified linear programming and class 0 bounds for graph pebbling. J. Combin. Optim. **34**(1), 114–132 (2017)
9. Czygrinow, A., Hurlbert, G., Kierstead, H., Trotter, W.: A note on graph pebbling. Graphs Combin. **18**(2), 219–225 (2002)
10. Hurlbert, G.: A survey of graph pebbling. Congr. Numer. **139**, 41–64 (1999)
11. Hurlbert, G.: Recent progress in graph pebbling. Graph Theor. Notes N.Y. **49**, 25–37 (2005)
12. Hurlbert, G.: The weight function lemma for graph pebbling. J. Combin. Optim. **34**(2), 343–361 (2017)
13. Lovász, L.: Combinatorial Problems and Exercises. North Holland, Amsterdam, New York, Oxford (1979)
14. Pachter, L., Snevily, H., Voxman, B.: On pebbling graphs. Congr. Numer. **107**, 65–80 (1995)
15. Valencia-Pabon, M., Vera, J.C.: On the diameter of Kneser graphs. Discret. Math. **305**(1–3), 383–385 (2005)

Structural and Combinatorial Properties of 2-Swap Word Permutation Graphs

Duncan Adamson[1] , Nathan Flaherty[1,2](✉) , Igor Potapov[2] ,
and Paul G. Spirakis[2]

[1] Leverhulme Research Centre for Functional Materials Design,
University of Liverpool, Liverpool, UK
{d.a.adamson,n.flaherty}@liverpool.ac.uk
[2] Department of Computer Science, University of Liverpool, Liverpool, UK
{potapov,spirakis}@liverpool.ac.uk

Abstract. In this paper, we study the graph induced by the 2-*swap* permutation on words with a fixed Parikh vector. A 2-swap is defined as a pair of positions $s = (i, j)$ where the word w induced by the swap s on v is $v[1]v[2] \ldots v[i-1]v[j]v[i+1] \ldots v[j-1]v[i]v[j+1] \ldots v[n]$. With these permutations, we define the *Configuration Graph*, $G(P)$ for a given Parikh vector. Each vertex in $G(P)$ corresponds to a unique word with the Parikh vector P, with an edge between any pair of words v and w if there exists a swap s such that $v \circ s = w$. We provide several key combinatorial properties of this graph, including the exact diameter of this graph, the clique number of the graph, and the relationships between subgraphs within this graph. Additionally, we show that for every vertex in the graph, there exists a Hamiltonian path starting at this vertex. Finally, we provide an algorithm enumerating these paths from a given input word of length n with a delay of at most $O(\log n)$ between outputting edges, requiring $O(n \log n)$ preprocessing.

1 Introduction

In information theory and computer science, there are several well-known edit distances between strings which are based on insertions, deletions and substitutions of single characters or various permutations of several characters, including swaps of adjacent or non-adjacent characters, shuffling, etc. [15,16,24].

These operations are well motivated by problems in physical science, for example, the biological swaps which occur at a gene level are non-adjacent swap operations of two symbols (mutation swap operator) representing gene mutations [9]. In recent work on Crystal Structure Prediction the swap operation on a pair of symbols in a given word representing layers of atomic structures was used to generate new permutations of those layers, with the aim of exploring the configuration space of crystal structures [14]. In computer science string-to-string correction has been studied for adjacent swaps [23] and also in the

This work is supported by the Leverhulme Research Centre for Functional Materials Design and EPSRC grants EP/P02002X/1, EP/R018472/1.

context of sorting networks [10], motion on graphs and diameter of permutation groups [22]. In group theory, the distance between two permutations (the Cayley distance) measures the minimum number of transpositions of elements needed to turn one into the other [21].

A *configuration graph* is a graph where words (also known as strings) are represented by vertices and operations by edges between the strings. For example, one may define the operations as the standard suite of edits (insertions, deletions, and substitutions), with each edge corresponding to a pair of words at an edit distance of one. In such a graph, the distance between any pair of words corresponds to the edit distance between these words. In this paper, we study the structural properties of such graphs defined by swap operations of two symbols on a given word (2-swap permutations), a permutation defined by a pair of indices (i, j) and changing a word w by substituting the symbol at position i with that at position j, and the symbol at position j with that at position i. As the number of occurrences of each symbol in a given word can not be changed under this operation, we restrict our work to only those words with a given Parikh vector[1]. We focus on studying several fundamental properties of the structure of these graphs, most notably the diameter, clique number, number of cliques, and the Hamiltonicity of the graph. Similar problems have been heavily studied for Cayley graphs [21], and permutation graphs [17]. It has been conjectured that the diameter of the symmetric group of degree n is polynomially bounded in n, where only recently the exponential upper bound [11] was replaced by a quasipolynomial upper bound [18]. The diameter problem has additionally been studied with respect to a random pair of generators for symmetric groups [19]. In general, finding the diameter of a Cayley graph of a permutation group is NP-hard and finding the distance between two permutations in directed Cayley graphs of permutation groups is PSPACE-hard [20].

To develop efficient exploration strategies for these graphs it is essential to investigate structural and combinatorial properties. As mentioned above the problem is motivated by problems arising in chemistry regarding Crystal Structure Prediction (CSP) which is computationally intractable in general [4,5]. In current tools [13,14], chemists rely on representing crystal structures as a multiset of discrete blocks, with optimisation performed via a series of permutations, corresponding to swapping blocks. Understanding reachability properties under the swap operations can help to evaluate and improve various heuristic space exploration tools and extend related combinatorial toolbox [6,8].

Our Results. We provide several key combinatorial properties of the graph defined by 2-swap permutations over a given word. First, we show that this graph is *locally isomorphic*, that is, the subgraph of radius r centred on any pair of vertices w and u are isomorphic. We strengthen this by providing an exact diameter on the graph for any given Parikh vector. Finally, we show that, for every vertex v in the graph, there is a Hamiltonian path starting at v. We build upon this by providing a novel algorithm for enumerating the Hamiltonian

---

[1] The Parikh vector of a word w denotes a vector with the number of occurrences of the symbols in the word w.

path starting at any given vertex v in a binary graph with at most $O(\log n)$ delay between outputting the swaps corresponding to the transitions made in the graph. Our enumeration results correlate well with the existing work on the enumeration of words. This includes work on explicitly outputting each word with linear delay [3,27], or outputting an implicit representation of each word with either constant or logarithmic delay relative to the length of the words [1,2, 25,28,29]. The surveys [26,30] provide a comprehensive overview of a wide range of enumeration results.

2 Preliminaries

Let $\mathbb{N} = \{1, 2, \dots\}$ denote the set of natural numbers, and $\mathbb{N}_0 = \mathbb{N} \cup \{0\}$. We denote by $[n]$ the set $\{1, 2, \dots, n\}$ and by $[i, n]$ the set $\{i, i+1, \dots, n\}$, for all $i, n \in \mathbb{N}_0, i \le n$. An *alphabet* Σ is an ordered, finite set of symbols. Tacitly assume that the alphabet $\Sigma = [\sigma] = \{1, 2, \dots, \sigma\}$, where $\sigma = |\Sigma|$. We treat each symbol in Σ both as a symbol and by the numeric value, i.e. $i \in \Sigma$ represents both the symbol i and the integer i. A *word* is a finite sequence of symbols from a given alphabet. The length of a word w, denoted $|w|$, is the number of symbols in the sequence. The notation Σ^n denotes the set of n-length words defined over the alphabet Σ, and the notation Σ^* denotes the set of all words defined over Σ.

For $i \in [|w|]$, the notation $w[i]$ is used to denote the ith symbol in w, and for the pair $i, j \in [|w|], w[i, j]$ is used to denote the sequence $w[i]w[i+1] \dots w[j]$, such a sequence is called a *factor* of w. We abuse this notation by defining, for any pair $i, j \in [|w|]$ such that $j < i$, $w[i, j] = \varepsilon$, where ε denotes the empty string.

Definition 1 (2-swap). *Given a word $w \in \Sigma^n$ and pair $i, j \in [n], i < j$ such that $w[i] \ne w[j]$, the 2-swap of w by (i, j), denoted $w \circ (i, j)$, returns the word*

$$w[1, i-1]w[j]w[i+1, j-1]w[i]w[j+1, n].$$

Example 1. Given the word $w = 11221122$ and pair $(2, 7)$, $w \circ (2, 7) = 12221112$.

Given a word $w \in \Sigma^n$, the *Parikh vector* of w, denoted $P(w)$ is the σ-length vector such that the ith entry of $P(w)$ contains the number of occurrences of symbol i in w, formally, for $i \in [\sigma]$ $P(w)[i] = |\{j \in [n] \mid w[j] = i\}|$, where $n = |w|$. For example, the word $w = 11221122$ has Parikh vector $(4, 4)$. The set of words with a given Parikh vector P over the alphabet Σ is denoted $\Sigma^{*|P}$, formally $\Sigma^{*|P} = \{w \in \Sigma^* \mid P(w) = P\}$. It is notable that $|\Sigma^{*|P}| = \frac{n!}{\prod_{i \in [\sigma]} P[i]!}$. Unless stated otherwise we define $n := \sum_{i \in [\sigma]} P[i]$.

Definition 2. *For a given alphabet Σ and Parikh vector P, the* configuration graph *of $\Sigma^{*|P}$ is the undirected graph $G(P) = \{V(P), E(P)\}$ where:*

- $V(P) = \{v_w \mid w \in \Sigma^{*|P}\}$.
- $E(P) = \{\{v_w, v_u\} \in V(P) \times V(P) \mid \exists i, j \in [n] \text{ s.t. } w \circ (i, j) = u\}$.

Informally, the configuration graph for a given Parikh vector P is the graph with each vertex corresponding to some word in $\Sigma^{*|P}$, and each edge connecting every pair of words $w, u \in \Sigma^{*|P}$ such that there exists some 2-swap transforming w into u.

A *path* (also called a *walk*) in a graph is an ordered set of edges such that the second vertex in the ith edge is the first vertex in the $(i+1)^{th}$ edge, i.e. $p = \{(v_1, v_2), (v_2, v_3), \dots, (v_{|p|}, v_{|p|}+1)\}$. Note that a path of length i visits $i+1$ vertices. A path p *visits* a vertex v if there exists some edge $e \in p$ such that $v \in e$. A *cycle* (also called a *circuit*) is a path such that the first vertex visited is the same as the last. A *Hamiltonian* path p is a path visiting each vertex exactly once, i.e. for every $v \in V$, there exists at most two edges $e_1, e_2 \in p$ such that $v \in e_1$ and $v \in e_2$. A cycle is Hamiltonian if it is a Hamiltonian path and a cycle. A path p covers a set of vertices V if, for every $v \in V$, there exists some $e \in p$ such that $v \in e$. Note that a Hamiltonian path is a path cover of every vertex in the graph. and a Hamiltonian cycle is a cycle cover of every vertex in the graph.

The *distance* between a pair of vertices $v, u \in V$, denoted $D(v, u)$ in the graph G is the smallest value $d \in \mathbb{N}_0$ for which there exists some path p of length d covering both v and u, i.e. the minimum number of edges needed to move from v to u. If $v = u$, then $D(v, u)$ is defined as 0. The *diameter* of a graph G is the maximum distance between any pair of vertices in the graph, i.e. $\max_{v, u \in V} D(v, u)$.

Given two graphs $G = (V, E)$ and $G' = (V', E')$, G is *isomorphic* to G' if there exists a bijective mapping $f : V \mapsto V'$ such that, for every $v, u \in V$, $(v, u) \in E$ if and only if $(f(v), f(u)) \in E'$. The notation $G \cong G'$ is used to denote that G is isomorphic to G', and $G \not\cong G'$ to denote that G is not isomorphic to G'. A *subgraph* of a graph $G = (V, E)$ is a graph $G' = (V', E')$ such that $V' \subseteq V$ and $E' \subseteq E$. A *clique* $G' = (V', E')$ is a subgraph, $G' \subseteq G$ which is complete (i.e. for all $u, v \in G'$, $(u, v) \in E'$). And the clique number ω of a graph G is the size of the largest clique in G.

3   Basic Properties of the Configuration Graph

In this section, we provide a set of combinatorial results on the configuration graph. We first show that every subgraph of the configuration graph $G(p) = (V, E)$ with the vertex set $V'(v) = \{u \in V \mid D(v, u) \leq \ell\}$, and edge set $E' = (V' \times V') \cap E$ are isomorphic. We build on this by providing a tight bound on the diameter of these graphs. We start by considering some local structures within the graph.

Lemma 1. *Given a Parikh vector P with associated configuration graph $G(P)$, each vertex $v \in V(P)$ belongs to $\sum_{j \in \Sigma} \prod_{i \in \Sigma \setminus \{j\}} P[i]$ maximal cliques, with the size of each such clique being in $\{P[i] + 1 \mid i \in \Sigma\}$.*

Proof. Consider first the words with Parikh vector $P = (k, 1)$. Note that every word in $\Sigma^{*|P}$ consists of k copies of the symbol 1, and one copy of the symbol 2. Therefore, given any pair of words $w, u \in \Sigma^P$ such that $w[i] = u[j] = 2$, the

2-swap (i,j) transforms w into u and hence there exists some edge between w and u. Hence $G(P)$ must be a complete graph, a clique, of size $k+1$.

In the general case, consider the word $w \in \Sigma^{*|P}$ where $P = (k_1, k_2, \ldots, k_\sigma)$, and $n = \sum_{k_i \in P} k_i$. Let $\text{Pos}(w, i) = \{j \in [n] \mid w[j] = i\}$. Given $i, j \in [\sigma], i \neq j$, let $i_1, i_2 \in \text{Pos}(w, i)$ and $j_1 \in \text{Pos}(w, j)$ be a set of indices. Let $v_1 = w \circ (i_1, j_1)$ and $v_2 = w \circ (i_2, j_1)$. Then, $v_1[i_1] = v_2[i_2], v_2[i_2] = v_1[i_1]$, and $v_1[j_1] = v_2[j_1]$. Further, for every $\ell \in [n]$ such that $\ell \notin \{i_1, i_2, j_1\}$, $v_1[\ell] = v_2[\ell]$ as these positions are unchanged by the swaps. Therefore, $v_1 = v_2 \circ (i_1, i_2)$, and hence these words are connected in $G(P)$. Further, as this holds for any $j_1 \in \text{Pos}(v, j)$, the set of words induced by the swaps (j, ℓ), for some fixed $\ell \in \text{Pos}(w, i)$ correspond to a clique of size $P[j]+1$. Therefore, there exists $\prod_{i \in \Sigma \setminus \{j\}} P[i]$ cliques of size $P[j]+1$ including w, for any $j \in \Sigma$.

We now show that the cliques induced by the set of swaps $S(i,j) = \{(i', j) \mid i' \in \text{Pos}(w, w[i])\}$ are maximal. Let $C(i,j,w) = \{w\} \cup \{w \circ (i', j) \mid (i', j) \in S(i,j)\}$, i.e. the clique induced by the set of swaps in $S(i,j)$. Consider a set of swaps, $(i_1, j_1), (i_2, j_1), (i_1, j_2)$ and (i_2, j_2), where $i_1, i_2 \in \text{Pos}(w, i)$ and $j_1, j_2 \in \text{Pos}(w, j)$. Let $v_{1,1} = w \circ (i_1, j_1), v_{2,1} = w \circ (i_2, j_1), v_{1,2} = w \circ (i_1, j_2)$ and $v_{2,2} = w \circ (i_2, j_2)$. Note that $\{w, v_{1,1}, v_{2,1}\} \subseteq C(i, j_1), \{w, v_{1,1}, v_{1,2}\} \subseteq C(j, i_1)$, $\{w, v_{2,1}, v_{2,2}\} \subseteq C(i, j_2)$ and $\{w, v_{2,1}, v_{2,2}\} \subseteq C(j, i_2)$.

We now claim that there exists no swap transforming $v_{1,1}$ in to $v_{2,2}$. Observe first that, for every $\ell \in [|w|]$ such that $\ell \notin \{i_1, i_2, j_1, j_2\}$, $v_{1,1}[\ell] = v_{2,2}[\ell]$. As $v_{1,1}[i_1] = v_{2,2}[j_1]$, and $v_{1,1}[j_1] = v_{2,2}[i_2]$, exactly two swaps are needed to transform $v_{1,1}$ into $v_{2,2}$. Therefore, for any pair of swaps $(i_1, j_1), (i_2, j_2) \in \text{Pos}(i, w) \times \text{Pos}(j, w)$, such that $i_1 \neq i_2$ and $j_1 \neq j_2$, the words $w \circ (i_1, j_1)$ and $w \circ (i_2, j_2)$ are not adjacent in $G(v)$. Similarly, given a set of indices $i' \in \text{Pos}(i, w), j' \in \text{Pos}(j, w)$ and $\ell' \in \text{Pos}(\ell, w)$ and swaps $(i', j'), (i', j')$, observe that as $w[j'] \neq w[\ell']$, the distance between $w \circ (i', j')$ and $w \circ (i', \ell')$ is 2. Therefore, every clique induced by the set of swaps $S(i,j) = \{(i', j) \mid i' \in \text{Pos}(w, w[i])\}$ is maximal.

Corollary 1. *Let $v \in V(P)$ be the vertex in $G(P)$ corresponding to the word $w \in \Sigma^{*|P}$. Then, v belongs only to the maximal cliques corresponding to the set of words $\{w \circ (i,j) \mid i \in \text{Pos}(w, x)\}$ for some fixed symbol $x \in \Sigma$ and position $j \in [|w|], w[j] \neq x$, where $\text{Pos}(w, x) = \{i \in [|w|], w[i] = x\}$.*

Now since the edges in each maximal clique only swap two types of symbols we have the following corollary for the number of cliques.

Corollary 2. *There are $\sum_{(i,j) \in \Sigma \times \Sigma} \frac{(\sum_{k \in \Sigma \setminus \{i,j\}} P[k])!}{\prod_{k \in \Sigma \setminus \{i,j\}} P[k]!}$ maximal cliques in $G(p)$.*

Corollary 3. *The clique number $\omega(G(P))$ is equal to $\max_{i \in [\sigma]} P[i] + 1$.*

Lemma 2. *Let $G_r(v)$ be the subgraph of $G(P)$ induced by all vertices of distance at most r away from a given vertex v. Then, for any pair of vertices $u, v \in V$ and given any $r \in \mathbb{Z}^+$, $G_r(u) \cong G_r(v)$.*

Proof. Let $\pi \in S_n$ be the permutation such that $u \circ \pi = v$. We use the permutation π to define an isomorphism $f : G_r(u) \to G_r(v)$ such that $f(w) = w \circ \pi$. In order to show that f is an isomorphism we need to show that it preserves adjacency. We start by showing that for every word, $w \in G_1(u)$, $f(w) \in G_1(v)$.

Let $\tau = (\tau_1, \tau_2)$ be the 2-swap such that $w = u \circ \tau$. We now have 3 cases for how π and τ interact, either none of the indices in τ are changed by π, just one of τ_1 or τ_2 are changed by π, or both τ_1 and τ_2 are changed by π. In the first case, $f(w)$ is adjacent to v as $v \circ \tau = f(w)$. In the second case, let (τ_1, τ_2) be a swap that that $\pi[\tau_1] = \tau_1$, i.e. τ_1 is not changed by the permutation π. We define a new swap τ' such that $v \circ \tau' = f(w)$. Let $x, y \in [n]$ be the positions in v such that $\pi[x] = \tau_2$ and $\pi[\tau_2] = y$. Now, let $\tau' = (\tau_1, y)$. Observe that $v[y] = w[\tau_2]$, and $v[\tau_1] = w[\tau_1]$. Therefore, the word $v \circ \tau' = u \circ \tau \circ \pi$. Note that as the ordering of the indices in the swap does not change the swap, the same argument holds for the case when $\pi[\tau_2] = \tau_2$. In the final case, let $\tau' = (\pi[\tau_1], \pi[\tau_2])$. Note that by arguments above, $u[\pi[\tau_1]] = v[\tau_1]$ and $u[\pi[\tau_2]] = v[\tau_2]$, and hence $v \circ \tau' = u \circ \tau \circ \pi$. Repeating this argument for each word at distance $\ell \in [1, r]$ proves this statement.

We now provide the exact value of the diameter of any configuration graph $G(P)$. Theorem 1 states the explicit diameter of the graph, with the remainder of the section dedicated to proving this result.

Theorem 1. *The diameter of the Configuration Graph, $G(P)$ for a given Parikh vector P is $n - \max_{i \in [\sigma]} P[i]$.*

Theorem 1 is proven by first showing that the upper bound matches $n - \max_{i \in [\sigma]} P[i]$ (Lemma 3). We then show that the lower bound on the diameter matches the upper bound (Lemma 5), concluding our proof of Theorem 1.

Lemma 3 (Upper Bound of Diameter). *The diameter of the Configuration Graph, $G(P)$ for a given Parikh vector P is at most $n - \max_{i \in [\sigma]} P[i]$.*

Proof (Proof of Upper Bound). This claim is proven by providing a procedure to determine a sequence of $n - \max_{i \in [\sigma]} P[i]$ swaps to transform any word $w \in \Sigma^{*|P}$ into some word $v \in \Sigma^{*|P}$. We assume, without loss of generality, that $P[1] \geq P[2] \geq \cdots \geq P[\sigma]$. The procedure described by Algorithm 1 operates by iterating over the set of symbols in Σ, and the set of occurrences of each symbol in the word. At each step, we have a symbol $x \in [2, \sigma]$ and index $k \in [1, P[x]]$. The procedure finds the position i of the kth appearance of symbol x in w, and the position j of the kth appearance of x in v. Formally, i is the value such that $w[i] = x$ and $|\{i' \in [1, i-1] \mid w[i'] = x\}| = k$ and j the value such that $v[j] = x$ and $|\{j' \in [1, j-1] \mid v[j'] = x\}| = k$. Finally, the algorithm adds the swap (i, j) to the set of swaps, and then moves to the next symbol.

This procedure requires one swap for each symbol in w other than 1, giving a total of $n - \max_{i \in [\sigma]} P[i]$ swaps. Note that after each swap, the symbol at position j of the word is the symbol $v[j]$. Therefore, after all swaps have been applied, the symbol at position $j \in \{i \in [1, |w|] \mid v[i] \in \Sigma \setminus \{1\}\}$ must equal $v[j]$. By extension, for any index i such that $v[i] = 1$, the symbol at position i must be 1, and thus equal $v[i]$. Therefore this procedure transforms w into v.

Algorithm 1. Procedure to select 2-swaps to generate a path from w to v.

1: $S \leftarrow \emptyset$ ▷ Set of 2-swaps
2: **for** $x \in \Sigma \setminus \{1\}$ **do**
3:      **for** $1 \leq k \leq P_x$ **do**
4:          $i \leftarrow$ index of kth occurrence of x in w
5:          $j \leftarrow$ index of kth occurrence of x in v
6:          $S \leftarrow S \cup (i, j)$
7:      **end for**
8:      Apply all 2-swaps in S to w and set $S \leftarrow \emptyset$
9: **end for**

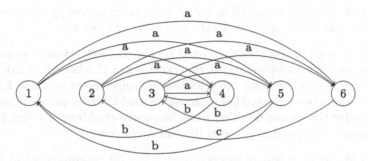

Fig. 1. The graph $G(aaabbc, bcbaaa)$ with the edge (i, j) labelled by $w[i] = v[j]$.

In order to prove the lower bound on the diameter (i.e. that $\mathtt{diam}(G) \geq n - \max_{i \in [\sigma]} P[i]$) we introduce a new auxiliary structure, the *2-swap graph*. Informally, the 2-swap graph, defined for a pair of words $w, v \in \Sigma^{*|P}$ and denoted $G(w, v) = \{V(w, v), E(w, v)\}$ is a directed graph such that the edge $(u_i, u_j) \in V(w, v) \times V(w, v)$ if and only if $w[i] = v[j]$. Note that this definition allows for self-loops.

Definition 3 (2-swap Graph). *Let $w, v \in \Sigma^{*|P}$ be a pair of words. The 2-swap graph $G(w, v) = \{V(w, v), E(w, v)\}$ contains the vertex set $V(w, v) = \{u_1, u_2, \ldots, u_{|w|}\}$ and edge set $E(v, w) = \{(v_i, v_j) \in V(w, v) \times V(w, v) \mid w[i] = v[j] \text{ or } v[i] = w[j]\}$. The edge set, E, is defined as follows, for all $i, j \in \Sigma$ there exists an edge $(i, j) \in E$ if and only if $w[i] = v[j]$.*

An example of the 2-swap graph is given in Fig. 1.

Lemma 4. *Let $G(w, v)$ be a graph constructed as above for transforming w into v using 2-swaps. Then, there exists a procedure to convert any cycle cover of $G(w, v)$, C, into a $w - v$ path in $G(P)$*

Proof. Let $C \in \mathcal{C}$ be a cycle where $C = (e^1, e^2, \ldots, e^{|C|})$ and $e_2^i = e_1^{i+1 \bmod |C|}$. The $w - v$ path (i.e. a sequence of 2-swaps) is constructed as follows. Starting with $i = 1$ in increasing value of $i \in [|C| - 1]$, the 2-swap (e_1^i, e_2^i) is added to the set of 2-swaps S. Where e_1^i and e_2^i are the endpoints of edge e^i for each i.

Assume, for the sake of contradiction, that S does not correspond to a proper set of 2-swaps converting w into v. Then, there must exist some symbol at position i such that the symbol $w[i]$ is placed at some position j such that $w[i] \neq v[j]$. As w_i must be placed at some position that is connected to node i by an edge, there must be an edge between i and j, hence $w_i = v_j$, contradicting the construction of $G(w,v)$. Therefore, S must correspond to a proper set of 2-swaps.

Corollary 4. *Let C be a cycle cover of $G(w,v)$. Then there exists a set of $\sum_{c \in C} |c| - 1$ 2-swaps transforming w in to v.*

Corollary 5. *Let S be the smallest set of 2-swaps transforming w in to v, then S must correspond to a vertex disjoint cycle cover of $G(w,v)$.*

Proof. For the sake of contradiction, let S be the smallest set of 2-swaps transforming w into v, corresponding to the cycle cover C where C is not vertex disjoint. Let $c_1, c_2 \in C$ be a pair of cycles sharing some vertex u. Then, following the construction above, the symbol w_u must be used in two separate positions in v, contradicting the assumption that v can be constructed from w using 2-swaps. Hence S must correspond to a vertex disjoint cycle cover.

Corollary 6. *Given a pair of words $w, v \in \Sigma^{*|_P}$, the minimum set of 2-swaps transforming w into v S corresponds to the vertex disjoint cycle cover of $G(w,v)$ maximising the number of cycles.*

Lemma 5 (Lower Bound). *The diameter of $G(p)$ is at least $n - \max_{i \in [\sigma]} P[i]$.*

Proof. We assume w.l.o.g. that $P[1] \geq P[2] \geq \cdots \geq P[\sigma]$. Let $w, v \in \Sigma^{*|_P}$ satisfy:

$$w = (123 \ldots \sigma)^{P[\sigma]} (123 \ldots \sigma - 1)^{P[\sigma-1]-P[\sigma]} \ldots 1^{P[1]-P[2]}$$

and

$$v = (23 \ldots \sigma 1)^{P[\sigma]} (23 \ldots (\sigma - 1)1)^{P[\sigma-1]-P[\sigma]} \ldots 1^{P[1]-P[2]},$$

i.e. w is made up of $P[1]$ subwords, each of which are of the form $12 \ldots k$, and v is made up of the same subwords as w but each of them has been cyclically shifted by one (for example when $P = (3, 2, 1)$ we have $w = 123121$ and $v = 231211$). Following Corollary 6, the minimum number of 2-swaps needed to convert w into v can be derived from a vertex disjoint cycle cover of $G(w,v)$ with the maximum number of cycles.

Observe that any occurrence of symbol σ must have an outgoing edge in $G(w,v)$ to symbol 1, and an incoming edge from symbol $\sigma - 1$. Repeating this logic, each instance of σ must be contained within a cycle of length σ. Removing each such cycles and repeating this argument gives a set of $P[1]$ cycles, with $P[\sigma]$ cycles of length σ, $P[\sigma - 1] - P[\sigma]$ cycles of length $\sigma - 1$, and generally $P[i] - P[i+1]$ cycles of length i. This gives the number of 2-swaps needed to transform w to v being a minimum of $n - P[1] = n - \max_{i \in [\sigma]} P[i]$.

Theorem 1 follows from Lemmas 3 and 5.

4   Hamiltonicity

In this section, we prove that the configuration graph contains Hamiltonian paths and, for binary alphabets, we provide an efficient algorithm for enumerating a Hamiltonian path. We first show that every configuration graph over a binary Parikh vector is Hamiltonian. This is then generalised to alphabets of size σ, using the binary case to build Hamiltonian paths with alphabets of size σ.

Binary Alphabets. For notational conciseness, given a symbol a the notation $\bar{a}$ is used to denote $\bar{a} \in \Sigma, a \neq \bar{a}$, i.e. if $a = 1$, then $\bar{a} = 2$. We prove Hamiltonicity via a recursive approach that forms the basis for our enumeration algorithm. Our proof works by taking an arbitrary word in the graph w, and constructing a path starting with w. At each step of the path, the idea is to find the shortest suffix of w such that both symbols in Σ appear in the suffix. Letting $w = ps$, the path is constructed by first forming a path containing every word ps', for every $s' \in \Sigma^{P(s)}$, i.e. a path from w transitioning through every word formed by maintaining the prefix p and permuting the suffix s. Once every such word has been added to the path, the algorithm repeats this process by performing some swap of the form $(|p|, i)$ where $i \in [|p| + 1, |w|]$, i.e. a swap taking the last symbol in the prefix p, and replacing it with the symbol $\overline{w[|p|]}$ from some position in the suffix.

This process is repeated, considering increasingly long suffixes, until every word has been covered by the path. Using this approach, we ensure that every word with the same prefix is added to the path first, before shortening the prefix. The algorithm HAMILTONENUMERATION outlines this logic within the context of the enumeration problem, where each transition is output while constructing the path.

Theorem 2. *For every Parikh vector $P \in \mathbb{N}_0^2$ and word $w \in \Sigma^{*|P}$, there exists a Hamiltonian path starting at w in the configuration graph $G(P) = (V(P), E(P))$.*

As stated in [12] the binary reflected grey code gives an ordering for the words over a Binary alphabet of a certain Parikh Vector $(k, n - k)$ with a single 2-swap between each subsequent word. This does indeed prove Theorem 2 by providing a Hamiltonian Circuit for a given binary Parikh vector (it is worth noting that following a Hamiltonian Circuit starting at w gives a Hamiltonian path from W.) However, we also present our own inductive proof for this case to provide an explanation of our enumeration algorithm.

Proof. We prove this statement in a recursive manner. As a base case, consider the three vectors of length 2, $(2, 0), (0, 2)$ and $(1, 1)$. Note that there exists only a single word with the Parikh vectors $(2, 0)$ or $(0, 2)$, and thus the graph must, trivially, be Hamiltonian. For the Parikh vector $(1, 1)$, there exists only the words 12 and 21, connected by the 2-swap $(1, 2)$ and therefore is also Hamiltonian path and it can be found stating at either word.

In the general case, assume that for every Parikh vector $P' = (P'_1, P'_2)$ with $P'_1 + P'_2 < \ell$, the graph $G(P')$ contains a Hamiltonian path, and further there exists such a path starting at every word in $\Sigma^{P'}$. Now, let $P = (P_1, P_2)$ be an arbitrary Parikh vector such that $P_1 + P_2 = \ell$. Given some word $w \in \Sigma^{*|P}$, observe that there must exist some Hamiltonian path starting at the word $w[2, \ell]$ in the subgraph $G'(P) = (V'(P), E'(P))$ where $V'(P) = \{u \in V(P) \mid u[1] = w[1]\}$ and $E'(P) = (V'(P) \times V'(P)) \cap E(P)$. Let w' be the last word visited by the Hamiltonian path in $G'(P)$, and let i be some position in w' such that $w'[i] = \overline{w[1]}$. Note that there must exist Hamiltonian path starting at $(w \circ (1, i))[2, \ell]$ in the subgraph $G''(P) = (V''(P), E''(P))$ where $V''(P) = \{u \in V(P) \mid u[1] = \overline{w[1]}\}$ and $E''(P) = (V''(P) \times V''(P)) \cap E(P)$. As every vertex in $G(P)$ is either in the subgraph $G'(P)$ or $G''(P)$, the Hamiltonian paths starting at w in $G'(P)$ and at $w' \circ (1, i)$ in $G''(P)$ cover the complete graph. Further, as these paths are connected, there exists a Hamiltonian path starting at the arbitrary word $w \in \Sigma^{*|P}$, and therefore the Theorem holds.

Enumeration. We now provide our enumeration algorithm. Rather than output each word completely, we instead maintain the current state of the word in memory and output the swaps taken at each step, corresponding to the transitions. This way, at any given step the algorithm may be paused and the current word fully output, while the full path can be reconstructed from only the output. There are two key challenges behind this algorithm. First is the problem of deciding the next swap to be taken to move from the current word in the graph to the next word. Second, is the problem of minimising the worst-case delay in the output of these swaps, keeping in mind that the output is of constant size.

High-Level Idea of HAMILTONENUMERATION
From a given word w with Parikh vector, P, the algorithm works by first finding the shortest suffix s of w such that there exists some pair of indices i, j for which $s[i] \neq s[j]$. Using this suffix and letting $w = us$, we find a path through every vertex in $G(P)$ with the prefix u. Note that following the same arguments as Theorem 2, such a path must exist. Once every word in $G(P)$ with the prefix u has been visited by the path, the algorithm then enumerates every word with the prefix $u[1, |u| - 1]$, extending the current path. When adding every word with the prefix $u[1, |u| - 1]$ to the path, note that every word with the prefix u has already been added, thus all that is left is to add those words with the prefix $u[1, |u| - 1]\overline{u[|u|]}$, which is achieved via the same process as before.

The swaps are determined as follows. From the initial word w, let R_1 be the last occurrence of the symbol 1 in w, and let R_2 be the last occurrence of 2 in w. The first swap is made between $\min(R_1, R_2)$ and $\min(R_1, R_2) + 1$, with the algorithm then iterating through every word with the Parikh vector $P[w[\min(R_1, R_2), |w|]] - P\left[\overline{w[\min(R_1, R_2)]}\right]$.

In the general case, a call is made to the algorithm with a Parikh vector $P = (P_1, P_2)$, with the current word w fixed, and the assumption that no word with the prefix $w[1, |w| - (P_1 + P_2)]$ has been added to the path other than w. The algorithm, therefore, is tasked with iterating through every word with the current prefix. Let R_1 be the last occurrence of the symbol 1, and R_2 be the last occurrence of the symbol 2 in the current word. The algorithm first enumerates every word with the prefix $w[1, \min(R_1, R_2) - 1]$. Noting that there exists only a single word with the prefix $w[1, \min(R_1, R_2)]$, it is sufficient to only enumerate through those words with the prefix $w[1, \min(R_1, R_2) - 1] \overline{w[\min(R_1, R_2)]}$. The first swap made by this algorithm is between $(\min(R_1, R_2), \min(R_1, R_2) + 1)$, allowing a single recursive call to be made to $HamiltonianEnumeration(P(w \circ (\min(R_1, R_2), \min(R_1, R_2) + 1))[\min(R_1, R_2) + 1, |w|],$ CURRENT CALL$)$, where CURRENT CALL denotes the pointer to the current call on the stack. From this call the algorithm enumerates every word with the prefix $w[1, \min(R_1, R_2) - 1] \overline{w[\min(R_1, R_2)]}$. As every word with the prefix $w[1, \min(R_1, R_2)]$ has already been output and added to the path, once this recursive call has been made, every word with the prefix $w[1, \min(R_1, R_2) - 1]$ will have been added to the path. Note that the word w is updated at each step, ending at the word w'.

After every word with the prefix $w'[1, \min(R_1, R_2) - 1]$ has been added to the path, the next step is to add every word with the prefix $w'[1, \min(R_1, R_2) - 2]]$ to the path. As every word with the prefix $w[1, \min(R_1, R_2) - 1]$ is already in the path, it is sufficient to add just those words with the prefix $w'[1, \min(R_1, R_2) - 2] \overline{w[\min(R_1, R_2)]}$ to the path. This is achieved by making the swap between $\min(R_1, R_2) - 2$, and the smallest value $i > \min(R_1, R_2) - 2$ such that $w[i] \neq w'[\min(R_1, R_2) - 1]$, then recursively enumerating every word with the prefix $w'[1, \min(R_1, R_2) - 2]$. This process is repeated in decreasing prefix length until every word has been enumerated.

To efficiently determine the last position in the current word w containing the symbols 1 and 2, a pair of balanced binary search trees are maintained. The tree T_1 corresponds to the positions of the symbol 1 in w, with each node in T_1 being labelled with an index and the tree sorted by the value of the labels. Analogously, tree T_2 corresponds to the positions of the symbol 2 in w. Using these trees, note that the last position in w at which either symbol appears can be determined in $O(\log n)$ time, and further each tree can be updated in $O(\log n)$ time after each swap.

Lemma 6. *Let P be a Parikh vector of length n, and let $w \in \Sigma^{*|P}$ be a word.* HAMILTONIANENUMERATION *outputs a path visiting every word in $\Sigma^{*|P}$ starting at w.*

Proof. This lemma is proven via the same tools as Theorem 2. Explicitly, we show first that the algorithm explores every suffix in increasing length, relying on the exploration of suffixes of length 2 as a base case, then provide an inductive proof of the remaining cases. We assume that the starting word has been fully output as part of the precomputation. With this in mind, note that there are two cases for length 2 prefixes, either the suffix contains two copies of the same symbol or one copy of each symbol. In the first case, as w has been output, so has every permutation of the length 2 prefix of w. Otherwise, the algorithm outputs the swap $(n - 1, n)$ and returns to the previous call.

In the general case, we assume that for some $\ell \in [n]$, every permutation of $w[n - \ell + 1, n]$ has been visited by the path. Further, we assume the algorithm can, given any word v, visit every word of the form $v[1, n - \ell]u$, for every $u \in \Sigma^{P(v[n - \ell + 1, n])}$, i.e. the algorithm is capable of taking any word v as an input, and visiting every word with the same Parikh vector $P(v)$ and prefix $v[1, n - \ell + 1]$. Note that in the case that $w[n-\ell, n] = w[n-\ell]^\ell$, the algorithm has already visited every word in $\Sigma^{P(w)}$ with the prefix $w[1, n - \ell]$. Otherwise, as the algorithm has, by this point, visited every word of the form $w[1, n - ell + 1]u$, for every $u \in \Sigma^{P(v[n-\ell-1,n])}$, it is sufficient to show that the algorithm visits every word of the form $w[1, \ell - 1]\overline{w[\ell]}u$, for every $u \in \Sigma^{P'}$, $P' = P(w[n - \ell, n]) - P(\overline{w}[\ell])$.

Let w' be the last word visited by the algorithm with the prefix $w[1, n-\ell+1]$. Note that the first step taken by the algorithm is to determine the first position j in $w'[n - \ell + 1, n]$ containing the symbol $\overline{w[\ell]}$. Therefore, by making the swap $(n - \ell, j)$, the algorithm moves to some word with a suffix in $\Sigma^{P'}$, where $P' = P(w[n - \ell, n]) - P(\overline{w[n - \ell]})$. As the algorithm can, by inductive assumption, visit every word with a suffix of length $\ell - 1$, the algorithm must also be able to visit every word with a suffix of length ℓ, completing the proof.

Lemma 7. *Let P be a Parikh vector, and let $w \in \Sigma^{*|P}$ be a word. The path output by* HAMILTONENUMERATION *does not visit any word in $w \in \Sigma^{*|P}$ more than once.*

Proof. Note that this property holds for length 2 words. By extension, the length at most 2 path visiting every word with the prefix $w[1, n - 2]$ does not visit the same word twice before returning to a previous call on the stack.

Assume now that, given any input word $v \in \Sigma^{*|P}$, the algorithm visits every word in $\Sigma^{*|P}$ with the prefix $v[1, n - l + 1]$ without repetition, and has only visited words with this prefix. Further, assume that $P(v[n - \ell, n]) \neq (0, \ell - 1)$ or $(\ell - 1, 0)$. Then, after every such word has been visited by the path, the algorithm returns to the previous state, with the goal of enumerating every word with the prefix $v[1, n - \ell]$. As every word in $\Sigma^{*|P}$ with the prefix $v[1, n - \ell + 1]$ has been visited, it is sufficient to show that only those words with the prefix $v[1, n - \ell]\overline{v[n\ell + 1]}$ are enumerated. The first swap made at this state is between ℓ and the smallest index $j \in [n - \ell + 1, n]$ such that $v[n\ell] \neq v[j]$, which, as the algorithm has only visited words with the prefix $v[1, n - l + 1]$, has not previously been visited. After this swap, the algorithm enumerates every word with the prefix $v[1, n-\ell+1]\overline{v[n - \ell]}$, which, by the inductive assumption, is done without visiting the same word. Therefore, by induction. every word with the prefix $v[1, n - \ell]$ is visited by the path output by HAMILTONIANENUMERATION exactly once.

Theorem 3. *Given a Parikh vector $P = (P_1, P_2)$ such that $P_1 + P_2 = n$, and word $w \in \Sigma^P$,* HAMILTONIANENUMERATION *outputs a Hamiltonian path with at most $O(\log n)$ delay between the output of each edge after $O(n \log n)$ preprocessing.*

Proof. Following Lemmas 6 and 7, the path outputted by HAMILTONENUMERATION is Hamiltonian. In the preprocessing step, the algorithm constructs two balanced binary search trees T_1 and T_2. Every node in T_1 is labelled by some index $i_1 \in [n]$ for which $w[i_1] = 1$, and sorted by the values of the labels. Similarly, every node in T_2 is labelled by some index $i_2 \in [n]$ for which $w[i_2] = 2$, and sorted by the values of the labels. As each of these constructions requires at most $O(n \log n)$ time, the total complexity of the preprocessing is $O(n \log n)$.

During each call, we have one of three cases. If either value of the Parikh vector is 0, then the algorithm immediately returns to the last state without any output. If the Parikh vector is $(1, 1)$, then the algorithm outputs a swap between the two symbols, updates the trees T_1 and T_2, requiring at most $O(\log n)$ time, then returns to the last state. In the third case, the Parikh vector (P_1, P_2) satisfies $P_1 > 0, P_2 > 0$. First, the algorithm determines the last position in the current state of the word w containing the symbol 1 and the last position containing the symbol 2, i.e. the values $R_1 = \max_{j \in [1,n]} w[i_1] = 1$ and $R_2 = \max_{j \in [1,n]} w[i_2] = 2$. These values can be determined in $O(\log n)$ time using the trees T_1 and T_2. Using these values, the algorithm iterates through every length from $\min(T_1, T_2)$ to $n - (P_1 + P_2 - 1)$, enumerating every word in $\Sigma^{P(w)}$ with the prefix $w[1, n - (P_1 + P_2 - 1)]$. For each $\ell \in [\min(T_1, T_2), n - (P_1 + P_2 - 1)]$, the algorithm outputs the swap (ℓ, j), where $j \in [n - \ell, n]$ is the largest value for which $w[j] = \overline{w[\ell]}$. After this output, the algorithm updates the trees T_1 and T_2. Note that both finding the value of j and updating the trees require $O(\log n)$ time. After this swap, the algorithm makes the next call to HAMILTONIANENUMERATION. Note that after this call, HAMILTONIANENUMERATION must either return immediately to the last state or output some swap before either returning or making the next recursive call. Therefore, ignoring the time complexity of returning to a previous state in the stack, the worst case delay between outputs is $O(\log n)$, corresponding to searching and updating the trees T_1 and T_2.

To avoid having to check each state in the stack after returning from a recursive call, the algorithm uses tail recursion. Explicitly, rather than returning to the state in the stack from which the algorithm was called, the algorithm is passed a pointer to the last state in the stack corresponding to a length ℓ such that some word with the prefix $w[1, n - \ell]$ has not been output. To do so, after the swap between $n - (P_1 + P_2 - 1)$ and j is made, for the value j as defined above, the algorithm passes the pointer it was initially given, denoted in the algorithm as *last_state* to the call to HAMILTONIANENUMERATION, allowing the algorithm to skip over the current state during the recursion process.

General Alphabets. We now show that the graph is Hamiltonian for any alphabet of size $\sigma > 2$. The main idea here is to build a cycle based on recursively grouping together sets of symbols. Given a Parikh vector $P = (P_1, P_2, \ldots, P_\sigma)$, our proof operates in a set of σ recursive phases, with the ith step corresponding to finding a Hamiltonian path in the graph $G(P_i, P_{i+1}, \ldots, P_\sigma)$, then mapping this path to one in $G(P)$. The paths in $G(P_i, P_{i+1}, \ldots, P_\sigma)$ are generated in turn by a recursive process. Starting with the word w, first, we consider the path visiting every vertex corresponding to a permutation of the symbols $i + 1, \ldots, \sigma$ in w. Explicitly, every word v in this path is of the form:

$$v[i] = \begin{cases} w[i] & w[i] \in \{1, 2, \ldots, i\} \\ x_i \in \{i+1, \ldots, \sigma\} & w[i] \notin \{1, 2, \ldots, i\} \end{cases},$$

where x_i is some arbitrary symbol $\{i, i+1, \ldots, \sigma\}$. Further, every such word is visited exactly once.

After this path is output, a single swap corresponding to the first swap in $G(P_i, (P_{i+1} + P_{i+2}, \ldots, P_\sigma))$ is made, ensuring that this swap must involve some position in w containing the symbol i. After this swap, another path visiting exactly once every word corresponding to a permutation of the symbols $i + 1, \ldots, \sigma$ in w can be output. By repeating this for every swap in $G(P_i, (P_{i+1} + P_{i+2}, \ldots, P_\sigma))$, inserting a path visiting exactly once every word corresponding to a permutation of the symbols $i + 1, \ldots, \sigma$ in w between each such swap, note that every permutation of the symbols $i, i+1, \ldots, \sigma$ in w is output exactly once. In other words, every word $v \in \Sigma^{*|P}$ of the form

$$v[i] = \begin{cases} w[i] & w[i] \in \{1, 2, \ldots, i-1\} \\ x_i \in \{i, i+1, \ldots, \sigma\} & w[i] \in \{i, i+1, \ldots, \sigma\} \end{cases},$$

where x_i is some symbol in $\{i, i+1, \ldots, \sigma\}$. Further, each such word is visited exactly once. Using the binary alphabet as a base case, this process provides an outline of the proof of the Hamiltonicity of $G(P)$. A full proof of Theorem 4 can be found in the full version of this paper [7].

Theorem 4. *Given an arbitrary Parikh vector $P \in \mathbb{N}^\sigma$, there exists a Hamiltonian path starting at every vertex v in the configuration graph $G(P)$.*

Conclusion: Following the work on 2-swap, the most natural step is to consider these problems for k-swap based on two variants with exactly k and less or equal to k. Note that a configuration graph for exactly k-swap permutation might not have a single component. We also would like to point to other attractive directions of permutations on multidimensional words [5] and important combinatorial objects such as necklaces and bracelets [6,8]. For the 2-swap graph specifically, we leave open the problem of determining the shortest path between two given words w and v. We conjecture that the simple greedy algorithm used to derive the upper bound in Lemma 3 can be used to find the shortest path between any pair of vertices.

References

1. Ackerman, M., Mäkinen, E.: Three new algorithms for regular language enumeration. In: Ngo, H.Q. (ed.) COCOON 2009. LNCS, vol. 5609, pp. 178–191. Springer, Heidelberg (2009). https://doi.org/10.1007/978-3-642-02882-3_19
2. Ackerman, M., Shallit, J.: Efficient enumeration of words in regular languages. Theoret. Comput. Sci. **410**(37), 3461–3470 (2009)
3. Adamson, D.: Ranking and unranking k-subsequence universal words. In: WORDS, p. 47–59. Springer Nature Switzerland (2023). https://doi.org/10.1007/978-3-031-33180-0_4
4. Adamson, D., Deligkas, A., Gusev, V., Potapov, I.: On the hardness of energy minimisation for crystal structure prediction. Fund. Inform. **184**(3), 181–203 (2021)

5. Adamson, D., Deligkas, A., Gusev, V.V., Potapov, I.: The complexity of periodic energy minimisation. In: MFCS 2022. LIPIcs, vol. 241, pp. 8:1–8:15 (2022)
6. Adamson, D., Deligkas, A., Gusev, V.V., Potapov, I.: The k-centre problem for classes of cyclic words. In SOFSEM 2023. LNCS, vol. 13878, pp. 385–400. Springer (2023). https://doi.org/10.1007/978-3-031-23101-8_26
7. Adamson, D., Flaherty, N., Potapov, I., Spirakis, P.: Structural and combinatorial properties of 2-swap word permutation graphs. arXiv:2307.01648 (2023)
8. Adamson, D., Gusev, V.V., Potapov, I., Deligkas, A.: Ranking Bracelets in Polynomial Time. In: CPM 2023, LIPIcs, vol. 191, pp. 4:1–4:17 (2021)
9. Amir, A., Paryenty, H., Roditty, L.: On the hardness of the consensus string problem. Inf. Process. Lett. **113**(10), 371–374 (2013)
10. Angel, O., Holroyd, A., Romik, D., Virág, B.: Random sorting networks. Adv. Math. **215**(2), 839–868 (2007)
11. Babai, L., Seress, A.: On the diameter of permutation groups. Eur. J. Comb. **13**(4), 231–243 (1992)
12. Buck, M., Wiedemann, D.: Gray codes with restricted density. Discret. Math. **48**(2–3), 163–171 (1984)
13. Collins, C., Darling, G.R., Rosseinsky, M.J.: The flexible unit structure engine (FUSE) for probe structure-based composition prediction. Faraday Discuss. **211**, 117–131 (2018)
14. Collins, C., et al.: Accelerated discovery of two crystal structure types in a complex inorganic phase field. Nature **546**(7657), 280–284 (2017)
15. Crochemore, M., Rytter, W.: Jewels of stringology. World Scientific (2003)
16. Ganczorz, M., Gawrychowski, P., Jez, A., Kociumaka, T.: Edit distance with block operations. In: ESA, Schloss Dagstuhl-Leibniz-Zentrum fuer Informatik (2018)
17. Goddard, W., Raines, M.E., Slater, P.J.: Distance and connectivity measures in permutation graphs. Discret. Math. **271**(1), 61–70 (2003)
18. Helfgott, H.A., Seress, Á.: On the diameter of permutation groups. Annals of Mathematics, pp. 611–658 (2014)
19. Helfgott, H.A., Seress, Á., Zuk, A.: Random generators of the symmetric group: diameter, mixing time and spectral gap. J. Algebra **421**, 349–368 (2015)
20. Jerrum, M.R.: The complexity of finding minimum-length generator sequences. Theoret. Comput. Sci. **36**, 265–289 (1985)
21. Konstantinova, E.: Some problems on cayley graphs. Linear Algebra Appli. **429**(11), 2754–2769 (2008); Special Issue devoted to selected papers presented at the first IPM Conference on Algebraic Graph Theory
22. Kornhauser, D., Miller, G.L., Spirakis, P.G.: Coordinating pebble motion on graphs, the diameter of permutation groups, and applications. In: FOCS 1984, pp. 241–250. IEEE Computer Society (1984)
23. Levy, A.: Exploiting pseudo-locality of interchange distance. In: SPIRE 2021, pp. 227–240. Springer (2021), https://doi.org/10.1007/978-3-030-86692-1_19
24. Maji, H., Izumi, T.: Listing center strings under the edit distance metric. In: Lu, Z., Kim, D., Wu, W., Li, W., Du, D.-Z. (eds.) COCOA 2015. LNCS, vol. 9486, pp. 771–782. Springer, Cham (2015). https://doi.org/10.1007/978-3-319-26626-8_57
25. Mäkinen, E.: On lexicographic enumeration of regular and context-free languages. Acta Cybernet. **13**(1), 55–61 (1997)
26. Mütze, T.: Combinatorial Gray codes - an updated survey. arXiv preprint arXiv:2202.01280 (2022)
27. Ruskey, F., Savage, C., Min Yih Wang, T.: Generating necklaces. J. Algorithms **13**(3), 414–430 (1992)

28. Schmid, M.L., Schweikardt, N.: Spanner evaluation over slp-compressed documents. In: PODS 2021, pp. 153–165. ACM (2021)
29. Schmid, M.L., Schweikardt, N.: Query evaluation over slp-represented document databases with complex document editing. In: PODS 2022, pp. 79–89. ACM (2022)
30. Wasa, K.: Enumeration of enumeration algorithms. arXiv e-prints, pp. 1605 (2016)

Directed Ear Anonymity

Marcelo Garlet Milani[✉] [ID]

National Institute of Informatics, Tokyo, Japan
mgmilani@nii.ac.jp

Abstract. We define and study a new structural parameter for directed graphs, which we call *ear anonymity*. Our parameter aims to generalize the useful properties of *funnels* to larger digraph classes. In particular, funnels are exactly the acyclic digraphs with ear anonymity one. We prove that computing the ear anonymity of a digraph is NP-hard and that it can be solved in $\mathcal{O}(m(n + m))$-time on acyclic digraphs (where n is the number of vertices and m is the number of arcs in the input digraph). It remains open where exactly in the polynomial hierarchy the problem of computing ear anonymity lies, however for a related problem we manage to show Σ_2^p-completeness.

Keywords: Digraphs · Algorithms · Ear anonymity · Computational complexity

1 Introduction

One approach for handling computationally hard problems is to design algorithms which are efficient if certain structural parameters of the input are small. In undirected graphs, *width* parameters such as *treewidth* [3,6,18] and *cliquewidth* [7] are very effective in handling a number of problems (see also [9]).

Width parameters for directed graphs, however, seem to be less powerful [11,12]. While *directed treewidth* helps when solving LINKAGE, where the task is to connect terminal pairs by disjoint paths, [13], the algorithm has a running time of the form $\mathcal{O}(n^{f(k,dtw)})$, where k is the number of terminals and dtw is the directed treewidth of the input digraph. At the same time, there is no $f(k)n^{g(dtw)}$-time algorithm for LINKAGE [20] under standard assumptions, and many further problems remain hard even if the directed treewidth of the input is a constant [11].

One of the shortcomings of directed treewidth is that it cannot explain the structural complexity of acyclic digraphs, as those digraphs have directed treewidth zero. Indeed, the digraph constructed in the hardness reduction for LINKAGE provided by [20] is acyclic. Since fundamental problems like LINKAGE remain NP-hard even if the input digraph is acyclic, it is natural to search for additional parameters which may help in the study of the structural of digraphs and also of acyclic digraphs.

© The Author(s), under exclusive license to Springer Nature Switzerland AG 2024
J. A. Soto and A. Wiese (Eds.): LATIN 2024, LNCS 14579, pp. 77–97, 2024.
https://doi.org/10.1007/978-3-031-55601-2_6

Funnels are an algorithmically useful subclass of acyclic digraphs [17]. For example, it is easy to solve LINKAGE in polynomial time on funnels. Further, FUNNEL ARC DELETION SET, the problem of deleting at most k arcs from a digraph in order to obtain a funnel, admits a polynomial kernel [16].

Funnels have found application in *RNA assembly* [15], modeling a class of digraph on which FLOW DECOMPOSITION is easy to solve. Additionally, [4] considers two generalization of funnels, namely k-*funnels* and a class called $\mathcal{ST}_k$, and then shows that STRING MATCHING TO LABELED GRAPH can be solved more efficiently on k-funnels and in digraphs of the class $\mathcal{ST}_k$ if k is small.

In this work, we generalize the properties of funnels by defining a parameter called *ear anonymity*. This parameter is defined in such a way that funnels are exactly the acyclic digraphs with ear anonymity one. We show that, while computing the ear anonymity of a digraph is NP-hard in general, it can be computed in $\mathcal{O}(m(n + m))$-time if the input digraph is acyclic.

We define ear anonymity together with three relevant computational problems in Sect. 3. In Sect. 4 we prove one of our main results, providing a polynomial-time algorithm for EAR ANONYMITY on acyclic digraphs. In Sect. 5, we show that all computational problems defined here regarding ear anonymity are NP-hard in the general setting. Further, in Sect. 6 we show another of our main results, namely that one of these problems is even Σ_2^p-complete, a class which is "above" NP in the polynomial hierarchy. To achieve this, we define two additional computational problems which we use to help us construct our reduction, proving that each of them is also Σ_2^p-hard.

Since the literature on hardness results on higher levels of the polynomial hierarchy is not as rich as for NP-hardness results, we consider the techniques used in Sect. 6 to be of independent interest and to be potentially useful in showing that further problems on digraphs are Σ_2^p-complete. In particular, the auxiliary problems considered are related to finding linkages in directed graphs, a fundamental problem often used in NP-hardness reductions. To the best of our knowledge (see [19] for a survey on related hardness results), none of the hard problems in the polynomial hierarchy "above" NP studied so far are related to linkages on digraphs.

Finally, we provide some concluding remarks and discuss future work in Sect. 7. Due to space constrains, proofs marked with $(\star)$ are deferred to full version of this manuscript.

2   Preliminaries

A *directed graph*, or *digraph*, is a tuple $D := (V, E)$ where V is the *vertex set* and $E \subseteq \{(v, u) \mid v, u \in V \text{ and } v \neq u\}$ is the *arc set*. We write $V(D)$ for the set V and $E(D)$ for the set E.

The inneighbors of a vertex v in a digraph D are denoted by $\mathsf{in}_D(v) = \{u \in V \mid (u, v) \in E\}$; its outneighbors are given by $\mathsf{out}_D(v) = \{u \in V \mid (v, u) \in E\}$. The indegree of v is written as $\mathsf{indeg}_D(v) = |\mathsf{in}_D(v)|$, and its outdegree as $\mathsf{outdeg}_D(v) = |\mathsf{out}_D(v)|$. A vertex v is a *source* if $\mathsf{indeg}_D(v) = 0$ and a *sink* if $\mathsf{outdeg}_D(v) = 0$. We omit the index D if the digraph is clear from the context.

We extend the definition of set operators for digraphs. Let $G = (V, E), H = (U, F)$ be two digraphs. We define $H \subseteq G \Leftrightarrow U \subseteq V$ and $F \subseteq E; H \cup G = (U \cup V, F \cup E); H \cap G = (U \cap V, F \cap E)$ and $H \setminus G = H - V(G)$. In particular, we say that H is a *subgraph* of G if $H \subseteq G$.

A *walk* of length ℓ in D is a vertex sequence $W := (v_0, v_1, \ldots, v_\ell)$ such that $(v_i, v_{i+1}) \subseteq E(D)$ holds for all $0 \le i < \ell$. We say that W is a v_0-v_ℓ-walk and write $\mathsf{start}(W)$ for v_0 and $\mathsf{end}(W)$ for v_ℓ.

A walk W is said to be a v_0-v_ℓ-*path* if no vertex appears twice along the walk; W is a *cycle* if $v_0 = v_\ell$ and $(v_0, v_1, \ldots, v_{\ell-1})$ is a path and $\ell \ge 2$; further, W is a *directed ear* if it is either a path or a cycle. Finally, D is *acyclic* if it does not contain any cycles.

Given two walks $W_1 = (x_1, x_2, \ldots, x_j), W_2 = (y_1, y_2, \ldots, y_k)$ with $\mathsf{end}(W_1) = \mathsf{start}(W_2)$, we make use of the concatenation notation for sequences and write $W_1 \cdot W_2$ for the walk $W_3 := (x_1, x_2, \ldots, x_j, y_2, y_3, \ldots, y_k)$. If W_1 or W_2 is an empty sequence, then the result of $W_1 \cdot W_2$ is the other walk (or the empty sequence if both walks are empty).

Let P be a path and X a set of vertices with $V(P) \cap X \ne \emptyset$. We consider the vertices $p_1, \ldots, p_m$ of P ordered by their occurrence on P. Let i be the highest index such that $p_i \in X$, we call p_i the *last vertex of P in X*. Similarly, for the smallest index j with $p_j \in X$ we call p_j the *first vertex of P in X*.

In digraphs, the vertices which can be reached from a vertex v are given by $\mathsf{out}^*(v)$. The vertices which can reach v are given by $\mathsf{in}^*(v)$. That is $u \in \mathsf{in}^*(v)$ if and only if there is a (u, v)-path, and $u \in \mathsf{out}^*(v)$ if and only if there is a (v, u)-path.

Given a digraph D and an arc $(v, u) \in E(D)$, we say that (v, u) is *butterfly contractible* if $\mathsf{outdeg}(v) = 1$ or $\mathsf{indeg}(u) = 1$. The *butterfly contraction* of (v, u) is the operation which consists of removing v, u from D, then adding a new vertex vu, together with the arcs $\{(w, vu) \mid w \in \mathsf{in}_D(v) \setminus \{u\}\}$ and $\{(vu, w) \mid w \in \mathsf{out}_D(u) \setminus \{v\}\}$. Note that, by definition of digraph, we *remove* duplicated arcs and arcs of the form (w, w). If there is a subgraph D' of D such that we can construct another digraph H from D' by means of butterfly contractions, then we say that H is a *butterfly minor of D*, or that D *contains H as a butterfly minor*.

A *subdivision* of an arc (v, u) is the operation of replacing (v, u) by a path v, w, u. We say that digraph H is a *topological minor* of a digraph D if some subdivision H' of H, obtained by iteratively subdiving arcs, is isomorphic to some subgraph of D. See [2, 8] for further information on digraphs.

Definition 1 ([17]). *A digraph D is a funnel if D is a DAG and for every path P from a source to a sink of D of length at least one there is some arc $a \in A(P)$ such that for any different path Q from a (possibly different) source to a (possibly different) sink we have $a \notin A(Q)$.*

Given two sets A, B of vertices in a digraph D, we say that a set of pairwise vertex-disjoint paths $\mathcal{L}$ is a *linkage* from A to B if all paths in $\mathcal{L}$ start in A and end in B.

Definition 2. *Let $i \geq 1$. A language L over an alphabet Γ is in Σ_i^p if there exists a polynomial $p : \mathbb{N} \to \mathbb{N}$ and a polynomial-time Turing Machine M such that for every word $x \in \Gamma^*$,*

$$x \in L \Leftrightarrow \exists u_1 \in \Gamma^{p(|x|)} \, \forall u_2 \in \Gamma^{p(|x|)} \ldots Q_i u_i \in \Gamma^{p(|x|)} \, M(x, u_1, \ldots, u_i) = 1,$$

where Q_i denotes $\forall$ or $\exists$ depending on whether i is even or odd, respectively.
The polynomial hierarchy *is the set* $PH = \bigcup_i \Sigma_i^p$.

It is easy to verify that $\Sigma_1^p = \mathsf{NP}$. We also define, for every $i \geq 1$, the class $\Pi_i^p = \mathsf{co}\Sigma_i^p = \{\overline{L} \mid L \in \Sigma_i^p\}$. The concept of reductions and hardness can be defined in a similar way as for NP-completeness. See [1] for further information on computational complexity.

3   Definition of Ear Anonymity

Funnels are characterized by how easy it is to uniquely identify a maximal path[1]: it suffices to take the private arc of the path. In acyclic digraphs, the maximal paths correspond exactly to the paths which start in a source and end in a sink. In general, this does not have to be case. Indeed, a cycle contains several distinct maximal paths, all of them overlapping. Hence, it is natural that, in general digraphs, we consider not only how to identify maximal paths, but also cycles, leading us to the well-known concept of ears.

We then come to the question of how to uniquely identify a maximal ear in a digraph. Clearly, a single arc does not always suffice, as it can be in several ears. If we take a set of arcs, ignoring their order on the ear, then some rather simple digraphs will require a large number of arcs to uniquely identify an ear, for example the digraph in Fig. 1.

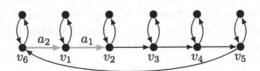

Fig. 1. For any subset of at most 5 arcs of the cycle $(v_1, v_2, v_3, v_4, v_5, v_6, v_1)$ we can find some path visiting such arcs which is distinct from the cycle considered.

Hence, we consider not only the arcs of the ear, but also their order along the ear. We also require the existence of at least one arc in the identifying sequence in order to ensure the parameter is closed under the subgraph relation.

Definition 3. *Let P be an ear. A sequence $(a_1, a_2, \ldots, a_k)$ of arcs of P is an identifying sequence for P if $k \geq 1$ and every ear Q containing $(a_1, a_2, \ldots, a_k)$ in this order is a subgraph of P.*

---

[1] Maximal with respect to the subgraph relation.

Note that the cycle $(v_1, v_2, v_3, v_4, v_5, v_6, v_1)$ in Fig. 1 admits an identifying sequence of length two, namely $\bar{a} = (a_1, a_2)$ where $a_1 = (v_1, v_2)$ and $a_2 = (v_6, v_1)$. We also note that using vertices instead of arcs does not lead to a well-defined parameter, as can be observed in the example given in Fig. 2. As every ear can be uniquely described by ordering its entire arc-set according to its occurrence along the ear, the parameter defined above is well-defined for all ears.

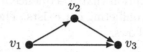

Fig. 2. A digraph with two maximal ears. While the ear (v_1, v_2, v_3) contains all vertices of the ear (v_1, v_3), both ears admit an identifying sequence of length 1.

Intuitively, the shorter the identifying sequence of an ear, the less information is necessary in order to uniquely identify or find such an ear. This leads to the following definition (we ignore maximal ears consisting of a single vertex, as they are seldom interesting and can be found in linear time).

Definition 4. *Let $P = (a_1, a_2, \ldots, a_k)$ be a maximal ear in a digraph D, given by its arc-sequence (in the case of a cycle, any arc of P can be chosen as a_1). The* ear anonymity *of P in D, denoted by $\mathsf{ea}_D(P)$, is the length of the shortest identifying sequence for P. If $k = 0$, we say that $\mathsf{ea}_D(P) = 0$.*

As we are often interested in the worst-case running time of an algorithm, if some ear of a digraph has high anonymity, then this digraph could be a difficult instance in the worst case.

Definition 5. *The* ear anonymity *of a digraph D, denoted by $\mathsf{ea}(D)$, is the maximum ear anonymity of the maximal ears of D.*

It is a simple exercise to compare Definition 5 and Definition 1 to verify the following observation.

Remark 1. An acyclic digraph D is a funnel if and only if $\mathsf{ea}(D) \leq 1$.

It is sometimes useful to know that a parameter is closed under certain operations. For ear anonymity, we can show the following.

Remark 2 (⋆). Let D, H be digraphs such that H is a butterfly minor or topological minor of D. Then $\mathsf{ea}(H) \leq \mathsf{ea}(D)$.

We will use Remark 2 later to draw a connection between ear anonymity and directed treewidth. We now investigate the complexity of computing the ear anonymity of a digraph. Definitions 3 to 5 naturally lead us to three related computational problems.

Since most of the literature on decision problems concerns itself with problems in NP, we formulate the question of our decision problems as an "existential" question (instead of a "for all" question). Hence, the question of whether an arc-sequence $\bar{a}$ is an identifying sequence for an ear P becomes the question of the existence of another ear as defined below.

Definition 6. *Let P be an ear and let $\bar{a}$ be a sequence of arcs of P, sorted according to their order on P. We say that an ear Q is a* conflicting *ear for $(P, \bar{a})$ if Q visits the arcs of $\bar{a}$ in the given order, yet Q is not a subgraph of P.*

It is immediate from definition that a sequence $\bar{a}$ is an identifying sequence for an ear if, and only if, no conflicting ear exists. The first problem we consider can then be formulated as follows.

CONFLICTING EAR
Input   A digraph D, a maximal ear P in D and a sequence $\bar{a}$ of arcs
         of P, sorted according to their occurrence on P.
Question Is there a conflicting ear for $(P, \bar{a})$?

From the above definition it is trivial to derive the following observation.

Remark 3. CONFLICTING EAR is in NP.

Note that the question "is $\bar{a}$ an identifying sequence for P?" is the complement of CONFLICTING EAR and thus, by Remark 3, a coNP question. We can also formulate this question as "for all ears Q, is Q not a conflicting ear for $(P, \bar{a})$?". When considering the problem of determining the ear anonymity of an ear, it seems thus unavoidable to have a quantifier alternation in the question: asking for the existence of an identifying sequence means chaining an existential question with a "for all" question.

EAR-IDENTIFYING SEQUENCE
Input   A digraph D, a maximal ear P in D and an integer k.
Question Is there an identifying sequence $\bar{a}$ for P of length at most k?

Unlike CONFLICTING EAR, it is not clear from the definition whether EAR-IDENTIFYING SEQUENCE is in NP, but one can easily verify containment in a class higher up in the polynomial hierarchy.

Remark 4. EAR-IDENTIFYING SEQUENCE is in Σ_2^p.

As before, asking if an ear has high anonymity is equivalent to asking if no short identifying sequence for that ear exists. It seems again unavoidable to add another quantifier alternation when deciding if a digraph has high ear anonymity: asking if a digraph has high ear anonymity means asking for the existence of an ear for which no short identifying sequence exists.

EAR ANONYMITY
Input   A digraph D and an integer k.
Question Is there a maximal ear P in D such that $ea_D(P) \geq k$?

While it is not clear from the definition whether EAR ANONYMITY is even in Σ_2^p, it is easy to verify that it is in Σ_3^p.

Remark 5. EAR ANONYMITY is in Σ_3^p.

In Sect. 4 we show that CONFLICTING EAR, EAR-IDENTIFYING SEQUENCE and EAR ANONYMITY are in P on DAGs. In Sect. 5 we show that the three previous decision problems are NP-hard in general using some of the results from Sect. 4. Finally, in Sect. 6, we show that EAR-IDENTIFYING SEQUENCE is Σ_2^p-complete.

4   EAR ANONYMITY on DAGs

We start by identifying certain substructures which increase the anonymity of an ear by enforcing certain arcs to be present in any identifying sequence. Two such substructures, called *deviations* and *bypasses*, are defined in Definitions 7 and 8 and illustrated in Figs. 4 and 3 below. Of particular interest are subpaths of an ear which must be *hit* by any identifying sequence. We call these subpaths *blocking subpaths* since they prevent a potential conflicting ear from containing the corresponding bypass or deviation as a subgraph.

Definition 7. *Let P be an ear and let Q be a path in a digraph D. We say that Q is a* deviation *for P if Q is internally disjoint from P and exactly one of* end(Q), start(Q) *lies in P. Additionally, the* start(P)-end(Q) *subpath of P is called a* blocking subpath *for Q if* end(Q) $\in V(P)$, *and the* start(Q)-end(P) *subpath of P is called a* blocking subpath *for Q if* start(Q) $\in V(P)$.

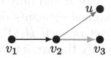

Fig. 3. The path (v_2, u) is a deviation for the path $P = (v_1, v_2, v_3)$. The unique identifying sequence of length one for P is $((v_2, v_3))$.

Definition 8. *Let P be a path in a digraph D. Let $v_1, v_2, \ldots v_n$ be the vertices of P sorted according to their order in P. A* bypass *for P is a path Q in D from some v_i to some v_j with $i < j$ such that $V(Q) \cap V(P) = \{v_i, v_j\}$ and Q is not a subpath of P. Further, the v_i-v_j subpath of P is called the* blocking subpath *for Q.*

If an ear contains many arc-disjoint blocking subpaths, then every identifying sequence must be long. If, on the other hand, the blocking subpaths overlap, then a short identifying sequence may still exist. In order to better analyze the relationship between the length of an identifying sequence and the blocking subpaths of an ear, we model this problem as a problem on intervals. Intuitively, we can consider each arc on an ear to be an integer, ordered naturally along the ear, and each blocking subpath as an interval over the integers. Hence, we are interested in finding a minimum set of integers which hit all the intervals. This naturally leads us to the definitions given below.

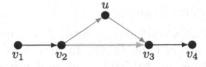

Fig. 4. The path (v_2, u, v_3) is a bypass for $P = (v_1, v_2, v_3, v_4)$. Note that there is exactly one identifying sequence of length 1 for P, namely $((v_2, v_3))$.

Definition 9. *Let $Q_1, Q_2, \ldots, Q_k$ be subpaths of an ear P. The* arc-interval set *of $Q_1, Q_2, \ldots, Q_k$ is the set of intervals $\mathcal{I} = \{I_1, I_2, \ldots, I_k\}$ with $I_i = E(Q_i)$ for all $1 \leq i \leq k$.*

Definition 10. *Let $\mathcal{I} = \{I_1, I_2, \ldots, I_n\}$ be a set of intervals over a finite (ordered) domain U. A set $X \subseteq U$ is a* hitting set *for $\mathcal{I}$ if $I_i \cap X \neq \emptyset$ for every $I_i \in \mathcal{I}$.*

Since an ear can have an exponential number of bypasses and deviations, we are interested in reducing the number of blocking subpaths we need to consider. In particular, if a blocking subpath is fully contained within another, then we can ignore the longer subpath.

Formally, we define a partial ordering $\preceq$ over the blocking paths of the bypasses and the deviations for an ear P as follows. For two blocking subpaths B_a, B_b set $B_a \preceq B_b$ if $\mathsf{start}(B_a)$ is not before $\mathsf{start}(B_b)$ in P and $\mathsf{end}(B_a)$ is not after $\mathsf{end}(B_b)$ in P. That is, $B_a \preceq B_b$ if and only if B_a is a subpath of B_b.

Let $B_1, B_2, \ldots B_k$ be the minimal elements of $\preceq$. Every set of intervals $\mathcal{I}$ which contains the arc-interval set of each $B_1, B_2, \ldots B_k$ is called the *blocking interval set* for P. If $\mathcal{I}$ contains only the arc-interval sets of $B_1, B_2, \ldots B_k$, then it is the minimum blocking interval set for P.

We now establish a connection between hitting sets for a blocking interval set for an ear and the identifying sequence for that ear.

Lemma 1. *Let P be a maximal ear in an acyclic digraph D and let $\mathcal{I}$ be a blocking interval set for P. Let $\bar{a} = (a_1, a_2, \ldots, a_k)$ be a hitting set for $\mathcal{I}$, sorted according to the occurrence of the arcs along P. Then $\bar{a}$ is an identifying sequence for P.*

Proof. Let Q be a maximal ear visiting $\bar{a}$ in this order. Let $(v_i^s, v_i^e) = a_i$ for each $1 \leq i \leq k$. Partition Q and P as follows. For each $2 \leq i \leq k$ let Q_i be the v_{i-1}^e-v_i^e subpath of Q and let P_i be the v_{i-1}^e-v_i^e subpath of P. Let Q_1 be the $\mathsf{start}(Q)$-v_1^e subpath of Q and let Q_{k+1} be the v_k^e-$\mathsf{end}(Q)$ subpath of Q. Similarly, let P_1 be the $\mathsf{start}(P)$-v_1^e subpath of P and let P_{k+1} be the v_k^e-$\mathsf{end}(P)$ subpath of P. Note that $P = P_1 \cdot P_2 \cdot \ldots \cdot P_{k+1}$ and $Q = Q_1 \cdot Q_2 \cdot \ldots \cdot Q_{k+1}$.

Assume towards a contradiction that $Q \neq P$. In particular, $Q_i \neq P_i$ holds for some $1 \leq i \leq k+1$.

If Q_i contains an arc v_i, v_j such that both v_i and v_j lie in P, but (v_i, v_j) is not an arc in P, then there is no arc of $\bar{a}$ between v_i and v_j along P, as D is acyclic.

However, (v_i, v_j) is a bypass for P, and its corresponding blocking subpath is not hit by $\bar{a}$, a contradiction to the choice of $\bar{a}$. Hence, Q_i must contain some arc (u_1, u_2) such that exactly of u_1, u_2 is in P.

We now distinguish between two cases.

Case 1: $i = 1$ or $i = k + 1$. Assume without loss of generality that $i = 1$. The case $i = k + 1$ follows analogously. Let (u_1, u_2) be the first arc along Q_1 such that $u_1 \notin V(P_1)$ and $u_2 \in V(P_1)$. Since (v_1^s, v_1^e) is both in Q_1 and in P_1 and $Q_1 \neq P_1$, such an arc (u_1, u_2) exists.

If u_2 comes after or at v_1^e along P_1, then Q_1 must contain a subpath Q' from u_2 to some u_3 such that u_3 comes before or at v_1^s along P_1. This however implies the existence of a cycle in D, a contradiction to the assumption that D is acyclic. Hence, u_2 lies before or at v_1^s along P_1.

By definition, (u_1, u_2) is a deviation for P_1. Hence, the $\mathsf{start}(P_1)$-u_2 subpath of P_1 contains a blocking path B which is not hit by $\bar{a}$, a contradiction to the assumption that $\bar{a}$ is a hitting set for $\mathcal{I}$.

Case 2: $2 \leq i \leq k$. Let (u_1, u_2) be the first arc along Q_i such that $u_1 \in V(P_i)$ and $u_2 \notin V(P_i)$. Since both Q_i and P_i contain v_{i-1}^e and v_i^s, such an arc (u_1, u_2) exists. As (v_i^s, v_i^e) is the last arc along Q_i, there must be a u_2-v_i^s path Q' in Q. If Q' intersects some vertex of P which comes at or after v_i^e along P, then there is a cycle in D, a contradiction. Hence, Q' must contain a bypass whose blocking subpath B does not contain any arc of $\bar{a}$, contradicting the assumption that $\bar{a}$ is a hitting set for $\mathcal{I}$.

As both **Case 1** and **Case 2** lead to a contradiction, we conclude that $Q = P$ and, hence, $\bar{a}$ is an identifying sequence for P, as desired. $\square$

Note that Lemma 1 is not true if we allow the digraph to contain cycles, with Fig. 5 being a counter-example.

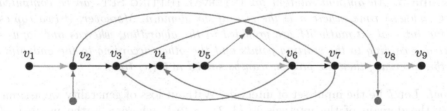

Fig. 5. The set $\{(v_1, v_2), (v_5, v_6), (v_8, v_9)\}$ is a hitting set of size 3 for the blocking interval set for $P = (v_1, v_2, \ldots, v_9)$, yet $\mathsf{ea}_D(P) = 4$, witnessed by the sequence $((v_1, v_2), (v_5, v_6), (v_7, v_8), (v_8, v_9))$.

The reverse direction of Lemma 1, however, does hold in general, and is proven below.

Lemma 2 $(\star)$. *Let P be a maximal path in a digraph D and let $\mathcal{I}$ be the set of blocking intervals for P. Let $\bar{a}$ be an identifying sequence for P. Then $\bar{a}$ is a hitting set for $\mathcal{I}$.*

Algorithm minimumHittingSet. Compute a minimum hitting set for a set of intervals.

1: **function** minimumHittingSet(set of intervals $\mathcal{I} = \{I_0, I_1, \ldots, I_{n-1}\}$)
2:     $\mathcal{I}_{start} \leftarrow$ sort $\mathcal{I}$ by starting points ; $\mathcal{I}_{end} \leftarrow$ sort $\mathcal{I}$ by endpoints
3:     $hit \leftarrow$ an array of length n, initialized with **false**
4:     $i_{start} \leftarrow 0$; $i_{end} \leftarrow 0$; $X \leftarrow \emptyset$
5:     **while** $i_{end} < n$ **do**
6:         $e \leftarrow$ end($\mathcal{I}_{end}[i_{end}]$) ; $X \leftarrow X \cup \{e\}$
7:         **while** $i_{start} < n$ **and** start($\mathcal{I}_{start}[i_{start}]$) $\leq e$ **do**
8:             $I_j \leftarrow \mathcal{I}_{start}[i_{start}]$
9:             $hit[j] \leftarrow$ **true**
10:            $i_{start} \leftarrow i_{start} + 1$
        ▷ The function id returns the index of the interval, that is, id(I_j) $= j$.
11:        **while** $i_{end} < n$ **and** $hit[\text{id}(\mathcal{I}_{end}[i_{end}])] =$ **true do**
12:            $i_{end} \leftarrow i_{end} + 1$
13:    **return** X

Together, Lemmas 1 and 2 allow us to reduce EAR-IDENTIFYING SEQUENCE on acyclic digraphs to a hitting set problem on intervals which can be solved efficiently, as shown below.

INTERVAL HITTING SET
Input   A set $\mathcal{I}$ of intervals over some finite domain U and an integer k.
Question Is there a hitting set $X \subseteq U$ for $\mathcal{I}$ of size at most k?

The algorithm in Lemma 3 uses standard techniques (see, for example, [5]) to greedily compute the desired hitting set.

Lemma 3. *An optimal solution for* INTERVAL HITTING SET *can be computed in $\mathcal{O}(n \log n)$ time, where n is the size of the domain. Moreover, if two copies of the interval set mathcalI are provided to the algorithm, whereas one copy is sorted according to the starting points and the other according to the endpoints of the intervals, then the problem can be solved in $\mathcal{O}(n)$ time.*

Proof. Let $\mathcal{I}$ be the input set of intervals. Without loss of generality we assume that the domain of the intervals is $\{1, 2, \ldots, 2n\}$, where n is the number of intervals. If this is not the case, we can compress the domain in $\mathcal{O}(n \log n)$ time by sorting the start and endpoints and then assigning each one of them a number from 1 to $2n$, preserving the original order.

We show that the minimumHittingSet computes a minimum hitting set for I in $\mathcal{O}(n \log n)$ time. Towards this end, let Y be a hitting set of $\mathcal{I}$ that is distinct from X. Let $e \in X \setminus Y$ be the earliest such element in the domain of $\mathcal{I}$. If no such e exists, then the choice of X was clearly optimal and there is nothing to show.

Otherwise, e was chosen as the endpoint of some interval I_j. Furthermore, e is the only element in X hitting I_j as all arcs added to X afterwards come after the end of I_j. Hence, there is some $e' \in Y \setminus X$ such that e' also hits I_j. Further, e' must come before e in the domain of $\mathcal{I}$, as e is the endpoint of I_j. Let Y' be the elements of Y coming before e' and let X' be the elements of X coming before e.

By assumption, $X' \subseteq Y'$. Hence, Y' hits all intervals hit by X' and potentially more. Let $\mathcal{I}'$ be the intervals hit by X'. Since I_j was chosen as the interval with the earliest endpoint in $\mathcal{I} \setminus \mathcal{I}'$, every interval in $\mathcal{I} \setminus \mathcal{I}'$ which is hit by e' is also hit by e. Thus, the set $Z = (Y \setminus \{e'\}) \cup \{e\}$ is a hitting set for $\mathcal{I}$ and $|Z| \leq |Y|$. By repeatedly applying the argument above, we obtain a hitting set Z' with $X \subseteq Z'$. By choosing Y as a minimum hitting set, we obtain that X must be a minimum hitting set as well.

We now analyze the running time of the algorithm above. Sorting the intervals can be done in $\mathcal{O}(n \log n)$ time. The while loops on lines 7 and 11 iterate at most n times each, as in each iteration the variable i_{start} or i_{end} is incremented by one. Further, each iteration takes $\mathcal{O}(1)$ time. Hence, the total running time is in $\mathcal{O}(n \log n)$, as desired.

If we the set $\mathcal{I}$ is already sorted both ways during at the input, then we can skip the steps which sort this set, obtaining a running time of $\mathcal{O}(n)$ instead. $\square$

In order to effectively use Lemma 3 when solving EAR-IDENTIFYING SEQUENCE on acyclic digraphs, we need to be able to efficiently compute the blocking interval set for an ear.

Lemma 4 ($\star$). *Let P be a maximal ear in a digraph D. Then a set $\mathcal{I}$ of blocking intervals for P can be computed in $\mathcal{O}(n + m)$ time, where $n = |V(D)|$ and $m = |E(D)|$. Further $|\mathcal{I}| \leq |V(P)| + 2$. Finally, two copies of $\mathcal{I}$ can be outputted simultaneously, one sorted according to the starting points of the intervals, and one sorted according to the end points.*

Combining the previous results, we can now conclude that EAR-IDENTIFYING SEQUENCE is in P if the input digraph is acyclic.

Theorem 1 ($\star$). *Given an acyclic digraph D and a maximal ear P in D, we can compute $ea_D(P)$ and find an identifying sequence for P of minimum length in $\mathcal{O}(n + m)$ time, where $n = |V(D)|$ and $m = |E(D)|$.*

Further, using similar methods as in Theorem 1, we can also solve CONFLICTING EAR in polynomial time if the input digraph is acyclic.

Theorem 2 ($\star$). CONFLICTING EAR *can be solved in $\mathcal{O}(n + m)$ time if the input digraph D is acyclic, where $n = |V(D)|$ and $m = |E(D)|$.*

In order to solve EAR ANONYMITY in polynomial time on DAGs, we compute for each vertex v a number anon$[v]$ which is a lower bound to the number of arcs required in the ear-identifying sequence of any maximal path containing v. We do this by following the topological ordering of the vertices and by keeping track of the bypasses and deviations found.

The algorithm relies on the property of DAGs that, given four distinct vertices v_1, v_2, v_3, v_4, sorted according to their topological ordering, if v_2 can reach

---

Algorithm DAGEarAnonymity. Compute the ear anonymity of a DAG.

1: **function** DAGEarAnonymity(DAG D)
2:     $V \leftarrow$ sort $V(D)$ by the topological ordering of D
3:     anon $\leftarrow$ empty array over V
4:     **for each** $v \in V$ **do**
5:         anon$[v] \leftarrow \max(\{\text{anon}[u] : u \in \text{in}(v)\} \cup \{0\})$
6:         **if** indeg$(v) > 1$ and anon$[v] = 0$ **then**
7:             anon$[v] \leftarrow 1$
8:         **else**
9:             **for each** $u \in \text{in}(v)$ **do**
10:                 $V_u \leftarrow$ vertices which can reach u
11:                 $U_u \leftarrow \{w \in V_u :$ there is a w-v path in $D - u$
12:                     which is internally disjoint from V_u $\}$
13:                 anon$[v] \leftarrow \max(\{\text{anon}[w] + 1 : w \in U_u\} \cup \{\text{anon}[v]\})$
14:         **if** outdeg$(v) = 0$ and indeg$(v) > 0$ **then**
15:             $P \leftarrow$ shortest path ending on v such that indeg(start$(P)) > 1$
16:             **if** no such P exists **then**
17:                 $P' \leftarrow$ the unique maximal path with end$(P') = v$
18:             **else if** there is some $u \in V(P)$ with outdeg$(u) > 1$ **then**
19:                 $P' \leftarrow$ shortest subpath of P ending on v such that
20:                     outdeg(start$(P')) > 1$
21:             **else**
22:                 **continue**
23:             **for each** $u \in V(P') \setminus \{\text{start}(P')\}$ **do**
24:                 anon$[u] \leftarrow$ anon$[u] + 1$
25:     **return** $\max(\{\text{anon}[v] : v \in V(D)\})$

---

v_3, then every v_1-v_2 path is disjoint from every v_3-v_4 path. This allows us to efficiently compute bypasses using breadth-first search. The pseudo-code is provided in DAGEarAnonymity.

Lemma 5. *At the end of the execution of DAGEarAnonymity, for every $v \in V(D)$ there is a path P which starts at some source of D and ends in v such that, for every path Q starting in v and ending in some sink, the path $R := P \cdot Q$ has $\text{ea}_D(R) \geq \text{anon}[v]$. Furthermore, there is an ear-identifying sequence $\bar{a}$ of minimum length for R such that at least $\text{anon}[v]$ arcs of $\bar{a}$ lie in P.*

Proof. We prove the statement by induction on the index of v in the topological ordering of D.

The statement is clearly true if v is a source. So assume that v is not a source.

Case 1: The last change in the value of anon$[v]$ was on line 5. Then there is some $u \in \text{in}(v)$ such that anon$[u] = \text{anon}[v]$.

By the induction hypothesis, there is some P which starts in a source and ends in u satisfying the additional conditions given at the statement. Let Q be some path starting in v and ending in some sink. Let $P' = P \cdot (u, v)$.

By assumption, there is some ear-identifying sequence $\bar{a}$ for $R := P \cdot (u, v) \cdot Q$ such that at least anon$[u] = \text{anon}[v]$ arcs of $\bar{a}$ lie in P and, hence, in P'.

Case 2: The last change in the value of anon$[v]$ was on line 7. Then anon$[u] =$ 0 for all inneighbors $u \in \text{in}(v)$ of v. Let u_1, u_2 be two distinct inneighbors of v. By the induction hypothesis, there is some path P starting at some source and ending in u_1 satisfying the properties given in the statement.

Let $P' = P \cdot (u_1, v)$. Let Q be some path starting in v and ending in some sink. Let $R' = P' \cdot Q$ and let $\bar{a}$ be some ear-identifying sequence for R' of minimum length. As there are at least two maximal paths (one coming from u_1 and the other from u_2) visiting all arcs of $\bar{a}$ lying on Q, some arc of P' must be on $\bar{a}$. Hence, P' contains at least anon$[v] = 1$ arcs of $\bar{a}$, as desired.

Case 3: The last change in the value of anon$[v]$ was on line 13. Then there are vertices w, u_1 such that u_1 is an inneighbor of v, anon$[w] = $ anon$[v] - 1$, there is a $w - v$ path P_1 which is disjoint from u_1, and w can reach u_1. By the induction hypothesis, there is a path P starting at a source and ending in w which satisfies the conditions given in the statement.

Let $P' = P \cdot P_1$ Let Q be some path starting at v and ending at some sink. By assumption, there is an ear-identifying sequence $\bar{a}$ for $R := P \cdot P_1 \cdot Q$ of minimum length such that at least anon$[w]$ arcs of $\bar{a}$ lie on P and hence on P'. However, there are at least two different paths from w to v (one visiting u_1 and one not). Hence, at least one arc of $\bar{a}$ must lie on P_1 and hence on P'.

Case 4: The last change in the value of anon$[v]$ was on line 24.

If the path P' was chosen on line 17, then there is exactly one maximal path Q containing v. In particular, every vertex in Q has indegree at most one. Hence, $\text{ea}_D(Q) = $ anon$[v] = 1$ and the statement follows.

Otherwise, there is a path P with $\text{indeg}(\text{start}(P)) > 1$ which ends on some sink t and contains v. Furthermore, there is exactly one v-t path in D. Let P' be the shortest subpath of P ending on t such that $\text{outdeg}(\text{start}(P')) > 1$. Note that P' contains v since anon$[v]$ was incremented on line 24. Further, $\text{indeg}(u) = 1$ holds for all $u \in V(P') \setminus \{\text{start}(P')\}$. In particular, $\text{indeg}(v) = 1$ and the value of anon$[v]$ was not modified on line 7 nor on 13.

Let $u \in \text{in}(v)$. By the induction hypothesis, there is a path P_u ending on u and satisfying the conditions in the statement. Note that anon$[v] \leq$ anon$[u] + 1$.

If $u \in V(P') \setminus \{\text{start}(P')\}$, then anon$[u] = $ anon$[v]$ and there is exactly one u-v path in D. The statement follows trivially.

Otherwise we have $u = \text{start}(P')$, $\text{outdeg}(u) > 1$ and $\text{indeg}(v) = 1$. Further, there is exactly one path Q starting in v and ending on some sink, and there is exactly one u-v path. Let $R = P_u \cdot (u, v) \cdot Q$ and let $w \in \text{outdeg}(u) \setminus \{v\}$.

By the induction hypothesis, there is an ear-identifying sequence $\bar{a}$ of minimum length for R such that at least anon$[u]$ arcs of $\bar{a}$ lie in P_u. Since (u, w) is a deviation for R, at least one arc of $\bar{a}$ must lie on $(u, v) \cdot Q$. As Q is the only maximal path starting in v, we choose $\bar{a}$ such that it contains the arc (u, v) and no arcs in Q. Hence, at least anon$[v] = $ anon$[u] + 1$ arcs of $\bar{a}$ lie in $P_u \cdot (u, v)$, as desired. □

Lemma 6. *At the end of the execution of DAGEarAnonymity, for every $v \in$ $V(D)$ and every maximal path $R := P \cdot v \cdot Q$ there is an ear-identifying sequence $\bar{a}$ of minimum length for R such that at most* anon$[v]$ *arcs of $\bar{a}$ lie in P.*

Proof. We prove the following slightly stronger statement.

Claim: for every $v \in V(D)$ and every maximal path $R := P \cdot v \cdot Q$ there is an ear-identifying sequence $\bar{a}$ of minimum length for R such that for every $u \in V(P \cdot v)$, at most anon$[u]$ arcs of $\bar{a}$ lie in P_u, where P_u is the subpath of P starting on start(P) and ending on u.

We prove the statement by induction on the index of v in the topological ordering of D.

The statement is clearly true if v is a source, as P then becomes empty. So assume that v is not a source and let $R := P \cdot (u, v) \cdot Q$ be a maximal path, where u is the predecessor of v along R. Let $P_v = P \cdot (u, v)$.

By the induction hypothesis, there is an ear-identifying sequence $\bar{a}$ of minimum length for R which satisfies the condition given in the claim above. In particular, at most anon$[u]$ arcs of $\bar{a}$ lie in P.

If anon$[v] >$ anon$[u]$, then clearly at most anon$[u] + 1 \leq$ anon$[v]$ arcs of $\bar{a}$ lie in P_v.

Now assume that anon$[v] =$ anon$[u]$. If anon$[v] = 0$, then indeg$(u') \leq 1$ holds for all u' which can reach v, as otherwise we would increment anon$[u']$ on line 7 and propagate this through line 5. In particular, there is exactly one path which starts at some source and ends in v. Since v is not a source and anon$[v]$ was not incremented on line 24, we know that v can reach some vertex w' with anon$[w'] \geq 1$. This means that v has at least one outneighbor w. Hence, any sequence containing some arc in Q satisfies the required condition. Thus, we can assume that anon$[v] \geq 1$.

If $\bar{a}$ does not contain (u, v), then there is nothing to show. Further, if less than anon$[u]$ arcs of $\bar{a}$ lie in P, then clearly at most anon$[u] =$ anon$[v]$ arcs of $\bar{a}$ lie in P_v.

Let P_a be the shortest subpath of P starting at start(P) which contains the same arcs of $\bar{a}$ as P, and let P_b be the rest of P, that is, $P = P_a \cdot P_b$. Observe that, by the induction hypothesis and by the assignment on line 5, anon$[u'] =$ anon$[u]$ holds for all $u' \in V(P_b)$.

If there is some path R' starting in $V(P_b)$ and ending on v without using (u, v), then anon$[v]$ is incremented on line 13, as either start(R') can reach u or u can reach the predecessor of v on R'. This contradicts, however, the equality anon$[v] =$ anon$[u]$ assumed previously.

Assume towards a contradiction that, if $\bar{b}$ is an ear-identifying sequence of minimum length for R satisfying the conditions in the claim above, then $\bar{b}$ contains (u, v) and exactly anon$[u]$ arcs of $\bar{b}$ lie in P. We consider the following cases.

Case 1: v is not a sink.

Let w be the successor of v in R and let $\bar{b}_1$ be the arc sequence obtained by replacing (u, v) with (v, w) in $\bar{b}$ while preserving the topological ordering of the arcs.

By assumption, there is a conflicting ear R' for $(R, \bar{b}_1)$. Further, R' does not contain (u, v) as D is acyclic and $\bar{b}$ is an ear-identifying sequence for R.

If R' is disjoint from P, then $\mathsf{anon}[u] = 0$, a contradiction.

Otherwise, $\mathsf{start}(R')$ must lie on P_b, as R' must visit all arcs of $\bar{a}$ which are in P_a. Since R' contains v, it also contains a path from $V(P_b)$ to v avoiding (u, v). This contradicts the argumentation above before the case distinction.

Case 1: v is a sink.

Let $\bar{b}_1$ be the subsequence of $\bar{b}$ obtained by removing (u, v) from $\bar{b}$. By assumption, there is a conflicting ear R' for $(R, \bar{b}_1)$. In particular, R' intersects P_a, as at least $\mathsf{anon}[u] \geq 1$ arcs of $\bar{a}$ lie in P_a. This implies that R' contains a path from $V(P_b)$ to v which avoids (u, v), a contradiction. $\qquad \square$

Lemmas 5 and 6 essentially prove that DAGEarAnonymity is correct. Hence, we now only need to provide a running-time analysis.

Theorem 3. Ear Anonymity *can be solved in* $\mathcal{O}(m(n + m))$-*time if the input digraph* D *is acyclic, where* $n = |V(D)|$ *and* $m = |E(D)|$.

Proof. We run DAGEarAnonymity on the input digraph, obtaining a value $k = \mathsf{anon}[v]$ for some sink $v \in V(D)$ for which $\mathsf{anon}[v]$ is maximum. By Lemma 5, $k \leq \mathsf{ea}(D)$. By Lemma 6, $k \geq \mathsf{ea}(D)$, and so $k = \mathsf{ea}(D)$.

We now analyze the running time of DAGEarAnonymity. Sorting $V(D)$ according to the topological ordering can be done in $\mathcal{O}(n + m)$-time using standard techniques.

The **for**-loop on line 4 is executed exactly n times. We then iterate over all inneighbors of v. Hence, each arc is considered a constant number of times. For each arc, we compute the sets V_u, U_u on lines 10 and 11 using breadth-first searches in $\mathcal{O}(n + m)$-time. For each sink, the paths P, P' on lines 15, 17 and 19 can also be computed with breadth-first search. Hence, the running time is dominated by computing a constant number of breadth-first searches for each arc, and so it lies in $\mathcal{O}(m(n + m))$. $\qquad \square$

5 NP-hardness

We consider the problems Conflicting Ear, Ear-Identifying Sequence and Ear Anonymity in the general setting without any restrictions on the input digraph. We show that Conflicting Ear is NP-hard, providing a reduction from the NP-complete problem Linkage, defined below.

Linkage

Input   A digraph D, an integer k and a set $S = \{(s_1, t_1), (s_2, t_2), \ldots, (s_k, t_k)\}$ of vertex pairs.

Question Is there a linkage $\mathcal{L}$ in D such that for each $(s_i, t_i) \in S$ there is some s_i-t_i path $L_i \in \mathcal{L}$?

Linkage remains NP-hard even if $k = 2$. [10]

Theorem 4. CONFLICTING EAR *is NP-complete even if $\bar{a}$ has length 3.*

Proof. From Remark 3 we know that CONFLICTING EAR is in NP. To show that it is NP-hard, we provide a reduction as follows. Let (D, S) be a LINKAGE instance where $k = |S| = 2$. Construct a digraph D' as follows (see Fig. 7 below for an illustration of the construction).

Start with D. Add the vertices $\{u_1, \ldots, u_3, v_1, \ldots, v_6\}$ and the following paths to D', where each path is given by its vertex-sequence:

$$P_1 = (u_1, v_1, v_2, u_2, v_3, v_4, u_3, v_5, v_6),$$
$$P_2 = (v_2, u_1, s_1), P_3 = (t_1, u_3, v_3), P_4 = (v_4, u_2, s_2), P_5 = (t_2, v_5).$$

Set $\bar{a} = ((v_1, v_2), (v_3, v_4), (v_5, v_6))$. Note that $\bar{a}$ is a sequence of arcs of P_1, sorted according to their occurrence on P_1. This completes the construction of the CONFLICTING EAR instance $(D', P_1, \bar{a})$.

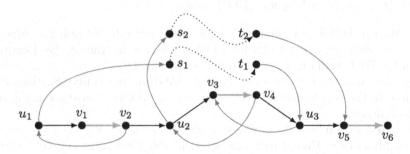

Fig. 6. The digraph H_1 for $n = 3$, used in the reduction of the proof of Theorem 6. Every $u_{1,0}$-$u_{1,12}$ path in H_1 contains $2n = 6$ arc-disjoint blocking subpaths.

We first show that, if the LINKAGE instance (D, S) is a "yes"-instance, then so is the CONFLICTING EAR instance $(D', P_1, \bar{a})$.

Let $\mathcal{L}$ be a solution for (D, S). Let L_i be the s_i-t_i path in $\mathcal{L}$ for $i \in \{1, 2\}$. We construct a conflicting ear Q for $(P_1, \bar{a})$ as follows. We set $Q = (v_1, v_2) \cdot P_2 \cdot L_1 \cdot P_3 \cdot (v_3, v_4) \cdot P_4 \cdot L_2 \cdot P_5 \cdot (v_5, v_6)$.

Clearly Q is not a subgraph of P_1 and Q visits the arcs of $\bar{a}$ in the given order. By assumption, L_1 and L_2 are disjoint paths, and so Q is a path. Hence, Q is a conflicting ear for $(P_1, \bar{a})$, as desired.

For the other direction, let Q be a conflicting ear for $(P_1, \bar{a})$. We first show that Q does not contain (v_2, u_2). Assume towards a contradiction that it does contain (v_2, u_2). Then Q does not contain (v_2, u_1) or (v_4, u_2). Since Q contains (v_3, v_4), it must also contain (v_4, u_3). However, (u_3, v_3) closes a cycle with the arcs (v_3, v_4) and (v_4, u_3). Hence, Q cannot contain (u_3, v_3). This implies that Q must contain (u_2, v_3) in order to reach v_3. Finally, Q must contain (u_3, v_5) and (v_5, v_6). However, we now have $Q = P_1$, a contradiction to the assumption that Q is a conflicting ear for $(P_1, \bar{a})$.

Since Q contains (v_1, v_2) but not (v_2, u_2), it must contain (v_2, u_1). As (u_1, v_1) closes a cycle, Q does not contain this arc and must contain (u_1, s_1) instead. Because Q contains (v_3, v_4), it must reach v_3 through u_3 or through u_2. However, if Q contains (u_2, v_3), then it also contains (v_4, u_2), which closes a cycle. Hence, Q does not contain (u_2, v_3) and must contain (u_3, v_3) instead. As before, Q cannot contain (v_4, u_3) as this would close a cycle, so Q must contain (v_4, u_2) and (u_2, s_2).

In order to reach u_3, Q must contain (t_1, u_3). Since Q contains both s_1 and s_2, it must also contain t_1 and t_2, as they are the only vertices of D which are reachable by s_1 and s_2 and have arcs to P_1. Hence, Q contains (t_2, v_5). Since v_5 can only reach v_6, Q must visit (t_1, u_3) before visiting (t_2, v_5). Further, Q visits t_1 before visiting s_2 and it visits s_1 before visiting t_1. Hence, Q must visit t_2 after s_2 and must also contain two paths L_1 and L_2, where L_1 is an s_1-t_1 path in D and L_2 is an s_2-t_2 path in D. Since Q is an ear, L_1 and L_2 must be disjoint. Thus, L_1 and L_2 are a solution to the LINKAGE instance (D, S), as desired. □

Using Theorem 4 and Lemma 2, it is simple to show that EAR-IDENTIFYING SEQUENCE is NP-hard as well.

Theorem 5 ($\star$). EAR-IDENTIFYING SEQUENCE *is NP-hard even if* $k \leq 3$.

For the next hardness result, we provide a reduction from the following NP-complete problem. [10]

u-v-w-PATH
Input   A digraph D and three vertices $u, v, w \in V(D)$.
Question Is there a path from u to w in D containing v?

Theorem 6 ($\star$). EAR ANONYMITY *is NP-hard.*

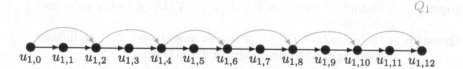

$u_{1,0} \quad u_{1,1} \quad u_{1,2} \quad u_{1,3} \quad u_{1,4} \quad u_{1,5} \quad u_{1,6} \quad u_{1,7} \quad u_{1,8} \quad u_{1,9} \quad u_{1,10} \quad u_{1,11} \quad u_{1,12}$

Fig. 7. Gadget for the reduction in the proof of Theorem 4. Bold, green arcs belong to $\bar{a}$, and the red arcs belong to the conflicting ear Q but not to P_1. The dotted lines correspond to disjoint paths in D. (Color figure online)

It remains open whether EAR ANONYMITY is complete for NP, but in the next section we show that EAR-IDENTIFYING SEQUENCE is Σ_2^p-complete.

6 Σ_2^p-Hardness for EAR-IDENTIFYING SEQUENCE

In order to show that EAR-IDENTIFYING SEQUENCE is Σ_2^p-hard, we define two auxiliary problems and show that each of them is Σ_2^p-hard. Using intermediate problems helps us reduce the complexity of our final reduction. We provide a reduction from SHORTEST IMPLICANT CORE, defined below.

Definition 11. *Let φ be a propositional formula and let I be a set of literals. We say that I is an* implicant *for φ if $(\bigwedge_{\ell \in I} \ell) \to \varphi$ is a tautology (that is, it evaluates to true under every assignment of the variables).*

SHORTEST IMPLICANT CORE
Input   A DNF formula , an implicant C of and an integer k.
Question Is there an implicant $C'C$ of of size at most k?

Lemma 7 ([21, **Theorem 3**]). SHORTEST IMPLICANT CORE *is Σ_2^p-complete.*

The first auxiliary problem is about linkages in a digraph. We want to decide if there is a subset of the terminals which cannot be linked, whereas we are forced to always connect some fixed pairs. The last restriction is useful when constructing our gadgets, since it allows us to adapt the reduction used to show NP-hardness for LINKAGE [10], reusing one of their gadgets.

RESTRICTED SUBSET LINKAGE
Input   A digraph D, two sets $T_0, T \subseteq V(D) \times V(D)$ of vertex pairs
             and an integer k.
Question Is there a subset $T' \subseteq T$ of size at most k such that no linkage
             connecting the terminal pairs of $T' \cup T_0$ in D exist?

Lemma 8 ($\star$). RESTRICTED SUBSET LINKAGE *is Σ_2^p-hard.*

For the hardness reduction for EAR-IDENTIFYING SEQUENCE, it is convenient to consider a variant of RESTRICTED SUBSET LINKAGE in which $T_0 = \emptyset$, because then we do not need to differentiate between T' and T_0.

SUBSET LINKAGE
Input   A digraph D, two sets $T \subseteq V(D) \times V(D)$ of vertex pairs and
             an integer k.
Question Is there a subset $T' \subseteq T$ of size at most k such that no linkage
             connecting the terminal pairs of T' in D exist?

Lemma 9 ($\star$). SUBSET LINKAGE *is Σ_2^p-hard.*

We can now provide our main reduction. One of the biggest challenges in the construction is to use a single conflicting ear Q to count how many pairs from T' were already taken. Further, it is not clear how to model taking an arc into an ear-identifying sequence $\bar{a}$ as a choice of some terminal (s_i, t_i), and we instead take terminal pairs based on arcs which are not taken into $\bar{a}$.

Theorem 7 ($\star$). EAR-IDENTIFYING SEQUENCE *is Σ_2^p-hard.*

7   Remarks

Using Remark 2 and Lemma 2 and the *directed grid theorem* below, it is possible to draw a connection between directed treewidth and ear anonymity.

Theorem 8 ([14]). *There is a computable function $f : \mathbb{N} \to \mathbb{N}$ such that every digraph D with $\mathsf{dtw}(D) \geq f(k)$ contains a cylindrical grid of order k as a butterfly minor, where $\mathsf{dtw}(D)$ is the directed tree-width of D.*

It is easy to verify that a cylindrical grid of order k has ear anonymity at least $2k$. Take any cycle C on the cylindrical grid which is neither the outermost nor the innermost cycle. Then, a subpath Q_i of C from row i to row $i + 1 \bmod 2k$ is a blocking subpath of a bypass for C. Since C has at least $2k$ internally disjoint blocking subpaths, by Lemma 2 we have $k \leq \mathsf{ea}_D(C) \leq \mathsf{ea}(D)$. Hence, we obtain the following inequality.

Remark 6. There is a computable function $f : \mathbb{N} \to \mathbb{N}$ such that $\mathsf{dtw}(D) \leq f(\mathsf{ea}(D))$.

Remark 6 naturally raises the following question.

Question 1. What is the smallest function f such that $\mathsf{dtw}(D) \leq f(\mathsf{ea}(D))$ holds for all digraphs D?

On the other hand, directed acyclic graphs have directed treewidth zero but can have arbitrarily high ear anonymity. For example, the digraph H_1 used in the reduction in the proof of Theorem 6 (see Fig. 6) is acyclic and $\mathsf{ea}(H_1) = 2n$. Thus, there is no function $f : \mathbb{N} \to \mathbb{N}$ for which $\mathsf{ea}(D) \leq f(\mathsf{dtw}(D))$ holds for all digraphs D.

Since EAR ANONYMITY is in P if the input digraph is acyclic, it is natural to ask what is the parameterized complexity of EAR ANONYMITY when parameterized by directed treewidth.

Question 2. Can EAR ANONYMITY be solved in $\mathcal{O}(n^{f(\mathsf{dtw}(D))})$ time, where $\mathsf{dtw}(D)$ is the directed treewidth of the input digraph D?

It is still unclear where exactly in the polynomial hierarchy EAR ANONYMITY lies. On the one hand, EAR-IDENTIFYING SEQUENCE looks like a subproblem of EAR ANONYMITY, yet if the digraph has very high ear anonymity, then there are many "correct" guesses for some ear of high anonymity, which could make the problem easier, and not harder, than EAR-IDENTIFYING SEQUENCE.

Question 3. Is EAR ANONYMITY in NP?

Finally, one could also ask if a phenomenon similar to the directed grid theorem also occurs with ear anonymity. That is, while a path with high ear anonymity witnesses that an acyclic digraph has high ear anonymity, is there also some witness which gives us an upper bound on the ear anonymity of the same digraph?

Question 4. Is there some "small" witness W and some function f which allow us to efficiently verify that $\mathrm{ea}(D) \leq f(W)$?

Question 5. Can we solve LINKAGE in $f(k)n^{g(\mathrm{ea}(d))}$-time? (In general?) (On DAGs?)

Acknowledgments. An anonymous reviewer provided useful insights which helped improve the running time of the algorithms in Lemma 4 and Theorems 1 and 2.

Disclosure of Interests. The author has no competing interests to declare that are relevant to the content of this article.

References

1. Arora, S., Barak, B.: Computational Complexity: A Modern Approach. Cambridge University Press (2009)
2. Bang-Jensen, J., Gutin, G.Z.: Digraphs: Theory, Algorithms and Applications. Springer Science & Business Media (2008)
3. Bodlaender, H.L.: Treewidth: structure and algorithms. In: Prencipe, G., Zaks, S. (eds.) SIROCCO 2007. LNCS, vol. 4474, pp. 11–25. Springer, Heidelberg (2007). https://doi.org/10.1007/978-3-540-72951-8_3
4. Cáceres, M.: Parameterized algorithms for string matching to dags: funnels and beyond. In: Bulteau, L., Lipták, Z. (eds.) 34th Annual Symposium on Combinatorial Pattern Matching, CPM 2023, 26–28 June 2023, Marne-la-Vallée, France. LIPIcs, vol. 259, pp. 7:1–7:19. Schloss Dagstuhl - Leibniz-Zentrum für Informatik (2023). https://doi.org/10.4230/LIPICS.CPM.2023.7
5. Cormen, T.H., Leiserson, C.E., Rivest, R.L., Stein, C.: Introduction to Algorithms, Second Edition, chap. 16, pp. 370–404. The MIT Press and McGraw-Hill Book Company (2001)
6. Courcelle, B.: The monadic second-order logic of graphs. i. recognizable sets of finite graphs. Inform. Comput. **85**(1), 12–75 (1990)
7. Courcelle, B., Olariu, S.: Upper bounds to the clique width of graphs. Discret. Appl. Math. **101**(1–3), 77–114 (2000). https://doi.org/10.1016/S0166-218X(99)00184-5
8. Diestel, R.: Graph Theory. GTM, vol. 173. Springer, Heidelberg (2017). https://doi.org/10.1007/978-3-662-53622-3
9. Downey, R.G., Fellows, M.R.: Fundamentals of Parameterized Complexity. TCS, Springer, London (2013). https://doi.org/10.1007/978-1-4471-5559-1
10. Fortune, S., Hopcroft, J., Wyllie, J.: The directed subgraph homeomorphism problem. Theoret. Comput. Sci. **10**(2), 111–121 (1980). https://doi.org/10.1016/0304-3975(80)90009-2
11. Ganian, R., Hliněný, P., Kneis, J., Langer, A., Obdržálek, J., Rossmanith, P.: Digraph width measures in parameterized algorithmics. Discret. Appl. Math. **168**, 88–107 (2014). https://doi.org/10.1016/j.dam.2013.10.038
12. Ganian, R., et al.: Are there any good digraph width measures? J. Comb. Theory, Ser. B **116**, 250–286 (2016). https://doi.org/10.1016/j.jctb.2015.09.001
13. Johnson, T., Robertson, N., Seymour, P.D., Thomas, R.: Directed tree-width. J. Combinat. Theory, Series B **82**(1), 138–154 (2001)
14. Kawarabayashi, K.I., Kreutzer, S.: The directed grid theorem. In: Proceedings of the Forty-Seventh Annual ACM Symposium on Theory of Computing, pp. 655–664 (2015)

15. Khan, S., Kortelainen, M., Cáceres, M., Williams, L., Tomescu, A.I.: Improving RNA assembly via safety and completeness in flow decompositions. J. Comput. Biol. **29**(12), 1270–1287 (2022). https://doi.org/10.1089/CMB.2022.0261
16. Milani, M.G.: A polynomial kernel for funnel arc deletion set. Algorithmica, pp. 1–21 (2022)
17. Milani, M.G., Molter, H., Niedermeier, R., Sorge, M.: Efficient algorithms for measuring the funnel-likeness of dags. J. Comb. Optim. **39**(1), 216–245 (2020)
18. Robertson, N., Seymour, P.D.: Graph minors. II. algorithmic aspects of treewidth. J. Algorithms **7**(3), 309–322 (1986). https://doi.org/10.1016/0196-6774(86)90023-4
19. Schaefer, M., Umans, C.: Completeness in the polynomial-time hierarchy: a compendium. SIGACT News **33**(3), 32–49 (2002)
20. Slivkins, A.: Parameterized tractability of edge-disjoint paths on directed acyclic graphs. SIAM J. Discret. Math. **24**(1), 146–157 (2010)
21. Umans, C.: The minimum equivalent DNF problem and shortest implicants. J. Comput. Syst. Sci. **63**(4), 597–611 (2001). https://doi.org/10.1006/jcss.2001.1775

Separating Path Systems in Complete Graphs

Cristina G. Fernandes[1] , Guilherme Oliveira Mota[1] ,
and Nicolás Sanhueza-Matamala[2](✉)

[1] Instituto de Matemática e Estatística, Universidade de São Paulo, São Paulo, Brazil
{cris,mota}@ime.usp.br
[2] Facultad de Ciencias Físicas y Matemáticas, Universidad de Concepción,
Concepción, Chile
nsanhuezam@udec.cl

Abstract. We prove that in any n-vertex complete graph there is a collection $\mathcal{P}$ of $(1 + o(1))n$ paths that *strongly separates* any pair of distinct edges e, f, meaning that there is a path in $\mathcal{P}$ which contains e but not f. Furthermore, for certain classes of n-vertex αn-regular graphs we find a collection of $(\sqrt{3\alpha + 1} - 1 + o(1))n$ paths that strongly separates any pair of edges. Both results are best-possible up to the $o(1)$ term.

Keywords: Paths · Separating systems · Complete graphs

1 Introduction

Let $\mathcal{P}$ be a family of paths in a graph G. We say that two edges e, f are *weakly separated by* $\mathcal{P}$ if there is a path in $\mathcal{P}$ which contains one of these edges but not both. We also say that they are *strongly separated by* $\mathcal{P}$ if there are two paths $P_e, P_f \in \mathcal{P}$ such that P_e contains e but not f, and P_f contains f but not e.

We are interested in the problem of finding "small" families of paths ("path systems") that separate any pair of edges in a given graph. A path system in a graph G is *weak-separating* (resp. *strong-separating*) if all pairs of edges in G are weakly (resp. strongly) separated by it. Let $\operatorname{wsp}(G)$ and $\operatorname{ssp}(G)$, respectively, denote the size of smallest such families of paths in a graph G. Since every strong-separating path system is also weak-separating, the inequality $\operatorname{wsp}(G) \leq \operatorname{ssp}(G)$ holds for any graph G, but equality is not true in general.

The study of general separating set systems was initiated by Rényi [11]. The variation which considers the separation of edges using subgraphs was considered many times in the computer science community, motivated by the application of efficiently detecting faulty links in networks [13]. The question got renewed interest in the combinatorics community after it was raised by Katona in a conference in 2013, and was considered simultaneously by Falgas-Ravry, Kittipassorn, Korándi, Letzter, and Narayanan [4] in its weak version, and by Balogh, Csaba, Martin, and Pluhár [1] in its strong version. Both teams conjectured that n-vertex graphs G admit (weak and strong) separating path systems of size linear in n, that is, $\operatorname{wsp}(G), \operatorname{ssp}(G) = O(n)$, and both also observed that an $O(n \log n)$

© The Author(s), under exclusive license to Springer Nature Switzerland AG 2024
J. A. Soto and A. Wiese (Eds.): LATIN 2024, LNCS 14579, pp. 98–113, 2024.
https://doi.org/10.1007/978-3-031-55601-2_7

bound holds. After a partial progress obtained by Letzter [10], the conjecture was settled by Bonamy, Botler, Dross, Naia, and Skokan [2], who proved that $ssp(G) \leq 19n$ holds for any n-vertex graph G.

1.1   Separating Cliques

An interesting open question is to replace the value '19' in $ssp(G) \leq 19n$ by the smallest possible number. Perhaps, it could be possible even to replace this value by $1 + o(1)$. The behaviour of $ssp(K_n)$ and $wsp(K_n)$ has been enquired repeatedly by many authors (e.g. [4, Sect. 7]), so studying separating path systems in complete graphs seems particularly relevant.

For the weak separation, we know that $wsp(K_n) \geq n - 1$ (see the remark before Conjecture 1.2 in [4]). One can also check that $ssp(K_n) \geq n$. Our first main result shows that this lower bound is asymptotically correct.

Theorem 1. $ssp(K_n) = (1 + o(1))n$.

Let us summarise the history of upper bounds for this problem. First we learned that $wsp(K_n) = O(n)$ [4, Theorem 1.3], and then that $ssp(K_n) \leq 2n + 4$ [1, Theorem 3]. Wickes [12] showed that $wsp(K_n) \leq n$ whenever n or $n - 1$ is a prime number, and that $wsp(K_n) \leq (21/16 + o(1))n$ in general.

The problem of estimating $ssp(K_n)$ is connected with the older problem of finding *orthogonal double covers* (ODC), which are collections $\mathcal{C}$ of subgraphs of K_n in which every edge appears in exactly two elements of $\mathcal{C}$, and the intersection of any two elements of $\mathcal{C}$ contains exactly one edge. If each graph of $\mathcal{C}$ is a Hamiltonian path, then it is easy to check that $\mathcal{C}$ must contain exactly n paths and that it forms a strong-separating path system. Thus, we know that $ssp(K_n) = n$ whenever an ODC with Hamiltonian paths exists. This is known to be false for $n = 4$, but is known to be true for infinitely many values of n. In particular, it holds if n can be written as a product of the numbers 5, 9, 13, 17, and 29 [8]. See the survey [7] for more results and details.

1.2   Separating Regular Graphs

Our main result for cliques (Theorem 1) follows from a more general result which works for "robustly connected" graphs which are *almost regular*, meaning that each vertex has approximately the same number of neighbours. For simplicity, we give the statement for regular graphs here. Let $\alpha \in [0, 1]$ and consider an αn-regular graph G on n vertices. A counting argument (Proposition 1) shows that $ssp(G) \geq (\sqrt{3\alpha + 1} - 1 - o(1))n$ must hold. Our second main result shows that this bound essentially holds with equality if we also assume some vertex-connectivity condition. We say an n-vertex graph G is (δ, L)-*robustly-connected* if, for every $x, y \in V(G)$, there exists $1 \leq \ell \leq L$ such that there are at least δn^ℓ (x, y)-paths with exactly ℓ inner vertices each.

Theorem 2. *Let* $\alpha, \delta \in (0, 1)$ *and* $L \geq 1$. *If* G *is an* n-*vertex graph which is* αn-*regular and* (δ, L)-*robustly-connected, then* $ssp(G) = (\sqrt{3\alpha + 1} - 1 + o(1))n$.

We note that at least some kind of connectivity is required in the previous result. Indeed, the graph G formed by two vertex-disjoint cliques with $n/2$ vertices is $(n/2-1)$-regular, but has clearly $\text{ssp}(G) = 2 \cdot \text{ssp}(K_{n/2}) \geq n - o(n)$, whereas Theorem 2 would give an incorrect upper bound around $(0.582 + o(1))n$.

The function $f(\alpha) = \sqrt{3\alpha + 1} - 1$ satisfies $\alpha < f(\alpha) < \sqrt{\alpha} < 1$ for $\alpha \in (0, 1)$, so in particular it shows that all n-vertex graphs G covered by Theorem 2 satisfy $\text{ssp}(G) \leq (1 + o(1))n$. From Theorem 2 we obtain as corollaries results for many interesting classes of graphs as balanced complete bipartite graphs, regular graphs with large minimum degree, regular robust expanders, etc. (see Sect. 8 for details).

2   Lower Bounds

In this short section we provide a lower bound for $\text{ssp}(G)$ for general dense graphs G. Our main results, Theorems 1 and 2, match this lower bound.

Given a path system $\mathcal{P}$ in a graph G and $e \in E(G)$, let $\mathcal{P}(e) \subseteq \mathcal{P}$ be the paths of $\mathcal{P}$ which contain e. Note that $\mathcal{P}$ is weak-separating if and only if the sets $\mathcal{P}(e)$ are different for all $e \in E(G)$; and $\mathcal{P}$ is strong-separating if and only if no set $\mathcal{P}(e)$ is contained in another $\mathcal{P}(f)$.

Proposition 1. *For any $\alpha, \varepsilon \in (0, 1]$, the following holds for all sufficiently large n. If G is an n-vertex graph with $\alpha\binom{n}{2}$ edges, then $\text{ssp}(G) \geq (\sqrt{3\alpha + 1} - 1 - \varepsilon)n$.*

Proof. Let $\mathcal{P}$ be a strong-separating path system of size $\text{ssp}(G)$, and suppose β is such that $|\mathcal{P}| = \beta n$ (we know that $\beta \leq 19$ by the result of [2]). Note that

$$\sum_{e \in E(G)} |\mathcal{P}(e)| = \sum_{P \in \mathcal{P}} |E(P)| \leq \beta n(n-1) = 2\beta \binom{n}{2}.$$

For $i \in \{1, 2\}$, let $E_i \subseteq E(G)$ be the set of edges e such that $|\mathcal{P}(e)| = i$. Then

$$2\beta \binom{n}{2} \geq \sum_{e \in E(G)} |\mathcal{P}(e)| \geq |E_1| + 2|E_2| + 3\left(\alpha\binom{n}{2} - |E_1| - |E_2|\right)$$

$$= 3\alpha\binom{n}{2} - 2|E_1| - |E_2|. \tag{1}$$

Since $\mathcal{P}$ is strong-separating, if $e \in E_2$, then the two paths of $\mathcal{P}$ that contain e cannot both contain any other $f \in E_2$. Thus, $|E_2| \leq \binom{|\mathcal{P}|}{2} \leq \binom{\beta n}{2} < \beta^2\binom{n}{2} + \beta^2 n$. Also, $|E_1| \leq |\mathcal{P}| \leq \beta n$. Applying these bounds on $|E_1|$ and $|E_2|$ in (1), we get

$$\beta^2\binom{n}{2} + 2\beta\binom{n}{2} \geq 3\alpha\binom{n}{2} - \beta^2 n - 2\beta n \geq 3\alpha\binom{n}{2} - 400n,$$

where in the last step we used $\beta \leq 19$ to get $\beta^2 n + 2\beta n \leq 400n$. Thus the inequality $\beta^2 + 2\beta \geq 3\alpha - 800/n$ holds. Since $\beta > 0$ and n is sufficiently large, solving the quadratic in terms of β gives $\beta \geq \sqrt{3\alpha + 1} - 1 - \varepsilon$, as desired. $\square$

3 Preliminaries

3.1 Hypergraph Matchings

We use a recent result by Glock, Joos, Kim, Kühn, and Lichev [6] (see similar results in [3]). This result allows us to find almost perfect matchings in hypergraphs $\mathcal{H}$ which avoid certain 'conflicts'. Each conflict is a subset of edges $X \subseteq E(\mathcal{H})$ which we do not want to appear together in the matching M, i.e., we want $X \not\subseteq M$ for all such conflicts X. We encode these conflicts using an auxiliary 'conflict hypergraph' $\mathcal{C}$ whose vertex set is $E(\mathcal{H})$ and each edge is a different conflict, i.e., each edge of $\mathcal{C}$ encodes a set of edges of $\mathcal{H}$.

Given a (not necessarily uniform) hypergraph $\mathcal{C}$ and $k \geq 1$, let $\mathcal{C}^{(k)}$ denote the subgraph of $\mathcal{C}$ consisting of all edges of size exactly k. If $\mathcal{C} = \mathcal{C}^{(k)}$, then $\mathcal{C}$ is a k-graph. For a hypergraph $\mathcal{H}$ and $j \geq 1$, let $\delta_j(\mathcal{H})$ (resp. $\Delta_j(\mathcal{H})$) be the minimum (resp. maximum), taken over all j-sets S of vertices, of the number of edges of $\mathcal{H}$ which contain S. A hypergraph $\mathcal{H}$ is $(x \pm y)$-regular if $x - y \leq \delta_1(\mathcal{H}) \leq \Delta_1(\mathcal{H}) \leq x + y$. Let $N_{\mathcal{H}}(v)$ denote the set of neighbours of v in $\mathcal{H}$. Given a hypergraph $\mathcal{C}$ with $V(\mathcal{C}) = E(\mathcal{H})$, we say $E \subseteq E(\mathcal{H})$ is $\mathcal{C}$-free if no C in $E(\mathcal{C})$ is a subset of E. Also, $\mathcal{C}$ is a (d, ℓ, ρ)-bounded conflict system for $\mathcal{H}$ if

(C1) $3 \leq |C| \leq \ell$ for each $C \in \mathcal{C}$;
(C2) $\Delta_1(\mathcal{C}^{(j)}) \leq \ell d^{j-1}$ for all $3 \leq j \leq \ell$; and
(C3) $\Delta_{j'}(\mathcal{C}^{(j)}) \leq \ell d^{j-j'-\rho}$ for all $3 \leq j \leq \ell$ and $2 \leq j' < j$.

We say that a set of edges $Z \subseteq E(\mathcal{H})$ is (d, ρ)-trackable[1] if $|Z| \geq d^{1+\rho}$.

Theorem 3 ([6, Theorem 3.2]). *For all $k, \ell \geq 2$, there exists $\rho_0 > 0$ such that for all $\rho \in (0, \rho_0)$ there exists d_0 so that the following holds for all $d \geq d_0$. Suppose $\mathcal{H}$ is a k-graph on $n \leq \exp(d^{\rho^3})$ vertices with $(1 - d^{-\rho})d \leq \delta_1(\mathcal{H}) \leq \Delta_1(\mathcal{H}) \leq d$ and $\Delta_2(\mathcal{H}) \leq d^{1-\rho}$ and suppose $\mathcal{C}$ is a (d, ℓ, ρ)-bounded conflict system for $\mathcal{H}$. Suppose $\mathcal{Z}$ is a set of (d, ρ)-trackable sets of edges in $\mathcal{H}$ with $|\mathcal{Z}| \leq \exp(d^{\rho^3})$. Then, there exists a $\mathcal{C}$-free matching $\mathcal{M} \subseteq \mathcal{H}$ of size at least $(1 - d^{-\rho^3})n/k$ with $|Z \cap \mathcal{M}| = (1 \pm d^{-\rho^3})|\mathcal{M}||Z|/|E(\mathcal{H})|$ for all $Z \in \mathcal{Z}$.*

3.2 Building a Base Hypergraph

The next lemma constructs an auxiliary hypergraph which we will use as base to apply Theorem 3 later.

Lemma 1. *For any $\alpha, \beta, \lambda > 0$ with $\beta = \sqrt{3\alpha + 1} - 1 < \lambda$, there exists n_0 such that the following holds for every $n \geq n_0$. There exists a 3-graph J such that*

(J1) *there is a partition $\{U_1, U_2\}$ of $V(J)$ with $|U_1| = \lambda n$ and $|U_2| = \lambda n\beta/2$;*

---

[1] This corresponds to the $j = 1$ and $\varepsilon = \rho$ case of the definition of $(d, \varepsilon, \mathcal{C})$-trackable *test systems* of [6, Sect. 3]. The original definition requires more properties but reduces to the definition we have given when $j = 1$. In particular, $\mathcal{C}$ does not play a role anymore, so we opted for removing it from the notation.

(J2) *there is a partition $\{J_1, J_2\}$ of $E(J)$ such that*
   - *$e \subseteq U_1$ for each $e \in J_1$, and*
   - *$|e \cap U_1| = 2$ for each $e \in J_2$;*

(J3) *every pair $\{i, j\} \in U_1$ is contained only in edges of J_1, or in at most one edge of J_2;*

(J4) *$\Delta_2(J) \leq \ln^2 n$;*

(J5) *J has $\alpha\binom{n}{2} \pm n^{2/3}$ edges in total; and*

(J6) *J is $(\beta n/\lambda, n^{2/3})$-regular.*

Proof. Note that $3\alpha = 2\beta + \beta^2$ from the definition of β. Defining $d_2 = \beta^2 n/\lambda$ and $d_3 = 3(\alpha - \beta^2)n/(2\lambda)$, we obtain $d_2 + d_3 = \beta n/\lambda$. A $\{2,3\}$-*graph* is a hypergraph whose edges have size either 2 or 3. We say it is an *antichain* if no edge is contained in another. We start by defining an antichain $\{2,3\}$-graph on a set U_1 of λn vertices and we claim that there is such an antichain with

(F1) each vertex is adjacent to $d_2 \pm n^{2/3}$ edges of size 2 in I;

(F2) each vertex is adjacent to $d_3 \pm n^{2/3}$ edges of size 3 in I.

Indeed, define a random graph $I^{(2)}$ on U_1 by including each edge independently with probability $p := \beta^2/\lambda^2 < 1$. Let $\bar{I}$ be the complement of $I^{(2)}$. In expectation, each vertex is contained in around $p|U_1| = d_2$ many edges. A standard application of Chernoff's inequality shows that, with overwhelmingly large probability, each vertex of U_1 is contained in $d_2 \pm n^{2/3}$ edges of $I^{(2)}$, and thus we can assume that a choice of $I^{(2)}$ is fixed and satisfies that property. Similarly, we can also assume that every vertex is contained in $(1-p)^3 \binom{\lambda n}{2} \pm n^{4/3}$ triangles in $\bar{I}$. Next, we form a 3-graph $I^{(3)}$ on U_1 by including each triple of vertices which forms a triangle in $\bar{I}$ with probability $q := 3(\alpha - \beta^2)/((1-p)^3 \lambda^3 n)$. If a vertex x is contained in $d := (1-p)^3 \binom{\lambda n}{2}$ triangles in $\bar{I}$, then in expectation it must be contained in $dq = d_3 \pm n^{1/2}$ many 3-edges in $I^{(3)}$. Thus, again by Chernoff's inequality, we can assume that each vertex in $I^{(3)}$ is contained in $d_3 \pm n^{2/3}$ many triangles in $I^{(3)}$. We conclude by taking $I = I^{(2)} \cup I^{(3)}$.

Now we transform I into a 3-graph. To achieve this, we will add a set U_2 of extra vertices to I, and extend each 2-edge of I to a 3-edge by including to it a vertex in U_2. Let U_2 have size $r := \lambda n \beta/2$ and vertices $\{v_1, \ldots, v_r\}$. Randomly partition the 2-edges of $I^{(2)}$ into r sets $F_1, \ldots, F_r$ by including each edge of $I^{(2)}$ in an F_i with probability $1/r$. Next, define sets of 3-edges $H_1, \ldots, H_r$ given by $H_i := \{xyv_i : xy \in F_i\}$. Let J be the 3-graph on vertex set $U_1 \cup U_2$ whose edges are $I^{(3)} \cup \bigcup_{i=1}^r H_i$. Note that, by construction, J satisfies **(J1)**–**(J3)**, so it only remains to verify **(J4)**, **(J5)** and **(J6)**.

We show that **(J4)** holds. Let x, y be a pair of vertices in $V(J)$ and let us consider the possible cases. If $x, y \in U_1$ and $xy \in I^{(2)}$, then $\deg_J(xy) = 1$, because its only neighbour is v_i (if $xy \in F_i$). If $x, y \in U_1$ and $xy \in \bar{I}$, then $\deg_J(xy)$ is precisely the number of triangles of $I^{(3)}$ that contain xy. This is a random variable with expected value at most $nq = O(1)$. So, by standard concentration inequalities, $\deg_J(xy) > \ln^2 n$ holds with probability at most $n^{-\ln n}$, and we can comfortably use the union bound to ensure that $\deg_J(xy) \leq \ln^2 n$ for every such pair $xy \in \bar{I}$. If $x \in U_1$ and $y \in U_2$, then $y = v_i$ for some $1 \leq i \leq r$, and $\deg_J(xy)$

is the number of triangles of the form $xzv_i \in H_i$. For a fixed x, there are at most $|U_1| = \lambda n$ choices for z to form an edge $xz \in I^{(2)}$. Recall that each such edge belongs to H_i with probability $1/r = O(1/n)$. So the expected value of $\deg_J(xy)$ is again of the form $O(1)$, and we can conclude the argument similarly. Finally, if $x, y \in U_2$ then $\deg_J(xy) = 0$ by construcion. This finishes the proof of **(J4)**.

From (F1), we deduce that $|E(I^{(2)})|$ is $\lambda n(d_2 \pm n^{2/3})/2 = \beta^2 n^2/2 \pm n^{2/3}$ and, from (F2), we deduce that $|E(I^{(3)})|$ is $\lambda n(d_3 \pm n^{2/3})/3 = (\alpha - \beta^2)n^2/2 \pm n^{2/3}$, hence $|E(J)| = |E(I^{(2)})| + |E(I^{(3)})| = \alpha n^2/2 \pm O(n^{2/3})$, which proves **(J5)**.

Now we prove **(J6)**. Let $i \in V(J)$. If $i \in U_1$, then $\deg_J(i) = \deg_I(i)$. Since $d_2 + d_3 = \beta n/\lambda$, we have that $\deg_I(i) = d_2 + d_3 + O(n^{2/3}) = \beta n/\lambda + O(n^{2/3})$. Assume now that $i \in U_2$. Recall that we defined J in a way that each vertex of U_2 belongs to $|E(I^{(2)})|/r$ edges, so we conclude observing that

$$\frac{|E(I^{(2)})|}{r} = \frac{\beta^2 \binom{n}{2} \pm O(n^{4/3})}{\lambda n \beta/2} = \frac{\beta n}{\lambda} \pm O(n^{1/3}).$$

$\square$

4 Separating Almost All Edges

In this section we show how to separate most pairs of edges of robustly-connected graphs by paths and cycles, guaranteeing additional structural properties.

In what follows, let $\varepsilon, \delta > 0$, let L be an integer and let G be an n-vertex graph. A *2-matching* in G is a collection of vertex-disjoint cycles and paths in G. We say a 2-matching Q in G is (δ, L)-*robustly-connected* if, for every $x, y \in V(Q)$, there exists $1 \le \ell \le L$ such that there are at least δn^ℓ (x, y)-paths with exactly ℓ inner vertices each, all in $V(G) \setminus V(Q)$. A 2-matching Q in G is ε-*compact* if each cycle in Q has length at least $1/\varepsilon$ and Q contains at most εn paths. A collection $\{Q_1, \ldots, Q_t\}$ of 2-matchings in G is ε-*compact* if each Q_i is ε-compact.

Let $\mathcal{Q} = \{Q_1, \ldots, Q_t\}$ be a collection of subgraphs of G. We use $E(\mathcal{Q})$ to denote the set $\bigcup_{i=1}^t E(Q_i)$. We say $\mathcal{Q}$ *separates an edge e from all other edges* of G if the set $\{i : e \in E(Q_i)\}$ is not contained in the set $\{j : f \in E(Q_j)\}$ for each $f \in E(G) \setminus \{e\}$. Clearly, if an edge e is separated from all other edges of G by $\mathcal{Q}$, then $e \in E(\mathcal{Q})$. We also say that $\mathcal{Q}$ *strongly separates* a set E' of edges if, for every distinct $e, f \in E'$, the sets $\{i : e \in E(Q_i)\}$ and $\{j : f \in E(Q_j)\}$ are not contained in each other.

Lemma 2. *Let $1/n \ll \varepsilon, \varepsilon', \alpha, \delta, 1/L, \rho$. Let $\beta = \sqrt{3\alpha + 1} - 1$. If G is an n-vertex $(\alpha n \pm n^{1-\rho})$-regular graph that is (δ, L)-robustly-connected, then there exists an ε'-compact collection $\mathcal{Q}$ of 2-matchings in G with $|\mathcal{Q}| = \beta n/(1-\varepsilon)$ such that the following holds.*

(Q1) $\Delta(G - E(\mathcal{Q})) \le \varepsilon' n$;

(Q2) $\mathcal{Q}$ *strongly separates $E(\mathcal{Q})$;*

(Q3) *each vertex in G is the endpoint of at most $\varepsilon' n$ paths among all $Q \in \mathcal{Q}$;*

(Q4) *each $e \in E(\mathcal{Q})$ is contained in two or three 2-matchings in $\mathcal{Q}$; and*

(Q5) *each $Q \in \mathcal{Q}$ is $(\varepsilon^\ell \delta/2, L)$-robustly-connected.*

Proof. Our proof has five steps. First, we will build an auxiliary hypergraph $\mathcal{H}$ such that a large matching $M \subseteq \mathcal{H}$, which avoids certain conflicts, yields a family of graphs with the desired properties. We wish to apply Theorem 3 to find such a matching. In the second step, we will verify that $\mathcal{H}$ satisfies the hypotheses of Theorem 3. In the third step, we will define our conflict hypergraph $\mathcal{C}$. In the fourth step, we will define some test sets and prove they are trackable. Having done this, we are ready to apply Theorem 3, which is done in the last step. Then we verify that the construction gives the desired graphs.

Step 1: Constructing the auxiliary hypergraph. Obtain an oriented graph D by orienting each edge of G uniformly at random. Each vertex v has expected in-degree and out-degree $d_G(v)/2 = (\alpha n \pm n^{1-\rho})/2$. So, by using a concentration inequality like the Chernoff bounds, we can assume that in D every vertex has in-degree and out-degree of the form $\alpha n/2 \pm 2n^{1-\rho}$.

Next, consider an auxiliary bipartite graph B whose clusters are copies V_1 and V_2 of $V(G)$, where each vertex $x \in V(G)$ is represented by two copies $x_1 \in V_1$ and $x_2 \in V_2$, and such that $x_1 y_2 \in E(B)$ if and only if $(x, y) \in E(D)$. Thus we have that $|E(B)| = |E(G)| = \alpha\binom{n}{2} \pm n^{2-\rho}$, because G is $(\alpha n \pm n^{1-\rho})$-regular. Finally, let $\lambda = \beta/(1 - \varepsilon)$. Apply Lemma 1 with α, β, γ to obtain a 3-graph J which satisfies **(J1)**–**(J6)** and assume that $U_1 = [\lambda n]$ and $V(J) = [|V(J)|]$.

Now we build an initial auxiliary 8-graph $\mathcal{H}'$ as follows. Let Z be the complete bipartite graph between clusters $V(B)$ and $V(J)$. The vertex set of $\mathcal{H}'$ is $E(B) \cup E(J) \cup E(Z)$. Each edge in $\mathcal{H}'$ is determined by a choice $x_1 y_2 \in E(B)$ and $ijk \in E(J)$, which form an edge together with the 6 edges in Z that join x_1 and y_2 to i, j, and k. More precisely, the edge determined by $x_1 y_2 \in E(B)$ and $ijk \in E(J)$ is $\Phi(x_1 y_2, ijk) := \{x_1 y_2, ijk, x_1 i, x_1 j, x_1 k, y_2 i, y_2 j, y_2 k\}$; and the edge set of $\mathcal{H}'$ is given by $E(\mathcal{H}') = \{\Phi(x_1 y_2, ijk) : x_1 y_2 \in E(B), ijk \in E(J)\}$.

The idea behind the construction of $\mathcal{H}'$ is as follows. Suppose M is a matching in $\mathcal{H}'$, that $x_1 y_2 \in E(B)$ is covered by M and appears together with $ijk \in E(J)$ in an edge of M. By **(J2)**, $\{i, j, k\} \cap U_1$ has size 2 or 3. Recall that we want to obtain an ε'-compact collection $\mathcal{Q} := \{Q_1, \ldots, Q_t\}$ of 2-matchings in G, where $t = \lambda n$, satisfying **(Q1)**–**(Q5)**. We will add edges $xy \in E(G)$ such that $\Phi(x_1 y_2, ijk) \in M$ or $\Phi(y_1 x_2, ijk) \in M$ to the graphs Q_a if $a \in \{i, j, k\} \cap U_1$.

By construction, and since M is a matching, at most one edge in B involving x_1 (resp. x_2) appears in an edge of M together with some $a \in U_1$. By considering the contributions of the two copies $x_1, x_2 \in V(B)$ of a vertex $x \in V(G)$, this means that the graphs $Q_a \subseteq G$ have maximum degree 2, and thus these graphs are 2-matchings in G, as we wanted. By construction and property **(J2)**, each edge in $E(G)$ belongs to either 0, 2 or 3 graphs Q_a. Importantly, property **(J3)** implies that, for two distinct edges $e, f \in E(G)$, no two non-empty sets of the type $\{a : e \in E(Q_a)\}$ and $\{b : f \in E(Q_b)\}$ can be contained in each other. Straightforward calculations reveal the degrees of the vertices in $V(\mathcal{H}')$.

Claim 1. *The following hold.*

(i) $\deg_{\mathcal{H}'}(x_1 y_2) = \alpha\binom{n}{2} \pm n^{2/3}$ *for each* $x_1 y_2 \in E(B)$;

(ii) $\deg_{\mathcal{H}'}(ijk) = \alpha\binom{n}{2} \pm n^{2-\rho}$ *for each* $ijk \in E(J)$; *and*

(iii) $\deg_{\mathcal{H}'}(x_a i) = \frac{\alpha\beta}{\lambda}\binom{n}{2} \pm 2n^{2-\rho}$ *for each* $x_a i \in E(Z)$.

Since $\mathcal{H}'$ is not quite regular, we will actually work with a carefully-chosen subgraph $\mathcal{H} \subseteq \mathcal{H}'$. Let $p := \beta/\lambda = 1 - \varepsilon$. For each $i \in V(J)$, select a subset $X_i \subseteq V(G)$ by including in X_i each vertex of G independently at random with probability p. This defines a family $\{X_i : i \in V(J)\}$ of subsets of $V(G)$. For each $x \in V(G)$, consider the random set $Y_x = \{i \in V(J) : x \in X_i\}$. Finally, let $\mathcal{H} \subseteq \mathcal{H}'$ be the induced subgraph of $\mathcal{H}'$ obtained after removing all vertices $x_1 i, x_2 i \in E(Z)$ whenever $x \notin X_i$ (or, equivalently, $i \notin Y_x$). Thus, we have that $E(\mathcal{H}) = \{\Phi(x_1 y_2, ijk) : x_1 y_2 \in E(B), ijk \in E(J), \{x, y\} \subseteq X_i \cap X_j \cap X_k\}$.

Claim 2. *The following hold simultaneously with positive probability.*

(i) X_i *has* $pn \pm n^{2/3}$ *vertices of* G *for each* $i \in V(J)$;

(ii) Y_x *has* $3\alpha n/2 \pm n^{2/3}$ *vertices of* J *for each* $x \in V(G)$;

(iii) $\mathcal{H}$ *is* $(p^6 \alpha\binom{n}{2} \pm 2n^{2-\rho})$*-regular; and*

(iv) *for each* $i \in V(J)$ *and each pair of distinct vertices* $x, y \in X_i$, *there exists* ℓ *with* $1 \le \ell \le L$ *such that there are at least* $\varepsilon^\ell \delta n^\ell / 2$ (x, y)*-paths in* G *with exactly* ℓ *inner vertices each, all in* $V(G) \setminus X_i$.

From now on, we assume that the sets $\{X_i : i \in V(J)\}$, $\{Y_x : x \in V(G)\}$, and the hypergraph $\mathcal{H}$ satisfy properties (i)–(iv) of Claim 2.

Step 2: Verifying properties of $\mathcal{H}$. We start by defining $d := \Delta_1(\mathcal{H})$. Note that from Claim 2(iii), we have $d = p^6 \alpha n^2 / 2 \pm 2n^{2-\rho}$. We will apply Theorem 3 to $\mathcal{H}$. The following claim guarantees that $\mathcal{H}$ satisfies the required hypotheses and can be checked with the information given by the previous claims.

Claim 3. *The following facts about* $\mathcal{H}$ *hold.*

(H1) $\mathcal{H}$ *has at most* $\exp(d^{\rho^3})$ *vertices;*

(H2) $d(1 - d^{-\rho/3}) \le \delta_1(\mathcal{H}) \le \Delta_1(\mathcal{H}) = d$; *and*

(H3) $\Delta_2(\mathcal{H}) \le d^{2/3}$.

Step 3: Setting the conflicts. We must ensure that the collection $\mathcal{Q}$ of 2-matchings we want to obtain is ε'-compact: each 2-matching in $\mathcal{Q}$ has at most $\varepsilon' n$ paths and each cycle in $\mathcal{Q}$ has length at least $1/\varepsilon'$. This condition on the cycle lengths will be encoded by using conflicts.

Recall that D is the oriented graph obtained by orienting each edge of G uniformly at random. In what follows, let $r := 1/\varepsilon'$. We define our conflict hypergraph $\mathcal{C}$ on vertex set $E(\mathcal{H})$ and edge set defined as follows. For each ℓ with $3 \le \ell \le r$, each ℓ-length directed cycle $C \subseteq D$ with vertices $\{v^1, \ldots, v^\ell\}$, each $i \in U_1$, and each $j_1 k_1, \ldots, j_\ell k_\ell \in N_J(i)$, we define the following edge:

$$F(C, i, j_1 k_1, \ldots, j_\ell k_\ell) = \{\Phi(v_1^a v_2^{a+1}, i j_a k_a) : 1 \le a \le \ell\},$$

where $v_2^{\ell+1} = v_2^1$. Note that $F(C, i, j_1 k_1, \ldots, j_\ell k_\ell)$ corresponds to a set of ℓ edges of $\mathcal{H}'$, associated to the triples $(i\,j_1 k_1), \ldots, (i\,j_\ell k_\ell)$ and the edges of the ℓ-length directed cycle C in D. In such a case, we say i is the *monochromatic colour* of the conflicting cycle C. The edges of the conflict hypergraph $\mathcal{C}$ consist of all edges of type $F(C, i, j_1 k_1, \ldots, j_\ell k_\ell)$ which are contained in $E(\mathcal{H})$. The next claim establishes that $\mathcal{C}$ is a (d, r, ρ)-bounded conflict system for $\mathcal{H}$.

Claim 4. *The following facts about $\mathcal{C}$ hold.*

(C1) $3 \le |F| \le r$ *for each* $F \in \mathcal{C}$; *and*
(C2) $\Delta_{j'}(\mathcal{C}^{(j)}) \le r d^{j-j'-\rho}$ *for each* $3 \le j \le r$ *and* $1 \le j' < j$.

Step 4: Setting the test sets. For each $x_1 \in V_1 \subseteq V(B)$ and each $y_2 \in V_2 \subseteq V(B)$, define $Z_{x_1} = \{\Phi(x_1 y_2, ijk) \in E(\mathcal{H}) : y_2 \in N_B(x_1), ijk \in E(J)\}$ and define Z_{y_2} in a similar manner. Furthermore, define $Z_i = \{\Phi(x_1 y_2, ijk) \in E(\mathcal{H}) : x, y \in X_i\}$ for each $i \in V(J)$. We claim that $\mathcal{Z} := \{Z_{x_1} : x_1 \in V_1\} \cup \{Z_{y_2} : y_2 \in V_2\} \cup \{Z_i : i \in V(J)\}$ is a suitable family of trackable sets. Specifically, the next claim shows that $\mathcal{Z}$ is not very large and has only (d, ρ)-trackable sets.

Claim 5. *The following facts about $\mathcal{Z}$ hold.*

(Z1) $|\mathcal{Z}| \le \exp(d^{\rho^3})$; *and*
(Z2) *each* $Z \in \mathcal{Z}$ *is* (d, ρ)-*trackable.*

Step 5: Finishing the proof. Recall that $d = \Delta_1(\mathcal{H})$. By Claims 3, 4 and 5, we can apply Theorem 3 to $\mathcal{H}$, using $\mathcal{C}$ as a conflict system and $\mathcal{Z}$ as a set of trackable sets, and $\rho/3$ in place of ρ. By doing so, we obtain a matching $\mathcal{M} \subseteq \mathcal{H}$ such that

   (i) $\mathcal{M}$ is $\mathcal{C}$-free,
   (ii) $\mathcal{M}$ has size at least $(1 - d^{-(\rho/3)^3})|V(\mathcal{H})|/8$, and
   (iii) $|Z_a \cap \mathcal{M}| = (1 \pm d^{-(\rho/3)^3})|\mathcal{M}||Z_a|/|E(\mathcal{H})|$ for each $a \in V(B) \cup V(J)$.

Using $\mathcal{M}$, we define the graphs $\{Q_i\}_{i=1}^t$ as follows. For an edge $x_1 y_2 \in E(B)$, suppose there exists $ijk \in E(J)$ such that $\Phi(x_1 y_2, ijk) \in \mathcal{M}$. In that case, we will add the edge $xy \in E(G)$ to the graph Q_a such that $a \in \{i, j, k\} \cap U_1$. Now we verify that $\mathcal{Q} = \{Q_i\}_{i=1}^t$ is ε'-compact and satisfies properties **(Q1)–(Q5)** with $\varepsilon^\ell \delta/2$, λ and ε' in the place of δ, β and ε, respectively.

We start by verifying that $\mathcal{Q}$ is a collection of 2-matchings. Note that, for each $1 \le i \le t$, the graph Q_i has maximum degree at most 2. Indeed, let $x \in V(G)$ be any vertex. Since $\mathcal{M}$ is a matching in $\mathcal{H}$, at most two edges in $\mathcal{M}$ can cover the vertices $x_1 i, x_2 i \in V(\mathcal{H})$; and this will yield at most two edges adjacent to x belonging to Q_i. Also, $\mathcal{Q}$ is ε'-compact, which means that each 2-matching in $\mathcal{Q}$ has at most $\varepsilon' n$ paths and each cycle in $\mathcal{Q}$ has length at least $1/\varepsilon'$. The latter holds because we avoided the conflicts in $\mathcal{C}$. More precisely, an ℓ-cycle in Q_i corresponds to a sequence of ℓ edges, all of which are in Q_i. This means the cycle was formed from a length-ℓ directed cycle in D, all of whose edges were joined (via $\mathcal{M}$) to triples in J, all containing vertex $i \in V(J)$. Recall that $r = 1/\varepsilon'$. If $\ell \le r$, this forms a conflict in $\mathcal{C}$, so, as $\mathcal{M}$ is $\mathcal{C}$-free, we deduce that $\ell > r$. To check that Q_i has few paths, from (iii), we have that

$|E(Q_i)| = |Z_i \cap \mathcal{M}| \geq (1 - \varepsilon'/2)|X_i| = (1 - \varepsilon'/2)|V(Q_i)|$, and then the number of degree-one vertices in Q_i is at most $2(|V(Q_i)| - |E(Q_i)|) \leq \varepsilon'|V(Q_i)| \leq \varepsilon'n$.

Property **(Q1)** follows from the properties of the chosen test sets. More precisely, we want to prove that $\Delta(G') \leq \varepsilon'n$ for $G' := G - E(Q)$. Since for any $x \in V(G)$ we have $\deg_{G'}(x) = \deg_G(x) - (|Z_{x_1} \cap \mathcal{M}| + |Z_{x_2} \cap \mathcal{M}|)$, it is enough to prove that $|Z_{x_1} \cap \mathcal{M}| + |Z_{x_2} \cap \mathcal{M}| \geq \deg_G(x) - \varepsilon'n$. For that, by using (ii) and (iii) and the facts that $|E(\mathcal{H})| \leq |V(\mathcal{H})|\Delta_1(\mathcal{H})/8$ and $|Z_{x_1}| + |Z_{x_2}| \geq \delta_1(\mathcal{H})(|N_B(x_1)| + |N_B(x_2)|) \geq d(1 - d^{-\rho/3})\deg_G(x)$, we have the following for any $x \in V(G)$:

$$|Z_{x_1} \cap \mathcal{M}| + |Z_{x_2} \cap \mathcal{M}| \geq \frac{(1 - d^{-(\rho/3)^3})^2(|Z_{x_1}| + |Z_{x_2}|)|V(\mathcal{H})|/8}{|V(\mathcal{H})|\Delta_1(\mathcal{H})/8}$$

$$\geq (1 - d^{-(\rho/3)^3})^2(1 - d^{-\rho/3})\deg_G(x)$$

$$\geq \deg_G(x) - \varepsilon'n/2, \tag{2}$$

where inequality (2) holds for sufficiently large n because $\deg_G(x) = \Theta(n)$ and $d = \Theta(n^2)$. Thus **(Q1)** holds.

Property **(Q2)** can be checked as follows. Let e, f be distinct edges of $E(Q)$. There are orientations $(x, y), (x', y') \in E(D)$ of e, f respectively, and edges $ijk, i'j'k' \in E(J)$ such that $\Phi(x_1y_2, ijk)$ and $\Phi(x_1'y_2', i'j'k')$ belong to $\mathcal{M}$. We have, respectively, that $A_e := \{a : e \in E(Q_a)\} = \{i, j, k\} \cap U_1$ and $A_f := \{a : f \in E(Q_a)\} = \{i', j', k'\} \cap U_1$. For a contradiction, suppose $A_e \subseteq A_f$. If $|A_e| = 3$, then we would have that $ijk = i'j'k'$, contradicting that $\mathcal{M}$ is a matching, so $|A_e| = 2$; say, $A_e = \{i, j\}$, and ij is a pair in V_1 (from the construction of J). We recall that by **(J3)** no pair ij is contained both in an edge with intersection 2 and 3 with V_1, so this rules out the case $|A_f| = 3$. Thus we can only have $A_e = A_f = \{i, j\}$. But again **(J3)** implies ij is contained in a unique edge in J, say, ijr. This implies that $ijk = i'j'k' = ijr$, contradicting the fact that $\mathcal{M}$ is a matching. Therefore Q strongly separates $E(Q)$, and **(Q2)** holds.

To prove **(Q3)**, let $x \in V(G)$. Recall that $Y_x \subseteq V(J)$ is the random set $Y_x = \{i : x \in X_i\}$ and from Claim 2(ii) we have that $|Y_x| = 3\alpha n/2 \pm n^{2/3}$. Note that if x is the end of a path in some 2-matching Q_i, then there is an edge $\Phi(x_1y_2, ijk)$ in $Z_{x_1} \cap \mathcal{M}$, but no edge $\Phi(x_2y_1, ijk)$ is in $Z_{x_2} \cap \mathcal{M}$; or there is an edge $\Phi(x_2y_1, ijk)$ in $Z_{x_2} \cap \mathcal{M}$, but no edge $\Phi(x_1y_2, ijk)$ is in $Z_{x_1} \cap \mathcal{M}$. This motivates the following definition: for each $x \in V(G)$, a set $F(x_1)$ of indexes $i \in V(J)$ such that there is an edge $\Phi(x_1y_2, ijk)$ in $\mathcal{M}$; and a set $F(x_2)$ of indexes $i \in V(J)$ such that there is $\Phi(x_2y_1, ijk)$ in $\mathcal{M}$. Note that, from the way we construct $\mathcal{H}$, we know that $F(x_1), F(x_2) \subseteq Y_x$. In view of the above discussion, the number of times x is the endpoint of a path in the 2-matchings of Q is the number of indexes $i \in Y_x$ such that $i \notin F(x_1) \cap F(x_2)$. Therefore, this number of indexes i such that x is the endpoint of a path in Q_i is at most

$$|Y_x \setminus F(x_1)| + |Y_x \setminus F(x_2)| \leq 3\alpha n + 2n^{2/3} - 3(|Z_{x_1} \cap \mathcal{M}| + |Z_{x_2} \cap \mathcal{M}|)$$
$$\leq 3\alpha n + 2n^{2/3} - 3(\deg_G(x) - \varepsilon' n/2)$$
$$\leq 3\alpha n + 2n^{2/3} - 3(\alpha n - n^{1-\rho} - \varepsilon' n/2)$$
$$\leq \varepsilon' n,$$

where in the first inequality we use the facts that $|F(x_1)| = 3|Z_{x_1} \cap \mathcal{M}|$ and $|F(x_2)| = 3|Z_{x_2} \cap \mathcal{M}|$, and also that $Y_x \leq 3\alpha n + n^{2/3}$; in the second inequality we use (2); the third inequality follows from $\deg_G(x) \geq \alpha n - n^{1-\rho}$; and since n is sufficiently large, the last inequality holds. This verifies **(Q3)**.

To see **(Q4)**, let $e \in E(\mathcal{Q})$ be arbitrary. As explained before, there exists an orientation $(x, y) \in E(D)$ of e and an edge $ijk \in E(J)$ such that $\Phi(x_1 y_2, ijk)$ belongs to $\mathcal{M}$, and $\{a : e \in E(Q_a)\} = \{i, j, k\} \cap U_1$. Since the latter set obviously has at most three elements, **(Q4)** follows.

Finally, to prove **(Q5)**, from Claim 2(iv), because $V(Q_i) \subseteq X_i$, we deduce that Q_i is $(\varepsilon^\ell \delta/2, L)$-robustly-connected, as required. Thus **(Q5)** holds.   □

5   Breaking Cycles and Connecting Paths

For a number $\varepsilon \geq 0$, a collection $\mathcal{P}$ of paths in G is an ε-*almost separating path system* if there is a set $E' \subseteq E(G)$ such that $\mathcal{P}$ separates every edge in E' from all other edges in G and $\Delta(G - E') \leq \varepsilon n$. Note that such $\mathcal{P}$ strongly separates E'.

The next lemma will be used to prove our main result. It takes as an input the 2-matchings from Lemma 2 and yields a collection of paths in return.

Lemma 3. *For each ε, δ, and L, there exist ε' and n_0 such that the following holds for every $n \geq n_0$ and every $\beta \in (0, 1)$. Let G be an n-vertex graph and $\mathcal{Q}$ be an ε'-compact collection of (δ, L)-robustly-connected 2-matchings in G. If*

(i) *$|\mathcal{Q}| = \beta n$ and $\mathcal{Q}$ strongly separates $E(\mathcal{Q})$,*
(ii) *each vertex in G is the endpoint of at most $\varepsilon' n$ paths among all $Q \in \mathcal{Q}$,*
(iii) *each $e \in E(\mathcal{Q})$ is contained in at most three of the 2-matchings in $\mathcal{Q}$, and*
(iv) *$\Delta(G - E(\mathcal{Q})) \leq \varepsilon' n$,*

then there exists an ε-almost separating path system with βn paths.

Proof. A 2-matching is *acyclic* if it has no cycle. The first step is to transform $\mathcal{Q}$ into a collection of acyclic 2-matchings. We do this by deleting one edge, chosen at random, in each cycle belonging to some 2-matching of $\mathcal{Q}$. By construction, after the deletion we obtain an acyclic and $2\varepsilon'$-compact collection, and it will still be (δ, L)-robustly-connected and satisfy (i) and (iii). Moreover, if we remove at most $4\varepsilon' n$ edges incident to each vertex of G, then (ii) will hold with $5\varepsilon' n$ in the place of $\varepsilon' n$, and the degree of u in $G - E_i$ will increase by at most $4\varepsilon' n$, which implies (iv) with $5\varepsilon' n$ in the place of $\varepsilon' n$. The random choice ensures that, with non-zero probability, (ii) and (iv) hold. We proceed the proof assuming $\mathcal{Q}$ satisfies the assumption of the statement with ε', but in addition is acyclic.

Let $t = \beta n$. We will describe a sequence $\mathcal{C}_0, \ldots, \mathcal{C}_t$ of collections of acyclic 2-matchings in G and sets $E_0, \ldots, E_t$ of edges of G, the idea being that $\mathcal{C}_i$ strongly separates E_i, and that each $\mathcal{C}_i$ will be obtained from $\mathcal{C}_{i-1}$ by replacing Q_i with a path P_i. Then $\mathcal{C}_t$ will be the desired path system.

For each vertex u and $0 \le i \le t$, let $d_i(u)$ be the total number of paths in the 2-matchings in $Q_{i+1}, \ldots, Q_t$ that have u as an endpoint. We will make sure the following invariants on $\mathcal{C}_i$ and E_i hold for each $0 \le i \le t$:

(I1) each $\mathcal{C}_i$ separates every edge in E_i from all other edges of G;
(I2) edges in more than three of the 2-matchings in $\mathcal{C}_i$ are in $E(\mathcal{C}_i) \setminus E(\mathcal{C}_0)$; and
(I3) the degree of each vertex u in $G - E_i$ is at most εn if $d_i(u) = 0$ and at most $\sqrt{\varepsilon'} n - 2d_i(u)$ if $d_i(u) > 0$.

Let $E_0 = E(\mathcal{Q})$ and $\mathcal{C}_0 = \mathcal{Q}$. Note that $\mathcal{Q}_0$ and E_0 satisfy the three invariants. We will define $\mathcal{C}_i = (\mathcal{C}_{i-1} \setminus \{Q_i\}) \cup \{P_i\}$ for $i = 1, \ldots, t$, where P_i is a path that contains all paths in Q_i. Therefore, if invariants **(I1)** and **(I3)** hold for $i = t$ and $\varepsilon' \le \varepsilon^2$, then $\mathcal{C}_t$ will be an ε-almost separating path system with t paths, and the proof of the lemma will be complete, as $t = \beta n$.

Suppose $i \ge 1$. To describe how we build P_i from Q_i, we need some definitions. Let f be an edge of G not in Q_i such that $Q_i + f$ is a 2-matching. Let E^f be f plus the set of edges of Q_i in E_{i-1} that are not separated from f by $(\mathcal{C}_{i-1} \setminus \{Q_i\}) \cup \{Q_i + f\}$. We can prove that if f is in at most three of the 2-matchings in $\mathcal{C}_{i-1}$, then $|E^f| \le 4$.

A vertex u is *tight* if its degree in $G - E_{i-1}$ is more than $\varepsilon n - 2$ if $d_{i-1}(u) = 0$, or more $\sqrt{\varepsilon'} n - 2d_{i-1}(u) - 2$ if $d_{i-1}(u) > 0$. An edge f is *available for* P_i if $f \notin E(\mathcal{C}_{i-1}) \setminus E(\mathcal{C}_0)$ and the extremes of the edges in E^f are not tight.

To transform Q_i into P_i, we will proceed as follows. Start with P_i' being one of the paths in Q_i and let $Q_i' = Q_i \setminus \{P_i'\}$. While Q_i' is non-empty, let P be one of the paths in Q_i'. Call y one of the ends of P and x one of the ends of P_i'. An (x, y)-path in G is *good* if it has length at most L, all its edges are available and its inner vertices are not in $V(P_i) \cup V(Q_i')$. If a good (x, y)-path exists, we extend P_i' by gluing P_i' and P; we remove P from Q_i', and repeat this process until Q_i' is empty. When Q_i' is empty, we let P_i be P_i'. Recall that $\mathcal{C}_i = (\mathcal{C}_{i-1} \setminus \{Q_i\}) \cup \{P_i\}$, and we let E_i be E_{i-1} minus all edges contained in more than three 2-matchings of $\mathcal{C}_i$ and all edges not separated by $\mathcal{C}_i$ from some other edge of G.

This process is well-defined if the required (x, y)-good paths exists at every point in the construction. We will show that, indeed, assuming that the invariants hold, there is always a good (x, y)-path to be chosen in the gluing process above. Then, to complete the proof, we will prove that the invariants hold even after Q_i is modified by the choice of any good path.

First, note that the number of vertices in P_i' not in Q_i is less than $L\varepsilon' n$. Indeed, each connecting path has at most L inner vertices and Q_i is ε'-compact, so Q_i has at most $\varepsilon' n$ paths and we use less than $\varepsilon' n$ connecting paths to get to P_i. If $\varepsilon' < \delta/(4L)$, then the number of vertices in P_i' not in Q_i is less than $\delta n/4$.

Second, let us consider the tight vertices. We start by arguing that x is not tight. This happens because $d_i(x) = d_{i-1}(x) - 1$ and, by invariant **(I3)**, the degree of x in $G - E_{i-1}$ is at most $\sqrt{\varepsilon'} n - 2d_{i-1}(x) = \sqrt{\varepsilon'} n - 2d_i(x) - 2$. For the same reasons, y is not tight. Now, note that $E_i \setminus E_{i-1} \subseteq \bigcup \{E^f : f \in E(P_i) \setminus E(Q_i)\}$. Hence, $|E_i \setminus E_{i-1}| \leq 4L\varepsilon'n$ because $|E^f| \leq 4$ for each f, and because Q_i consists of at most $\varepsilon'n$ paths. This, $\Delta(G - E(Q)) \leq \varepsilon'n$, and $d_i(G) \leq \varepsilon'n$ imply that the maximum number of tight vertices is at most $(\varepsilon'n + 4L\varepsilon'\beta n)/(\sqrt{\varepsilon'} - 2\varepsilon') = \varepsilon'(1+4L\beta)n/(\sqrt{\varepsilon'} - 2\varepsilon')$. As long as $2\varepsilon' < \sqrt{\varepsilon'}/2$, that is, $\varepsilon' < 1/16$, we have that this number is less than $2\sqrt{\varepsilon'}(1 + 4L\beta)n$. If additionally $\varepsilon' < (\delta/(8(1+4L\beta)))^2$, we have that the number of tight vertices is less than $2\sqrt{\varepsilon'}(1 + 4L\beta)n < \delta n/4$.

Third, $|E(C_{i-1}) \setminus E(C_0)| < 4L\varepsilon'n(i-1) < 4L\varepsilon'\beta n^2$ because $i \leq \beta n$. Hence, by invariant **(I2)**, at most $4L\varepsilon'\beta n^2$ edges are used more than three times by C_{i-1}. Let $e \in E(C_{i-1}) \setminus E(C_0)$. Because Q_i is (δ, L)-robustly-connected, there exist $\ell \leq L$ and δn^ℓ (x, y)-paths in G, each with ℓ internal vertices, all in $V(G) \setminus V(Q_i)$. If e is not incident to x or y, then the number of (x, y)-paths in G with ℓ internal vertices and containing e is at most $n^{\ell-2}$. Hence, the number of (x, y)-paths in G with ℓ internal vertices, containing an edge in $E(C_{i-1}) \setminus E(C_0)$ not incident to x or y, is less than $4L\varepsilon'\beta n^\ell$. If e is incident to x or y, then the number of (x, y)-paths in G with ℓ internal vertices and containing e is at most $n^{\ell-1}$. But, there are less than $\sqrt{\varepsilon'}n$ edges incident to x and less than $\sqrt{\varepsilon'}n$ edges incident to y contained in more than three 2-matchings in C_{i-1}, by invariant **(I3)**. Thus, the number of (x, y)-paths in G of length ℓ containing an edge in $E(C_{i-1}) \setminus E(C_0)$ incident to x or y is less than $2\sqrt{\varepsilon'}n^\ell$. Since ε' is small enough, we have $4L\varepsilon'\beta + 2\sqrt{\varepsilon'} < \delta/4$, and thus at most $\delta n^\ell/4$ (x, y)-paths of length ℓ contain some edge of $E(C_{i-1}) \setminus E(C_0)$.

Summarising, for ε' small enough, the number of vertices in P_i' not in Q_i is less than $\delta n/4 \leq \delta n^\ell/4$, the number of tight vertices is also less than $\delta n/4 \leq \delta n^\ell/4$, and the number of (x, y)-paths containing some edge in $E(C_{i-1}) \setminus E(C_0)$ is less than $\delta n^\ell/4$. This means that at least $\delta n^\ell/4$ of the δn^ℓ (x, y)-paths in G, each with ℓ internal vertices, all in $V(G) \setminus V(Q_i)$, are good. As long as n_0 is such that $\delta n_0^\ell/4 \geq \delta n_0/4 \geq 1$, there is a good (x, y)-path.

Now let us verify the invariants. By the definition of E_i, invariant **(I1)** holds for i as C_i separates every edge in E_i from all other edges of G. Invariant **(I2)** holds because edges in more than three 2-matchings in C_i lie in used connecting paths, i.e., lie in $E(P_j) \setminus E(Q_j)$ for some j with $1 \leq j \leq i$. For invariant **(I3)**, note that $E_i \setminus E_{i-1} \subseteq E(P_i)$, so the degree of v from $G - E_{i-1}$ to $G - E_i$ decreases only for untight vertices, and by at most two. As the degree of an untight vertex u in $G - E_{i-1}$ is at most $\varepsilon n - 2$ if $d_{i-1}(u) = 0$ and at most $\sqrt{\varepsilon'}n - 2d_{i-1}(u) - 2$ if $d_{i-1}(u) > 0$, every vertex u in $G - E_i$ has degree at most εn if $d_i(u) = 0$ and at most $\sqrt{\varepsilon'}n - 2d_i(u)$ if $d_i(u) > 0$, also as $d_i(u) \leq d_{i-1}(u)$. So **(I3)** holds. $\qquad \square$

6   Separating the Last Few Edges

In this section we deal with a subgraph H of G, of small maximum degree, whose edges are not separated by the path family obtained in the previous sections. The main result we prove now is the following.

Lemma 4. *Let $\varepsilon, \delta, L > 0$ and let G and H be n-vertex graphs with $H \subseteq G$ such that $\Delta(H) \leq \varepsilon n$ and G is (δ, L)-robustly-connected. Then there exist paths*

$\{P_i\}_{i=1}^r$, $\{Q_i\}_{i=1}^r$ in G, with $r \leq 600L\delta^{-1}\sqrt{\varepsilon}n$, such that, for each $e \in E(H)$, there exist distinct $1 \leq i < j \leq r$ such that $\{e\} = E(P_i) \cap E(P_j) \cap E(Q_i) \cap E(Q_j)$.

We only sketch the proof of the above result[2]. The first step of the proof is to find a family of matchings which separates the edges of H.

Lemma 5. *Let $\Delta \geq 0$ and let H be an n-vertex graph with $\Delta(H) \leq \Delta$. Then there is a collection of $t \leq 300\sqrt{\Delta n}$ matchings $M_1, \ldots, M_t \subseteq H$ such that*

(M1) *each edge in H belongs to exactly two matchings M_i, M_j; and*

(M2) *matchings M_i, M_j have at most one edge in common for each $1 \leq i < j \leq t$.*

Lemma 5 is proven by assigning a 'label' $\{i, j\} \subseteq \{1, \ldots, t\}$ at random, with $i \neq j$, to each edge in $E(H)$ and letting each M_i be the set of edges of H with i in their labels. By construction, **(M1)** holds. From Lovász Local Lemma, the graphs M_i are matchings and satisfy **(M2)** with non-zero probability.

Given $M_1, \ldots, M_t$, we can prove Lemma 4 as follows. First (by partitioning each M_i into smaller matchings, and relabeling) we can assume that each M_i has at most $\delta n/(4L)$ edges and we have $r \leq 600L\delta^{-1}\sqrt{\varepsilon}n$ matchings in total. Next, for each i, we find two paths P_i, Q_i such that $E(P_i) \cap E(Q_i) = M_i$, which is enough to conclude. The paths P_i, Q_i are found by using the (δ, L)-robust-connectivity of G greedily, incorporating one edge of M_i at a time.

7   Proof of the Main Result

Now we have the tools to prove our main result, from which Theorem 1 and Theorem 2 immediately follow (in combination with Proposition 1).

Theorem 4. *Let $\alpha, \rho, \varepsilon, \delta \in (0, 1)$, $L > 0$ and n be sufficiently large. Let G be an n-vertex $(\alpha n \pm n^{1-\rho})$-regular graph which is (δ, L)-robustly-connected. Then $\mathrm{ssp}(G) \leq (\sqrt{3\alpha + 1} - 1 + \varepsilon)n$.*

Proof. Let $\varepsilon_2 := 1 - 1/(1 + \varepsilon/2)$ and $\delta' := \varepsilon_2^\ell \delta/2$. Choose ε' and n_0 so that Lemma 3 holds with $(\varepsilon\delta/(2400L))^2$, L and δ' in the roles of ε, L and δ. Apply Lemma 2 with $1 - 1/(1+\varepsilon/2)$ as ε to obtain a family $\mathcal{Q}$ of $t := (\sqrt{3\alpha+1}-1+\varepsilon/2)n$ many 2-matchings $Q_1, \ldots, Q_t$ satisfying **(Q1)**–**(Q5)**. Now apply Lemma 3 to G and $\mathcal{Q}$ to obtain an $(\varepsilon\delta/4800L)^2$-almost separating path system $\mathcal{P}$ in G of size t.

Let $E' \subseteq E(G)$ be the subset of edges which are strongly separated by $\mathcal{P}$ from every other edge. Since $\mathcal{P}$ is $(\varepsilon\delta/(2400L))^2$-almost separating, the subgraph $J := G - E'$ satisfies $\Delta(J) \leq (\varepsilon\delta/(2400L))^2 n$. We apply Lemma 4 with J and $(\varepsilon\delta/(2400L))^2$ playing the roles of H and ε to obtain two families $\mathcal{R}_1, \mathcal{R}_2$ of paths, with at most $\varepsilon n/4$ paths each, such that, for each $e \in E(J)$, there exist two paths $P_i, P_j \in \mathcal{R}_1$ and $Q_i, Q_j \in \mathcal{R}_2$ such that $\{e\} = E(P_i) \cap E(P_j) \cap E(Q_i) \cap E(Q_j)$.

Let $\mathcal{P}' := \mathcal{P} \cup \mathcal{R}_1 \cup \mathcal{R}_2$. Note that $\mathcal{P}'$ has at most $t + \varepsilon n/2 = (\sqrt{3\alpha+1}-1+\varepsilon)n$ many paths. We claim that $\mathcal{P}'$ is a strong-separating path system for G. Indeed,

---

[2] for a full proof of Lemma 4 and also a more detailed proof of the main result, see [5].

let e, f be distinct edges in $E(G)$; we need to show that there exists a path in $\mathcal{P}'$ which contains e and not f. If $e \in E'$, then such a path is contained in $\mathcal{P}$, so we can assume that $e \in E(J)$. There exist four paths $P_i, P_j, Q_i, Q_j \in \mathcal{P}'$ such that $\{e\} = E(P_i) \cap E(P_j) \cap E(Q_i) \cap E(Q_j)$, which in particular implies that one of these paths does not contain f. □

8 Concluding Remarks

Now we apply Theorem 4 to bound $\mathrm{ssp}(G)$ for graphs G belonging to certain families of graphs. In all cases, we just need to check that the corresponding graphs are (δ, L)-robustly-connected for suitable parameters.

Corollary 1. $\mathrm{ssp}(K_{n/2,n/2}) = (\sqrt{5/2} - 1 + o(1))n$.

Let us now describe a well-known family of graphs which satisfies the connectivity assumptions of Theorem 4. Given $0 < \nu \leq \tau \leq 1$, a graph G on n vertices, and a set $S \subseteq V(G)$, the ν-*robust neighbourhood of* S is the set $\mathrm{RN}_{\nu,G}(S) \subseteq V(G)$ of all vertices with at least νn neighbours in S. We say that G is a *robust* (ν, τ)-*expander* if, for every $S \subseteq V(G)$ with $\tau n \leq |S| \leq (1 - \tau)n$, we have $|\mathrm{RN}_{\nu,G}(S)| \geq |S| + \nu n$. Many families of graphs are robust (ν, τ)-expanders for suitable values of ν, τ, including large graphs with $\delta(G) \geq dn$ for fixed $d > 1/2$, dense random graphs, dense regular quasirandom graphs [9, Lemma 5.8], etc.

Corollary 2. *For each $\varepsilon, \alpha, \tau, \nu, \rho > 0$ with $\alpha \geq \tau + \nu$, there exists n_0 such that the following holds for each $n \geq n_0$. Let G be an n-vertex $(\alpha n \pm n^{1-\rho})$-regular robust (ν, τ)-expander. Then $\mathrm{ssp}(G) \leq (\sqrt{3\alpha + 1} - 1 + \varepsilon)n$.*

Acknowledgments. This research was partly supported by a joint project FAPESP and ANID (2019/13364-7) and by CAPES (Finance Code 001). C. G. Fernandes was supported by CNPq (310979/2020-0), G. O. Mota by CNPq (306620/2020-0, 406248/2021-4) and N. Sanhueza-Matamala by ANID FONDECYT Iniciación Nº11220269 grant. ANID is the Chilean National Agency for Research and Development. CAPES is the Coordenação de Aperfeiçoamento de Pessoal de Nível Superior. CNPq is the National Council for Scientific and Technological Development of Brazil. FAPESP is the São Paulo Research Foundation.

References

1. Balogh, J., Csaba, B., Martin, R.R., Pluhár, A.: On the path separation number of graphs. Discret. Appl. Math. **213**, 26–33 (2016)
2. Bonamy, M., Botler, F., Dross, F., Naia, T., Skokan, J.: Separating the edges of a graph by a linear number of paths. Adv. Comb., 1–7 (2023)
3. Delcourt, M., Postle, L.: Finding an almost perfect matching in a hypergraph avoiding forbidden submatchings (2022). https://arxiv.org/abs/2204.08981
4. Falgas-Ravry, V., Kittipassorn, T., Korándi, D., Letzter, S., Narayanan, B.P.: Separating path systems. J. Comb. **5**(3), 335–354 (2014)

5. Fernandes, C.G., Mota, G.O., Sanhueza-Matamala, N.: Separating path systems in complete graphs (2023). https://doi.org/10.48550/arXiv.2312.14879
6. Glock, S., Joos, F., Kim, J., Kühn, M., Lichev, L.: Conflict-free hypergraph matchings. In: Proceedings of the Annual ACM-SIAM Symposium on Discrete Algorithms (SODA), pp. 2991–3005 (2023)
7. Gronau, HD.O.F., Grüttmüller, M., Hartmann, S., Leck, U., Leck, V.: On orthogonal double covers of graphs. Des. Codes Cryptogr. **27**(1–2), 49–91 (2002). https://doi.org/10.1023/A:1016546402248
8. Horton, J.D., Nonay, G.M.: Self-orthogonal Hamilton path decompositions. Discret. Math. **97**(1–3), 251–264 (1991)
9. Kühn, D., Osthus, D.: Hamilton decompositions of regular expanders: applications. J. Combin. Theor. Ser. B **104**, 1–27 (2014)
10. Letzter, S.: Separating paths systems of almost linear size. arXiv arXiv:2211.07732 (2022)
11. Rényi, A.: On random generating elements of a finite Boolean algebra. Acta Sci. Math. (Szeged) **22**, 75–81 (1961)
12. Wickes, B.: Separating path systems for the complete graph. Discret. Math. **347**(3), 113784 (2024)
13. Zakrevski, L., Karpovsky, M.: Fault-tolerant message routing for multiprocessors. In: Rolim, J. (ed.) IPPS 1998. LNCS, vol. 1388, pp. 714–730. Springer, Heidelberg (1998). https://doi.org/10.1007/3-540-64359-1_737

Infinite Separation Between General and Chromatic Memory

Alexander Kozachinskiy$^{(\boxtimes)}$ (iD)

IMFD and CENIA, Santiago, Chile
alexander.kozachinskyi@cenia.cl

Abstract. In this paper, we construct a winning condition W over a finite set of colors such that, first, every finite arena has a strategy with 2 states of general memory which is optimal w.r.t. W, and second, there exists no k such that every finite arena has a strategy with k states of chromatic memory which is optimal w.r.t. W.

Keywords: Games on graphs · Finite-memory strategies · Chromatic memory

1 Introduction

Memory requirements for games on graphs have been studied for decades. Initially, these studies were motivated by applications to automata theory and the decidability of logical theories. For example, the memoryless determinacy of parity games is a key ingredient for the complementation of tree automata and leads to the decidability of the monadic second-order theory of trees [15]. Recently, games on graphs have become an important tool in reactive synthesis [1]. They serve there as a model of the interaction between a reactive system and the environment. One question studied in games on graphs is which winning conditions admit "simple" winning strategies. The prevailing measure of the complexity of strategies in the literature is memory. In this note, we study two kinds of memory – *general* (a.k.a. *chaotic*) memory and *chromatic* memory. The relationship between them was first addressed in the Ph.D. thesis of Kopczyński [12], followed by several recent works [3,6,7].

We focus on deterministic games, infinite-duration and turn-based. We call our players Protagonist and Antagonist. They play over a finite[1] directed graph called an *arena*. Its set of nodes has to be partitioned into ones controlled by Protagonist and ones controlled by Antagonist. Players move a token over the

---

[1] There are papers that study these games over infinite graphs, but in this note we only work with finite graphs.

The author is funded by ANID - Millennium Science Initiative Program - Code ICN17002, and the National Center for Artificial Intelligence CENIA FB210017, Basal ANID.

J. A. Soto and A. Wiese (Eds.): LATIN 2024, LNCS 14579, pp. 114–128, 2024.
https://doi.org/10.1007/978-3-031-55601-2_8

nodes of the graph along its edges. In each turn, the token is moved by the player controlling the current node.

After infinitely many turns, this process produces an infinite path in our graph. A *winning condition* is a set of infinite paths that are winning for Protagonist. In the literature, a standard way of defining winning conditions assumes that arenas are edge-colored by elements of some set of colors C. Then any subset $W \subseteq C^\omega$ is associated with a winning condition, consisting of all infinite paths whose sequence of colors belongs to W.

In this paper, we seek simple winning strategies of Protagonist, while the complexity of Antagonist's strategies is mostly irrelevant to us. Such asymmetry is motivated by reactive synthesis, where Protagonist represents a system and Antagonist represents the environment. Now, the main measure of the complexity of Protagonist's strategies for us is memory. Qualitatively, we distinguish between *finite-memory* strategies and *infinite-memory* strategies. In turn, among finite-memory strategies, we prefer those that have fewer states of memory.

Finite-memory strategies are defined through so-called *memory structures*. Intuitively, a memory structure plays the role of a "hard disk" of a strategy. Formally, a *general* memory structure $\mathcal{M}$ is a deterministic finite automaton whose input alphabet is the set of edges of an arena. During the game, edges over which the token moves are fed to $\mathcal{M}$ one by one. Correspondingly, the state of $\mathcal{M}$ is updated after each move. Now, a strategy *built on top of a memory structure* $\mathcal{M}$ (or simply an $\mathcal{M}$-strategy) is a strategy whose moves at any moment depend solely on two things: first, the current arena node, and second, the current state of $\mathcal{M}$. A strategy is finite-memory if it can be built on top of some memory structure. More precisely, if this memory structure has k states, then strategies built on top of it are strategies *with k states of general memory*. Of course, some strategies cannot be built on top of any memory structure. Such strategies are infinite-memory strategies.

We also consider a special class of general memory structures called *chromatic* memory structures. A memory structure is chromatic if its transition function does not distinguish edges of the same color. In other words, chromatic memory structures only reads colors of edges that are fed into them. Alternatively, a chromatic memory structure can be viewed as a finite automaton whose input alphabet is not the set of edges, but the set of colors. Correspondingly, strategies that are built on top of a chromatic memory structure with k states are called strategies with k *states of chromatic memory*.

Around a Kopczyński's question

Complexity of strategies brings us to complexity of winning conditions. For a given winning condition, we want to determine the minimal amount of memory which is sufficient to win whenever it is possible to win. More specifically, the **general memory complexity** of a winning condition W, denoted by $\mathsf{GenMem}(W)$, is the minimal $k \in \mathbb{N}$ such that in every arena there exists a Protagonist's strategy S with k states of general memory which is optimal w.r.t. W. If no such k exists, we set $\mathsf{GenMem}(W) = +\infty$. Now, "$S$ is optimal w.r.t. W"

means that there exists no node v such that some Protagonist's strategy is winning from v w.r.t. W and S is not. Substituting "general memory" by "chromatic memory", we obtain a definition of the **chromatic memory complexity** of W, which is denoted by ChrMem(W).

For any W, we have GenMem(W) $\leq$ ChrMem(W). Our paper revolves around a question from the Ph.D. thesis of Kopczyński [12].

Question 1. Is this true that GenMem(W) $=$ ChrMem(W) for every winning condition W?

To understand Kopczyński's motivation, we first have to go back to 1969, when Büchi and Landweber [4] established that ChrMem(W) is finite for all ω-regular W. An obvious corollary of this is that GenMem(W) is also finite for all ω-regular W. Since then, there is an unfinished quest of *exactly characterizing* ChrMem(W) and GenMem(W) for ω-regular W. In particular, it is open whether ChrMem(W) and GenMem(W) are computable given an ω-regular W as an input (assuming W is given, say, in a form of a non-deterministic Büchi automaton recognizing W).

In his Ph.D. thesis, Kopczyński contributed to this question by giving an algorithm computing ChrMem(W) for prefix-independent ω-regular W (a winning condition is called prefix-independent if it is invariant under adding and removing finite prefixes). Prior to that, he published a weaker version of this result in [11]. He asked Question 1 to find out, whether his algorithm also computes GenMem(W) for prefix-independent ω-regular W. His other motivation was that the same chromatic memory structure can be used in different arenas. Indeed, transition functions of chromatic memory structures can be defined over colors so that we do not have to specify them individually for each arena.

Question 1 was recently answered by Casares in [6]. Namely, for every $n \in \mathbb{N}$ he gave a *Muller* condition W over n colors with GenMem(W) $= 2$ and ChrMem(W) $= n$.

Definition 1. *A winning condition* $W \subseteq C^\omega$ *is* **Muller** *if* C *is finite and* $\alpha \in W \iff \beta \in W$ *for any two* $\alpha, \beta \in C^\omega$ *that have the same sets of colors occurring infinitely often in them.*

Every Muller condition is prefix-independent and ω-regular. Hence, we now know that Kopczyński's algorithm does not always compute GenMem(W) for prefix-independent ω-regular W. It is still open whether some other algorithm does this job.

In a follow-up work, Casares, Colcombet and Lehtinen [7] achieve a larger gap between GenMem(W) and ChrMem(W). Namely, they construct a Muller W over n colors such that GenMem(W) is linear in n and ChrMem(W) is exponential in n.

It is worth mentioning that Casares, Colcombet and Lehtinen derive these examples from their new automata-theoretic characterizations of ChrMem(W) and GenMem(W) for Muller W. First, Casares [6] showed that ChrMem(W) equals the minimal size of a deterministic Rabin automaton, recognizing W,

for every Muller W. Second, Casares, Colcombet and Lehtinen [7] showed that
GenMem(W) equals the minimal size of a good-for-games Rabin automaton,
recognizing W, for every Muller W. The latter result complements an earlier work by Dziembowski, Jurdziński and Walukiewicz [10], who characterized
GenMem(W) for Muller W in terms of their Zielonka's trees [15].

These examples, however, do not answer a natural follow-up question – can
the gap between GenMem(W) and ChrMem(W) be infinite? To answer it, we
have to go beyond Muller and even ω-regular conditions (because ChrMem(W)
is finite for them).

Question 2. Is it true that for every **finite** set of colors C and for every winning
condition $W \subseteq C^\omega$ we have GenMem(W) < $+\infty$ $\implies$ ChrMem(W) < $+\infty$?

Remark 1. If we do not insist on finiteness of C, a negative answer to Question 2
follows from the example of Casares. Namely, for every n he defines a winning
condition $W_n \subseteq \{1, 2, \ldots n\}^\omega$, consisting of all $\alpha \in \{1, 2, \ldots n\}^\omega$ such that there
are exactly two numbers from 1 to n that occur infinitely often in α. He then
shows that GenMem(W_n) = 2 and ChrMem(W_n) = n for every n. We can now
consider the union of these winning conditions $\cup_{n\geq 2} W_n$, which is a winning
condition over $C = \mathbb{N}$. On one hand, GenMem($\cup_{n\geq 2} W_n$) = 2 because every arena
has only finitely many natural numbers as colors, and hence $\cup_{n\geq 2} W_n$ coincides
with W_n for some n there. On the other hand, we have ChrMem($\cup_{n\geq 2} W_n$) $\geq$
ChrMem(W_n) = n for every n, which means that ChrMem($\cup_{n\geq 2} W_n$) = $+\infty$.

In this paper, we answer negatively to Question 2.

Theorem 1. *There exists a finite set of colors C and a winning condition $W \subseteq$
C^ω such that* GenMem(W) = 2 *and* ChrMem(W) = $+\infty$.

Topologically, our W belongs to the Σ_2^0-level of the Borel hierarchy. Next, the
size of C in our example is 5, and there is a chance that it can be reduced. In turn,
GenMem(W) is optimal because GenMem(W) = 1 implies ChrMem(W) = 1 (one
state of general memory is equally useless as one state of chromatic memory).

We call our W the "Rope Ladder" condition. We define it in Sect. 3. The upper
bound on GenMem(W) is given in Sect. 4. Before that, we give Preliminaries in
Sect. 2. The proof of the lower bound on ChrMem(W) is omitted due to the space
constraints, but it can be found in the arXiv version of this paper.

Further Open Questions

Still, some intriguing variations of Question 2 remain open. For example, it is
interesting to obtain Theorem 1 for a closed condition, i.e. a condition in the
Π_1^0-level of the Borel hierarchy, or equivalently, a condition given by a set of
prohibited finite prefixes. In the game-theoretic literature, such conditions are
usually called safety conditions. Our W is an infinite union of safety conditions.
In [9], Colcombet, Fijalkow and Horn give a characterization GenMem(W) for
safety W. Recently, Bouyer, Fijalkow, Randour, and Vandenhove [5] obtained a
characterization of ChrMem(W) for safety W.

Problem 1. Construct a finite set of colors C and a safety winning condition $W \subseteq C^\omega$ such that $\mathsf{GenMem}(W) < \infty$ and $\mathsf{ChrMem}(W) = +\infty$.

It is equally interesting to obtain Theorem 1 for a prefix-independent W, as our W is not prefix-independent.

Problem 2. Construct a finite set of colors C and a prefix-independent winning condition $W \subseteq C^\omega$ such that $\mathsf{GenMem}(W) < \infty$ and $\mathsf{ChrMem}(W) = +\infty$.

There is also a variation of Question 2 related to a paper of Bouyer, Le Roux, Oualhadj, Randour, and Vandenhove [3]. In this paper, they introduce and study the class of arena-independent finite-memory determined winning conditions. When the set of colors C is finite, this class can be defined as the class of W such that both $\mathsf{ChrMem}(W)$ and $\mathsf{ChrMem}(C^\omega \setminus W)$ are finite[2] (meaning that both Protagonist and Antagonist can play optimally w.r.t. W using some constant number of states of chromatic memory).

First, Bouyer et al. obtain an automata-theoretic characterization of arena-independent finite-memory determinacy. Second, they deduce a one-to-two-player lifting theorem from it. Namely, they show that as long as both $\mathsf{ChrMem}(W)$ and $\mathsf{ChrMem}(\neg W)$ are finite in arenas without the Antagonist's nodes, the same is true for all arenas.

A natural step forward would be to study conditions W for which both $\mathsf{GenMem}(W)$ and $\mathsf{GenMem}(\neg W)$ are finite. Unfortunately, it is even unknown whether this is a larger class of conditions. This raises the following problem.

Problem 3. Construct a finite set of colors C and a winning condition $W \subseteq C^\omega$ such that $\mathsf{GenMem}(W)$ and $\mathsf{GenMem}(\neg W)$ are finite, but $\mathsf{ChrMem}(W)$ is infinite.

In fact, it is not clear if our W from Theorem 1 solves this problem. We do not know whether $\mathsf{GenMem}(\neg W)$ is finite for this W.

Question 2 is also open over infinite arenas. There is a relevant result due to Bouyer, Randour and Vandenhove [2], who showed that the class of W for which $\mathsf{ChrMem}(W)$ and $\mathsf{ChrMem}(\neg W)$ are both finite in infinite arenas coincides with the class of ω-regular W. Thus, it would be sufficient to give a non-ω-regular W for which both $\mathsf{GenMem}(W)$ and $\mathsf{GenMem}(\neg W)$ are finite in infinite arenas.

Finally, let us mention a line of work which studied the relationship between chromatic and general memory in the non-uniform setting. Namely, fix a single arena $\mathcal{A}$ and some winning condition W, and then consider two quantities: first, the minimal k_{gen} such that $\mathcal{A}$ has an optimal strategy with k_{gen} states of general memory, and second, the minimal k_{chr} such that $\mathcal{A}$ has an optimal strategy with k_{chr} states of chromatic memory. In [14], Le Roux showed that if k_{gen} is finite, then k_{chr} is also finite. There is no contradiction with Theorem 1 because k_{chr} depends not only on k_{gen}, but also on $\mathcal{A}$. A tight bound on k_{chr} in terms of k_{gen} and the number of nodes of $\mathcal{A}$ was obtained in [13].

---

[2] In their original definition, the "memory structure" (see Preliminaries) must be the same in all arenas. When C is finite, this definition is equivalent, because there are just finitely many chromatic memory structures up to a certain size. If none of them works for all arenas, one can construct a finite arena where none of them works.

2   Preliminaries

Notation. For a set A, we let A^* and A^ω stand for the set of all finite and the set of all infinite sequences of elements of A, respectively. For $x \in A^*$, we let $|x|$ denote the length of x. We also set $|x| = +\infty$ for $x \in A^\omega$. We let $\circ$ denote the function composition. The set of positive integral numbers is denoted by $\mathbb{Z}^+$.

2.1   Arenas

Definition 2. *Let C be a non-empty set. A tuple $\mathcal{A} = \langle V_P, V_A, E \rangle$ is called an* **arena over the set of colors** *C if the following conditions hold:*

- V_P, V_A, E *are finite sets such that $V_P \cap V_A = \varnothing$, $V_P \cup V_A \neq \varnothing$ and $E \subseteq (V_P \cup V_A) \times C \times (V_P \cup V_A)$;*
- *for every $s \in V_P \cup V_A$ there exist $c \in C$ and $t \in V_P \cup V_A$ such that $(s, c, t) \in E$.*

Elements of the set $V = V_P \cup V_A$ will be called nodes of $\mathcal{A}$. Elements of V_P will be called nodes controlled by Protagonist (or simply Protagonist's nodes). Similarly, elements of V_A will be called nodes controlled by Antagonist (or simply Antagonist's nodes). Elements of E will be called edges of $\mathcal{A}$. For an edge $e = (s, c, t) \in E$ we define $\mathsf{source}(e) = s, \mathsf{col}(e) = c$ and $\mathsf{target}(e) = t$. We imagine $e \in E$ as an arrow which is drawn from the node $\mathsf{source}(e)$ to the node $\mathsf{target}(e)$ and which is colored into $\mathsf{col}(e)$. Note that the second condition in the definition of an arena means that every node has at least one out-going edge.

We extend the domain of col to the set $E^* \cup E^\omega$ by

$$\mathsf{col}(e_1 e_2 e_3 \ldots) = \mathsf{col}(e_1)\mathsf{col}(e_2)\mathsf{col}(e_3)\ldots, \qquad e_1, e_2, e_3, \ldots \in E.$$

A non-empty sequence $p = e_1 e_2 e_3 \ldots \in E^* \cup E^\omega$ is called a path if for any $1 \leq i < |p|$ we have $\mathsf{target}(e_i) = \mathsf{source}(e_{i+1})$. We set $\mathsf{source}(p) = \mathsf{source}(e_1)$ and, if p is finite, $\mathsf{target}(p) = \mathsf{target}(e_{|p|})$. For technical convenience, every node $v \in V$ is assigned a 0-length path λ_v, for which we set $\mathsf{source}(\lambda_v) = \mathsf{target}(\lambda_v) = v$ and $\mathsf{col}(\lambda_v) = $ empty string.

Paths are sequences of edges, so we will say that some paths are prefixes of the others. However, we have to define this for 0-length paths. Namely, we say that λ_v is a prefix of a path p if and only if $\mathsf{source}(p) = v$.

2.2   Strategies

Let $\mathcal{A} = \langle V_P, V_A, E \rangle$ be an arena over the set of colors C. A Protagonist's strategy in $\mathcal{A}$ is any function

$$S\colon \{p \mid p \text{ is a finite path in } \mathcal{A} \text{ with } \mathsf{target}(p) \in V_P\} \to E,$$

such that for every p from the domain of S we have $\mathsf{source}(S(p)) = \mathsf{target}(p)$. In this paper, we do not mention Antagonist's strategies, but, of course, they can be defined similarly.

The set of finite paths in $\mathcal{A}$ is the set of positions of the game. Possible starting positions are 0-length paths $\lambda_s, s \in V$. When the starting position[3] is λ_s, we say that the game starts at s. Now, consider any finite path p. Protagonist is the one to move after p if and only if $t = \mathsf{target}(p)$ is a Protagonist's node. In this situation, Protagonist must choose some edge starting at t. A Protagonist's strategy fixes this choice for every p with $\mathsf{target}(p) \in V_P$. We then append this edge to p and get the next position in the game. Antagonist acts the same for those p such that $\mathsf{target}(p)$ is an Antagonist's node.

Let us define paths that are consistent with a Protagonist's strategy S. First, any 0-length path λ_v is consistent with S. Now, a non-empty path $p = e_1 e_2 e_3 \ldots$ (which may be finite or infinite) is consistent with S if the following holds:

- if $\mathsf{source}(p) \in V_P$, then $e_1 = S(\lambda_{\mathsf{source}(p)})$;
- for every $1 \leq i < |p|$, if $\mathsf{target}(e_i) \in V_P$, then $e_{i+1} = S(e_1 e_2 \ldots e_i)$.

For brevity, paths that are consistent with S will also be called *plays with S*. For a node v, we let $\mathsf{FinitePlays}(S, v)$ and $\mathsf{InfinitePlays}(S, v)$ denote the set of finite plays with S that start at v and the set of infinite plays with S that start at v, respectively. For $U \subseteq V$, we define $\mathsf{FinitePlays}(S, U) = \bigcup_{v \in U} \mathsf{FinitePlays}(S, v)$ and $\mathsf{InfinitePlays}(S, U) = \bigcup_{v \in U} \mathsf{InfinitePlays}(S, v)$.

2.3   Memory Structures

Let $\mathcal{A} = \langle V_P, V_A, E \rangle$ be an arena over the set of colors C. A memory structure in $\mathcal{A}$ is a tuple $\mathcal{M} = \langle M, m_{init}, \delta \rangle$, where M is a finite set, $m_{init} \in M$ and $\delta \colon M \times E \to M$. Elements of M are called states of $\mathcal{M}$, m_{init} is called the initial state of $\mathcal{M}$ and δ is called the transition function of $\mathcal{M}$. Given $m \in M$, we inductively define the function $\delta(m, \cdot)$ over arbitrary finite sequences of edges:

$$\delta(m, \text{empty sequence}) = m,$$
$$\delta(m, se) = \delta(\delta(m, s), e), \qquad s \in E^*, e \in E.$$

In other words, $\delta(m, s)$ is the state of $\mathcal{M}$ it transits to from the state m if we fed s to it.

A memory structure $\mathcal{M} = \langle M, m_{init}, \delta \rangle$ is called chromatic if $\delta(m, e_1) = \delta(m, e_2)$ for every $m \in M$ and for every $e_1, e_2 \in E$ with $\mathsf{col}(e_1) = \mathsf{col}(e_2)$. In this case, there exists $\sigma \colon M \times C \to M$ such that $\delta(m, e) = \sigma(m, \mathsf{col}(e))$. In other words, we can view $\mathcal{M}$ as a deterministic finite automaton over C, with σ being its transition function.

A strategy S is built on top of a memory structure $\mathcal{M}$ if we have $S(p_1) = S(p_2)$ for any two paths p_1, p_2 with $\mathsf{target}(p_1) = \mathsf{target}(p_2)$ and $\delta(m_{init}, p_1) = \delta(m_{init}, p_2)$. In this case, we sometimes simply say that S is an $\mathcal{M}$-strategy. To define an $\mathcal{M}$-strategy S, it is sufficient to give its *next-move function* $n_S \colon V_P \times$

---

[3] We do not have to redefine S for every starting position. The same S can be played from any of them.

$M \to E$. For $v \in V_P$ and $m \in M$, the value of $n_S(v, m)$ determines what S does for paths that end at v and bring $\mathcal{M}$ to m from m_{init}.

A strategy S built on top of a memory structure $\mathcal{M}$ with k states is called a strategy with k states of general memory. If $\mathcal{M}$ is chromatic, then S is a strategy with k states of chromatic memory.

For brevity, if S is an $\mathcal{M}$-strategy and p is a finite path, we say that $\delta(m_{init}, p)$ is the state of S after p.

2.4   Winning Conditions and Their Memory Complexity

A winning condition is any set $W \subseteq C^\omega$. We say that a Protagonist's strategy S is winning from a node u w.r.t. W if the image of InfinitePlays(S, u) under col is a subset of W. In other words, any infinite play from u against S must give a sequence of colors belonging to W. Now, a Protagonist's strategy S is called optimal w.r.t. W if there exists no node u such that some Protagonist's strategy is winning from u w.r.t. W and S is not.

We let GenMem(W) be the minimal $k \in \mathbb{Z}^+$ such that in every arena $\mathcal{A}$ over C there exists a Protagonist's strategy with k states of general memory which is optimal w.r.t. W. If no such k exists, we set GenMem$(W) = +\infty$. Likewise, we let ChrMem(W) be the minimal $k \in \mathbb{Z}^+$ such that in every arena $\mathcal{A}$ over C there exists a Protagonist's strategy with k states of general memory which is optimal w.r.t. W. Again, if no such k exists, we set ChrMem$(W) = +\infty$.

3   The "Rope Ladder" Condition

Consider the partially ordered set $\Omega = (\mathbb{N} \times \{0, 1\}, \preceq)$, where $\preceq$ is defined by

$$\forall (n, a), (m, b) \in \mathbb{N} \times \{0, 1\} \qquad (n, a) \preceq (m, b) \iff (n, a) = (m, b) \text{ or } n < m.$$

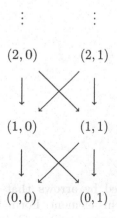

Above is its Hasse diagram, with arrows representing $\preceq$ (they are directed from bigger elements to smaller elements):

We will use an abbreviation $\overline{0} = (0,0)$. Next, we let $\mathbb{M}$ be the set of all functions $f: \Omega \to \Omega$ that are monotone w.r.t. $\preceq$. Being monotone w.r.t. $\preceq$ means that $x \preceq y \implies f(x) \preceq f(y)$ for all $x, y \in \Omega$.

Definition 3. *The Rope Ladder condition is a set* $\mathsf{RL} \subseteq \mathbb{M}^\omega$, *consisting of all infinite sequences* $(f_1, f_2, f_3, \ldots) \in \mathbb{M}^\omega$ *for which there exists* $(N, b) \in \Omega$ *such that* $f_n \circ \ldots \circ f_2 \circ f_1(\overline{0}) \preceq (N, b)$ *for all* $n \geq 1$.

We will use the following informal terminology with regard to RL. Imagine that there is an ant which can move over the elements of Ω. Initially, it sits at $\overline{0}$. Next, take any sequence $(f_1, f_2, f_3, \ldots) \in \mathbb{M}^\omega$. We start moving the ant by applying functions from the sequence to the position of the ant. Namely, we first move the ant from $\overline{0}$ to $f_1(\overline{0})$, then from $f_1(\overline{0})$ to $f_2 \circ f_1(\overline{0})$, and so on. Now, $(f_1, f_2, f_3, \ldots) \in \mathsf{RL}$ if and only if there exists a "layer" in Ω which is never exceeded by the ant.

Remark 2. RL is defined over infinitely many colors, but for our lower bound on its chromatic memory complexity we will consider its restriction to some finite subset of $\mathbb{M}$.

To illustrate these definitions, we establish the following fact. It can also be considered as a warm-up for our lower bound.

Fact 1 $\mathsf{ChrMem}(\mathsf{RL}) > 1$.

Proof. First, consider $u, v: \Omega \to \Omega$, depicted below:

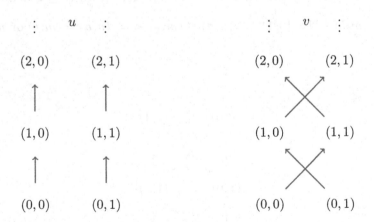

These functions are defined by arrows that direct each element of Ω to the value of the function on this element. Formally, $u((n, a)) = (n + 1, a)$ and $v((n, a)) = (n + 1, 1 - a)$ for every $(n, a) \in \Omega$. It holds that $u, v \in \mathbb{M}$ because they both always increase the first coordinate by 1.

We also consider the following two functions $f_0, f_1 \colon \Omega \to \Omega$:

$$f_b((n,a)) = \begin{cases} (n,a) & (n,a) = (0,0), (0,1) \text{ or } (1,b), \\ (n+1,a) & \text{otherwise,} \end{cases} \qquad b \in \{0,1\} \qquad (1)$$

For the reader's convenience, we depict them as well.

$$\vdots \quad f_0 \quad \vdots \qquad\qquad\qquad \vdots \quad f_1 \quad \vdots$$

$(2,0)$	$(2,1)$		$(2,0)$	$(2,1)$
$\circlearrowleft$	$\uparrow$		$\uparrow$	$\circlearrowright$
$(1,0)$	$(1,1)$		$(1,0)$	$(1,1)$
$\circlearrowleft$	$\circlearrowleft$		$\circlearrowleft$	$\circlearrowleft$
$(0,0)$	$(0,1)$		$(0,0)$	$(0,1)$

To see that $f_0, f_1 \in \mathbb{M}$, observe that both functions have 3 fixed points, and at remaining points, they act by increasing the first coordinate by 1. There could be a problem with monotonicity if below some fixed point there were a point which is not a fixed point. However, the sets of fixed points of f_0 and f_1 are downwards-closed w.r.t. $\preceq$.

Consider the following arena.

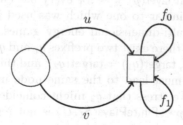

The circle is controlled by Antagonist and the square is controlled by Protagonist. Assume that the game starts in the circle. We first show that Protagonist has a winning strategy w.r.t. RL. Then we show that Protagonist does not have a positional strategy which is winning w.r.t. RL. This implies that ChrMem(RL) > 1.

Let us start with the first claim. After the first move of Antagonist, the ant moves either to $u(\overline{\mathbf{0}}) = (1,0)$ or to $v(\overline{\mathbf{0}}) = (1,1)$. In the first case, Protagonist wins by forever using the f_0-edge (the ant will always stay at $(1,0)$). In the second case, Protagonist wins by always using the f_1-edge (the ant will always stay at $(1, 1)$).

Now we show that every positional strategy of Protagonist is not winning w.r.t. RL. In fact, there are just 2 Protagonist's positional strategies – one which always uses the f_0-edge and the other which always uses the f_1-edge. The first one loses if Antagonist goes by the v-edge. Then the ant moves to $v(\overline{\mathbf{0}}) = (1, 1)$. If we start applying f_0 to the ant's position, the first coordinate of the ant will get arbitrarily large. Similarly, the second Protagonist's positional strategy loses if Antagonist goes by the u-edge.   □

4   Upper Bound on the General Memory

In this section, we establish

Proposition 1. GenMem(RL) = 2.

By Fact 1, we only have to show an upper bound GenMem(RL) $\leq$ 2. For that, for every arena $\mathcal{A}$ over $\mathbb{M}$ and for every Protagonist's strategy S_1 in $\mathcal{A}$ we construct a Protagonist's strategy S_2 with 2 states of general memory for which the following holds: for every node v of $\mathcal{A}$, if S_1 is winning w.r.t. RL from v, then so is S_2.

We will use the following notation. Take any finite path $p = e_1 \ldots e_m$ in $\mathcal{A}$. Define $\mathsf{ant}(p) = \mathsf{col}(e_m) \circ \ldots \circ \mathsf{col}(e_2) \circ \mathsf{col}(e_1)(\overline{\mathbf{0}})$. In other words, $\mathsf{ant}(p)$ is the position of the ant after the path p. In case when p is a 0-length path, we set $\mathsf{ant}(p) = \overline{\mathbf{0}}$. We also write $\mathsf{layer}(p)$ for the first coordinate of $\mathsf{ant}(p)$.

Let U be the set of nodes of $\mathcal{A}$ from which S_1 is winning w.r.t. RL. By definition of RL, for every $P \in \mathsf{InfinitePlays}(S_1, U)$ there exists $N \in \mathbb{N}$ such that $\mathsf{layer}(p) \leq N$ for every finite prefix p of P. The first step of our argument is to change the quantifiers here. That is, we obtain a strategy S_1' for which there exists some $N \in \mathbb{N}$ such that $\mathsf{layer}(p) \leq N$ for every $p \in \mathsf{FinitePlays}(S_1', U)$.

We use an argument, similar to one which was used in [8] to show finite-memory determinacy of multi-dimensional energy games. We call a play $p \in \mathsf{FinitePlays}(S_1, U)$ *regular* if there exist two prefixes q_1 and q_2 of p such that, first, q_1 is shorter than q_2, second, $\mathsf{target}(q_1) = \mathsf{target}(q_2)$, and third, $\mathsf{ant}(q_1) = \mathsf{ant}(q_2)$. In other words, q_1 and q_2 must lead to the same node in $\mathcal{A}$ and to the same position of the ant in Ω. We stress that q_2 might coincide with p, but q_1 must be a proper prefix of p. If $p \in \mathsf{FinitePlays}(S_1, U)$ is not regular, then we call it *irregular*.

First, we show that there are only finitely many irregular plays in the set $\mathsf{FinitePlays}(S_1, U)$. Note that any prefix of an irregular play is irregular. Thus, irregular plays form a collection of trees with finite branching (for each $u \in U$ there is a tree of irregular plays that start at u). Assume for contradiction that there are infinitely many irregular plays. Then, by Kőnig's lemma, there exists an infinite branch in one of our trees. It gives some $P \in \mathsf{InfinitePlays}(S_1, U)$ whose finite prefixes are all irregular. However, P must be winning for Protagonist w.r.t. RL. In other words, there exists $N \in \mathbb{N}$ such that $\mathsf{layer}(p) \leq N$ for every finite prefix p of P. So, if p ranges over finite prefixes of P, then $\mathsf{ant}(p)$ takes only finitely many values. Hence, there exist a node v of $\mathcal{A}$ and some $(n, b) \in \Omega$ such that $v = \mathsf{target}(p)$ and $(n, b) = \mathsf{ant}(p)$ for infinitely many prefixes p of P. Consider any two such prefixes. A longer one is regular because the shorter one is its prefix and leads to the same node in $\mathcal{A}$ and to the same position of the ant. This is a contradiction.

We now define S_1'. It will maintain the following invariant for plays that start at U: if p_{cur} is the current play, then there exists an irregular $p \in \mathsf{FinitePlays}(S_1, U)$ such that $\mathsf{target}(p_{cur}) = \mathsf{target}(p)$ and $\mathsf{ant}(p_{cur}) = \mathsf{ant}(p)$. Since there are only finitely many irregular plays, this invariant implies that $\mathsf{ant}(p_{cur})$ takes only finitely many values over $p_{cur} \in \mathsf{FinitePlays}(S_1', U)$, as required from S_1'.

To maintain the invariant, S_1' plays as follows. In the beginning, $p_{cur} = \lambda_w$ for some $w \in U$. Hence, we can set $p = \lambda_w$ also. Indeed, $\lambda_w \in \mathsf{FinitePlays}(S_1, U)$ and it is irregular as it has no proper prefixes. Let us now show how to maintain the invariant. Consider any play p_{cur} with S_1' for which there exists an irregular $p \in \mathsf{FinitePlays}(S_1, U)$ such that $\mathsf{target}(p_{cur}) = \mathsf{target}(p)$ and $\mathsf{ant}(p_{cur}) = \mathsf{ant}(p)$. In this position, if its Protagonist's turn to move, S_1' makes the same move as S_1 from p. As a result, some edge e is played. Observe that $pe \in \mathsf{FinitePlays}(S_1, U)$. In turn, our new current play with S_1' is $p_{cur}e$. We have that $\mathsf{target}(p_{cur}e) = \mathsf{target}(pe) = \mathsf{target}(e)$ and $\mathsf{ant}(p_{cur}e) = \mathsf{col}(e)(\mathsf{ant}(p_{cur})) = \mathsf{col}(e)(\mathsf{ant}(p)) = \mathsf{ant}(pe)$. So, if pe is irregular, then the invariant is maintained. Now, assume that pe is regular. Then there are two prefixes q_1 and q_2 of pe such that, first, q_1 is shorter than q_2, second, $\mathsf{target}(q_1) = \mathsf{target}(q_2)$, and third, $\mathsf{ant}(q_1) = \mathsf{ant}(q_2)$. Since p is irregular, q_2 cannot be a prefix of p. Hence, $q_2 = pe$. By the same reason, q_1 is irregular. Thus, invariant is maintained if we set the new value of p be q_1. Indeed, $\mathsf{target}(p_{cur}e) = \mathsf{target}(pe) = \mathsf{target}(q_2) = \mathsf{target}(q_1)$ and $\mathsf{ant}(p_{cur}e) = \mathsf{ant}(pe) = \mathsf{ant}(q_2) = \mathsf{ant}(q_1)$.

We now turn S_1' into a strategy S_2 with 2 states of general memory which is winning w.r.t. RL from every node of U.

Preliminary Definitions. Let X be the set of nodes reachable from U by plays with S_1'. Next, for $v \in X$, define $\Omega_v \subseteq \Omega$ as the set of all $(n, b) \in \Omega$ such that $(n, b) = \mathsf{ant}(p)$ for some $p \in \mathsf{FinitePlays}(S_1', W)$ with $v = \mathsf{target}(p)$. In other words, Ω_v is the set of all possible positions of the ant that can arise at v if we play according to S_1' from a node of U.

Now, take any $v \in X$. The set Ω_v is non-empty and, by our requirements on S_1', finite. Hence, it has 1 or 2 maximal elements w.r.t. $\preceq$. We will denote them by M_0^v and M_1^v. If Ω_v has just a single maximal element, then $M_0^v = M_1^v$. If Ω_v has two different maxima, then let M_0^v be the one having 0 as the second coordinate. Finally, for every $v \in X$ and for every $b \in \{0, 1\}$ fix some $p_b^v \in \mathsf{FinitePlays}(S_1', U)$ such that $\mathsf{target}(p_b^v) = v$ and $\mathsf{ant}(p_b^v) = M_b^v$.

Description of S_2. Two states of S_2 will be denoted by 0 and 1. The initial state of S_2 is 0. The next-move function of S_2 is defined as follows. Assume that the state of S_2 is $I \in \{0, 1\}$ and it has to make a move from a node v. If $v \notin X$, it makes an arbitrary move (this case does not matter for the argument below). Now, assume that $v \in X$. Then S_2 make the same move as S_1' after p_I^v.

We now describe the memory structure of S_2. Assume that it receives an edge e when its state is $I \in \{0, 1\}$. The new state $J \in \{0, 1\}$ is computed as follows. Denote $u = \mathsf{source}(e)$ and $v = \mathsf{target}(e)$. If $u \notin X$ or $v \notin X$, then $J = 0$ (again, this case is irrelevant for the rest of the argument). Assume now that $u, v \in X$. If $\mathsf{col}(e)\big(M_I^u\big) \in \Omega_v$, then we find some $b \in \{0, 1\}$ such that $\mathsf{col}(e)\big(M_I^u\big) \preceq M_b^v$ and set $J = b$. Otherwise, we set $J = 0$.

Showing that S_2 is Winning from W. First, we observe that $\mathsf{target}(p) \in X$ for every $p \in \mathsf{FinitePlays}(S_2, U)$ (in other words, S_2 cannot leave X if we start somewhere in U). Indeed, assume for contradiction that some play with S_2 leaves X from some node $v \in X$. Let I be the state of S_2 at the moment just before leaving X. If it is a Protagonist's turn to move, then it moves as S_1' after p_I^v. Recall that $p_I^v \in \mathsf{FinitePlays}(S_1', U)$. Thus, we obtain a continuation of p_I^v which is consistent with S_1' and leads outside X. This contradicts the definition of X. Now, if it is an Antagonist's turn to move from v, then any continuation of p_I^v by one edge is consistent with S_1', so we obtain the same contradiction.

Next, we show that for any play $p \in \mathsf{FinitePlays}(S_2, U)$ we have $\mathsf{ant}(p) \preceq M_I^{\mathsf{target}(p)}$, where I is the state of S_2 after p. This statement implies that S_2 is winning w.r.t. RL from every node of U. Note that $M_I^{\mathsf{target}(p)}$ is well-defined thanks to the previous paragraph.

We prove this statement by induction on the length of p. Let us start with the induction base. Assume that $|p| = 0$ (then $p = \lambda_w$ for some $w \in U$). The state of S_2 after p is the initial state, that is, 0. Thus, we have to show that $\mathsf{ant}(p) \preceq M_0^{\mathsf{target}(p)}$. Note that p has length 0 and hence is consistent with any strategy. In particular, $p \in \mathsf{FinitePlays}(S_1', U)$. Hence, $\mathsf{ant}(p) \in \Omega_{\mathsf{target}(p)}$. If $\Omega_{\mathsf{target}(p)}$ has just a single maximum, then $\mathsf{ant}(p)$ does not exceed this maximum, as required. Now, if $M_0^{\mathsf{target}(p)} \neq M_1^{\mathsf{target}(p)}$, then the second coordinate of $M_0^{\mathsf{target}(p)}$ is 0, so we have $\mathsf{ant}(p) \preceq M_0^{\mathsf{target}(p)}$ just because $\mathsf{ant}(p) = \overline{\mathbf{0}}$.

Next, we establish the induction step. Consider any $p \in \mathsf{FinitePlays}(S_2, U)$ of positive length and assume that for all paths from $\mathsf{FinitePlays}(S_2, U)$ of smaller length the statement is already proved. We prove our statement for p. Let e be the last edge of p. Correspondingly, let q be the part of p preceding e. Denote
$$u = \mathsf{target}(q) = \mathsf{source}(e) \quad \text{and} \quad v = \mathsf{target}(p) = \mathsf{target}(e)$$

Any prefix of p is also in $\mathsf{FinitePlays}(S_2, U)$, so $q \in \mathsf{FinitePlays}(S_2, U)$. Therefore, our statement holds for q. Namely, if I is the state of S_2 after q, then $\mathsf{ant}(q) \preceq M_I^u$.

Let J be the state of S_2 after p. Our goal is to show that $\mathsf{ant}(p) \preceq M_J^v$. Note that $\mathsf{ant}(p) = \mathsf{col}(e)\big(\mathsf{ant}(q)\big)$ by definition of ant. Since $\mathsf{col}(e) \in \mathbb{M}$ is monotone and $\mathsf{ant}(q) \preceq M_I^u$, we have that $\mathsf{ant}(p) = \mathsf{col}(e)\big(\mathsf{ant}(q)\big) \preceq \mathsf{col}(e)\big(M_I^u\big)$. It remains to show that $\mathsf{col}(e)\big(M_I^u\big) \preceq M_J^v$. Note that J is the state into which S_2 transits from the state I after receiving e. By definition of the memory structure of S_2, it is sufficient to show that $\mathsf{col}(e)\big(M_I^u\big) \in \Omega_v$.

By definition of p_I^u, we have that $M_I^u = \mathsf{ant}(p_I^u)$. Hence, $\mathsf{col}(e)\big(M_I^u\big) = \mathsf{ant}(p_I^u e)$. The path $p_I^u e$ starts at some node of U and ends in $\mathsf{target}(e) = v$. Thus, to establish $\mathsf{ant}(p_I^u e) \in \Omega_v$, it remains to show consistency of $p_I^u e$ with

S_1'. We have $p_I^u \in \mathsf{FinitePlays}(S_1', U)$ by definition of p_I^u. In turn, if Protagonist is the one to move from $u = \mathsf{target}(p_I^u)$, then $e = S_1'(p_I^u)$. Indeed, e is the edge played by S_2 from u when its state is I. Hence, $e = S_1'(p_I^u)$, by the definition of the next-move function of S_2.

References

1. Bloem, R., Chatterjee, K., Jobstmann, B.: Graph Games and Reactive Synthesis. In: Handbook of Model Checking, pp. 921–962. Springer, Cham (2018). https://doi.org/10.1007/978-3-319-10575-8_27
2. Bouyer, P., Randour, M., Vandenhove, P.: Characterizing omega-regularity through finite-memory determinacy of games on infinite graphs. Theoretics **2**, 1–48 (2023)
3. Bouyer, P., Le Roux, S., Oualhadj, Y., Randour, M., Vandenhove, P.: Games where you can play optimally with arena-independent finite memory. Logical Methods Comput. Sci. **18**.11:1–11:44 (2022)
4. Büchi, J.R., Landweber, L.H.: Solving sequential conditions by finite-state strategies. Trans. Am. Math. Soc. **138**, 295–311 (1969)
5. Bouyer, P., Fijalkow, N., Randour, M., and Vandenhove, P.: How to play optimally for regular objectives? In: 50th International Colloquium on Automata, Languages, and Programming (ICALP 2023). Leibniz International Proceedings in Informatics (LIPIcs), vol. 261, pp. 118:1–118:18. Schloss Dagstuhl - Leibniz-Zentrum für Informatik,
6. Casares, A.: On the minimisation of transition-based rabin automata and the chromatic memory requirements of muller conditions. In: Manea, F., Simpson, A. (eds.) 30th Eacsl Annual Conference on Computer Science Logic (CSL 2022), Dagstuhl, Germany. Leibniz International Proceedings in Informatics (LIPIcs), vol. 216, pp. 12:1–12:17. Schloss Dagstuhl - Leibniz-Zentrum für Informatik (2022)
7. Casares, A., Colcombet, T., Lehtinen, K.: On the size of good-for-games Rabin automata and its link with the memory in Muller games. In: 349th International Colloquium on Automata, Languages, and Programming (ICALP). Leibniz International Proceedings in Informatics (LIPIcs), vol. 229, pp. 117:1–117:20. Schloss Dagstuhl - Leibniz-Zentrum für Informatik (2022)
8. Chatterjee, K., Doyen, L., Henzinger, T.A., Raskin, J.-F.: Generalized mean-payoff and energy games. In: IARCS Annual Conference on Foundations of Software Technology and Theoretical Computer Science (FSTTCS 2010). Schloss Dagstuhl-Leibniz-Zentrum fuer Informatik (2010)
9. Colcombet, T., Fijalkow, N., Horn, F.: Playing safe. In: 34th International Conference on Foundation of Software Technology and Theoretical Computer Science (FSTTCS 2014) Schloss Dagstuhl-Leibniz-Zentrum fuer Informatik (2014)
10. Dziembowski, S., Jurdzinski, M., Walukiewicz, I.: How much memory is needed to win infinite games? In: Proceedings of Twelfth Annual IEEE Symposium on Logic in Computer Science, pp. 99–110. IEEE (1997)
11. Kopczyński, E.: Omega-regular half-positional winning conditions. In: Duparc, J., Henzinger, T.A. (eds.) CSL 2007. LNCS, vol. 4646, pp. 41–53. Springer, Heidelberg (2007). https://doi.org/10.1007/978-3-540-74915-8_7
12. Kopczyński, E.: Half-positional determinacy of infinite games. PhD thesis, Warsaw University (2008)
13. Kozachinskiy, A.: State complexity of chromatic memory in infinite-duration games. arXiv preprint arXiv:2201.09297 (2022)

14. Le Roux, S.: Time-aware uniformization of winning strategies. In: Anselmo, M., Della Vedova, G., Manea, F., Pauly, A. (eds.) CiE 2020. LNCS, vol. 12098, pp. 193–204. Springer, Cham (2020). https://doi.org/10.1007/978-3-030-51466-2_17
15. Zielonka, W.: Infinite games on finitely coloured graphs with applications to automata on infinite trees. Theoret. Comput. Sci. **200**(1–2), 135–183 (1998)

Parameterized Algorithms

Sparsity in Covering Solutions

Pallavi Jain and Manveer Singh Rathore[(✉)]

Indian Institute of Technology Jodhpur, Jodhpur, India
pallavi@iitj.ac.in, manveersingh520420@gmail.com

Abstract. In the classical covering problems, the goal is to find a subset of vertices/edges that "covers" a specific structure of the graph. In this work, we initiate the study of the covering problems where given a graph G, in addition to the covering, the solution needs to be sparse, i.e., the number of edges with both the endpoints in the solution are minimized. We consider two well-studied covering problems, namely VERTEX COVER and FEEDBACK VERTEX SET. In SPARSE VERTEX COVER, given a graph G, and integers k, t, the goal is to find a minimal vertex cover S of size at most k such that the number of edges in $G[S]$ is at most t. Analogously, we can define SPARSE FEEDBACK VERTEX SET. Both the problems are NP-hard. We studied these problems in the realm of parameterized complexity. Our results are as follows:
1. SPARSE VERTEX COVER admits an $\mathcal{O}(k^2)$ vertex kernel and an algorithm that runs in $\mathcal{O}(1.3953^k \cdot n^{O(1)})$ time.
2. SPARSE FEEDBACK VERTEX SET admits an $\mathcal{O}(k^4)$ vertex kernel and an algorithm that runs in $\mathcal{O}(5^k \cdot n^{O(1)})$ time.

Keywords: Parameterized Complexity · Kernelization · Vertex Cover · Feedback Vertex Set · Sparsity

1 Introduction

In last few years, classical covering problems with an additional constraint of *independence* on the solution - no edge between any pair of vertices in the solution - has gained a lot of attention [2–5,12,13,15,16]. One of the well-studied problems in this stream is INDEPENDENT FEEDBACK VERTEX SET(IFVS) [1,6–8,10,19, 20], in which given a graph G and an integer k, we aim for a set $S \subseteq V(G)$ of size k such that $G[S]$ is edgeless as well as acyclic, respectively. Even though finding an independent set or vertex cover of size k are NP-hard problems, INDEPENDENT VERTEX COVER can be solved in polynomial time as it reduces to testing if the given graph is a bipartite graph with one part of size k. However, IFVS remains NP-hard [23]. Note that an independent vertex cover or independent feedback vertex set need not exist.

In this paper, we generalise the notion of independence and study the sparsity in the solution. In particular, we study the following questions.

© The Author(s), under exclusive license to Springer Nature Switzerland AG 2024
J. A. Soto and A. Wiese (Eds.): LATIN 2024, LNCS 14579, pp. 131–146, 2024.
https://doi.org/10.1007/978-3-031-55601-2_9

SPARSE VERTEX COVER (SVC)   **Parameter:** k

Input: A graph G, and two non-negative integers k and t.

Question: Does there exist a minimal subset $S \subseteq V(G)$ of size at most k such that for every edge $uv \in E(G)$, $S \cap \{u, v\} \neq \emptyset$, and the number of edges in $G\langle S \rangle$ is at most t?

Given a graph $G = (V, E)$ and a set $S \subseteq V(G)$, $G\langle S \rangle$ is a subgraph of G with vertex set S and edge set $\{uv \in E(G) : u \neq v \text{ and } u, v \in S\}$. It is basically an induced subgraph on S where we ignore the self-loops and the multiplicity of an edge.

Similarly, we can define SPARSE FEEDBACK VERTEX SET.

SPARSE FEEDBACK VERTEX SET (SFVS)   **Parameter:** k

Input: A graph G, and two non-negative integers k and t

Question: Does there exist a minimal subset $S \subseteq V(G)$ of size at most k such that $G - S$ is acyclic, and the number of edges in $G\langle S \rangle$ is at most t?

Note that for $t = 0$, the sparsity constraint on the solution of a problem is same as the independence, while for $t = \binom{k}{2}$, it is the same as the problem without any further constraint. Thus, both the problems are NP-hard.

Our Results: Note that, due to the NP-hardness of IFVS, SFVS is NP-hard even for $t = 0$. Thus, in this paper we focus on the parameterized complexity with respect to k. We note there is a trivial XP algorithm for SVC with respect to t: we guess the set of edges in the solution, after that, we are reduced to solving INDEPENDENT VERTEX COVER. An FPT algorithm for SVC with respect to t elude us so far, however, we design a kernel and an FPT algorithm with respect to k. Next, we discuss our results and methodologies, in detail.

1. **A Quadratic Kernel for SVC.** We present an $\mathcal{O}(k^2)$ vertex kernel. Our kernel is similar to the VERTEX COVER kernel using Buss rule, however, we cannot delete high degree vertices from the graph, as we also need to take care of the edges incident to it in the solution. Thus, we add some dummy vertices, to ensue that high degree vertices are always in the solutions. A linear kernel elude us so far. The crown decomposition does not help any longer to design the linear kernel as we might need to take some vertices from "crown" as well, in the interest of sparsity.

2. **An FPT algorithm for SVC.** We design an algorithm using a branching technique. The simple $\mathcal{O}^{\star}(2^k)$[1] algorithm for VERTEX COVER works for our case as well. In fact, the algorithm that branches on the vertex of degree at least three also works as we can solve the graph of degree at most two in polynomial time (Theorem 1). Thus, using the ideas of known algorithms for VERTEX COVER, we obtain an algorithm that runs in $\mathcal{O}^{\star}(1.4656^k)$ time. However, the known algorithms with better running time does not extend to our case due to same difficulty as discussed for the kernel. We design

---

[1] The notation $\mathcal{O}^{\star}(\cdot)$ suppresses the polynomial factor in (n, m).

new branching rules and develop an algorithm that runs in $\mathcal{O}^*(1.3953^k)$ time (Sect. 3).

3. **An $\mathcal{O}(k^4)$ Kernel for SFVS.** We present an $\mathcal{O}(k^4)$ vertex kernel in Sect. 4.1. As we cannot delete high degree vertex in SVC, here we cannot delete a vertex v that is the only common vertex in $k + 1$ cycles, which is one of the crucial rules in the kernelization algorithm of FVS. Furthermore, we cannot short-circuit degree two vertices as these might belong to a solution to control the sparsity in the solution. Furthermore, due to the same reason, we cannot apply the Expansion Lemma here, which is crucially used to design the quadratic kernel of FVS. But, our idea is inspired from the idea of FVS kernel, i.e., try to bound the maximum degree and minimum degree of the graph. The same approach has been used for the other variants of FVS as well [2,17]. The intuitive description of our algorithm is as follows. Let M be the set of vertices in G that is the core vertex of $(k + 1)$-flower. Then, we delete degree one vertices in $G - M$. Furthermore, we apply a reduction rule to ensure that two degree two vertices are not adjacent. Then, using a marking scheme similar to the one in [17] and a reduction rule, we bounded the maximum degree of a vertex in $G - M$ by $\mathcal{O}(k^3)$, and eventually bound the number of vertices in $G - M$ by $\mathcal{O}(k^4)$. Since the size of M can be at most k in a yes-instance, the number of vertices in G is bounded by $\mathcal{O}(k^4)$.

4. **An FPT algorithm for SFVS.** For this algorithm also, we use the branching technique. The details are in Sect. 4.2.

Related Work (not Discussed Above). A problem related to the sparsity in the solution of feedback vertex set is 2-induced forests, where the solution of the FEEDBACK VERTEX SET problem also needs to be a forest, which has been extensively studied in combinatorics [14,22,24]. However, to the best of our knowledge, the problem has not been studied from the viewpoint of parameterized complexity. In fact, the problem of partitioning the graph into an independent set and a forest (i.e., IFVS) can be viewed as a sparsity concept in the vertex cover solution.

2 Preliminaries

Notations and Terminologies: We use the standard graph notations and terminologies in [11]. Throughout the paper, we follow the following notions. Let G be a graph, $V(G)$ and $E(G)$ denote the vertex set and the edge set of graph G, respectively. Let n and m denote the number of vertices and the number of edges of G, respectively. For an edge $uv \in E(G)$, we call u and v endpoints of the edge uv. Let G be a graph and $X \subseteq V(G)$, then $G[X]$ is the graph induced on X and $G - X$ is graph G induced on $V(G) \setminus X$. The notation $N_G(v)$ denote the neighbourhood of a vertex v in G. For set $S \subseteq V(G)$, $N_S(v)$ refers to $N(v) \cap S$. We skip the subscript from the notation when it is clear from the context. For two graphs G_1 and G_2, $G_1 \uplus G_2$ denote the disjoint union of these two graphs. That is, $V(G_1 \uplus G_2) = V(G_1) \uplus V(G_2)$ and $E(G_1 \uplus G_2) = E(G_1) \uplus E(G_2)$.

A cycle in the graph G is denoted by $(v_1, \ldots, v_\ell)$, where for each $i \in [\ell - 1]$, $v_i v_{i+1}$ is an edge in G and $v_1 v_\ell$ is an edge in G. A cycle of length three in a graph is called a *triangle*. In an undirected graph G, a component C is a connected subgraph of G which is not a subgraph of any larger connected subgraph. For an integer n, the notation $[n]$ denote the set $\{1, \ldots, n\}$

3 An $\mathcal{O}(1.3953^k \cdot n^{\mathcal{O}(1)})$ Algorithm for SVC

In this section, we design an algorithm that runs in $\mathcal{O}^\star(1.3953^k)$ time. Let (G, k, t) be an instance of SVC. As argued for the kernel, we cannot delete the vertices that are part of the solution. We also note that G might be a disconnected graph. Indeed, we can find a minimum vertex cover in each component independently, but here, we need to keep track of k and t. Thus, we solve each component independently, and then try to merge the solutions. Hence, we define the following variant of the problem. In the following definition, G need not be a connected graph. We will use connectivity, when required. Let $\mathcal{Z} = V(G)$. Note that $\mathcal{Z}$ will remain same throughout the algorithm.

CONSTRAINED SPARSE VERTEX COVER (CSVC)   **Parameter:** k
Input: a graph G, a set $X \subseteq V(G)$, and a non-negative integer k.
Question: find a set $\mathcal{S} = \{S_0, S_1, \ldots, S_k\}$ such that for each $i \in \{0, \ldots, k\}$, if $G - X$ does not have a vertex cover of size exactly i, then $S_i = \mathcal{Z}$, otherwise S_i is a vertex cover of $G - X$ of size i that maximises sparsity, i.e, $|E(G\langle S_i \cup X\rangle)| = \min\{|E(G\langle S \cup X\rangle)| : S$ is an i-sized vertex cover of $G - X\}$.

For an "invalid" instance of CSVC, we return $\mathcal{Z}$. During the execution of the algorithm, an instance may become invalid when k drops below 0.

We will solve SVC by designing an algorithm for CSVC. Let (G, k, t) be an instance of SVC. Let $(G, \emptyset, k)$ be an instance of CSVC. Clearly, (G, k, t) is a YES-instance if and only if solution $\mathcal{S}$ to $(G, \emptyset, k)$ has a set $S_i \neq \mathcal{Z}$ such that $|E(G\langle S_i\rangle)| \leq t$, which is captured from minimality. Thus, we next focus on designing an algorithm for CSVC.

We begin with applying the following simple reduction rule exhaustively.

Reduction Rule 1 (♣[2]) *Let (G, X, k) be an instance of* CSVC.

1. *If $k < 0$, then return $S := \{\mathcal{Z}\}$.*
2. *If $k \geq 0$, and there is no edge in $G - X$, return $S = \{S_0, \ldots, S_k\}$, where*
$$S_i = \arg \min_{\substack{U \subseteq V(G) \\ |U|=i}} \left(\sum_{v \in U} |N_X(v)| \right).$$

We begin with defining some family of graphs according to their unique properties such as degree or vertex count. We will further show that these family of graphs are solvable in polynomial time in Theorem 1.

---

[2] The correctness of Reduction Rules, Lemma, Theorem marked by ♣ are in appendix.

1. $\mathcal{G}_{1HD}$: family of graphs that contains at most one vertex of degree at least three in every connected component.
2. $\mathcal{G}_c$: family of graphs whose each connected component is of size at most c.
3. $\mathcal{G}_{c+1HD}$: family of graphs whose each connected component either belongs to $\mathcal{G}_{1HD}$ or $\mathcal{G}_c$.

For the above mentioned families of graphs with $c = \log(n)$, where $n = |V(G)|$, we have a polynomial time algorithm as stated in the following theorem . The algorithm uses dynamic programming.

Theorem 1. *An instance (G, X, k) of CSVC can be solved in polynomial time, if $G \in \mathcal{G}_{c+1HD}$, where c is a constant or $\log n$.*

We use the branching technique to solve our problem. Towards this, we first define some terminologies.

Branch into $\{A_1, A_2, .., A_z\}$: This means that we create z new instances of CSVC: $(G, X \cup A_1, k - |A_1|)$, $(G, X \cup A_2, k - |A_2|)$, $\ldots$, $(G, X \cup A_z, k - |A_z|)$, and solve them recursively.

We always apply a branching rule on a component in $G - X$ that does not belong to the family $\mathcal{G}_{\log(n)+1HD}$.

After each branching step, we reuse the notation (G, X, k), for convenience. Algorithm 1 to 7 present our branching rules. To show the correctness of our algorithm, we show the safeness of each branching step as follows. Let (G, X, k) be an instance of CSVC and we branch into $\{A_1, A_2, .., A_z\}$. We will show that for $i \in [k]$, if $A_j \subseteq S_i$, where $j \in [z]$, then $S_i \setminus A_j$ is a vertex cover of $G - (X \cup A_j)$ of size $i - |A_j|$ that maximises sparsity. Furthermore, let $\mathcal{S}_j = \{S_{j,0}, \ldots, S_{j,(k-|A_j|)}\}$ be a solution to $(G, X \cup A_j, k - |A_j|)$, where $j \in [z]$. For $i \in [k]$, let $\mathcal{X}_i = \{S_{j,(i-|A_j|)} : j \in [z]$ and $|\mathcal{S}_j| \geq i - |A_j|\}$. That is, $\mathcal{X}_i$ contains all those vertex covers in the solutions of subproblems that can be extended to i-sized vertex covers of $G - X$. If $\mathcal{X}_i \neq \emptyset$, then let $S_j = \arg\min_{S_j \in \mathcal{X}_i} |E(G\langle S_j \cup X \cup A_j\rangle)|$. We will show that $S' = S_j \cup A_j$ is a vertex cover of $G - X$ of size i that maximises sparsity. In our branching steps, we also try to branch on the vertices that lead to the graph class that is polynomial-time solvable. This helps in further decreasing the branching factor, and hence improves the running time.

We prove the following generic lemma that will be used to prove the safeness of our branching rules.

Lemma 1. *Let (G, X, k) be an instance of CSVC. Let v be a vertex in $G - X$ of degree at least one. Then, Branch into $\{v, N(v)\}$ is safe.*

We begin with the branching rule in Algorithm 1.

Lemma 2. *Algorithm 1 is correct and has a branching vector $(1, 4)$.*

Algorithm 1. Branching Rule 1: C is a component in $G - X$ that does not belong to $\mathcal{G}_{\log n + 1HD}$ and has a vertex of degree at least four.

1: **if** there exists a vertex v in C such that $|N_{G-X}(v)| \geq 4$ **then**
2:      Branch into $\{\{v\}, N_{G-X}(v)\}$.

After applying Algorithm 1 exhaustively, the maximum degree of a vertex in $G - X$ is at most three. Let (G, X, k) be an instance of CSFVS such that degree of the graph $G - X$ is at most three. Let $G_1, \ldots, G_\ell$ be the components in $G - X$. We will solve ℓ instances of CSVC independently: for each $i \in [\ell]$, we solve CSVC for the instance (G_i, X, k). We obtain the solution for (G, X, k) due to the following lemma.

Lemma 3. *Let (G, X, k) be an instance of CSVC. Let $G_1, \ldots, G_\ell$ be the components in $G - X$. Let $\mathcal{S}_i = \{S_{i,0}, \ldots, S_{i,k}\}$ be a solution to (G_i, X, k). There exists a polynomial time algorithm that constructs a solution of (G, X, k).*

Let (G, X, k) be an instance CSVC. Due to Lemma 3, we safely assume that $G - X$ is a connected graph. Furthermore, since Branching Rule 1 is not applicable, each component in $G - X$ that does not belong to the family $\mathcal{G}_{c+1HD}$ is of degree at most three. Note that we care about connectivity only at this point, and do not care later if the graph becomes disconnected. If the graph $G - X$ does not belong to the family $\mathcal{G}_{c+1HD}$ and 3-regular, then we apply the branching rule in Algorithm 2.

Algorithm 2. Branching Rule 2: $G - X$ is a 3-regular connected graph and does not belong to the family $\mathcal{G}_{c+1HD}$

1: let $v \in G - X$
2: Branch into $\{\{v\}, N_{G-X}(v)\}$.

The branching step in Algorithm 2 is safe due to Lemma 1. Note that this branching step is applicable only once as once we branch on a vertex, the graph is no longer a 3-regular graph as we do not add edges at any step of the algorithm. Thus, this step contributes only two additional nodes in the branching tree.

From now onwards, we assume that (G, X, k) is an instance CSVC such that the degree of $G - X$ is at most three and it is not a 3-regular graph. Next, we consider the case when there exists a component in $G - X$ that does not belong to the family $\mathcal{G}_{c+1HD}$ and has a degree one vertex. For this case, we apply the branching rules in Algorithm 3. Let v be a vertex in $G - X$. We say that u is a *nearest degree three vertex to v* if the length of a shortest path between v and u is smallest among the shortest paths between v and any other degree three vertex.

Lemma 4. *Algorithm 3 is correct and has a branching vector $(3, 4, 3)$.*

Algorithm 3. Branching Rule 3: C is a component in $G - X$ that does not belong to $\mathcal{G}_{\log(n)+1HD}$ and has a vertex of degree one.

1: **if** there exists a vertex $v \in C$ such that $N_C(v) = \{u\}$. **then**
2:     **if** $|N_C(u)| = 2$ **then**
3:         let x be the nearest degree 3 vertex to u in C $\triangleright$ x exists as $C \notin \mathcal{G}_{1HD}$
4:         Branch into $\{\{x\}, N_C(x)\}$
5:     **else**
6:         let $N_C(u) = \{v, x, y\}$
7:         **if** $|N_C(x)| = 3$ (or $|N_C(y)| = 3$) **then**
8:             Branch into $\{\{x\}, N_C(x)\}$ (or $\{\{y\}, N_C(y)\}$).
9:         **else if** $|N_C(x)| = 1$ (or $|N_C(y)| = 1$) **then**
10:             let $z \neq u$ is the nearest degree 3 vertex to y (or x)
11:             Branch into $\{\{z\}, N_C(z)\}$
12:         **else**
13:             let $a \neq u$ is the nearest degree 3 vertex to x
14:             **if** there exists $b \notin \{u, a\}$ which is the nearest degree 3 vertex to y **then**
15:                 Branch into $\{N_C(a), \{a \cup N_C(b)\}, \{a, b\}\}$.
16:             **else**
17:                 Branch into $\{\{a\}, N_C(a)\}$

After the exhaustive application of Algorithm 3, if a component C in $G - X$ does not belong to the family $\mathcal{G}_{c+1HD}$, then the vertices in C have degree either two or three in $G - X$. We proceed to our next branching rule in Algorithm 4 that is applicable when the graph has a triangle.

Algorithm 4. Branching Rule 4: C is a component in $G - X$ that does not belong to $\mathcal{G}_{c+1HD}$ and has a triangle

---

1: **if** there exists a triangle (u, v, w) in C **then**.
2:     **if** $|N_C(u)| = |N_C(v)| = 2$ and $|N_C(w)| = 3$ **then**
3:         Branch on $\{\{w\}, N_C(w)\}$.
4:     **else if** $|N_C(u)| = 2$ and $|N_C(v)| = |N_C(w)| = 3$ **then**
5:         Branch on $\{\{w\}, N_C(w)\}$.
6:     **else**
7:         let $|N_X(w)| \leq |N_X(v)|$ and $|N_X(w)| \leq |N_x(u)|$
8:         let $p \in N(u), q \in N(v)$, and $s \in N(w)$
9:         **if** $N(p) \subseteq N(w)$ and $N(q) \subseteq N(w)$ **then**
10:             Branch on $\{\{s\}, N(s)\}$
11:         **else**
12:             let $N(p) \nsubseteq N(w)$
13:             Branch on $\{\{w\}, N_C(w) \cup N_C(p)\}$.

Lemma 5. *Algorithm 4 is correct and has a branching vector* $(3, 4, 3)$.

After the exhaustive application of Algorithm 4, if a component C in $G - X$ does not belong to the family $\mathcal{G}_{c+1HD}$, then we know that there is no triangle in

C. Our next branching rule in Algorithm 5 branch on degree two vertices with some specific structure.

Algorithm 5. Branching Rule 5: C is a component in $G - X$ that does not belong to $\mathcal{G}_{c+1HD}$ and has a vertex of degree two

---
1: **if** there exists a vertex v of degree two in C **then**
2:     let $N_C(v) = \{u, w\}$
3:     **if** $|N_C(u)| = 2$ and $|N_C(w)| = 3$ **then**
4:         Branch on $\{\{w\}, N_C(w)\}$.
5:     **else if** $|N_C(u)| = 3$ and $|N_C(w)| = 3$ and $N_X(u) \leq N_X(v) - 1$ **then**
6:         Branch into $\{N_C(w), \{u, w\}\}$

Lemma 6. *Algorithm 5 is correct and has a branching vector* $(3, 4, 3)$.

After the exhaustive application of Algorithm 5, two degree two vertices are not adjacent. However, a degree two vertex might still be adjacent to two degree three vertices if it does not satisfy the condition in line 5 of Algorithm 5. Algorithm 6 and 7 consider degree three vertices for the branching.

Algorithm 6. Branching Rule 6: C is a component in $G - X$ that does not belong to $\mathcal{G}_{c+1HD}$ and has a degree three vertex which is adjacent to at least one degree three vertex and at most two degree three vertices

---
1: **if** there exists a vertex $v \in C$ such that $N_C(v) = \{x, y, z\}$ **then**
2:     **if** $N_C(x) = \{v, p\}, N_C(y) = \{v, p\}$ and $N_C(z) = \{v, s, t\}$. **then**
3:         Branch into $\{\{z, p\}, \{z\} \cup N(p), N(z)\}$
4:     **else if** $N_C(x) = \{v, p\}, N_C(y) = \{v, q\}$ and $N_C(z) = \{v, s, t\}$. **then**
5:         Branch into $\{N_C(v), N_C(z), \{v, z, p, q\}\}$
6:     **else if** $N_C(x) = \{v, p\}, N_C(y) = \{v, q, r\}$ and $N_C(z) = \{v, s, t\}$ **then**
7:         Branch into $\{N_C(y), \{y\} \cup N_C(z), \{x, y, z\}\}$

Lemma 7. *Algorithm 6 is correct and has a branching vector* $(3, 4, 3)$.

After exhaustive application of branching rules in Algorithm 6, if C is a component in $G - X$ that does not belong to $\mathcal{G}_{c+1HD}$, then every degree three vertex in C is adjacent to all degree two vertices or all degree three vertices. Recall that C is not three regular.

Lemma 8. *Algorithm 7 is correct and has a branching vector* $(3, 4, 3)$.

Lemma 9. *Let* (G, X, k) *be an instance of* CSVC. *Let* C *be a component in* $G - X$ *such that* C *is a graph of degree at most three and it is not a 3-regular graph. Then, after the exhaustive application of branching rules in Algorithm 3 to 7, the resulting component* C *belongs to* $\mathcal{G}_{c+1HD}$.

Algorithm 7. Branching Rule 7: C is a component in $G - X$ that does not belong to $\mathcal{G}_{c+1HD}$ and has a degree three vertex whose all the neighbors are degree two vertices

---

1: **if** there exists a vertex $v \in C$ such that $N_C(v) = \{a, b, s\}$ **then**
2:     **if** there exists $N_C(u) = \{a, b, d\}$ & $N_C(s) = \{v, x\}$ **then**
3:         Branch into $\{N_C(x), \{x\} \cup N_C(u), \{x, u\}\}$
4:     **else if** there exists $N_C(u) = \{a, d, e\}$ & $N_C(w) = \{b, d, f\}$ **then**
5:         **if** there exists $x \in C$ such that $N_C(x) = \{s, e, g\}$ **then**
6:             Branch into $\{N_C(x), N_C(w) \cup \{x\}, \{w, x\}\}$
7:         **else if** $N_C(s) = \{x, v\}$, $N_C(e) = \{y, u\}$ and $N_C(f) = \{z, w\}$ **then**
8:             Branch into $\{N_C(v), \{v\} \cup N_C(x), \{v, x\} \cup N_C(z), \{v, x, z\}\}$
9:     **else if** $N_C(a) = \{x, v\}$, $N_C(b) = \{y, v\}$ and $N_C(s) = \{z, v\}$ **then**
10:        Branch on $\{\{N_C(x)\}, \{x, y, z\}, \{\{x, z\} \cup N_C(y)\}, \{\{x, y\} \cup N_C(z)\}, \{\{x\} \cup N_C(y) \cup N_C(z)\}\}$

Table 1. Description of various cases in Algorithm 6. a) refers to branching in line 3. b) refers to branching in line 5. c) refers to branching in line 7.

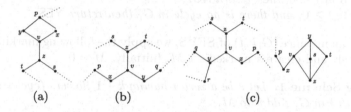

(a)   (b)   (c)

Table 2. Description of various cases in Algorithm 7. a refers to polynomial time solvable graph where each degree two vertex is connected to degree three vertex and vice-versa. b refers to branching in line 3.c refers to branching in line 6.d refers to branching in line 8.e refers to branching in line 10.

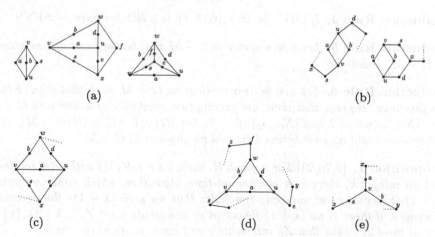

(a)   (b)

(c)   (d)   (e)

Thus, when none of the branching rules are applicable, we can solve the problem in polynomial time using Theorem 1.

Due to Algorithm 1 to 6 we have worst case branching vector of $(3, 4, 3)$. Thus, the running time of the algorithm is $\mathcal{O}^{\star}(1.3953^k)$.

4   SPARSE FEEDBACK VERTEX SET (SFVS)

In this section, we design an $\mathcal{O}(k^4)$ vertex kernel and an FPT algorithm that runs in $\mathcal{O}^{\star}(5^k)$ time.

4.1   An $\mathcal{O}(k^4)$ Vertex Kernel

We begin with the following simple reduction rule.

Reduction Rule 2. *Let (G, k, t) be an instance of SFVS.*

1. *If $k < 0$ or $t < 0$, then return NO.*
2. *If $k \geq 0$, $t \geq 0$, and there is no cycle in G, then return YES.*

Given an instance (G, k, t) of SFVS, we apply the following marking schemes exhaustively and create a marked set M. Initially, $M = \emptyset$.

Marking Scheme 1. *Let v be a vertex having $k + 1$ flowers (cycles only intersecting at v) in G. Add v to M.*

After applying the Marking Scheme 1, exhaustively, there is no vertex in $G - M$ that has $k + 1$ flower.

Reduction Rule 3. *For every vertex $v \in M$, add a self-loop to v.*

Reduction Rule 4. *If $|M| > k$, then (G, k, t) is a NO-instance of SFVS.*

Reduction Rule 5. *Let v be a vertex in $G - M$ that has at most one neighbour in $G - M$. Delete v.*

Reduction Rule 6. *Let u, v be two vertices in $G - M$ such that $uv \in E(G)$. Furthermore, suppose that there are exactly two neighbors of u and v in $G - M$ i.e., $|N_{G-M}(u)| = 2$ and $|N_{G-M}(v)| = 2$. Let $|N(u) \cap M| \geq |N(v) \cap M|$, then delete u and add an edge between its two neighbours in $G - M$.*

Proposition 1. [9,20,21] *For a graph H, a vertex $v \in V(H)$ without a self-loop, and an integer k, there is a polynomial-time algorithm, which either outputs a $(k + 1)$-flower at v, or correctly concludes that no such $(k + 1)$- flower exists. Moreover, if there is no $(k + 1)$-flower at v, it outputs a set $Z_v \subseteq V(H) \backslash \{v\}$ of size at most $2k$, such that Z_v intersects every cycle containing v in H.*

For $G' = G - M$, an integer k, and a vertex v without a self-loop, we invoke Proposition 1. Due to Marking Scheme 1, $G - M$ does not contain a $(k+1)$-flower at v. Hence, we obtained a vertex set Z_v. Let G'_v contain all the components of $G' \setminus (Z_v \cup \{v\})$ that have at least one neighbor of v. That is, $G'_v = C_1, \ldots, C_\ell$, where C_i is a component in $G' \setminus \{Z_v \cup \{v\}\}$ and has a neighbor of v. We can also say that all C_i will have exactly one neighbour of v, because it has at least one neighbor by construction, and if it has more than one neighbor then $G[v \cup C_i]$ has a cycle containing v that is not hit by any vertex of Z_v, a contradiction.

Reduction Rule 7. *If more than k components in G'_v contains a cycle, then return NO.*

Given an instance (G, k, t) of SFVS, after applying Reduction Rules 2 to 7, we apply the following marking schemes exhaustively and create a marked set M'. Initially, $M' = \emptyset$.

Marking Scheme 2. *Let v be a vertex in G without a self-loop. We add the following vertices to M'.*

- *Add the vertex v and all the vertices in Z_v to M'.*
- *Let C be a component in G'_v that contains a cycle. Add all the vertices of C to M'.*
- *Let n'_x be the number of components in G'_v that has at least two neighbor of x. For each $x \in Z_v$, add $\min\{k+2, n'_x\}$ components of G'_v that has at least two neighbor of x to M'.*
- *Let n_x be the number of components in G'_v that has a neighbor of x. For each $x \in Z_v$, add $\min\{k+2, n_x\}$ components of G'_v that has a neighbor of x to M'.*
- *For a pair of vertices x, y, let n_{xy} denote the number of components in G'_v that has a neighbour of x as well as a neighbour of y. For each pair of vertices $\{x, y\} \subseteq Z_v$, add $\min\{k+2, n_{xy}\}$ components of G'_v that has a neighbor of x as well as a neighbor of y to M'.*

Reduction Rule 8. *Let C be an unmarked component in G'_v after applying marking scheme 2. Let $uv \in E(G)$, where $u \in C$. Delete the edge uv.*

Let G be the graph obtained after exhaustive application of Reduction Rules 2 to 8. Then, we have the following bound on the vertices in $G - M$.

Lemma 10. *Let v be a vertex in $G - M$. Then the degree of v is bounded by $\mathcal{O}(k^3)$.*

We will use the following known result to obtain the bound on the number of vertices and edges in G.

Proposition 2. *[9] Let H be a graph with minimum degree at least 3, maximum degree at most δ, and a feedback vertex set of size at most k. Then, H has at most $(\delta + 1)k$ vertices and $2\delta k$ edges.*

Lemma 11. *Let G be the graph obtained after exhaustive application of Reduction Rules 2 to 8. The number of vertices and edges in G are bounded by $\mathcal{O}(k^4)$ and $\mathcal{O}(k^5)$, respectively.*

Due to Lemma 11, we have the following.

Theorem 2. SFVS *admits a kernel with $\mathcal{O}(k^4)$ vertices and $\mathcal{O}(k^5)$ edges.*

4.2 An $\mathcal{O}^\star(5^k)$ Algorithm

Let (G, k, t) be a given instance of SFVS. We begin with applying Reduction Rule 2. Let us assume that this reduction rule is no longer applicable. We first check whether G has a feedback vertex set of size at most k using the algorithm in [18]. Then, we apply the following reduction rule.

Reduction Rule 9. *If there does not exist a feedback vertex set of G of size at most k, then return* NO.

Towards designing an algorithm for SFVS, we define another problem, CONSTRAINED SPARSE FEEDBACK VERTEX SET (CSFVS), as follows.

CONSTRAINED SPARSE FEEDBACK VERTEX SET (CSFVS)   **Parameter: k**
Input: A graph G, two integers k, t, two subsets $X \subseteq V(G), Y \subseteq V(G)$ such that the number of components in $G[Y]$ is at most k and $G - (X \cup Y)$ is acyclic.
Question: Does there exist a set $S \subseteq V(G - (X \cup Y))$ of size at most $k - |X|$ such that $S \cup X$ is a minimal feedback vertex set of G and the number of edges in $G\langle S \cup X \rangle$ is at most t?

Note that if a component in $G[Y]$ is not a tree, then (G, X, Y, k, t) is a NO-instance of CSFVS. Thus, from now onwards, we assume that each component in $G[Y]$ is a tree. Clearly, if an instance (G, X, Y, k, t) is a YES-instance of CSFVS, then (G, k, t) is also a YES-instance of SFVS. However, the mapping of NO-instances is not clear. We will design an algorithm for CSFVS and use it as a blackbox for designing an algorithm for SFVS. Suppose that $\mathcal{A}$ is an algorithm for CSFVS that runs in $\mathcal{O}^\star(\alpha^k)$ time. Then, we design an algorithm for SFVS as follows. Let Z be a feedback vertex set of G of size at most k. Since Reduction Rule 9 is not applicable, Z exists and we can obtain it using the algorithm in [18]. Now, for every set $X \subseteq Z$ (X can also be an empty set), we call the algorithm $\mathcal{A}$ for the instance $\mathcal{I}_X = (G, X, Y = Z \setminus X, k, t)$, and return yes if the algorithm $\mathcal{A}$ return yes for any $X \subseteq Z$. Towards the correctness, we prove the following lemma.

Lemma 12. (G, k, t) *is a yes-instance of* SFVS *if and only if* $\mathcal{I}_X$ *is a* YES-*instance of* CSFVS, *for a set* $X \subseteq Z$.

Since we try all possible subsets of Z, the algorithm for SFVS runs in $\mathcal{O}^\star(\sum_{i=0}^{k} \binom{k}{i} \alpha^{k-i}) = \mathcal{O}^\star((\alpha + 1)^k)$.

Next, we focus on designing an algorithm for CSFVS.

An $\mathcal{O}^*(4^k)$ Algorithm for CSFVS. Our algorithm uses *branching* technique. Let $\mathcal{I} = (G, X, Y, k, t)$ be an instance of CSFVS. Let $Q_{\mathcal{I}} \subseteq V(G - (X \cup Y))$ be the set of vertices that does not belong to any cycle in $G - X$. Let $H_{\mathcal{I}} = G - (X \cup Y \cup Q_{\mathcal{I}})$.

We use the following measure to bound the depth of the branching: $\mu = b + c - g$, where the following holds.

1. b is the budget - the number of additional vertices that can be added to the solution being constructed. Initially, $b = k - |X| \leq k$.
2. c is the number of components(trees) in $G[Y]$. Initially, $1 \leq c \leq k$.
3. g is the number of *good* vertices in $H_{\mathcal{I}}$. We say that a vertex $v \in V(H_{\mathcal{I}})$ is good if it does not have any neighbor in $H_{\mathcal{I}}$ and has exactly two neighbors in Y. Mathematically, $N_{G-(X \cup Q_{\mathcal{I}})}(v) \subseteq Y$ and $|N_{G-(X \cup Q_{\mathcal{I}})}(v)| = 2$.

Observation 3. *If $g > b + c$ then $\mathcal{I} = (G, X, Y, k, t)$ is a* NO-*instance of* CSFVS

We begin with applying the following reduction rule exhaustively.

Reduction Rule 10. *Let $\mathcal{I} = (G, X, Y, k, t)$ be an instance of* CSFVS.

1. *If $b < 0$ or $\mu < 0$ return* NO.
2. *If $H_{\mathcal{I}} = \emptyset$ and $|E(G\langle X \rangle)| \leq t$ and $|X| \leq k$ and X is minimal feedback vertex set of G, then return* YES.
3. *If $H_{\mathcal{I}} = \emptyset$ and the conditions in the previous case does not hold, then return* NO.

Reduction Rule 11. *Suppose that uv is an edge in $H_{\mathcal{I}}$. Let $|N_X(u)| \geq |N_X(v)|$ and $|N_{Y \cup V(H_{\mathcal{I}})}(v)| = |N_{Y \cup V(H_{\mathcal{I}})}(u)| = 2$, i.e., the neighborhhod of u is larger than v in X and u and v both have exactly one more neighbor in $Y \cup V(H_{\mathcal{I}})$ (need not be a common neighbor). Furthermore, either u or v has at least one neighbor in Y. We delete u from $H_{\mathcal{I}}$ and add it to Y.*

Note that the above reduction rule is correct even without the constraint that "either u or v has at least one neighbor in Y". However, it is required to argue the running time, which will be cleared later. Without this constraint, we increase the number of components in Y, which is not desirable.

Reduction Rule 12. *Let C be a component in Y. If there exists a vertex $v \in H_{\mathcal{I}}$ such that v has at least two neighbours in C, then we add v to X and delete from $H_{\mathcal{I}}$.*

We apply all the above reduction rules exhaustively. When the reduction rules are not applicable, we move to our branching steps. We begin with defining the following terminology.

Branch on v : This means that we solve $(G, X \cup \{v\}, Y, k, t)$ and $(G, X, Y \cup \{v\}, k, t)$ recursively.

We will branch only on the vertices of $H_{\mathcal{I}}$. The correctness of branching step is due to the following lemma.

Lemma 13. (G, X, Y, k, t) *is a* YES-*instance of* CSFVS *if and only if either* $(G, X \cup \{v\}, Y, k, t)$ *is a* YES-*instance of* CSFVS *or* $(G, X, Y \cup \{v\}, k, t)$ *is a* YES-*instance of* CSFVS.

Our branching steps are in Algorithm 8. Recall that $G - (X \cup Y)$ is acyclic. Thus, due to the construction, $H_{\mathcal{I}}$ is also acyclic. We root every tree in $H_{\mathcal{I}}$.

Algorithm 8. Branching Procedure

1: **if** there exists a vertex v in $H_{\mathcal{I}}$ such that $|N_Y(v)| \geq 2$ and $|N_{Y \cup V(H_{\mathcal{I}})}(v)| \geq 3$. **then**
2:     Branch on v
3: **else if** there exists a leaf vertex (vertex of degree one) in $H_{\mathcal{I}}$ **then**
4:     let v be a leaf vertex in $H_{\mathcal{I}}$ that is at the farthest distance from its root and p be its parent. Branch on p.

The correctness of Algorithm 8 follows from Lemma 13. We prove the following lemma which will be used to argue the running time of the algorithm.

Lemma 14. *The branching vector of Algorithm 8 is* $(1, 1)$.

Next, we analyse the structure of the graph after exhaustive application of above reduction rules and the branching rules in Algorithm 8. The non-applicability of the reduction rules is used to argue the branching vector.

Lemma 15. *If none of the branching rule in Algorithm 8 is applicable, then, every vertex in* $H_{\mathcal{I}}$ *is a good vertex.*

After the exhaustive application of all the reduction rules and branching rules stated above, we know that every vertex in $H_{\mathcal{I}}$ is a good vertex vertex. Furthermore, it has neighbors in two distinct components of Y. Next, we apply the following reduction rule.

Reduction Rule 13. *Let* v *be the vertex in* $H_{\mathcal{I}}$ *such that* $|N_X(v)| = \max_{u \in H_{\mathcal{I}}} |N_X(u)|$, *i.e.,* v *has largest neighborhood in* X *among all the vertices in* $H_{\mathcal{I}}$. *We add* v *to* Y *and delete it from* $H_{\mathcal{I}}$.

Theorem 4. *There exists an algorithm that solves* CSFVS *in* $\mathcal{O}^*(4^k)$ *time.*

Theorem 5. *There exists an algorithm that solves* SFVS *in* $\mathcal{O}^*(5^k)$ *time.*

References

1. Agrawal, A., Gupta, S., Saurabh, S., Sharma, R.: Improved algorithms and combinatorial bounds for independent feedback vertex set. In: Guo, J., Hermelin, D. (eds.) IPEC, pp. 2:1–2:14 (2016)
2. Agrawal, A., Jain, P., Kanesh, L., Lokshtanov, D., Saurabh, S.: Conflict free feedback vertex set: a parameterized dichotomy. In: Potapov, I., Spirakis, P.G., Worrell, J. (eds.) MFCS, pp. 53:1–53:15 (2018)
3. Agrawal, A., Jain, P., Kanesh, L., Saurabh, S.: Parameterized complexity of conflict-free matchings and paths. Algorithmica $82(7)$, 1939–1965 (2020)
4. Arkin, E.M., et al.: Conflict-free covering. In: CCCG (2015)
5. Banik, A., Panolan, F., Raman, V., Sahlot, V., Saurabh, S.: Parameterized complexity of geometric covering problems having conflicts. Algorithmica $82(1)$, 1–19 (2020)
6. Bonamy, M., Dabrowski, K.K., Feghali, C., Johnson, M., Paulusma, D.: Independent feedback vertex sets for graphs of bounded diameter. Inf. Process. Lett. 131, 26–32 (2018)
7. Bonamy, M., Dabrowski, K.K., Feghali, C., Johnson, M., Paulusma, D.: Independent feedback vertex set for p 5-free graphs. Algorithmica, pp. 1342–1369 (2019)
8. Cranston, D.W., Yancey, M.P.: Vertex partitions into an independent set and a forest with each component small. SIAM J. Discret. Math. $35(3)$, 1769–1791 (2021)
9. Cygan, M., et al.: Parameterized Algorithms. Springer, Cham (2015). https://doi.org/10.1007/978-3-319-21275-3
10. Dabrowski, K.K., Johnson, M., Paesani, G., Paulusma, D., Zamaraev, V.: On the price of independence for vertex cover, feedback vertex set and odd cycle transversal. In: MFCS, pp. 63:1–63:15 (2018)
11. Diestel, R.: Graph theory. Springer, New York (2000)
12. Goddard, W., Henning, M.A.: Independent domination in graphs: a survey and recent results. Discret. Math. $313(7)$, 839–854 (2013)
13. Goddard, W., Henning, M.A.: Independent domination in outerplanar graphs. Discret. Appl. Math. 325, 52–57 (2023)
14. Hakimi, S.L., Schmeichel, E.F.: A note on the vertex arboricity of a graph. SIAM J. Discret. Math. $2(1)$, 64–67 (1989)
15. Jacob, A., Majumdar, D., Raman, V.: Parameterized complexity of conflict-free set cover. Theory Comput. Syst. $65(3)$, 515–540 (2021)
16. Jain, P., Kanesh, L., Misra, P.: Conflict free version of covering problems on graphs: classical and parameterized. Theory Comput. Syst. $64(6)$, 1067–1093 (2020)
17. Jain, P., Kanesh, L., Roy, S.K., Saurabh, S., Sharma, R.: Circumventing connectivity for kernelization. In: CIAC 2021, vol. 12701, pp. 300–313 (2021)
18. Kociumaka, T., Pilipczuk, M.: Faster deterministic feedback vertex set. Inf. Process. Lett. $114(10)$, 556–560 (2014)
19. Li, S., Pilipczuk, M.: An improved fpt algorithm for independent feedback vertex set. Theory Comput. Syst. 64, 1317–1330 (2020)
20. Misra, N., Philip, G., Raman, V., Saurabh, S.: On parameterized independent feedback vertex set. Theor. Comput. Sci. 461, 65–75 (2012)
21. Thomassé, S.: A $4k^2$ kernel for feedback vertex set. ACM Trans. Algorithms $6(2)$, 32:1–32:8 (2010)
22. Wu, Y., Yuan, J., Zhao, Y.: Partition a graph into two induced forests. J. Math. Study $1(01)$ (1996)

23. Yang, A., Yuan, J.: Partition the vertices of a graph into one independent set and one acyclic set. Discret. Math. **306**(12), 1207–1216 (2006)
24. Yang, A., Yuan, J.: On the vertex arboricity of planar graphs of diameter two. Discret. Math. **307**(19–20), 2438–2447 (2007)

Induced Tree Covering
and the Generalized Yutsis Property

Luís Cunha, Gabriel Duarte, Fábio Protti, Loana Nogueira,
and Uéverton Souza[✉] [iD]

Universidade Federal Fluminense, Niterói, Brazil
{lfignacio,fabio,loana,ueverton}@ic.uff.br, gabriel@id.uff.br

Abstract. The Yutsis property of a simple, connected, and undirected graph is the property of partitioning its vertex set into two induced trees. Although the first impression is that such a property is quite particular, it is more general than Hamiltonicity on planar graphs since a planar graph satisfies the Yutsis property if and only if its dual is Hamiltonian. Despite the fact that recognizing Yutsis graphs is NP-complete even on planar graphs, it is still possible to consider two even more challenging problems: (i) the recognition of k-Yutsis graphs, which are graphs that have their vertex sets partitioned into k induced trees, for a fixed $k \geq 2$; (ii) to find the minimum number of vertex-disjoint induced trees that cover all vertices of a graph G, which is called the tree cover number of G. The studies on Yutsis graphs emerge from the quantum theory of angular momenta since it appears as a graphical representation of general recoupling coefficients, and the studies on the tree cover number are motivated by its equality with the maximum positive semidefinite nullity on multigraphs with treewidth at most two.

Given the interest in the tree cover number on graphs with bounded treewidth, we investigate the parameterized complexity of the tree cover number computation. We prove that the tree cover number can be determined in $2^{\mathcal{O}(tw \log tw)} \cdot n^{\mathcal{O}(1)}$, where tw is the treewidth of the input graph, but it cannot be solved in $2^{o(tw \log tw)} \cdot n^{\mathcal{O}(1)}$ time unless ETH fails. Similarly, we conclude that recognizing k-Yutsis graphs can be done in $k^{\mathcal{O}(tw)} \cdot n^{\mathcal{O}(1)}$ time, but it cannot be done in $(k - \epsilon)^{tw} \cdot n^{\mathcal{O}(1)}$ time assuming SETH. We also show that the problem of determining the tree cover number of a graph G is polynomial-time solvable on graphs with bounded clique-width, but it is W[1]-hard considering clique-width parameterization while recognizing k-Yutsis graphs can be done in FPT time. Furthermore, contrasting with the polynomial-time recognition of k-Yutsis chordal graphs, for split graphs G having a partition $V(G) = (S, K)$ where S is an independent set and K is a clique, we prove that determining the tree cover number of G is NP-hard even when S has only vertices of degree 2 or 4, but it is polynomial-time solvable when each vertex of S has either odd degree or degree two in G. We also provide some characterizations for chordal k-Yutsis subclasses.

Keywords: Induced tree cover · Yutsis graphs · clique-width ·
treewidth · split graphs

Supported by CNPq and FAPERJ.

J. A. Soto and A. Wiese (Eds.): LATIN 2024, LNCS 14579, pp. 147–161, 2024.
https://doi.org/10.1007/978-3-031-55601-2_10

1   Introduction

The *Yutsis property* of a simple, connected, and undirected graph $G = (V, E)$ is the property of having a partition of $V(G)$ into two induced trees. A graph $G = (V, E)$ is a *Yutsis graph* if it satisfies the Yutsis property. According to Aldred, Van Dyck, Brinkmann, Fack, and McKay [1], cubic Yutsis graphs appear in the quantum theory of angular momenta as a graphical representation of general recoupling coefficients. They can be manipulated following certain rules in order to generate the so-called summation formulae for the general recoupling coefficient (see [3, 13, 27]).

Besides, the Yutsis property of graphs is also related to the notion of *bonds* in graphs. A *bond* of a connected graph $G = (V, E)$ is an inclusion-wise minimal disconnecting set F of edges of G. A bond F determines a cut $[S, V \setminus S]$ of G such that $G[S]$ and $G[V \setminus S]$ are both connected and F is the set of edges from S to $V \setminus S$. Yutsis graphs form exactly the class of graphs having a bond of size $|E(G)| - |V(G)| + 2$, which is the largest possible value of a bond. Regarding bonds on planar graphs, a folklore theorem states that if G is a connected planar graph, then a set of edges is a cycle in G if and only if it corresponds to a bond in the dual graph of G [18]. Note that each cycle separates the faces of G into the faces in the interior of the cycle and the faces of the exterior of the cycle, and the duals of the cycle edges are precisely the edges that cross from the interior to the exterior [24]. From this observation, it holds that the set of planar Yutsis graphs is precisely the dual class of Hamiltonian planar graphs. Since Garey, Johnson, and Tarjan [17] proved that the problem of establishing whether a 3-vertex-connected planar graph is Hamiltonian is NP-complete, it follows that recognizing planar Yutsis graphs is NP-complete as well. In 2008, Aldred, Van Dyck, Brinkmann, Fack, and McKay [1] showed that for any cubic Yutsis graph the partition of the vertex set into two trees results in trees of equal size.

Although the problem of recognizing Yutsis graphs is challenging since it generalizes Hamiltonicity in planar graphs, it can also be generalized from two different perspectives: viewed as a cutting problem, Yutsis graphs are extremal instances of the Largest Bond problem, as discussed earlier; in addition, observed as a partition problem, recognizing Yutsis graphs is a particular case of finding the smallest number of vertex-disjoint induced trees to cover the vertex set of an input graph G.

The (parameterized) complexity of computing the largest bond of a graph was recently studied in [10, 11, 15]. On the other hand, there are very few results about the minimum number of vertex-disjoint trees being induced subgraphs that cover all the vertices of G. Such an invariant, denoted as the *tree cover number* of a graph, was introduced in [2] as a tool for studying the maximum positive semidefinite nullity of a multigraph, and it was conjectured that the tree cover number of a graph G is upper bounded by the maximum positive semidefinite nullity of G. In [2], it was shown that the tree cover number of an outerplanar graph is equal to its maximum positive semidefinite nullity. Such a result was generalized in [14] showing that this equality also holds for multigraphs G with treewidth at most 2. In addition, bounds for the tree cover number of graphs were established in [4]. To the best of our knowledge, a multivariate investigation of the (parameterized) complexity of computing the tree cover number of a graph is missing in the literature. Therefore, our goal is to map sources of computational intractability for the problem and consequently identify features that make it tractable through the lenses of graph classes and parameterized complexity. Some tree covering problems where the trees do not need to be vertex-disjoint have been studied in [20] and references therein.

First, we consider the following generalized Yutsis property: the k-*Yutsis property* of a simple, connected, and undirected graph $G = (V, E)$ is the property of having a partition of $V(G)$ into k induced trees. A graph $G = (V, E)$ is a k-*Yutsis graph* if it satisfies the k-Yutsis property. Then, we consider the following related problems:

k-YUTSIS GRAPH RECOGNITION

Instance: A simple, undirected, and connected graph $G = (V, E)$.
Question: Is G a k-Yutsis graph?

Note that regarding k-YUTSIS GRAPH RECOGNITION the integer k is a fixed constant. Clearly, 1-Yutsis graphs are trees, so they are recognizable in polynomial time. On the other hand, 2-Yutsis graphs form the so-called Yutsis graph class, whose recognition is NP-complete even for planar graphs, as discussed earlier. In addition, it is easy to see that there is a monotonicity implying that recognizing k-Yutsis graphs is NP-complete for each $k \geq 2$. Besides, we still may consider a more general problem:

INDUCED TREE COVER

Instance: A simple, undirected, and connected graph $G = (V, E)$.
Goal: Determine the tree cover number of G.

We consider k-YUTSIS GRAPH RECOGNITION and INDUCED TREE COVER explicitly as distinct problems in the same way as k-COLORING and CHROMATIC NUMBER deserve particular attention, mainly from the point of view of parameterized complexity.

First, we investigate some subclasses of chordal graphs. We provide a polynomial-time algorithm for INDUCED TREE COVER on split graphs G having a partition $V(G) = (S, K)$ where S is an independent set, K is a clique, and all vertices of S have either odd degree or degree two in G. On the other hand, we show that INDUCED TREE COVER is NP-hard even on split graphs having all vertices of the independent part S with degree equal to either 2 or 4. In order to prove this hardness result, we introduce the MAXIMUM MATCHING WITH RESTRICTION ON SUBMATCHINGS problem (MMRS), which aims to find a maximum matching M of the input graph G satisfying that M cannot contain a perfect matching of some subgraphs of G. Surprisingly, we prove that MMRS is NP-hard even for graphs isomorphic to a disjoint union of K_2's. Continuing the investigation on chordal subclasses, we present forbidden induced subgraph characterizations for some k-tree Yutsis graphs. Also, we remark that k-YUTSIS GRAPH RECOGNITION on chordal graphs is reducible to the bounded treewidth case. Besides, the complexity of computing the tree cover number on graphs with bounded treewidth is also a natural question since the studies regarding this invariant are motivated by the equality with the maximum positive semidefinite nullity on graphs with treewidth equal to 2 [14].

Concerning treewidth, an algorithmic meta-theorem due to Courcelle [6] states that any problem expressible in the monadic second-order logic of graphs with edge set quantifications (MSO$_2$) can be solved in $f(tw) \cdot n$ time, where tw is the treewidth of the input graph. This meta-theorem shows that several problems can be efficiently solved when restricted to graphs of bounded treewidth. That is the case of k-YUTSIS GRAPH RECOGNITION for any constant k. Since k is a constant, one can describe an MSO formula expressing a partition into k parts such that each part is connected and acyclic (such a formula doesn't need edge set quantification). Conversely, it seems that INDUCED TREE COVER parameterized by treewidth cannot be solved in FPT time using Courcelle's framework based on model checking because the tree cover number of the input graph (number of parts) may be unbounded. Therefore, one of the contributions of this paper is to show that INDUCED TREE COVER parameterized by treewidth is in FPT. Our FPT-time algorithm is performed in $2^{\mathcal{O}(tw \log tw)} \cdot n^{\mathcal{O}(1)}$, where tw is the treewidth of the input graph. In addition, we also show that this running time is tight, assuming ETH. More precisely, we show that there is no $2^{o(tw \log tw)} \cdot n^{\mathcal{O}(1)}$ time algorithm to solve INDUCED TREE COVER unless ETH fails. Similarly, we also conclude that k-YUTSIS GRAPH RECOGNITION can be solved in $k^{\mathcal{O}(tw)} \cdot n^{\mathcal{O}(1)}$ time, but it cannot be solved in $(k - \epsilon)^{tw} \cdot n^{\mathcal{O}(1)}$ time assuming SETH.

In 2000, Courcelle, Makowsky, and Rotics [5] presented a "stronger" algorithmic meta-theorem. They stated that any problem expressible in the monadic second-order logic on graphs without edge set quantifications (MSO$_1$) can be solved in FPT time when parameterized by the clique-width of the input graph, which is a parameter stronger than treewidth in the sense that bounded treewidth implies bounded clique-width. Using such a meta-theorem, we can also conclude that k-YUTSIS GRAPH RECOGNITION is FPT concerning clique-width parameterization. On the other hand, we show that INDUCED TREE COVER is W[1]-hard when parameterized by clique-width. Besides that, we present a dynamic programming algorithm using expression-tree decompositions to solve INDUCED TREE COVER parameterized by clique-width in XP-time, implying that the problem is polynomial-time solvable for several interesting graph classes such as cographs and distance-hereditary graphs.

2   Induced Tree Cover on Split Graphs

A natural lower bound for the tree cover number of G is $\lceil \frac{\omega(G)}{2} \rceil$, where $\omega(G)$ is the size of the largest clique of G (any tree contains at most two vertices of a clique). From this remark, it holds that k-YUTSIS GRAPH RECOGNITION is "trivial" on chordal graphs G: if G has a clique of size $2k + 1$, then the answer is no, otherwise G has bounded treewidth, and we can decide the problem using an algorithm for bounded treewidth graphs (for instance, Courcelle Theorem).

Proposition 1. *Every k-Yutsis graph does not contain K_{2k+1} as a subgraph.*

The previous observation motivates the discussion on the complexity of INDUCED TREE COVER on chordal graphs. Next, we show that INDUCED TREE COVER is NP-hard even restricted to split graphs (a particular subclass of chordal graphs). Interestingly, this hardness depends on vertices of even degree greater than two: if a split graph G has a partition $V(G) = (S, K)$ such that S is an independent set, K is a clique, and any vertex of S has either degree odd or degree two in G, then we can determine the tree cover number of G in polynomial time.

Let $N_G(v)$ be the neighborhood of v in the graph G, and $d_G(v) = |N_G(v)|$.

Proposition 2. *Let G be a k-Yutsis graph that contains a clique K of size at most $2k$. Any graph G' obtained from G and successive additions of new vertices v_i, $i = 1, 2, \cdots$, where $N_{G'}(v_i) \subseteq K$ and $d_{G'}(v_i)$ is odd is a k-Yutsis graph.*

Proof. As G is a k-Yutsis graph, take a k-induced tree covering $\mathcal{C}$ of G. Recall that any tree of $\mathcal{C}$ contains at most 2 vertices of K. Thus, since $d(v_i)$ is odd ($i = 1, 2, \cdots$), there is a tree T_i of $\mathcal{C}$ such that v_i is adjacent to a unique vertex of T_i. Hence, each v_i can be added in its respective tree T_i, and G' is also covered by k-vertex disjoint induced trees.   □

At this point, we recall that one can enumerate all partitions of a split graph into a clique and an independent set in polynomial time. Therefore, we have Corollary 1 as a direct consequence of Proposition 2.

Corollary 1. *Let G be a split graph whose vertex set can be partitioned into a clique K and an independent set S such that all vertices of S have odd degree in G. It holds that the tree cover number of G can be determined in polynomial time.*

Theorem 1. *Let G be a split graph and $V(G) = (S, K)$ be a partition of G into a clique K and an independent set S. If every vertex $v_i \in S$ satisfies that either $d(v_i)$ is odd or $d(v_i) = 2$, then the tree cover number of G can be determined in polynomial time.*

Proof. By Proposition 2, we can forget about vertices in the independent set whose degree is odd since in any induced tree cover for the remaining graph, there will be a tree containing exactly one of their neighbors. Therefore, it is enough to consider only the case where all vertices of S have degree equal to 2.

It is easy to see that there is an optimal solution that never contains the two neighbors of a degree-2 vertex $v \in S$ in the same tree because it would force v to be a singleton, while we could add one of the neighbors of v to the singleton tree with v instead. Then, the problem of covering the graph with as few trees as possible means to take as many trees as possible that cover all vertices of the clique K without any such tree containing both neighbors of a degree-2 vertex of S. This can done by finding a maximum matching in the graph K' obtained from K by removing every edge whose both endpoints have a common neighbor $v_i \in S$ such that $d(v_i) = 2$. Given a maximum matching M of K', all vertices of K' that are not covered by M turn into additional trees of size one. Thus, the tree cover number of G is $|M| + (|K'| - 2 \cdot |M|) = |K'| - |M|$. □

Unlike the case where the vertices of the independent set of a split graph have odd degree or degree equal to 2, we prove that INDUCED TREE COVER is NP-hard when a split graph has vertices of the independent set with even degree greater than or equal to 4. Note that when G is a split graph having vertices with even degree greater than or equal to 4 belonging to S, the following property holds for any induced tree associated with G:

– For each $v \in S$ there is a clique formed by $\{v\} \cup N(v)$. Thus, at most one vertex $u \in N(v)$ must be chosen to be adjacent to v in an induced tree of G and the other vertices of $N(v) \setminus \{u\}$ cannot be adjacent to u in such a tree.

The above property holds because otherwise, v would belong to the same tree of two neighbors, which forms a K_3 and contradicts the construction of a tree of G. Thus, if one aims, for each $v \in S$, to avoid the graph $(\{v\}, \emptyset)$ as a trivial tree, one must be able to partition K into trees (K_1's or K_2's) such that there is at least one tree having only a single vertex of $N(v)$, for each $v \in S$. Therefore, the INDUCED TREE COVER problem on split graphs motivates a special kind of maximum matching problem.

Before we show that the INDUCED TREE COVER problem is NP-hard on split graphs when vertices of the independent set have even degree at least 4, we introduce a problem called MAXIMUM MATCHING WITH RESTRICTION ON SUBMATCHING, defined as follows:

MAXIMUM MATCHING WITH RESTRICTION ON SUBMATCHING (MMRS)

Instance: A graph H and a set $\mathcal{F}$ of subsets of $V(H)$.
Goal: Determine a maximum matching M with $H' = (V(H), M)$, such that for each $F_i \in \mathcal{F}$ the induced subgraph $H'[F_i]$ is not a matching.

Each set F_i is called a *conflict set* of H. In other words, in the MMRS problem, we are trying to find a maximum matching M of H where for each conflict set F_i of $\mathcal{F}$ the vertices of F_i may contain some edges of M but at least one isolated vertex with respect to $H[F_i]$. Next, we show that MMRS is NP-hard from a polynomial reduction from the INDEPENDENT SET problem.

Construction 1. *Given a simple graph G, we create an instance $(H, \mathcal{F})$ of MMRS from G as follows: for each vertex u of G create in H two adjacent vertices u_1 and u_2; for each edge uv of G create in $\mathcal{F}$ a conflict set $F_i = \{u_1, u_2, v_1, v_2\}$.*

Theorem 2. *MMRS is NP-hard for graphs isomorphic to a perfect matching, even when each conflict set $F_i \in \mathcal{F}$ satisfies $|F_i| = 4$.*

Proof. Let G be an instance of the INDEPENDENT SET problem. We construct an instance $(H, \mathcal{F})$ from G as described in Construction 1. Based on this construction, H is a disjoint union of $|V(G)|$ edges, and $\mathcal{F}$ is a set with $|E(G)|$ subsets of H, each of them with size equal to 4. We show that G has an independent set of size k if and only if H has an MMRS solution of size k.

Consider an independent set I of size k of G. By construction, this set I corresponds to a set M_I of k edges of H. Since each $F_i \in \mathcal{F}$ corresponds to an edge of G and M_I arise from an independent set of G, it holds that M_I is an MMRS solution (conflict-free matching) for $(H, \mathcal{F})$.

Now, consider an MMRS solution M_I of $(H, \mathcal{F})$ with k edges. By construction, this edges correspond to a set I of k vertices of G. As M_I is conflict-free, i.e., the induced subgraph $H'[F_i]$ is not a matching for each F_i, by construction, it follows that for each F_i the set M_I contain at most one edge of $H[F_i]$, implying that the corresponding set I is an independent set of G.   □

Now, we are able to prove that INDUCED TREE COVER is NP-hard for split graphs by a polynomial time reduction from MMRS on graphs isomorphic to a perfect matching and with conflict sets of size 4.

Theorem 3. INDUCED TREE COVER *is NP-hard for split graphs.*

Proof. Let $(H, \mathcal{F})$ be an instance obtained from a simple graph G with $|V(G)| = n$ according to Construction 1. In the following, we create a split graph Q, whose $V(Q) = K \cup S$, K is a clique and S is an independent set. i) Create the complete graph K among all vertices of H; ii) For each edge $uv \in E(K) \setminus E(H)$ create $2n - k + 1$ vertices, each of them adjacent to u and v; iii) For each $F_i \in \mathcal{F}$ create $2n - k + 1$ vertices, each of them adjacent to all vertices of F_i. Based on this construction, Q is a split graph whose complete part is K created in step i, and the independent set is S with the vertices created in steps ii and iii. Moreover, vertices of step ii have degree 2 and vertices of step iii have degree 4 (see Construction 1). We show that H has a solution of MMRS of size at least k if and only if Q has tree cover number at most $2n - k$.

Consider I' a solution of MMRS of size k for H. Recall that $|V(H)| = 2n$. We create a partition of Q into at most $2n - k$ trees as follows: each edge of I' is added into a distinct tree of the covering of Q, and each vertex of H that is not in I' is added into a new tree of this covering. Since $|V(H)| = 2n$ and $|I'| = k$, the resulting covering has size $2n - k$. By construction, as I' is a solution of MMRS, the vertices of S can be added into trees of the covering containing exactly one of their neighbors, resulting in an induced tree cover of Q.

Now, consider Q having tree cover number equal to $2n - k$, where $2n$ is the size of the clique part K of the split graph Q. Note that each tree of the covering has at most 2 vertices of K. As the tree cover number of Q is equal to $2n - k$, at least k trees have 2 vertices of K; otherwise, the tree cover number would be greater. By construction, each conflict set and non-edge of $(H, \mathcal{F})$ is represented by $2n - k + 1$ vertices of S. Given that the tree cover number of Q is equal to $2n - k$, it holds that the set of edges of K in trees of the covering form an MMRS solution of size at least k for $(H, \mathcal{F})$.   □

Corollary 2 follows as an immediate consequence from Theorem 3.

Corollary 2. INDUCED TREE COVER *is NP-hard even for split graphs* G *having a partition* $V(G) = (S, K)$ *such that* S *is an independent set,* K *is a clique and all vertices of* S *have even degree at most 4 in* G.

Since INDUCED TREE COVER is NP-hard even on split graphs, one could ask about the complexity of this problem on other special graph classes such as threshold graphs, cographs, and distance hereditary graphs, for example. Next, we present a stronger result that allows us to claim that, among others, all these classes are polynomial-time solvable concerning INDUCED TREE COVER.

3   Taking the Clique-Width as Parameter

Since we have proved that determining the tree cover number is NP-complete for split graphs, which have unbounded clique-width, now we consider graphs having bounded clique-width.

The *clique-width* of a graph G, denoted by $cw(G)$, is defined as the minimum number of labels needed to construct G, using the following four operations:

- create a single vertex v with an integer label l (denoted by $l(v)$);
- take the disjoint union (i.e., co-join) of two graphs (denoted by $\oplus$);
- join by an (arc) edge every vertex labeled i to every vertex labeled j for $i \neq j$ (denoted by $\eta(i, j)$);
- relabel all vertices with label i by label j (denoted by $\rho(i, j)$).

An algebraic term representing such a construction of G and using at most k labels is a k-*expression* of G. The clique-width of G is the minimum k for which G has a k-expression.

An expression tree $\mathcal{T}$ is *irredundant* if, for any join node, no vertex labeled i is adjacent to any vertex labeled j in the graph associated with its child.

Oum and Seymour [23] gave an algorithm that finds a $2^{\mathcal{O}(cw)}$-approximation of an optimal clique-width expression. In addition, Courcelle and Olariu [7] showed that every expression tree $\mathcal{T}$ of G can be transformed into an irredundant expression tree $\mathcal{T}$ of the same width in time linear in the size of $\mathcal{T}$. Therefore, without loss of generality, we can assume that we are given an irredundant expression tree $\mathcal{T}$ of G.

Let cw be the number of labels used in $\mathcal{T}$. Now, for every node ℓ of the irredundant expression tree $\mathcal{T}$, denote by G_ℓ the labeled graph associated with this node. First, we consider the decision version of INDUCED TREE COVER asking if the tree cover number is at most k (k is unbounded). Given integers ℓ and k, our dynamic programming algorithm is based on deciding if G_ℓ can be partitioned into at most k induced forests, such that each forest must be of a predetermined type.

The type of a tree T encodes for each label i if T has 0,1, or at least 2 vertices with label i. So, there are at most 3^{cw} types of trees. The type of a forest F encodes for each type of tree if F has 0,1, or at least 2 trees with such a type as a component. Thus, we have at most $3^{3^{cw}}$ types of forests. Note that information about at least two is needed to avoid cycles when doing join operations. Let $r = 3^{3^{cw}}$.

We define a table of Boolean values where each entry is of the form

$$c[\ell, k, \tau_1, ..., \tau_r],$$

representing if the labeled graph G_ℓ can be partitioned into at most k forests such that τ_j of them are of type j, for each $j \in [1..r]$. That is, τ_j is an integer between 0 and k representing the number of occurrences of forests of type j in the partition. Clearly, $\tau_1 + \tau_2 + ... + \tau_r \leq k$. Thus, the answer to the problem is **true** if and only if there is a **true** entry of the root node where each used forest type has only one tree.

This table has size bounded by $k^{3^{3^{cw}}} \cdot n$, where n is the number of nodes of the decomposition.

At this point, it is not hard to see that to compute $c[\ell, k, \tau_1, ..., \tau_r]$ from the entries of its children, it is enough to analyse the entries of its children compatible with the operation in question (relabel, disjoint union, and join) to obtain the entry to be filled in node ℓ.

Since each node ℓ' has at most $k^{3^{3^{cw}}}$ entries, we can compute each $c[\ell, k, \tau_1, ..., \tau_r]$ in XP time. To solve the optimization version, it is enough to check each possible $k \in [1..n]$ and take the least k for which the answer is **true**. Hence, the following holds.

Theorem 4. INDUCED TREE COVER *can be solved in* XP *time, considering the clique-width of the input graph as parameter.*

Besides that, Theorem 4 directly implies that INDUCED TREE COVER can be solved in polynomial time for some well-known graph classes, such as cographs or distance-hereditary graphs (which is a superclass of cographs), since they have cliquewidth equal to $cw = 2$ and $cw = 3$, respectively [19]. Also, for k-YUTSIS GRAPH RECOGNITION, as k is a constant, the algorithm performs in FPT time, being a simpler solver for the problem compared to the Courcelle framework.

Next, we show that the XP-solvability obtained by our algorithm cannot be "drastically improved" in the sense that INDUCED TREE COVER cannot be solved in FPT time, assuming $W[1] \neq FPT$.

Theorem 5. INDUCED TREE COVER *is W[1]-hard, considering the clique-width of the input graph as parameter.*

Proof. Fomin et al. [16] showed that CHROMATIC NUMBER is W[1]-hard when parameterized by clique-width. In such a problem, we are given a graph G and a positive integer ℓ, and asked whether $\chi(G) \leq \ell$, where $\chi(G)$ is the minimum number of colors to properly color the vertex set of G. Next, we present a (polynomial) reduction from CHROMATIC NUMBER to INDUCED TREE COVER where the resulting graph preserves the clique-width of the original graph (assuming it is at least 2).

Let G be an instance of CHROMATIC NUMBER. We create a graph H from G as follows. Make a disjoint union between G and a complete graph K' with ℓ vertices. Let H' be the resulting graph. Now, let H be the join of H' with a clique K'' with ℓ vertices. Note that if G can be constructed using $cw > 2$ labels, one can construct H with cw labels as well. Also, $K' \cup K''$ is a clique of size 2ℓ.

At this point, it is easy to see that G has an ℓ-coloring if and only if the resulting graph H has tree cover number ℓ. If G has a ℓ-coloring, then adding for each color class a distinct vertex of K' and K'', we obtain a partition of H into ℓ induced trees. Conversely, if H is partitioned into ℓ induced trees, then each tree has exactly two vertices of $K' \cup K''$, and the vertices of G must be in parts having exactly one vertex of K'' (due to connectivity and acyclicity), thus, $V(G)$ is partitioned into at most k independent sets, because any vertex of K'' is universal.   □

Due to the W[1]-hardness of INDUCED TREE COVER parameterized by clique-width, it is natural to ask whether INDUCED TREE COVER is fixed-parameter tractable when parameterized by treewidth.

4   Taking the Treewidth as Parameter

In the following, given a tree decomposition $\mathcal{T}$, we denote by ℓ one node of $\mathcal{T}$ and by X_ℓ the vertices contained in the *bag* of ℓ. We assume w.l.o.g that $\mathcal{T}$ is an extended version of a *nice* tree decomposition (see [8]), that is, we assume that there is a special root node r such that $X_\ell = \emptyset$ and all edges of the tree are directed towards r and each node ℓ has one of the following five types: *Leaf*; *Introduce vertex*; *Introduce edge*; *Forget vertex*; and *Join*. Moreover, define G_ℓ to be the subgraph of G which contains only vertices and edges that have been introduced in ℓ or in a descendant of ℓ.

Theorem 6. *Given a nice tree decomposition of G with width tw, one can find the tree cover number of G in $2^{\mathcal{O}(tw \log tw)} \cdot n$ time.*

Proof. Let $\mathcal{T}$ be a nice tree decomposition of G with width tw and $\partial(G)$ be a partition of $V(G)$ into induced trees such that $|\partial(G)|$ is the tree cover number of G.

Note that, for each node ℓ of $\mathcal{T}$, $\partial(G)$ defines a partition of X_ℓ, denoted by ρ^ℓ, where each part of ρ^ℓ corresponds to a maximal set of vertices of X_ℓ belonging to the same induced forest of G_ℓ that is a subgraph of some tree of $\partial(G)$. In addition, $\partial(G)$ also defines, for each set S_i in ρ^ℓ, a partition ρ_i^ℓ of S_i, where each part represents the vertices of S_i contained in the same tree of such an induced forest of G_ℓ.

Given a node ℓ of $\mathcal{T}$, assume ρ to be any partition of X_ℓ, and ρ_i to be a partition of $S_i \in \rho$. We define a table c indexed by $\ell \in V(\mathcal{T})$, a partition ρ of X_ℓ, and partitions $\rho_1 \ldots \rho_h$, where each ρ_i is corresponding to a set $S_i \in \rho$ and $h = |\rho|$. For simplicity, we denote $S_i = V(\rho_i)$. Note that given $\rho_1 \ldots \rho_h$ the partition ρ is defined.

An entry $c[\ell, \rho_1 \ldots \rho_h]$ of table c must store the size of a minimum covering into induced forest of the subgraph G_ℓ (partial solution) such that:

- each $V(\rho_i)$ is contained in the vertex set of a distinct forest F_i of this covering;
- each part of some ρ_i is contained in the vertex set of a distinct tree of F_i;
- each component of a disconnected forest of the covering contains vertices of X_ℓ.

Observe that each forest that does not contain vertices of X_ℓ is a tree. Hence, the disconnected forests are the elements that can be made connected through vertices of X_ℓ with vertices and edges to be inserted by ancestor nodes of the decomposition.

The following lemma is sufficient to show that a correct computation of table c is enough to determine the tree cover number of G.

Lemma 1. Let $\partial(G)$, $\rho^\ell = (V(\rho_1^\ell), \ldots, V(\rho_h^\ell))$ be a minimum partition into induced trees of G and its partition defined over X_ℓ, respectively. Let $\partial(G_\ell)$ be a minimum covering into induced forests of G_ℓ in accordance with $c[\ell, \rho_1^\ell \ldots \rho_h^\ell]$. It holds that $\partial(G_\ell)$ can be extended to a covering into induced trees of G with size at most $|\partial(G)|$.

Proof. Both $\partial(G_\ell)$ and $\partial(G)$ define the same partitions $\rho_1^\ell, \ldots, \rho_h^\ell$, by definition. Take a partition ρ_i^ℓ, the forest F_i^ℓ of $\partial(G_\ell)$ that contains $V(\rho_i^\ell)$ and the tree T_i of $\partial(G)$ that contains $V(\rho_i^\ell)$.

First, we observe that $G[V(F_i^\ell) \cup (V(T_i) \setminus V(G_\ell))]$ is a acyclic. If there is a cycle in $G[V(F_i^\ell) \cup (V(T_i) \setminus V(G_\ell))]$, it must contain vertices of both $V(F_i^\ell) \setminus X_\ell$ and $V(T_i) \setminus V(G_\ell)$, and at least two vertices of X_ℓ, because F_i^ℓ and T_i are acyclic and X_ℓ is a separator of G (assuming $V(G_\ell) \setminus X_\ell \neq \emptyset$ and $V(G) \setminus V(G_\ell) \neq \emptyset$). However, a path between any pair of vertices of $V(\rho_i^\ell)$ in F_i^ℓ implies that both vertices are in the same tree of F_i^ℓ, and then they are also connected in the graph induced by $V(T_i) \cap V(G_\ell)$ because $\partial(G_\ell)$ and $\partial(G)$ define the same partitions $\rho_1^\ell, \ldots, \rho_h^\ell$. Hence, the existence of a cycle in $G[V(F_i^\ell) \cup (V(T_i) \setminus V(G_\ell))]$ implies the existence of a cycle in T_i, which is a contradiction.

Now, observe that $G[V(F_i^\ell) \cup (V(T_i) \setminus V(G_\ell))]$ is connected. Recall that F_i^ℓ and $G[V(T_i) \cap V(G_\ell)]$ are forests having $|\rho_i^\ell|$ trees, where each one contains the vertices of a distinct set in ρ_i^ℓ. Since the paths between different trees of $G[V(T_i) \cap V(G_\ell)]$ is in $G[(V(T_i) \setminus V(G_\ell)) \cup X_\ell]$, such connections are preserved in $G[V(F_i^\ell) \cup (V(T_i) \setminus V(G_\ell))]$, so, it is connected.

At this point, we extend $\partial(G_\ell)$ to a partition into induced trees of G, denoted by $\partial^*(G)$, as follows: start with $\partial^*(G) = \partial(G_\ell)$; for each partition ρ_i^ℓ replace its forest F_i^ℓ with $G[V(F_i^\ell) \cup (V(T_i) \setminus V(G_\ell))]$; add the trees of $\partial(G)$ containing no vertex of G_ℓ. Since $\partial(G_\ell)$ covers the vertices of G_ℓ, $V(G) \setminus V(G_\ell)$ is covered by trees of $\partial(G_\ell)$ containing either no vertex of G_ℓ or some vertices of X_ℓ, and $G[V(F_i^\ell) \cup (V(T_i) \setminus V(G_\ell))]$ is a tree (for any ρ_i^ℓ), it holds that $\partial^*(G)$ is a covering of G into induced trees.

Finally, we claim that $|\partial^*(G)| \leq |\partial(G)|$. Both $\partial^*(G)$ and $\partial(G)$ have the same set of trees having no vertex of G_ℓ, and exactly one tree for each ρ_i^ℓ. Since $|\partial(G_\ell)| = c[\ell, \rho_1^\ell \ldots \rho_h^\ell]$, the number of trees of $\partial^*(G)$ containing only vertices of $V(G_\ell) \setminus X_\ell$ is minimum. Therefore, $|\partial^*(G)| \leq |\partial(G)|$.   □

Lemma 1 shows that working with table c we are dealing with a representative set that is enough to compute in order to determine the tree cover number of G.

The number of partitions of a set of k elements is at most $k!$. Since for each node ℓ, we consider any partition of X_ℓ, and for each part of such a partition we also consider its partition, it holds that the size of table c is upper bounded by $2^{O(tw \log tw)} \cdot n$.

We omit a complete description of how to compute the table c in a bottom-up manner using the extended nice tree decomposition due to space constraints.   □

Now, a natural question to ask is whether the dependence on tw appearing in the running time of the presented algorithm, i.e., $2^{O(tw \log tw)}$, is optimal for the problem. Next, assuming ETH, we show that for INDUCED TREE COVER, the dependence on tw in the running time cannot be improved to single exponential.

Theorem 7. *There is no $2^{o(tw \log tw)} \cdot n^{O(1)}$ time algorithm to solve* INDUCED TREE COVER *unless the ETH fails.*

Proof. In 2018, Lokshtanov, Marx, and Saurabh [22] proved that CHROMATIC NUMBER cannot be solved in $2^{o(tw \log tw)} \cdot n^{O(1)}$ unless the ETH fails. Next, we present a reduction from CHROMATIC NUMBER to INDUCED TREE COVER where the treewidth of the resulting graph is at most twice the treewidth of the original graph. This reduction together with the lower bound for CHROMATIC NUMBER presented in [22] is enough to claim that INDUCED TREE COVER also cannot be solved in $2^{o(tw \log tw)} \cdot n^{O(1)}$ unless the ETH fails.

Regarding CHROMATIC NUMBER parameterized by treewidth, it holds that $\ell \leq tw$, otherwise are dealing with a "yes" instance ($\chi(G) < \ell$ since $\chi(G) \leq tw$). Now, let G be an instance of CHROMATIC NUMBER and H be the graph constructed as in Theorem 5. From an optimal tree decomposition $\mathcal{T}$ of G, we can obtain an optimal tree decomposition $\mathcal{T}^*$ of H by adding the ℓ vertices of the K'' in all bags of $\mathcal{T}$, create a new node with a bag containing all vertices of the $K' \cup K''$, and then adding an edge from this new node to some other node. Since $\ell \leq tw$, H has treewdith at most $2tw$, where tw is the treewidth of G. As G has an ℓ-coloring if and only if H has tree cover number at ℓ, the proof is complete.   □

Returning to the discussion about the dependency on tw of our algorithm, we have two reasons for the $2^{O(tw \log tw)}$ dependency. The first is that the number of trees of the covering may be unbounded, so we should consider partitions of all sizes inside the bags. The second one is because we enumerate all partitions of a set to check connectivity throughout the algorithm. Nowadays, this connectivity checking can be improved using rank-based approach or Cut & Count. Thus, the $2^{\mathcal{O}(tw \log tw)}$ dependency is tight due to the need to consider partitions of all sizes inside a bag.

Regarding k-YUTSIS GRAPH RECOGNITION, we only need to consider partitions into at most k parts (k is a constant), which are $\mathcal{O}(k^{tw}) = 2^{\mathcal{O}(tw)}$. Thus, using the rank-based approach to improve the connectivity checking of each part, we can solve k-YUTSIS GRAPH RECOGNITION in $2^{\mathcal{O}(tw)} \cdot n^{\mathcal{O}(1)}$ time. Finally, Lokshtanov, Marx, and Saurabh [21] showed that for any $k \geq 3$, k-COLORING cannot be solved in time $(k - \epsilon)^{tw} \cdot n^{\mathcal{O}(1)}$. Again, using the same construction presented in Theorem 7 and assuming that $\ell = k$ is a *constant independent of the input* (we are dealing with k-YUTSIS GRAPH RECOGNITION), the treewidth of H is increased only by an additive constant from the treewidth of G; therefore, it follows that the same lower bound holds for k-YUTSIS GRAPH RECOGNITION.

Theorem 8. k-YUTSIS GRAPH RECOGNITION *can be solved in* $k^{\mathcal{O}(tw)} \cdot n^{\mathcal{O}(1)}$ *time, but it cannot be solved in* $(k - \epsilon)^{tw} \cdot n^{\mathcal{O}(1)}$ *time, unless SETH fails.*

5   On Some k-Tree Yutsis Graphs

We close our discussion with some structural characterization for k-trees. The proofs are omitted due to space constraints.

A *n-sun graph* is graph with $2n$ vertices consisting of a central complete graph K_n with an outer ring of n vertices, each of which is joined to both endpoints of the closest outer edge of the central core. 3-sun graph is denoted by S_3.

Observation 1. *A graph isomorphic to a* S_3 *is not a Yutsis graph.*

Observation 2. *Let G be a k-tree graph. If G is a Yutsis graph then $k \leq 3$.*

Theorem 9. *A 2-tree graph G is a Yutsis graph if and only if it is S_3-free.*

Observation 3. *Let G be a Yutsis graph with a clique K of size 3. Any graph G' obtained from G and adding a new vertex v where $N_{G'}(v) = K$ is also Yutsis.*

Observation 3 describes a closed operation for the property of having a Yutsis partition. Hence, a possible pre-process of any input graph G is to check whether G has a vertex v of degree 3 adjacent to a clique of size 3. In such a case, we can remove v from G and continue this process until there is no vertex satisfying these restrictions, or we know that the resulting graph satisfies the Yutsis property. Restricting to 3-tree graph, Theorem 10 follows straightforward.

Theorem 10. *Any 3-tree graph is a Yutsis graph.*

Now, we investigate the value k that a 2-tree graph is a k-Yutsis graph.

Theorem 11. *Let G be a connected 2-tree graph. If G has at most $k - 2$ subgraphs isomorphic to S_3 that share at most one face, then G is a k-Yutsis graph.*

Theorem 12. *Let G be a 2-tree graph. G is a k-Yutsis graph if and only if G does not contain $k - 1$ subgraphs isomorphic to a S_3 graph such that any pairwise of S_3 shares at most one face.*

6   Conclusion

This work investigates the complexity of INDUCED TREE COVER and k-YUTSIS GRAPH RECOGNITION. The reduction presented in Theorem 5 allows us to obtain many hardness results for INDUCED TREE COVER from CHROMATIC NUMBER such as W[1]-hardness concerning clique-width parameterization, ETH lower bounds for treewidth parameterization, and also NP-hardness for graphs classes NP-hard for CHROMATIC NUMBER and closed under the construction presented. In addition, in Theorem 3, we show that INDUCED TREE COVER is also NP-hard on split graphs. Continuing the investigation of INDUCED TREE COVER on special graph classes, it seems interesting to ask about it on interval and permutation graphs. It is also interesting to analyze the complexity of k-YUTSIS GRAPH RECOGNITION on classes of graphs where bounding the size of the largest clique does not imply restricting the size of its treewidth. It is worth mentioning that the recognition of Yutsis bipartite graphs is NP-complete, since the dual of an Eulerian planar graph is bipartite [26], and Hamiltonicity is NP-complete on 4-regular planar graphs [25] (Yutsis planar equals dual-Hamiltonian).

Finally, given the W[1]-hardness considering clique-width and the FPT tractability for treewidth, it is interesting to ask about the complexity of INDUCED TREE COVER considering structural parameterizations that are weaker or incomparable with clique-width and stronger or incomparable with treewidth such as distance to cluster, neighborhood diversity, twin-cover, co-treewidth, and co-degeneracy (c.f. [9,12]).

References

1. Aldred, R.E., Van Dyck, D., Brinkmann, G., Fack, V., McKay, B.D.: Graph structural properties of non-Yutsis graphs allowing fast recognition. Discret. Appl. Math., **157**(2), 377–386 (2009)
2. Barioli, F., Fallat, S., Mitchell, L., Narayan, S.: Minimum semidefinite rank of outerplanar graphs and the tree cover number. Electron. J. Linear Algebra **22**, 10–21 (2011)
3. Lawrence Christian Biedenharn and James D Louck. The Racah-Wigner algebra in quantum theory. Addison-Wesley (1981)
4. Bozeman, C., Catral, M., Cook, B., González, O., Reinhart, C.: On the tree cover number of a graph. Involve, a J. Math. **10**(5), 767–779 (2017)
5. Courcelle, B., Makowsky, J.A., Rotics, U.: Linear time solvable optimization problems on graphs of bounded clique-width. Theory Comput. Syst. **33**(2), 125–150 (2000)
6. Courcelle, B.: The monadic second-order logic of graphs. I. recognizable sets of finite graphs. Inform. Comput. **85**(1), 12–75 (1990)
7. Courcelle, B., Olariu, S.: Upper bounds to the clique width of graphs. Discret. Appl. Math. **101**(1–3), 77–114 (2000)
8. Cygan, M., et al.: Parameterized Algorithms. Springer, Cham (2015). https://doi.org/10.1007/978-3-319-21275-3

9. Duarte, G., Oliveira, M.D.O., Souza, U.S.: Co-degeneracy and co-treewidth: using the complement to solve dense instances. In: 46th International Symposium on Mathematical Foundations of Computer Science, MFCS, volume 202 of LIPIcs, pp. 42:1–42:17 (2021)

10. Durate, G.L., et al.: Computing the largest bond and the maximum connected cut of a graph. Algorithmica **83**(5), 1421–1458 (2021). https://doi.org/10.1007/s00453-020-00789-1

11. Duarte, G.L., Lokshtanov, D., Pedrosa, L.L., Schouery, R., Souza, U.S.: Computing the largest bond of a graph. In 14th International Symposium on Parameterized and Exact Computation, IPEC, volume 148 of LIPIcs, pp. 12:1–12:15 (2019)

12. Duarte, G.L., Souza, U.S.: On the minimum cycle cover problem on graphs with bounded co-degeneracy. In: 48th International Workshop on Graph-Theoretic Concepts in Computer Science, WG, vol. 13453 of LNCS, pp. 187–200 (2022). https://doi.org/10.1007/978-3-031-15914-5_14

13. Dyck, D.V., Fack, V.: On the reduction of Yutsis graphs. Discret. Math. **307**(11), 1506–1515 (2007)

14. Ekstrand, J., Erickson, C., Hay, D., Hogben, L., Roat, J.: Note on positive semidefinite maximum nullity and positive semidefinite zero forcing number of partial 2-trees. Electron. J. Linear Algebra **23**, 79–87 (2012)

15. Eto, H., Hanaka, T., Kobayashi, Y., Kobayashi, Y.: Parameterized algorithms for maximum cut with connectivity constraints. In: 14th International Symposium on Parameterized and Exact Computation, IPEC, vol. 148 of LIPIcs, pp. 13:1–13:15 (2019)

16. Fomin, F., Golovach, P., Lokshtanov, D., Saurabh, S.: Intractability of clique-width parameterizations. SIAM J. Comput. **39**(5), 1941–1956 (2010)

17. Garey, M.R., Johnson, D.S., Tarjan, R.E.: The planar hamiltonian circuit problem is NP-complete. SIAM J. Comput. **5**(4), 704–714 (1976)

18. Godsil, C., Royle, G.F.: Algebraic Graph Theory. Graduate texts in mathematics. Springer (2001)

19. Golumbic, M.C., Rotics, U.: On the clique—width of perfect graph classes. In: Widmayer, P., Neyer, G., Eidenbenz, S. (eds.) WG 1999. LNCS, vol. 1665, pp. 135–147. Springer, Heidelberg (1999). https://doi.org/10.1007/3-540-46784-X_14

20. Khani, M.R., Salavatipour, M.R.: Improved approximation algorithms for the min-max tree cover and bounded tree cover problems. Algorithmica **69**(2), 443–460 (2013). https://doi.org/10.1007/s00453-012-9740-5

21. Lokshtanov, D., Marx, D., Saurabh, S.: Known algorithms on graphs of bounded treewidth are probably optimal. ACM Trans. Algorithms, **14**(2) (2018)

22. Lokshtanov, D., Marx, D., Saurabh, S.: Slightly superexponential parameterized problems. SIAM J. Comput. **47**(3), 675–702 (2018)

23. Oum, S.I., Seymour, P.: Approximating clique-width and branch-width. J. Comb. Theory, B, **96**(4), 514–528 (2006)

24. Oxley, J.G.: Matroid theory, vol. 3. Oxford University Press, USA (2006)

25. Picouleau, C.: Complexity of the hamiltonian cycle in regular graph problem. Theoret. Comput. Sci. **131**(2), 463–473 (1994)

26. Welsh, D.J.A.: Euler and bipartite matroids. J. Combinatorial Theory **6**(4), 375–377 (1969)

27. Yutsis, A.P., Vanagas, V.V., Levinson, I.B.: Mathematical apparatus of the theory of angular momentum. Israel program for scientific translations (1962)

Knapsack: Connectedness, Path, and Shortest-Path

Palash Dey[✉], Sudeshna Kolay, and Sipra Singh[✉]

Indian Institute of Technology Kharagpur, Kharagpur, India
{palash.dey,skolay}@cse.iitkgp.ac.in, sipra.singh@iitkgp.ac.in

Abstract. We study the KNAPSACK problem with graph-theoretic constraints. That is, there exists a graph structure on the input set of items of KNAPSACK and the solution also needs to satisfy certain graph theoretic properties on top of the KNAPSACK constraints. In particular, we study CONNECTED KNAPSACK where the solution must be a connected subset of items which has maximum value and satisfies the size constraint of the knapsack. We show that this problem is strongly NP-complete even for graphs of maximum degree four and NP-complete even for star graphs. On the other hand, we develop an algorithm running in time $\mathcal{O}\left(2^{\mathcal{O}(\text{tw} \log \text{tw})} \cdot \text{poly}(n) \min\{s^2, d^2\}\right)$ where tw, s, d, n are respectively treewidth of the graph, the size of the knapsack, the target value of the knapsack, and the number of items. We also exhibit a $(1-\varepsilon)$ factor approximation algorithm running in time $\mathcal{O}\left(2^{\mathcal{O}(\text{tw} \log \text{tw})} \cdot \text{poly}(n, 1/\varepsilon)\right)$ for every $\varepsilon > 0$. We show similar results for PATH KNAPSACK and SHORTEST PATH KNAPSACK, where the solution must also induce a path and shortest path, respectively. Our results suggest that CONNECTED KNAPSACK is computationally the hardest, followed by PATH KNAPSACK and then SHORTEST PATH KNAPSACK.

Keywords: Knapsack · Graph Algorithms · Parameterised Complexity · Approximation algorithm

1 Introduction

The KNAPSACK problem is one of the most well-studied problems in computer science [5, 6, 19, 22]. Here, we are given a set of n items with corresponding sizes $w_1, \ldots, w_n$ and values $\alpha_1, \ldots, \alpha_n$, the size (aka budget) s of the bag, and the goal is to compute a subset of items whose total value is as high as possible and total size is at most s.

Often there exist natural graph structures on the set of items in the KNAPSACK problem, and we want the solution to satisfy some graph theoretic constraint also. For example, suppose a company wants to lease a set of mobile towers, each of which comes at a specific cost with a specific number of customers. The company has a budget and wants to serve the maximum number of customers. However, every mobile tower is in the range of some but not all

J. A. Soto and A. Wiese (Eds.): LATIN 2024, LNCS 14579, pp. 162–176, 2024.
https://doi.org/10.1007/978-3-031-55601-2_11

the other towers. The company wants to provide fast connection between every pair of its customers, which is possible only if they are connected via other intermediary towers of the company. Here we have a natural graph structure — the vertices are the towers, and we have an edge between two vertices if they are in the range of one another. In this situation, the company wants to lease a connected subset of towers which has the maximum total number of customers (to maximize its earning) subject to its budget constraint. We call this problem CONNECTED KNAPSACK.

Now, let us consider another application scenario where there is a railway network and a company wants to lease a railway path between station A and station B, maybe via other stations, to operate train services between them. Suppose the cost model puts a price tag on every station, and the links between two stations are complementary if the company owns both the stations. Each station also allows the owner to earn revenues from advertisement, etc. In this scenario, the company would like to lease from the entire railway network a path between stations A and B whose total revenues are as high as possible, subject to a budget constraint. We call this problem PATH KNAPSACK. If the company's primary goal is to provide the fastest connectivity between A and B, then the company wants to lease a shortest path between A and B, whose total revenues are as high as possible, subject to a budget constraint. We study this problem also under the name SHORTEST PATH KNAPSACK. The formal definitions of the above problems are in Sect. 2.

1.1 Contribution

We study the computational complexity of CONNECTED KNAPSACK, PATH KNAPSACK, and SHORTEST PATH KNAPSACK under the lens of parameterized complexity as well as approximation algorithms. We consider the treewidth of the underlying graph and the vertex cover size of the solution as our parameters. We note that both the parameters are well known to be small for many graphs appearing in various real-world applications. We summarize our results in Table 1.

We observe that all our problems admit fixed-parameter pseudo-polynomial time algorithms with respect to the treewidth of the graph. Further, all the problems admit an FPTAS for graphs with small (at most $\mathcal{O}(\log n)$) treewidth. Our results seem to indicate that SHORTEST PATH KNAPSACK is computationally the easiest, followed by PATH KNAPSACK and CONNECTED KNAPSACK.

1.2 Related Work

To the best of our knowledge, Yamada et al. [25] initiated the study of the KNAPSACK problem with a graph structure on the items (which they called conflict graph), where the goal is to compute an independent set which has the maximum total value and satisfies the budget constraint. The paper proposes a set of heuristics and upper bounds based on Lagrangean relaxation.

Table 1. Summary of results. † : tw is the treewidth of the input graph; ⋆ : s and d are respectively target size and target profit; ‡ : vcs is the size of the minimum vertex cover of the subgraph induced by the solution; ⋆⋆ : k is the number of vertices in the solution; $\alpha(\mathcal{V}) = \sum_{v \in \mathcal{V}} \alpha(v)$ where $\alpha(v)$ is the value of the vertex v in $\mathcal{V}$; m is the number of edges in the input graph.

CONNECTED KNAPSACK	Strongly NP-complete even if max degree is 4 [Theorem 1]
	NP-complete even for stars [Theorem 2]
	$2^{\mathcal{O}(\mathrm{tw}\log\mathrm{tw})} \cdot \mathrm{poly}(n) \min\{s^2, d^2\}^{\dagger\star}$ [Theorem 3]
	$2^{\mathcal{O}(\mathrm{tw}\log\mathrm{tw})} \cdot \mathrm{poly}(n, 1/\varepsilon)^{\dagger}$ time, $(1 - \varepsilon)$ approximation [Theorem 4]
	No $\mathcal{O}(f(vcs)).\mathrm{poly}(n, s, d)$ algorithm unless ETH fails‡ [Theorem 5]
PATH KNAPSACK	Strongly NP-complete even if max degree is 3 [Theorem 6]
	NP-complete for graphs with pathwidth 2 [Theorem 7]
	Polynomial-time algorithm for trees [Observation 1]
	$2^{\mathcal{O}(\mathrm{tw}\log\mathrm{tw})} \cdot \mathrm{poly}(\min\{s^2, d^2\})^{\dagger\star}$ [Theorem 8]
	$2^{\mathcal{O}(\mathrm{tw}\log\mathrm{tw})} \cdot \mathrm{poly}(n, 1/\varepsilon)^{\dagger}$ time, $(1 - \varepsilon)$ approximation [Theorem 9]
	$\mathcal{O}\left((2e)^{2vcs}vcs^{\mathcal{O}(\log vcs)}n^{\mathcal{O}(1)}\right)^{\ddagger}$ time algorithm [Corollary 2]
SHORTEST PATH KNAPSACK	NP-complete for graphs with pathwidth 2 [Corollary 3]
	$\mathcal{O}((m + n\log n) \cdot \min\{s^2, (\alpha(\mathcal{V}))^2\})$ time algorithm [Theorem 11]
	Polynomial-time algorithm for trees [Observation 2]
	$(1 - \varepsilon)$ approximation in $\mathrm{poly}(n, 1/\varepsilon)$ time [Theorem 12]

Later, Hifi and Michrafy [16,17] designed a meta-heuristic approach using reactive local search techniques for the problem. Pferschy and Schauer [23] showed that this problem is strongly NP-hard and presented FPTAS for some special graph classes. Bettinelli et al. [2] proposed a dynamic programming based pruning in a branching based algorithm using the upper bounds proposed by Held et al. [15]. Coniglio et al. [7] presented a branch-and-bound type algorithm for the problem. Finally, Luiz et al. [20] proposed a cutting plane based algorithm. Pferschy and Schauer [24], Gurski and Rehs [14], and Goebbels et al. [13] showed approximation algorithms for the KNAPSACK problem with more sophisticated neighborhood constraints. This problem (and also the variant where the solution should be a clique instead of an independent set) also admits an algorithm running in time $n^{\mathcal{O}(k)}$ where k is the thinness of the input graph [3,21].

Ito et al. [18] studied FPTASes for a generalized version of our problems, called the MAXIMUM PARTITION problem but on a specialized graph class, namely the class of series-parallel graphs. Although their work mentions that their results can be extended to bounded treewidth graphs, the authors state that the algorithmic techniques for that is more complex but do not explain the techniques. To the best of our knowledge, there is no follow-up work where the techniques for bounded treewidth graphs are explained.

Bonomo-Braberman and Gonzalez [4] studied a general framework for, which they called "locally checkable problems", which can be used to obtain FPT algorithms for all our problems parameterized by treewidth. However, the running time of these algorithms obtained by their framework is worse than our algorithms parameterized by treewidth.

There are many other generalizations of the basic KNAPSACK problem which have been studied extensively in the literature. We refer to [5,6,19,22] for an overview.

2   Preliminaries

We denote the set $\{0, 1, 2, \ldots\}$ of natural numbers with $\mathbb{N}$. For any integer ℓ, we denote the sets $\{1, \ldots, \ell\}$ and $\{0, 1, \ldots, \ell\}$ by $[\ell]$ and $[\ell]_0$ respectively. Given a graph $G = (V, E)$ the distance between two vertices $u, v \in V$ is denoted by $\mathrm{dist}_G(u, v)$. We now formally define our problems. We frame all our problems as decision problems so that we can directly use the framework of NP-completeness.

Definition 1 (CONNECTED KNAPSACK). *Given an undirected graph $\mathcal{G} = (\mathcal{V}, \mathcal{E})$, non-negative weights of vertices $(w(u))_{u \in \mathcal{V}}$, non-negative values of vertices $(\alpha(u))_{u \in \mathcal{V}}$, size s of the knapsack, and target value d, compute if there exists a subset $\mathcal{U} \subseteq \mathcal{V}$ of vertices such that: (i) $\mathcal{U}$ is connected, (ii) $w(\mathcal{U}) = \sum_{u \in \mathcal{U}} w(u) \leqslant s$, and (iii) $\alpha(\mathcal{U}) = \sum_{u \in \mathcal{U}} \alpha(u) \geqslant d$.*
We denote an arbitrary instance of CONNECTED KNAPSACK by $(\mathcal{G}, (w(u))_{u \in \mathcal{V}}, (\alpha(u))_{u \in \mathcal{V}}, s, d)$.

If not mentioned otherwise, for any subset $\mathcal{V}' \subseteq \mathcal{V}$, $\alpha(\mathcal{V}')$ and $w(\mathcal{V}')$, we denote $\sum_{u \in \mathcal{V}'} \alpha(u)$ and $\sum_{u \in \mathcal{V}'} w(u)$, respectively.

Definition 2 (PATH KNAPSACK). *Given an undirected graph $\mathcal{G} = (\mathcal{V}, \mathcal{E})$, non-negative weights of vertices $(w(u))_{u \in \mathcal{V}}$, non-negative values of vertices $(\alpha(u))_{u \in \mathcal{V}}$, size s of the knapsack, target value d, two vertices x and y, compute if there exists a subset $\mathcal{U} \subseteq \mathcal{V}$ of vertices such that: (i) $\mathcal{U}$ is a path between x and y in $\mathcal{G}$, (ii) $w(\mathcal{U}) = \sum_{u \in \mathcal{U}} w(u) \leqslant s$, and $\alpha(\mathcal{U}) = \sum_{u \in \mathcal{U}} \alpha(u) \geqslant d$. We denote an arbitrary instance of PATH KNAPSACK by $(\mathcal{G}, (w(u))_{u \in \mathcal{V}}, (\alpha(u))_{u \in \mathcal{V}}, s, d, x, y)$.*

In SHORTEST PATH KNAPSACK, the knapsack subset of items must be a shortest path between two given vertices. Thus, we have a function $(c(e))_{e \in \mathcal{E}}$ describing the positive weights of the edges of the graph as part of the input, and everything else remains the same as PATH KNAPSACK, resulting in an arbitrary instance being denoted as $(\mathcal{G}, (c(e))_{e \in \mathcal{E}}, (w(u))_{u \in \mathcal{V}}, (\alpha(u))_{u \in \mathcal{V}}, s, d, x, y)$.

If not mentioned otherwise, we use $n, s, d, \mathcal{G}$ and $\mathcal{V}$ to denote respectively the number of vertices, the size of the knapsack, the target value, the input graph, and the set of vertices in the input graph.

Definition 3 (Treewidth). *Let $G = (V_G, E_G)$ be a graph. A tree-decomposition of a graph G is a pair $(\mathbb{T} = (V_{\mathbb{T}}, E_{\mathbb{T}}), \mathcal{X} = \{X_t\}_{t \in V_{\mathbb{T}}})$, where $\mathbb{T}$ is a tree where every node $t \in V_{\mathbb{T}}$ is assigned a subset $X_t \subseteq V_G$, called a bag, such that the following conditions hold.*

▷ $\bigcup_{t \in V_{\mathbb{T}}} X_t = V_G$,
▷ for every edge $\{x, y\} \in E_G$ there is a $t \in V_{\mathbb{T}}$ such that $x, y \in X_t$, and
▷ for any $v \in V_G$ the subgraph of $\mathbb{T}$ induced by the set $\{t \mid v \in X_t\}$ is connected.

The width of a tree decomposition is $\max_{t \in V_{\mathbb{T}}} |X_t| - 1$. The treewidth of G is the minimum width over all tree decompositions of G and is denoted by $tw(G)$.

A tree decomposition $(\mathbb{T}, \mathcal{X})$ is called a nice edge tree decomposition if $\mathbb{T}$ is a tree rooted at some node r where $X_r = \emptyset$, each node of $\mathbb{T}$ has at most two children, and each node is of one of the following kinds:

▷ **Introduce node**: a node t that has only one child t' where $X_t \supset X_{t'}$ and $|X_t| = |X_{t'}| + 1$.
▷ **Introduce edge node** a node t labeled with an edge between u and v, with only one child t' such that $\{u, v\} \subseteq X_{t'} = X_t$. This bag is said to introduce uv.
▷ **Forget vertex node**: a node t that has only one child t' where $X_t \subset X_{t'}$ and $|X_t| = |X_{t'}| - 1$.
▷ **Join node**: a node t with two children t_1 and t_2 such that $X_t = X_{t_1} = X_{t_2}$.
▷ **Leaf node**: a node t that is a leaf of $\mathbb{T}$, and $X_t = \emptyset$.

We additionally require that every edge is introduced exactly once. One can show that a tree decomposition of width t can be transformed into a nice tree decomposition of the same width t and with $\mathcal{O}(t|V_G|)$ nodes, see e.g. [1, 9]. For a node $t \in \mathbb{T}$, let $\mathbb{T}_t$ be the subtree of $\mathbb{T}$ rooted at t, and $V(\mathbb{T}_t)$ denote the vertex set in that subtree. Then $\beta(t)$ is the subgraph of G where the vertex set is $\bigcup_{t' \in V(\mathbb{T}_t)} X_{t'}$ and the edge set is the union of the set of edges introduced in each $t', t' \in V(\mathbb{T}_t)$. We denote by $V(\beta(t))$ the set of vertices in that subgraph, and by $E(\beta(t))$ the set of edges of the subgraph.

In this paper, we sometimes fix a vertex $v \in V_G$ and include it in every bag of a nice edge tree decomposition $(\mathbb{T}, \mathcal{X})$ of G, with the effect of the root bag and each leaf bag containing v. For the sake of brevity, we also call such a modified tree decomposition a nice tree decomposition. Given the tree $\mathbb{T}$ rooted at the node r, for any nodes $t_1, t_2 \in V_{\mathbb{T}}$, the distance between the two nodes in $\mathbb{T}$ is denoted by $dist_{\mathbb{T}}(t_1, t_2)$.

We present our results in the next section. In the interest of space, we omit proofs of some of our results; they are marked by $\star$. They are available in the full version of the paper is [10].

3   CONNECTED KNAPSACK

We present our results for CONNECTED KNAPSACK in this section. First, we show that CONNECTED KNAPSACK is strongly NP-complete by reducing it from VERTEX COVER, which is known to be NP-complete even for 3-regular graphs [11, folklore]. Hence, we do not expect a pseudo-polynomial time algorithm for CONNECTED KNAPSACK, unlike KNAPSACK.

Theorem 1. CONNECTED KNAPSACK *is strongly* NP-*complete even when the maximum degree of the input graph is four.*

Proof: Clearly, CONNECTED KNAPSACK $\in$ NP. We reduce VERTEX COVER to CONNECTED KNAPSACK to prove NP-hardness. Let $(\mathcal{G} = (\mathcal{V} = \{v_i : i \in [n]\}, \mathcal{E}), k)$ be an arbitrary instance of VERTEX COVER where $\mathcal{G}$ is 3-regular. We construct the following instance $(\mathcal{G}' = (\mathcal{V}', \mathcal{E}'), (w(u))_{u \in \mathcal{V}'}, (\alpha(u))_{u \in \mathcal{V}'}, s, d)$ of CONNECTED KNAPSACK.

$$\mathcal{V}' = \{u_i, g_i : i \in [n]\} \cup \{h_e : e \in \mathcal{E}\}$$
$$\mathcal{E}' = \{\{u_i, h_e\} : i \in [n], e \in \mathcal{E}, e \text{ is incident on } v_i \text{ in } \mathcal{G}\}$$
$$\cup \{\{u_i, g_i\} : i \in [n]\} \cup \{\{g_i, g_{i+1}\} : i \in [n-1]\}$$
$$w(u_i) = 1, \alpha(u_i) = 0, w(g_i) = 0, \alpha(g_i) = 0, \forall i \in [n]$$
$$w(h_e) = 0, \alpha(h_e) = 1, \forall e \in \mathcal{E}, s = k, d = |\mathcal{E}|$$

We observe that the maximum degree of $\mathcal{G}'$ is at most four — (i) the degree of u_i is four for every $i \in [n]$, since $\mathcal{G}$ is 3-regular and u_i has an edge to g_i, (ii) the degree of h_e is two for every $e \in \mathcal{E}$, and (iii) the degree of g_i is at most three for every $i \in [n]$ since the set $\{g_i : i \in [n]\}$ induces a path. We claim that the two instances are equivalent.

In one direction, let us suppose that the VERTEX COVER instance is a YES instance. Let $\mathcal{W} \subseteq \mathcal{V}$ be a vertex cover of $\mathcal{G}$ with $|\mathcal{W}| \leqslant k$. We consider $\mathcal{U} = \{u_i : i \in [n], v_i \in \mathcal{W}\} \cup \{g_i : i \in [n]\} \cup \{h_e : e \in \mathcal{E}\} \subseteq \mathcal{V}'$. We claim that $\mathcal{G}'[\mathcal{U}]$ is connected. Since $\{g_i : i \in [n]\}$ induces a path and there is an edge between u_i and g_i for every $i \in [n]$, the induced subgraph $\mathcal{G}'[\{u_i : i \in [n], v_i \in \mathcal{W}\} \cup \{g_i : i \in [n]\}]$ is connected. Since $\mathcal{W}$ is a vertex cover of $\mathcal{G}$, every edge $e \in \mathcal{E}$ is incident on at least one vertex in $\mathcal{W}$. Hence, every vertex $h_e, e \in \mathcal{E}$, has an edge with at least one vertex in $\{u_i : i \in [n], v_i \in \mathcal{W}\}$ in the graph $\mathcal{G}'$. Hence, the induced subgraph $\mathcal{G}'[\mathcal{U}]$ is connected. Now we have $w(\mathcal{U}) = \sum_{i=1}^n w(u_i)\mathbb{1}(u_i \in \mathcal{U}) + \sum_{i=1}^n w(g_i) + \sum_{e \in \mathcal{E}} w(h_e) = |\mathcal{U}| \leqslant k$. We also have $\alpha(\mathcal{U}) = \sum_{i=1}^n \alpha(u_i)\mathbb{1}(u_i \in \mathcal{U}) + \sum_{i=1}^n \alpha(g_i) + \sum_{e \in \mathcal{E}} \alpha(h_e) = |\mathcal{E}|$. Hence, the CONNECTED KNAPSACK instance is also a YES instance.

In the other direction, let us assume that the CONNECTED KNAPSACK instance is a YES instance. Let $\mathcal{U} \subseteq \mathcal{V}'$ be a solution of the CONNECTED KNAPSACK instance. We consider $\mathcal{W} = \{v_i : i \in [n] : u_i \in \mathcal{U}\}$. Since $s = k$, we have $|\mathcal{W}| \leqslant k$. Also, since $d = |\mathcal{E}|$, we have $\{h_e : e \in \mathcal{E}\} \subseteq \mathcal{U}$. We claim that $\mathcal{W}$ is a vertex cover of $\mathcal{G}$. Suppose not, then there exists an edge $e \in \mathcal{E}$ which is not covered by $\mathcal{W}$. Then none of the neighbors of h_e belongs to $\mathcal{U}$ contradicting our assumption that $\mathcal{U}$ is a solution and thus $\mathcal{G}'[\mathcal{U}]$ should be connected. Hence, $\mathcal{W}$ is a vertex cover of $\mathcal{G}$ and thus the VERTEX COVER instance is a YES instance.

We observe that all the numbers in our reduced CONNECTED KNAPSACK instance are at most the number of edges of the graph. Hence, our reduction shows that CONNECTED KNAPSACK is strongly NP-complete. $\square$

Theorem 1 also implies the following corollary in the framework of parameterized complexity.

Corollary 1. CONNECTED KNAPSACK *is para-NP-hard parameterized by the maximum degree of the input graph.*

We next show that CONNECTED KNAPSACK is NP-complete even for trees. For that, we reduce from the NP-complete problem KNAPSACK.

Theorem 2 ($\star$). CONNECTED KNAPSACK *is* NP-*complete even for star graphs.*

We complement the hardness result in Theorem 2 by designing a pseudo-polynomial-time algorithm for CONNECTED KNAPSACK for trees. In fact, we have designed an algorithm with running time $2^{\mathcal{O}(\text{tw} \log \text{tw})} \cdot n^{\mathcal{O}(1)} \cdot \min\{s^2, d^2\}$ where the treewidth of the input graph is tw. We present this algorithm next.

3.1   Treewidth as a Parameter

Theorem 3. *There is an algorithm for* CONNECTED KNAPSACK *with running time* $2^{\mathcal{O}(\text{tw} \log \text{tw})} \cdot n \cdot \min\{s^2, d^2\}$ *where n is the number of vertices in the input graph, tw is the treewidth of the input graph, s is the input size of the knapsack and d is the input target value.*

Proof sketch: In the interest of space, we present the main ideas of our algorithm. We refer to [10] for the detailed algorithm with proof of correctness and the analysis of its running time. Let $(G = (V_G, E_G), (w(u))_{u \in V_G}, (\alpha(u))_{u \in V_G}, s, d)$ be an input instance of CONNECTED KNAPSACK such that tw $= tw(G)$. Let $\mathcal{U} \subseteq V_G$ be a solution subset for the input instance. For technical purposes, we guess a vertex $v \in \mathcal{U}$ — once the guess is fixed, we are only interested in finding solution subsets $\mathcal{U}'$ that contain v and $\mathcal{U}$ is one such candidate. We also consider a nice edge tree decomposition $(\mathbb{T} = (V_{\mathbb{T}}, E_{\mathbb{T}}), \mathcal{X})$ of G that is rooted at a node r, and where v has been added to all bags of the decomposition. Therefore, $X_r = \{v\}$ and each leaf bag is the singleton set $\{v\}$.

We define a function $\ell : V_{\mathbb{T}} \to \mathbb{N}$. For a vertex $t \in V_{\mathbb{T}}$, $\ell(t) = \text{dist}_{\mathbb{T}}(t, r)$, where r is the root. Note that this implies that $\ell(r) = 0$. Let us assume that the values that ℓ takes over the nodes of $\mathbb{T}$ are between 0 and L. Now, we describe a dynamic programming algorithm over $(\mathbb{T}, \mathcal{X})$ for CONNECTED KNAPSACK.

States. We maintain a DP table D where a state has the following components:

1. t represents a node in $V_{\mathbb{T}}$.
2. $\mathbb{P} = (P_0, \dots, P_m), m \leqslant \text{tw} + 1$ represents a partition of the vertex subset X_t.

Interpretation of States. For a state $[t, \mathbb{P}]$, if there is a solution subset $\mathcal{U}$ let $\mathcal{U}' = \mathcal{U} \cap V(\beta(t))$. Let $\beta_{\mathcal{U}'}$ be the graph induced on $\mathcal{U}'$ in $\beta(t)$. Let $C_1, C_2, \dots, C_m$ be the connected components of $\beta_{\mathcal{U}'}$. Note that $m \leqslant \text{tw}+1$. Then in the partition $\mathbb{P} = (P_0, P_1, \dots, P_m)$, $P_i = C_i \cap X_t, 1 \leqslant i \leqslant m$. Also, $P_0 = X_t \setminus \mathcal{U}'$.

Given a node $t \in V_{\mathbb{T}}$, a subgraph H of $\beta(t)$ is said to be a $\mathbb{P}$-subgraph if (i) the connected components $C_1, C_2, \dots, C_m$ of H, $m \leqslant \text{tw} + 1$ are such that $P_i = C_i \cap X_t, 1 \leqslant i \leqslant m$, (ii) $P_0 = X_t \setminus H$. For each state $[t, \mathbb{P}]$, a pair (w, α)

with $w \leqslant s$ is said to be feasible if there is a $\mathbb{P}$-subgraph of $\beta(t)$ whose total weight is w and total value is α. Moreover, a feasible pair (w, α) is said to be undominated if there is no other $\mathbb{P}$-subgraph with weight w' and value α' such that $w' \leqslant w$ and $\alpha' \geqslant \alpha$. Please note that by default, an empty $\mathbb{P}$-subgraph has total weight 0 and total value 0.

For each state $[t, \mathbb{P}]$, we initialize $D[t, \mathbb{P}]$ to the list $\{(0, 0)\}$. Our computation shall be such that in the end each $D[t, \mathbb{P}]$ stores the set of all undominated feasible pairs (w, α) for the state $[t, \mathbb{P}]$.

Dynamic Programming on D. We describe the following procedure to update the table D. We start updating the table for states with nodes $t \in V_{\mathbb{T}}$ such that $\ell(t) = L$. When all such states are updated, then we move to update states where the node t has $\ell(t) = L - 1$, and so on till we finally update states with r as the node — note that $\ell(r) = 0$. For a particular $i, 0 \leqslant i < L$ and a state $[t, \mathbb{P}]$ such that $\ell(t) = i$, we can assume that $D[t', \mathbb{P}']$ have been evaluated for all t' such that $\ell(t') > i$ and all partitions $\mathbb{P}'$ of $X_{t'}$. Now we consider several cases by which $D[t, \mathbb{P}]$ is updated based on the nature of t in $\mathbb{T}$:

1. Suppose t is a leaf node. Note that by our modification, $X_t = \{v\}$. There can be only 2 partitions for this singleton set — $\mathbb{P}_t^1 = (P_0 = \emptyset, P_1 = \{v\})$ and $\mathbb{P}_t^2 = (P_0 = \{v\}, P_1 = \emptyset)$. If $\mathbb{P} = \mathbb{P}_t^1$ then $D[t, \mathbb{P}]$ stores the pair $(w(v), \alpha(v))$ if $w(v) \leqslant s$ and otherwise no modification is made. If $\mathbb{P} = \mathbb{P}_t^2$ then $D[t, \mathbb{P}]$ is not modified.

2. Suppose t is a forget vertex node. Then it has an only child t' where $X_t \subset X_{t'}$ and there is exactly one vertex $u \neq v$ that belongs to $X_{t'}$ but not to X_t. Let $\mathbb{P}' = (P_0', P_1', \ldots, P_{m'}')$ be a partition of $X_{t'}$ such that when restricted to X_t we obtain the partition $\mathbb{P} = (P_0, P_1, \ldots, P_m)$. For each such partition, we shall do the following.

   Suppose $\mathbb{P}'$ has $u \in P_0'$, then each feasible undominated pair stored in $D[t', \mathbb{P}']$ is copied to $D[t, \mathbb{P}]$.

   Alternatively, suppose $\mathbb{P}'$ has $u \in P_i', i > 0$ and $|P_i'| > 1$. Then, each feasible undominated pair stored in $D[t', \mathbb{P}']$ is copied to $D[t, \mathbb{P}]$.

   Finally, suppose $\mathbb{P}'$ has $u \in P_i', i > 0$ and $P_i' = \{u\}$. Then we do not make any changes to $D[t, \mathbb{P}]$.

When t is an introduce node or an introduce edge node or a join node, then also we can analogously update the table entries.

Finally, in the last step of updating $D[t, \mathbb{P}]$, we go through the list saved in $D[t, \mathbb{P}]$ and only keep undominated pairs. The output of the algorithm is a pair (w, α) stored in $D[r, \mathbb{P} = (P_0 = \emptyset, P_1 = \{v\})]$ such that $w \leqslant s$ and α is the maximum value over all pairs in $D[r, \mathbb{P}]$.   □

3.2   A Fixed Parameter Fully Pseudo-polynomial Time Approximation Scheme

We now use the algorithm in Theorem 3 as a black-box to design an $(1 - \varepsilon)$-factor approximation algorithm for optimizing the value of the solution and running in time $2^{\mathcal{O}(\text{tw} \log \text{tw})} \cdot \text{poly}(n, 1/\varepsilon)$.

Theorem 4. *There is an $(1 - \varepsilon)$-factor approximation algorithm for* CON-NECTED KNAPSACK *for optimizing the value of the solution and running in time* $2^{\mathcal{O}(tw \log tw)} \cdot poly(n, 1/\varepsilon)$ *where* tw *is the treewidth of the input graph.*

Proof: Let $\mathcal{I} = (\mathcal{G} = (\mathcal{V}, \mathcal{E}), (w(u))_{u \in \mathcal{V}}, (\alpha(u))_{u \in \mathcal{V}}, s)$ be an arbitrary instance of CONNECTED KNAPSACK where the goal is to output a connected subgraph $\mathcal{U}$ of maximum $\alpha(\mathcal{U})$ subject to the constraint that $w(\mathcal{U}) \leqslant s$. Without loss of generality, we can assume that $w(u) \leqslant s$ for every $u \in \mathcal{V}$. If not, then we can remove every $u \in \mathcal{V}$ whose $w(u) > s$; this does not affect any solution since any vertex deleted can never be part of any solution. Let $\alpha_{\max} = \max\{\alpha(u) : u \in \mathcal{V}\}$. We construct another instance $\mathcal{I}' = \left(\mathcal{G} = (\mathcal{V}, \mathcal{E}), (w(u))_{u \in \mathcal{V}}, (\alpha'(u) = \left\lfloor \frac{n\alpha(u)}{\varepsilon\alpha_{\max}} \right\rfloor)_{u \in \mathcal{V}}, s\right)$ of CONNECTED KNAPSACK. We compute the optimal solution $\mathcal{W}' \subseteq \mathcal{V}$ of $\mathcal{I}'$ using the algorithm in Theorem 3 and output $\mathcal{W}'$. Let $\mathcal{W} \subseteq \mathcal{V}$ be an optimal solution of $\mathcal{I}$. Clearly $\mathcal{W}'$ is a valid (may not be optimal) solution of $\mathcal{I}$ also, since $w(\mathcal{W}') \leqslant s$ by the correctness of the algorithm in Theorem 3. We now prove the approximation factor of our algorithm.

$$\sum_{u \in \mathcal{W}'} \alpha(u) \geqslant \frac{\varepsilon\alpha_{\max}}{n} \sum_{u \in \mathcal{W}'} \left\lfloor \frac{n\alpha(u)}{\varepsilon\alpha_{\max}} \right\rfloor$$

$$\geqslant \frac{\varepsilon\alpha_{\max}}{n} \sum_{u \in \mathcal{W}} \left\lfloor \frac{n\alpha(u)}{\varepsilon\alpha_{\max}} \right\rfloor \qquad \text{[since } \mathcal{W}' \text{ is an optimal solution of } \mathcal{I}']$$

$$\geqslant \frac{\varepsilon\alpha_{\max}}{n} \sum_{u \in \mathcal{W}} \left(\frac{n\alpha(u)}{\varepsilon\alpha_{\max}} - 1 \right)$$

$$\geqslant \left(\sum_{u \in \mathcal{W}} \alpha(u) \right) - \varepsilon\alpha_{\max}$$

$$\geqslant \mathsf{OPT}(\mathcal{I}) - \varepsilon\mathsf{OPT}(\mathcal{I}) \qquad\qquad [\alpha_{\max} \leqslant \mathsf{OPT}(\mathcal{I})]$$

$$= (1 - \varepsilon)\mathsf{OPT}(\mathcal{I})$$

Hence, the approximation factor of our algorithm is $(1 - \varepsilon)$. We now analyze the running time of our algorithm.

The value of any optimal solution of $\mathcal{I}'$ is at most

$$\sum_{u \in \mathcal{V}} \alpha'(u) \leqslant \frac{n}{\varepsilon\alpha_{\max}} \sum_{u \in \mathcal{V}} \alpha(u) \leqslant \frac{n}{\varepsilon\alpha_{\max}} \sum_{u \in \mathcal{V}} \alpha_{\max} = \frac{n^2}{\varepsilon}.$$

Hence, the running time of our algorithm is $2^{tw \log tw} \cdot poly(n, 1/\varepsilon)$.   □

Other Parameters. We next consider *vcs*, the maximum size of a minimum vertex cover of the subgraph induced by any solution of CONNECTED KNAPSACK, as our parameter. That is, $vcs(\mathcal{I} = (\mathcal{G}, (w(u))_{u \in \mathcal{V}}, (\alpha(u))_{u \in \mathcal{V}}, s, d)) = \max\{\text{size of minimum vertex cover of } W : W \subseteq \mathcal{V} \text{ is a solution of } \mathcal{I}\}$. We already know from Theorem 2 that CONNECTED KNAPSACK is NP-complete for star graphs. We note that vcs is one for star graphs. Hence, CONNECTED KNAPSACK is para-NP-hard with respect to vcs, that is, there is no algorithm for CONNECTED KNAPSACK which runs in polynomial time even for constant values of vsc. However, whether there exists any algorithm with running time $\mathcal{O}(f(vcs).poly(n, s, d))$, remains a valid question. We answer this question negatively in Theorem 5. For that, we exhibit an FPT-reduction from PARTIAL VERTEX COVER which is known to be W[1]-hard parameterized by the size of the partial vertex cover we are looking for.

Theorem 5 ($\star$). *There is no algorithm for CONNECTED KNAPSACK running in time $\mathcal{O}(f(vcs).poly(n, s, d))$ unless ETH fails.*

4   PATH KNAPSACK

We now present the results of PATH KNAPSACK. We first show that PATH KNAPSACK is strongly NP-complete by reducing from HAMILTONIAN PATH which is defined as follows.

Definition 4 (HAMILTONIAN PATH). *Given a graph $\mathcal{G}(\mathcal{V}, \mathcal{E})$ and two vertices x and y, compute if there exists a path between x and y which visits every other vertex in $\mathcal{G}$. We denote an arbitrary instance of HAMILTONIAN PATH by $(\mathcal{G}, x, y)$.*

HAMILTONIAN PATH is known to be NP-complete even for graphs with maximum degree three [12].

Theorem 6 ($\star$). PATH KNAPSACK *is strongly NP-complete even for graphs with maximum degree three.*

However, PATH KNAPSACK is clearly polynomial-time solvable for trees, since there exists only one path between every two vertices in any tree.

Observation 1. PATH KNAPSACK *is polynomial-time solvable for trees.*

One immediate natural question is if Observation 1 can be generalized to graphs of bounded treewidth. The following result refutes the existence of any such algorithm.

Theorem 7 ($\star$). PATH KNAPSACK *is NP-complete even for graphs of pathwidth at most two. In particular, PATH KNAPSACK is para-NP-hard parameterized by pathwidth.*

Theorem 7 leaves the following question open: does there exist an algorithm for PATH KNAPSACK which runs in time $\mathcal{O}(f(tw) \cdot poly(n, s, d))$? We answer this question affirmatively in the following result. We omit its proof from this shorter version, since the dynamic programming algorithm is very similar to Theorem 3.

Theorem 8. *There is an algorithm for* PATH KNAPSACK *with running time* $2^{\mathcal{O}(tw \log tw)} \cdot n \cdot \min\{s^2, d^2\}$ *where tw is the treewidth of the input graph, s is the input size of the knapsack and d is the input target value.*

Using the technique in Theorem 4, we use Theorem 8 as a black-box to obtain the following approximation algorithm. We again omit its proof due to its similarity with Theorem 4.

Theorem 9. *There is an* $(1 - \varepsilon)$ *factor approximation algorithm for* PATH KNAPSACK *for optimizing the value of the solution running in time* $2^{\mathcal{O}(tw \log tw)} \cdot poly(n, 1/\varepsilon)$ *where tw is the treewidth of the input graph.*

We next consider the size of the minimum vertex cover vcs of the subgraph induced by a solution $\mathcal{W} \subseteq \mathcal{V}[\mathcal{G}]$. We observe that the size of the minimum vertex cover of $\mathcal{G}[\mathcal{W}]$ is at least half of $|\mathcal{W}|$ since there exists a Hamiltonian path in $\mathcal{G}[\mathcal{W}]$. Hence, it is enough to design an FPT algorithm with parameter $|\mathcal{W}|$. Our algorithm is based on the color coding technique [9].

Theorem 10 (⋆). *There is an algorithm for* PATH KNAPSACK *running in time* $\mathcal{O}\left((2e)^k k^{\mathcal{O}(\log k)} n^{\mathcal{O}(1)}\right)$ *where k is the number of vertices in the solution.*

Corollary 2. *There is an algorithm for* PATH KNAPSACK *running in time* $\mathcal{O}\left((2e)^{2vcs} vcs^{\mathcal{O}(\log vcs)} n^{\mathcal{O}(1)}\right)$ *where vcs is the size of the minimum vertex cover of the subgraph induced by the solution.*

5 SHORTEST PATH KNAPSACK

We now consider SHORTEST PATH KNAPSACK. We observe that all the u to v paths in the reduced instance of PATH KNAPSACK in the proof of Theorem 7 are of the same length. Hence, we immediately obtain the following result as a corollary of Theorem 7.

Corollary 3. SHORTEST PATH KNAPSACK *is* NP-*complete even for graphs of pathwidth at most two and the weight of every edge is one. In particular,* SHORTEST PATH KNAPSACK *is para-*NP-*hard parameterized by pathwidth.*

Interestingly, SHORTEST PATH KNAPSACK admits a pseudo-polynomial-time algorithm for any graph unlike CONNECTED KNAPSACK and PATH KNAPSACK.

Theorem 11. *There is an algorithm for* SHORTEST PATH KNAPSACK *running in time* $\mathcal{O}((m + n \log n) \cdot \min\{s^2, (\alpha(\mathcal{V}))^2\})$, *where m is the number of edges in the input graph.*

Proof sketch: Let $(\mathcal{G} = (\mathcal{V}, \mathcal{E}, (c(e))_{e \in \mathcal{E}}), (w(u))_{u \in \mathcal{V}}, (\alpha(u))_{u \in \mathcal{V}}, s, d, x, y)$ be an arbitrary instance of SHORTEST PATH KNAPSACK. We design a greedy and dynamic-programming based algorithm. For every vertex $v \in \mathcal{V}$, we store a boolean marker b_v, the distance δ_v of v from x, and a set $D_v = \{(w, \alpha) : \exists$ an x to v shortest-path $\mathcal{P}$ such that $w(\mathcal{P}) = w, \alpha(\mathcal{P}) = \alpha$, and for every other x to v shortest-path $\mathcal{Q}$, we have either $w(\mathcal{Q}) > w$ or $\alpha(\mathcal{Q}) < \alpha$ (or both)$\}$. That is, we store undominated weight-value pairs of all shortest x to v paths in D_v. We initialize $b_x = $ FALSE, $\delta_x = 0, D_x = \{(w(x), \alpha(x))\}, b_u = $ FALSE, $\delta_u = \infty$, and $D_u = \emptyset$ for every $u \in \mathcal{V} \setminus \{x\}$.

Updating DP table: We pick a vertex $z = \arg\min_{v \in \mathcal{V} : b_v = \text{FALSE}} \delta_v$. We set $b_z = $ TRUE. For every neighbor u of z, if $\delta_u > \delta_z + c(\{z, u\})$, then we reset $D_u = \emptyset$ and set $\delta_u = \delta_z + c(\{z, u\})$. If $\delta_u = \delta_z + c(\{z, u\})$, then update D_u as follows: for every $(w, \alpha) \in D_z$, we update D_u to $(D_u \cup \{(w + w(u), \alpha + \alpha(u))\})$ if $w + w(u) \leqslant s$. We remove all dominated pairs from D_u just before finishing each iteration. If we have $b_v = $ TRUE for every vertex, then we output YES if there exists a pair $(w, \alpha) \in D_y$ such that $w \leqslant s$ and $\alpha \geqslant d$. Else, we output NO.

We now argue the correctness of our algorithm. Following the proof of correctness of the classical Dijkstra's shortest path algorithm, we observe that if b_v is TRUE for any vertex $v \in \mathcal{V}$, its distance from x is δ_v [8]. We claim that at the end of updating a table entry in every iteration, the following invariant holds: for every vertex $v \in \mathcal{V}$ such that $b_v = $ TRUE, $(k_1, k_2) \in D_v$ if and only if there exists an x to v undominated shortest path $\mathcal{P}$ using only vertices marked TRUE such that $w(\mathcal{P}) = k_1$ and $\alpha(\mathcal{P}) = k_2$.

The invariant clearly holds after the first iteration. Let us assume that the invariant holds after i (>1) iterations; $\mathcal{V}_T$ be the set of vertices which are marked TRUE after i iterations. We have $|\mathcal{V}_T| = i > 1$. Suppose the algorithm picks the vertex z_{i+1} in the $(i + 1)$-th iteration; that is, we have $z_{i+1} = \arg\min_{v \in \mathcal{V} : b_v = \text{FALSE}} \delta_v$ when we start the $(i + 1)$-th iteration. Let $\mathcal{P}^* = x, \ldots, z, z_{i+1}$ be an undominated x to z_{i+1} shortest path using the vertices marked TRUE only. Then we need to show that $(w(\mathcal{P}^*), \alpha(\mathcal{P}^*)) \in D_{z_{i+1}}$. We claim that the prefix of the path $\mathcal{P}^*$ from x to z, let us call it $\mathcal{Q} = x, \ldots, z$, is an undominated x to z shortest path using the vertices marked TRUE only. It follows from the standard proof of correctness of Dijkstra's algorithm [8] that $\mathcal{Q}$ is a shortest path from x to z. Now, to show that $\mathcal{Q}$ is undominated, let us assume that another shortest path $\mathcal{R}$ from x to z dominates $\mathcal{Q}$. Then the shortest path $\mathcal{R}'$ from x to z_{i+1} which is $\mathcal{R}$ followed by z_{i+1} also dominates $\mathcal{P}^*$ contradicting our assumption that $\mathcal{P}^*$ is an undominated shortest path from x to z_{i+1}. We now observe that the iteration j when the vertex z was marked TRUE must be less than $(i+1)$ since z is already marked TRUE in the $(i+1)$-th iteration. Now, applying induction hypothesis after the j-th iteration, we have $(w(\mathcal{Q}), \alpha(\mathcal{Q})) \in D_z$ and $(w(\mathcal{P}^*), \alpha(\mathcal{P}^*)) \in D_{z_{i+1}}$. Also, at the end of the j-th iteration, the δ_z and $\delta_{z_{i+1}}$ values are set to the distances of z and z_{i+1} from x respectively. Thus, the D_z and $D_{z_{i+1}}$ are never reset to $\emptyset$ after j-th iteration. Also, we never remove any undominated pairs from DP tables. Since $\mathcal{P}^*$ is an undominated x to z_{i+1} path, we always have $(w(\mathcal{P}^*), \alpha(\mathcal{P}^*)) \in D_{z_{i+1}}$ from the end of j-th iteration and thus, in particular, after $(i + 1)$-th iteration. Hence, invariant (i) holds for z_{i+1} after $(i + 1)$-th iteration. Now consider any vertex z' other than z_{i+1} which is marked TRUE before $(i + 1)$-th iteration. Let $\mathcal{P}_1 = x, \ldots, z'$ be an undominated x to z' shortest path using the vertices marked TRUE only. If $\mathcal{P}_1$ does not pass through the vertex z_{i+1}, then we have $(w(\mathcal{P}_1), \alpha(\mathcal{P}_1)) \in D_{z_{z'}}$ by induction hypothesis after i-iterations. If $\mathcal{P}_1$ passes through z_{i+1}, we have $\delta_{z_{i+1}} < \delta_{z'}$ since all the edge weights are positive. However, this contradicts our assumption that z' had already been picked by the algorithm and marked TRUE before z_{i+1} was picked. Hence, $\mathcal{P}_1$ cannot use z_{i+1} as an intermediate vertex. Hence, after $(i + 1)$ iterations, for every vertex $v \in V$ such that $b_v =$ TRUE, for every x to v shortest path $\mathcal{P}$ using only vertices marked TRUE, we have $(w(\mathcal{P}), \alpha(\mathcal{P})) \in D_v$.

In the interest of space, we omit the other direction of the invariant and the analysis of its running time. They are available in [10]. $\qquad\square$

Clearly, SHORTEST PATH KNAPSACK admits a polynomial-time algorithm for trees since only one path exists between every two vertices.

Observation 2. SHORTEST PATH KNAPSACK *is in* P *for trees.*

Using the technique in Theorem 4, we use Theorem 11 as a black-box to obtain the following approximation algorithm. We again omit its proof due to its similarity with Theorem 4.

Theorem 12. *There is a poly$(n, 1/\varepsilon)$ time, $(1 - \varepsilon)$ factor approximation algorithm for* SHORTEST PATH KNAPSACK *for optimizing the value of the solution.*

6   Conclusion

We study the classical KNAPSACK problem with various graph theoretic constraints, namely connectedness, path and shortest path. We show that CONNECTED KNAPSACK and PATH KNAPSACK are strongly NP-complete whereas SHORTEST PATH KNAPSACK admits a pseudo-polynomial time algorithm. All the three problems admit FPTASes for bounded treewidth graphs; only SHORTEST PATH KNAPSACK admits an FPTAS for arbitrary graphs. It would be interesting to explore if meta-theorems can be proven in this general theme of knapsack on graphs.

References

1. A faster parameterized algorithm for pseudoforest deletion. Discrete Applied Mathematics **236**, 42–56 (2018)
2. Bettinelli, A., Cacchiani, V., Malaguti, E.: A branch-and-bound algorithm for the knapsack problem with conflict graph. INFORMS J. Comput. **29**(3), 457–473 (2017). https://doi.org/10.1287/ijoc.2016.0742
3. Bonomo, F., de Estrada, D.: On the thinness and proper thinness of a graph. Discret. Appl. Math. **261**, 78–92 (2019). https://doi.org/10.1016/J.DAM.2018.03.072

4. Bonomo-Braberman, F., Gonzalez, C.L.: A new approach on locally checkable problems. Discret. Appl. Math. **314**, 53–80 (2022). https://doi.org/10.1016/J.DAM.2022.01.019

5. Cacchiani, V., Iori, M., Locatelli, A., Martello, S.: Knapsack problems - an overview of recent advances. part I: single knapsack problems. Comput. Oper. Res. **143**, 105692 (2022). https://doi.org/10.1016/j.cor.2021.105692,

6. Cacchiani, V., Iori, M., Locatelli, A., Martello, S.: Knapsack problems - an overview of recent advances. part II: multiple, multidimensional, and quadratic knapsack problems. Comput. Oper. Res. **143**, 105693 (2022). https://doi.org/10.1016/j.cor.2021.105693

7. Coniglio, S., Furini, F., Segundo, P.S.: A new combinatorial branch-and-bound algorithm for the knapsack problem with conflicts. Eur. J. Oper. Res. **289**(2), 435–455 (2021). https://doi.org/10.1016/j.ejor.2020.07.023

8. Cormen, T.H., Leiserson, C.E., Rivest, R.L., Stein, C.: Introduction to Algorithms, 3rd Edition. MIT Press (2009). http://mitpress.mit.edu/books/introduction-algorithms

9. Cygan, M., et al.: Parameterized Algorithms. Springer, Cham (2015). https://doi.org/10.1007/978-3-319-21275-3

10. Dey, P., Kolay, S., Singh, S.: Knapsack: Connectedness, path, and shortest-path. CoRR abs/2307.12547 (2023)

11. Fleischner, H., Sabidussi, G., Sarvanov, V.I.: Maximum independent sets in 3- and 4-regular hamiltonian graphs. Discret. Math. **310**(20), 2742–2749 (2010). https://doi.org/10.1016/j.disc.2010.05.028

12. Garey, M.R., Johnson, D.S., Stockmeyer, L.J.: Some simplified np-complete problems. In: Constable, R.L., Ritchie, R.W., Carlyle, J.W., Harrison, M.A. (eds.) Proc. 6th Annual ACM Symposium on Theory of Computing, April 30 - May 2, 1974, Seattle, Washington, USA, pp. 47–63. ACM (1974). https://doi.org/10.1145/800119.803884

13. Goebbels, S., Gurski, F., Komander, D.: The knapsack problem with special neighbor constraints. Math. Methods Oper. Res. **95**(1), 1–34 (2022). https://doi.org/10.1007/s00186-021-00767-5

14. Gurski, F., Rehs, C.: Solutions for the knapsack problem with conflict and forcing graphs of bounded clique-width. Math. Methods Oper. Res. **89**(3), 411–432 (2019). https://doi.org/10.1007/s00186-019-00664-y

15. Held, S., Cook, W.J., Sewell, E.C.: Maximum-weight stable sets and safe lower bounds for graph coloring. Math. Program. Comput. **4**(4), 363–381 (2012). https://doi.org/10.1007/s12532-012-0042-3

16. Hifi, M., Michrafy, M.: A reactive local search-based algorithm for the disjunctively constrained knapsack problem. J. Oper. Res. Society **57**(6), 718–726 (2006)

17. Hifi, M., Michrafy, M.: Reduction strategies and exact algorithms for the disjunctively constrained knapsack problem. Comput. Oper. Res. **34**(9), 2657–2673 (2007)

18. Ito, T., Demaine, E.D., Zhou, X., Nishizeki, T.: Approximability of partitioning graphs with supply and demand. J. Discr. Algorithms **6**(4), 627–650 (2008)

19. Kellerer, H., Pferschy, U., Pisinger, D., Kellerer, H., Pferschy, U., Pisinger, D.: Multidimensional knapsack problems. Springer (2004). https://doi.org/10.1007/978-3-540-24777-7_9

20. Luiz, T.A., Santos, H.G., Uchoa, E.: Cover by disjoint cliques cuts for the knapsack problem with conflicting items. Oper. Res. Lett. **49**(6), 844–850 (2021). https://doi.org/10.1016/j.orl.2021.10.001

21. Mannino, C., Oriolo, G., Ricci-Tersenghi, F., Chandran, L.S.: The stable set problem and the thinness of a graph. Oper. Res. Lett. **35**(1), 1–9 (2007). https://doi.org/10.1016/J.ORL.2006.01.009
22. Martello, S., Toth, P.: Knapsack problems: algorithms and computer implementations. John Wiley & Sons, Inc. (1990)
23. Pferschy, U., Schauer, J.: The knapsack problem with conflict graphs. J. Graph Algorithms Appl. **13**(2), 233–249 (2009). https://doi.org/10.7155/jgaa.00186
24. Pferschy, U., Schauer, J.: Approximation of knapsack problems with conflict and forcing graphs. J. Comb. Optim. **33**(4), 1300–1323 (2017). https://doi.org/10.1007/s10878-016-0035-7
25. Yamada, T., Kataoka, S., Watanabe, K.: Heuristic and exact algorithms for the disjunctively constrained knapsack problem. Inform. Process. Society Japan J. **43**(9) (2002)

Parameterized Approximation Algorithms for Weighted Vertex Cover

Soumen Mandal[1($\boxtimes$)], Pranabendu Misra[2], Ashutosh Rai[1],
and Saket Saurabh[3,4]

[1] Department of Mathematics, IIT Delhi, New Delhi, India
{soumen.mandal,ashutosh.rai}@maths.iitd.ac.in
[2] Chennai Mathematical Institute, Chennai, India
pranabendu@cmi.ac.in
[3] Institute of Mathematical Sciences, Chennai, India
saket@imsc.res.in
[4] University of Bergen, Bergen, Norway

Abstract. A *vertex cover* of a graph is a set of vertices of the graph such that every edge has at least one endpoint in it. In this work, we study WEIGHTED VERTEX COVER with solution size as a parameter. Formally, in the (k, W)-VERTEX COVER problem, given a graph G, an integer k, a positive rational W, and a weight function $w : V(G) \to \mathbb{Q}^+$, the question is whether G has a vertex cover of size at most k of weight at most W, with k being the parameter. An (a, b)-bi-criteria approximation algorithm for (k, W)-VERTEX COVER either produces a vertex cover S such that $|S| \leq ak$ and $w(S) \leq bW$, or decides that there is no vertex cover of size at most k of weight at most W. We obtain the following results.

- A simple $(2, 2)$-bi-criteria approximation algorithm for (k, W)-VERTEX COVER in polynomial time by modifying the standard LP-rounding algorithm.
- A simple exact parameterized algorithm for (k, W)-VERTEX COVER running in $\mathcal{O}^*(1.4656^k)$ time (Here, the $\mathcal{O}^*$ notation hides factors polynomial in n.).
- A $(1+\epsilon, 2)$-approximation algorithm for (k, W)-VERTEX COVER running in $\mathcal{O}^*(1.4656^{(1-\epsilon)k})$ time.
- A $(1.5, 1.5)$-approximation algorithm for (k, W)-VERTEX COVER running in $\mathcal{O}^*(1.414^k)$ time.
- A $(2 - \delta, 2 - \delta)$-approximation algorithm for (k, W)-VERTEX COVER running in $\mathcal{O}^*\left(\sum_{i=\frac{\delta k(1-2\delta)}{1+2\delta}}^{\frac{\delta k(1-2\delta)}{2\delta}} \binom{\delta k+i}{\delta k-\frac{2i\delta}{1-2\delta}}\right)$ time for any $\delta < 0.5$. For example, for $(1.75, 1.75)$ and $(1.9, 1.9)$-approximation algorithms, we get running times of $\mathcal{O}^*(1.272^k)$ and $\mathcal{O}^*(1.151^k)$ respectively.

S. Mandal: Supported by Council of Scientific and Industrial Research, India.
Ashutosh Rai: Supported by Science and Engineering Research Board (SERB) Grant SRG/2021/002412.
S. Saurabh: Supported by the European Research Council (ERC) under the European Union's Horizon 2020 research and innovation programme (grant agreement No. 819416), and Swarnajayanti Fellowship (No. DST/SJF/MSA01/2017-18).

J. A. Soto and A. Wiese (Eds.): LATIN 2024, LNCS 14579, pp. 177–192, 2024.
https://doi.org/10.1007/978-3-031-55601-2_12

Our algorithms (expectedly) do not improve upon the running times of the existing algorithms for the unweighted version of VERTEX COVER. When compared to algorithms for the weighted version, our algorithms are the first ones to the best of our knowledge which work with arbitrary weights, and they perform well when the solution size is much smaller than the total weight of the desired solution.

Keywords: Weighted Vertex Cover · Parameterized Algorithms · Approximation Algorithms · Parameterized Approximation

1   Introduction

In the VERTEX COVER problem, we are given a graph G, and are asked to find a smallest set $S \subseteq V(G)$ of vertices such that every edge in the graph has at least one endpoint in S. This problem is known to be NP-complete and is in fact one of the 21 NP-complete problems listed by Karp [10]. This makes it unlikely that a polynomial time algorithm exists for the problem, and hence it has been studied extensively in the fields of *approximation algorithms* and *parameterized algorithms*.

For VERTEX COVER, there are several polynomial time 2-approximation algorithms [15]. On the other hand, it is known that VERTEX COVER is APX-complete, and hence cannot have a Polynomial Time Approximation Scheme (PTAS) unless P = NP [5]. In fact, under the assumption that P $\neq$ NP, there cannot be a $(\sqrt{2} - \epsilon)$-approximation algorithm for VERTEX COVER. Further, assuming the Unique Games Conjecture, there cannot be a $(2-\epsilon)$-approximation algorithm for VERTEX COVER [11]. This result basically shows that we cannot do better than the known approximation algorithms in polynomial time, unless the Unique Games Conjecture is false.

VERTEX COVER is also one of the most well-studied problems in parameterized algorithms, with the size of the vertex cover being the parameter. Formally, given a graph G and an integer k, it asks whether G has a vertex cover of size at most k. A branching algorithm for VERTEX COVER runs in time $\mathcal{O}^*(2^k)$, and there have been several results improving the running time to $\mathcal{O}^*(1.2738^k)$ [4].

There is another emerging field, *parameterized approximation algorithms*, that combines the fields of approximation algorithms and parameterized algorithms to get *better approximations* for problems in FPT time, than what is achievable in polynomial time. Naturally, the running time of a parameterized approximation algorithm, if one exists, is smaller than the running time of the exact FPT algorithm for the same problem. For VERTEX COVER, a parameterized $(1 + \epsilon)$-approximation algorithm is known to run in $\mathcal{O}^*(c^{(1-\epsilon)k})$ time [1,7], where the constant c is the base of the exponent of the best known FPT algorithm for vertex cover, running in time $\mathcal{O}^*(c^k)$. The running time was further improved by Brankovic and Fernau [2] for any $\epsilon \in (0, 0.5]$, and in particular, they gave a 1.5-approximation algorithm running in time $\mathcal{O}^*(1.0883^k)$. Recently, Kulik and Shachnai [14] introduced randomized branching as a tool for parameterized approximation and develop the mathematical machinery for its analysis. Using this they obtained the best known running times of parameterized approximation algorithms for VERTEX COVER and 3-HITTING SET for a wide range of approximation ratios. We refer to the survey [6] for other developments in the field of parameterized approximation algorithms.

While parameterized approximation algorithms for the unweighted version of VERTEX COVER have been extensively studied, the same cannot be said about the weighted version of the problem. In the weighted version of the VERTEX COVER problem, we are asked to find a vertex cover of minimum weight in the graph. In the decision version of the problem, we are given a graph G, weight function $w : V(G) \to \mathbb{Q}^+$, and a positive rational W; and the question is whether G has a vertex cover of weight at most W. We can get a 2-approximation algorithm for weighted vertex cover problem by doing a simple LP-rounding, and the negative results for approximation of the unweighted version carry over to weighted version of the problem as well. From the point of view of parameterized complexity, the weighted version is less studied than the unweighted version, but still a number of results have progressively brought the running time down to $\mathcal{O}^*(1.347^W)$, when parameterized by the weight W [8,16,18]. Shachnai and Zehavi [18] also designed an algorithm, that as an input to WEIGHTED VERTEX COVER, takes an integer parameter k instead of W, and in time $\mathcal{O}^*(1.363^k)$, either returns a vertex cover of weight at most W (with no guarantee on the size) or returns that there is no vertex cover of size at most k, where W is the weight of a minimum weight vertex cover of size at most k. One issue with all these algorithms is that they require all the vertex weights to be at least 1, because it can be seen through a simple reduction that otherwise the problem is not likely to be FPT when parameterized by W. Also, all these algorithms do not give any *guarantee on the number of vertices in the solution*, which can be at most W, but can be much smaller, since all the weights are at least 1.

This gives rise to the natural question of finding parameterized (approximation) algorithms for WEIGHTED VERTEX COVER, where we do not want too many vertices in the solution, with the size of the solution being the parameter. Formally, given a graph G, weight function $w : V(G) \to \mathbb{Q}^+$, integer k and positive rational W, the question is whether G has a vertex cover of size at most k and weight at most W. We call the problem (k, W)-VERTEX COVER, where the parameter is k. In [18], the authors define this problem, and show that it is NP-complete even on bipartite graphs.

Such a parameterization was considered in [3], where the authors gave exact parameterized algorithm for WEIGHTED FEEDBACK VERTEX SET parameterized by the solution size. Later, Saurabh [17] asked a similar question for the weighted version of DIRECTED FEEDBACK VERTEX SET, which was resolved by Kim et al. [12] recently using the technique of flow-augmentation. There is some other recent work being done with similar parameterizations for cut problems [9, 13], but somehow no work has been done on the WEIGHTED VERTEX COVER problem with this parameterization to the best of our knowledge. We fill this gap by studying (k, W)-VERTEX COVER from the perspective of parameterized (approximation) algorithms.

1.1 Our Problem, Results and Methods

We investigate (k, W)-VERTEX COVER from the point of view of approximation algorithms, parameterized algorithms, and parameterized approximation algorithms, with the size of the vertex cover k being the parameter for the latter two. This parameter is arguably better than the parameter W, since parameterization by W requires the weights to be at least 1 for the problem to be FPT, and in those cases the value of k is at most W, but it can be much smaller.

For approximation, we look at the notion of bi-criteria approximation, where we approximate both the size and weight of the vertex cover, within some factors a and b respectively. More formally, given an instance (G, w, k, W) of (k, W)-VERTEX COVER, an (a, b)-approximation algorithm either correctly determines that there is no vertex cover of size at most k and weight at most W or outputs a vertex cover S of G, such that $|S| \leq ak$ and $w(S) \leq bW$. It is easy to see that the $(2 - \epsilon)$ hardness for VERTEX COVER rules out any $(c, 2 - \epsilon)$- and $(2 - \epsilon, c)$-approximation algorithms running in polynomial time for any constants $c, \epsilon > 0$. In particular, it rules out $(2 - \epsilon, 2 - \epsilon)$-approximation algorithms running in polynomial time. We complement this negative result by observing a simple $(2, 2)$-approximation algorithm. This is done by modifying the known factor 2-approximation LP-rounding algorithm for VERTEX COVER. After this we move to parameterized algorithms and give the following results.

- A simple exact parameterized algorithm for (k, W)-VERTEX COVER running in $\mathcal{O}^*(1.4656^k)$ time.
- A $(1 + \epsilon, 2)$-approximation algorithm for (k, W)-VERTEX COVER running in $\mathcal{O}^*(1.4656^{(1-\epsilon)k})$ time.
- A $(1.5, 1.5)$-approximation algorithm for (k, W)-VERTEX COVER in $\mathcal{O}^*(1.414^k)$ time.
- A general $(2 - \delta, 2 - \delta)$-approximation algorithm for (k, W)-VERTEX COVER in
$$\mathcal{O}^*\left(\sum_{i=\frac{\delta k(1-2\delta)}{1+2\delta}}^{\frac{\delta k(1-2\delta)}{2\delta}} \binom{\delta k + i}{\delta k - \frac{2i\delta}{1-2\delta}}\right)$$ time for any $\delta < 0.5$. For example, for $(1.75, 1.75)$ and $(1.9, 1.9)$-approximation algorithms, the running times are $\mathcal{O}^*(1.272^k)$ and $\mathcal{O}^*(1.151^k)$ respectively.

The exact parameterized algorithm does a simple two-way branching on a vertex of degree at least 3, and solves (k, W)-VERTEX COVER in polynomial time on graphs of maximum degree at most 2 in the base case using the fact that such graphs have treewidth at most 2.

Then we use polynomial time $(2, 2)$-approximation algorithm along with a simple branching to get a $(1 + \epsilon, 2)$-approximation algorithm running in time $\mathcal{O}^*(1.4656^{(1-\epsilon)k})$. This algorithm follows along the ideas of [1] and [7].

Then we turn our attention to bringing both the approximation factors below 2. This is the main technical contribution of the paper. The main idea behind our algorithm is that if we do the branching on the vertices of large weight initially, applying the $(2, 2)$-approximation algorithm on the remaining instance cannot make the solution much worse. Branching on a maximum weight vertex first till the budget has dropped sufficiently, and then combining that with

$(2, 2)$-approximation algorithm gives us a $(1.5, 1.5)$-approximation algorithm running in time $\mathcal{O}^*(2^{\frac{k}{2}})$.

Finally, we generalize this algorithm and give a $(2 - \delta, 2 - \delta)$-approximation algorithm running in $\mathcal{O}^*\left(\sum_{i=\frac{\delta k(1-2\delta)}{1+2\delta}}^{\frac{\delta k(1-2\delta)}{2\delta}} \binom{\delta k + i}{\delta k - \frac{2i\delta}{1-2\delta}}\right)$ time for any $\delta < 0.5$. For example, putting the value of δ to be 0.25 and 0.1, this gives $(1.75, 1.75)$ and $(1.9, 1.9)$-approximation algorithms with running times of $\mathcal{O}^*(1.272^k)$ and $\mathcal{O}^*(1.151^k)$ respectively (see Fig. 1 for a graph of running times for different values of δ). This generalization requires different stopping criteria and a more involved analysis as compared to the previous algorithm to achieve the desired approximation factor.

Unlike the existing FPT algorithms for weighted version of the problem parameterized by W, which work for the weights of vertices being at least one, our algorithms work for all positive rational weights.

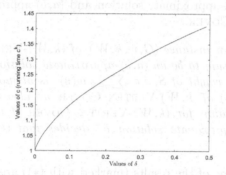

Fig. 1. Running time of $(2 - \delta, 2 - \delta)$-approximation algorithm for different values of δ

2   Preliminaries

In this section, we give the notations and definitions, along with preliminaries and some known results which are used in the paper.

For a graph $G = (V, E)$, we denote the set of vertices of the graph by $V(G)$ and the set of edges of the graph by $E(G)$. We denote $|V(G)|$ by n, where the graph is clear from the context. For a set $S \subseteq V(G)$, the subgraph of G induced by S is denoted by $G[S]$ and it is defined as the subgraph of G with vertex set S and edge set $\{\{u, v\} \in E(G) : u, v \in S\}$. The subgraph obtained after deleting S (and the edges incident to the vertices in S) is denoted as $G - S$. For ease of notation, we will use uv to denote an edge of a graph instead of $\{u, v\}$ where u and v are the endpoints of the edge. For any graph $G = (V, E)$ and $v \in V$, a vertex $u \in V$ is said to be a neighbour of v if $vu \in E(G)$. The open neighbourhood of v is denoted by $N(v)$ and it is defined as $N(v) = \{u \in V : vu \in E(G)\}$. We denote degree of a vertex v by $d(v)$ where $d(v) = |N(v)|$. The closed neighbourhood of a vertex v is $N[v] = N(v) \cup \{v\}$. By $\Delta(G)$, we denote the maximum degree of a vertex in G. We denote a polynomial function of n as $poly(n)$.

In a branching algorithm, if a branching rule has l possible branching options on an instance and the parameter drops by a_i in the ith option, then the vector $(a_1, a_2, ..., a_l)$ is called the *branching vector* for the rule. The *branching number* of a branching vector $(a_1, a_2, ..., a_l)$ is the unique positive root of the equation $\sum_{i=1}^{l} x^{-a_i} - 1 = 0$. If c is the largest branching number for all possible branching vectors in the branching algorithm, then the number of nodes in that branching tree is at most $\mathcal{O}(c^k)$ where k is the initial parameter. Also, if the algorithm takes only polynomial time on every node of the branching tree, then the total running time of the algorithm will be $\mathcal{O}^*(c^k)$.

For a (k, W)-VERTEX COVER instance (G, w, k, W), we say that the instance is a Yes instance if there is a vertex cover S of G of size at most k and weight at most W, otherwise, we say that it is a No instance. Let (G', w, k', W') be the instance after applying a reduction rule to an instance (G, w, k, W) of (k, W)-VERTEX COVER, then we say that the reduction rule is correct when (G, w, k, W) is a Yes instance if and only if (G', w, k', W') is a Yes instance. Now, we define (a, b)-approximate solution and (a, b)-approximation algorithm for (k, W)-VERTEX COVER.

Definition 1. *For an instance (G, w, k, W) of (k, W)-VERTEX COVER a vertex cover S of G is said to be an (a, b)-approximate solution if the size of S is at most ak and the weight of S, i.e. $\sum_{v \in S} w(v)$, is at most bW. For a given instance, (G, w, k, W) of (k, W)-VERTEX COVER an algorithm is called (a, b)-approximation algorithm for (k, W)-VERTEX COVER if the algorithm either returns an (a, b)-approximate solution or decides that (G, w, k, W) is a No instance.*

Full proofs of some of the results (marked with (*)) are deferred to the full version due to space constraints. We now state a general lemma which will be used later.

Lemma 1. (*) *If v is a vertex of a graph G with $d(v) \geq 1$, then an instance (G, w, k, W) of (k, W)-VERTEX COVER is a No instance if and only if both $(G - v, w, k-1, W-w(v))$ and $(G - N[v], w, k-|N(v)|, W-w(N(v)))$ are No instances.*

3   A $(2, 2)$-Approximation Algorithm for (k, W)-VERTEX COVER

In this section we will show a $(2, 2)$-approximation algorithm for (k, W)-VERTEX COVER.

Theorem 1. (*) *There exists a polynomial-time $(2, 2)$-approximation algorithm for (k, W)-VERTEX COVER.*

Proof. (sketch.) The standard Linear Program (LP) for VERTEX COVER is the following.

$$\text{minimize} \quad \sum_{j=1}^{n} w_j x_j \tag{1}$$

$$\text{subject to} \quad x_i + x_j \geq 1 \qquad \forall v_i v_j \in E(G) \tag{2}$$

$$0 \leq x_j \leq 1 \qquad \forall j \tag{3}$$

We add the cardinality constraint $\sum_{j=1}^{n} x_j \leq k$ to the above LP. Given any optimal solution to this new LP, it is easy to see that the standard rounding giving 2-approximate solution for weights also picks at most $2k$ vertices.   □

4   Parameterized Algorithm for (k, W)-VERTEX COVER

In this section, we will show a parameterized algorithm for (k, W)-VERTEX COVER running in $\mathcal{O}^*(1.4656^k)$ time. For that, we first make the following observation that we can solve (k, W)-VERTEX COVER in $poly(n)$ time for any graph G with $\Delta(G) \leq 2$. We know that for such a graph G, treewidth of G is bounded by 2, and following the standard dynamic programming on bounded treewidth graphs, we obtain a polynomial time algorithm for (k, W)-VERTEX COVER on such graphs.

Theorem 2. (∗) (k, W)–VERTEX COVER *can be solved in* $poly(n)$ *time for every graph G with* $\Delta(G) \leq 2$.

Given an instance (G, w, k, W) of (k, W)-VERTEX COVER, we first see if $\Delta(G) \leq 2$. If so, we solve (k, W)-VERTEX COVER in polynomial time using Theorem 2. Otherwise, we pick a vertex v of degree at least 3 from G and branch on including it or its neighbours in the solution. That is, we recursively solve the problems $(G - v, w, k - 1, W - w(v))$ and $(G - N(v), w, k - |N(v)|, W - w(N(v)))$, and return Yes if and only if at least one of the recursive calls returns Yes.

Theorem 3. (k, W)-VERTEX COVER *can be solved in time* $\mathcal{O}^*(1.4656^k)$.

Proof. We use the above described algorithm to solve an instance (G, w, k, W) of (k, W)-VERTEX COVER. The correctness of the algorithm follows from the correctness of the algorithm in Theorem 2 and from Lemma 1 that at least one of the recursive calls return Yes if and only if the original instance was a Yes instance. Since branch on a vertex of degree at least 3 and all the leaf nodes of the algorithm take polynomial time, this gives a branching vector of $(1, d)$, where $d \geq 3$, which results in running time being $\mathcal{O}^*(1.4656^k)$ as desired.   □

5   A Simple $(1 + \epsilon, 2)$-Approximation Algorithm

In this section, we give a $(1 + \epsilon, 2)$-approximation algorithm for (k, W)-Vertex Cover by combining the algorithms of Theorem 1 and Theorem 3. As a result, the running time of the algorithm scales between the running times of Theorem 1 and Theorem 3 with the value of ϵ. The algorithm either outputs No, which means that there is no vertex cover of size at most k with weight at most W, or outputs a vertex cover of size at most $(1 + \epsilon)k$ and weight at most $2W$. It is easy to see that the algorithm of Theorem 3 can be made to be constructive, so in this section, we assume that it returns a vertex cover of size at most k and weight at most W, if one exists.

Now we describe the algorithm. The algorithm checks at every step if the graph has maximum degree at most two, and if so, finds a vertex cover of weight at most W and size at most k in polynomial time using Theorem 2. Otherwise, the algorithm does a $(1, d)$-branching with $d \geq 3$ by picking a vertex v of degree at least 3, and picking either v or $N(v)$ into the solution, same as the algorithm of Theorem 3, while the size-budget is at least ϵk (the algorithm starts with a size-budget k and decreases it by i whenever it picks i vertices in the solution). When the size-budget drops below ϵk, the algorithm uses Theorem 1 to get a $(2, 2)$- approximate solution of the remaining instance in polynomial time and combines the solution with the vertices picked in the branching steps to get a vertex cover of G. The algorithm returns No if every recursive call returns No. Now we argue the correctness and running time of this algorithm. Observe that at the leaf nodes of the branching tree, either we solve the instance exactly in polynomial time using Theorem 2 or we apply $(2, 2)$-approximation algorithm that runs in polynomial time (Theorem 1).

Theorem 4. (∗) *For any $\epsilon \geq 0$, there exists a $(1 + \epsilon, 2)$-approximation algorithm for (k, W)-Vertex Cover with running time $\mathcal{O}^*(1.4656^{(1-\epsilon)k})$.*

The correctness of the proof follows from the correctness of branching, Theorem 1 and Theorem 3. For the running time analysis, we can think of this algorithm as a branching algorithm where the branching is the same as in the algorithm of Theorem 3 and every node of the branching tree including the leaf nodes (where we may apply Theorem 1 or Theorem 3) takes polynomial time only. We continue branching till the size-budget drops by $(1 - \epsilon)k$ which gives the running time of the algorithm to be $\mathcal{O}^*(1.4656^{(1-\epsilon)k})$. The detailed proof of this theorem is deferred to the full version of the paper.

6   A $(1.5, 1.5)$-Approximation Algorithm for (k, W)-Vertex Cover

In this section, we give a $(1.5, 1.5)$-approximation algorithm for (k, W)-Vertex Cover running in time $\mathcal{O}^*(2^{\frac{k}{2}})$. For that, we first make use of the following reduction rule.

Reduction Rule 1. *For any instance* (G, w, k, W) *of* (k, W)-VERTEX COVER, *if* u *is a maximum weight vertex of* G *and* $d(u) \leq 1$ *then delete* $N[u]$. *The new instance will be* $(G - N[u], w, k - |N(u)|, W - w(N(u)))$.

Lemma 2. (*) *Reduction Rule 1 is correct.*

The proof of the lemma follows from the fact that there always exists a solution of size at most k of minimum weight that does not include u. Now we describe the $(1.5, 1.5)$-approximation algorithm.

Given an instance (G, w, k, W), the algorithm first applies Reduction Rule 1 exhaustively, so that in the resulting graph, the degree of every maximum weight vertex is at least 2. Then the algorithm does branching on a maximum weight vertex till the depth $k/2$ in the branching tree. That is, branching is done on a node of the branching tree only if its distance from the root is less than $k/2$. Note that here we assume every internal node of the branching tree to have exactly two children, i.e., we assume that recursive call to the reduction rule is done at the same node of the branching tree in polynomial time. Any call to the algorithm returns No only if both the child nodes in the branching tree return No. At depth $k/2$, if the instance is $I = (G', w, k', W')$, then the algorithm applies the algorithm of Theorem 1 to get a $(2, 2)$-approximate solution for I, and combines it with the vertices picked in the solution while applying the reduction rule and branching, and outputs that as the solution. We describe the pseudo-code of the algorithm in Algorithm 1. To keep track of the depth of the branching tree at which we are working and for analysis, we pass the algorithm additional parameters α, β and x, where $x = k/2$ if (G, w, k, W) was the original instance. For the branching steps, we increase α in the branch where the maximum weight vertex is taken into the solution, and β is increased in the branch when its neighbourhood is taken into the solution. Hence, if $\alpha + \beta = \ell$, then we are at depth ℓ of the branching tree. We stop branching when $\alpha + \beta = k/2$ and use the $(2, 2)$-approximation algorithm. In the remainder of this section, we prove the correctness of Algorithm 1 and do the running time analysis.

Lemma 3. (*) *If Algorithm 1 returns No for the instance* (G, w, k, W), *then* (G, w, k, W) *is indeed a No instance.*

Lemma 4. (*) *If Algorithm 1 does not return No on an instance* (G, w, k, W) *of* (k, W)-VERTEX COVER, *then it returns a vertex cover* S *of* G.

Next, we move to the following lemma, which says that at the leaf nodes of the branching tree, where we apply the algorithm of Theorem 1, the value of the parameter is bounded.

Lemma 5. (*) *If Proc1, on input* $(G', w, k', W', \alpha, \beta, x)$, *calls the algorithm of Theorem 1 on an instance* (G', w, k', W') *in line 13, then* $k' \leq \alpha$.

Now, before we show the size and weight bounds on the vertex set returned by the algorithm, we define some notions that will help us with the analysis. We classify the instances on which we call the $(2, 2)$-approximation algorithm in line 13 into two types.

Let (G, w, k, W) be the original instance passed to Proc1. We say that an instance (G', w, k', W') on which $(2, 2)$-approximation algorithm is called in line 13 is Type-I, if $W' \leq W/2$ and is Type-II otherwise. We also partition the set S returned by Algorithm 1 into two parts. Let S_2 be the set returned by the $(2, 2)$-approximation algorithm and let $S_1 = S \setminus S_2$ be the set of vertices in the returned vertex cover which were picked during the branching or during the application of the reduction rule. The next two lemmas give the size and weight bounds of the returned solution.

Algorithm 1: $(1.5, 1.5)$-approximation algorithm for (k, W)-VERTEX COVER

Input   : An instance (G, w, k, W) of (k, W)-VERTEX COVER.

Output: A vertex cover S of size at most $1.5k$ with weight at most $1.5W$ or returns No.

1 *Fix $x = k/2$;*
2 *Initialize, $\alpha \leftarrow 0$ and $\beta \leftarrow 0$;*
3 return Proc1$(G, w, k, W, \alpha, \beta, x)$;
4 ;   /* We pass the parameters x, α, β to the algorithm with the instance (G, w, k, W) */
5 Proc1$(G, w, k, W, \alpha, \beta, x)$:
6 if $\min\{k, W\} < 0$ then
7 | return No;
8 end
9 if $E(G) = \emptyset$ then
10 | return $\emptyset$;
11 end
12 if $\alpha + \beta = x$ then
13 | return the output of Algorithm of Theorem 1 for the instance (G, w, k, W);
14 end
15 Let u be a maximum weight vertex in G;
16 if $d(u) \leq 1$ then
17 | return $N(u) \cup$ Proc1$(G - N[u], w, k - |N(u)|, W - w(N(u)), \alpha, \beta, x)$;
18 else
19 | recursively call ;   /* Branching step */
   | (i) Proc1$(G - u, w, k - 1, W - w(u), \alpha + 1, \beta, x)$;
   | (ii) Proc1$(G - N[u], w, k - |N(u)|, W - w(N(u)), \alpha, \beta + 1, x)$;

20 | if *both calls return No* then
21 | | return No ;
22 | else if Proc1$(G - u, w, k - 1, W - w(u), \alpha + 1, \beta, x)$ *returns S* then
23 | | return $S \cup \{u\}$;
24 | else if Proc1$(G - N[u], w, k - |N(u)|, W - w(N(u)), \alpha, \beta + 1, x)$ *returns S then*
25 | | return $S \cup N(u)$;
26 | end
27 end

Lemma 6. (∗) *If Algorithm 1 returns a set S for an input instance (G, w, k, W) after calling $(2, 2)$-approximation algorithm on a Type-I instance (G_2, w, k_2, W_2) in line 13, then $|S| \leq 1.5k$ and $w(S) \leq 1.5W$.*

Lemma 7. (∗) *If Algorithm 1 returns a set S for an input instance (G, w, k, W) after calling $(2, 2)$-approximation algorithm on a Type-II instance (G_3, w, k_3, W_3) in line 13, then $|S| \leq 1.5k$ and $w(S) \leq 1.5W$.*

Theorem 5. *There exists a $(1.5, 1.5)$-approximation algorithm for (k, W)-VERTEX COVER with running time $\mathcal{O}^*(2^{\frac{k}{2}})$.*

Proof. Given an instance (G, w, k, W) of (k, W)-VERTEX COVER we call Algorithm 1 on it. In Lemmas 3-7, we have already argued that if the algorithm returns No, then (G, w, k, W) is indeed a No instance, and if the algorithm returns a set S, then S is a vertex cover of G with $|S| \leq 1.5k$ and $w(S) \leq 1.5W$. So all we need to show is that Algorithm 1 takes time $\mathcal{O}^*(2^{\frac{k}{2}})$ on an instance (G, w, k, W). Our algorithm branches till depth $k/2$ in the branching tree, so the branching tree has $\mathcal{O}(2^{\frac{k}{2}})$ nodes, and on each of the nodes we spend at most polynomial time. Hence the running time follows. □

7 A $(2 - \delta, 2 - \delta)$-Approximation Algorithm for $\delta < 0.5$

In this section, we will present a $(2-\delta, 2-\delta)$-approximation algorithm for (k, W)-VERTEX COVER for any $\delta < 0.5$. We first give a description of this algorithm, that we call Algorithm 2.

Given an instance (G, w, k, W) of (k, W)-VERTEX COVER to Algorithm 2, it first fixes $Y := W - \delta W$ and $x := k - \delta k$. Then the algorithm calls Proc2 on instance (G, w, k, W, x, Y) and returns the solution returned by Proc2 (x and Y are never changed by Proc2). Proc2 first applies Reduction Rule 1 exhaustively, so that in the resulting graph, every maximum weight vertex has degree at least 2. Then it branches on a maximum weight vertex u, calling itself on $(G - u, k - 1, W - w(u), x, Y)$ and $(G - N[u], k - |N(u)|, W - w(N(u)), x, Y)$ recursively. Proc2 is very similar to Proc1 and the main difference is the stopping criteria. Proc2 stops branching on an instance (G', w, k', W', x, Y) in the branching tree when either i) G' becomes empty, or ii) $k' \leq x$ and either $W' \leq Y$ or there is no vertex of weight $\frac{Y}{k'}$ in G'. Note that here we assume every internal node of the branching tree to have exactly two children, i.e., we assume that the recursive call after the reduction rule is applied is made at the same node of the branching tree in polynomial time. Any recursive call to Proc2 returns No only if both the child nodes in the branching tree return No. If we stop branching on an instance (G', w, k', W', x, Y) and G' is empty then Proc2 returns the vertices picked in the solution during branching and applying Reduction Rule 1 to obtain (G', w, k', W', x, Y). Otherwise, Proc2 applies the algorithm of Theorem 1 (that is, $(2, 2)$-approximation algorithm) to the instance (G', w, k', W') to get a $(2, 2)$-approximate solution for (G', w, k', W') and combines it with the vertices added to the solution while applying the reduction rule and branching, and outputs that as a solution. Due to space constraints, we defer the pseudo-code of Algorithm 2 which describes this algorithm more formally to the full version. The following two lemmas give the correctness of Algorithm 2.

Lemma 8. (∗) *If Algorithm 2 returns* No *for the instance* (G, w, k, W), *then* (G, w, k, W) *is indeed a* No *instance.*

Lemma 9. (∗) *If Algorithm 2 does not return* No *on an instance* (G, w, k, W) *of* (k, W)-VERTEX COVER, *then it returns a vertex cover* S *of* G *such that* $|S| \leq (2 - \delta)k$ *and* $w(S) \leq (2 - \delta)W$.

Now, before we argue the running time of the algorithm, we first define some notations and prove some lemmas about the types of instances that can be generated during the course of the algorithm.

Due to Reduction Rule 1, the branching we are doing in Proc2 is a $(1, d)$-branching (where $d \geq 2$) but here we have additional stopping criteria for any instance (G', w, k', W', x, Y) with $E(G') \neq \emptyset$. The criteria are i) $W' \leq Y$ together with $k' \leq x$, or ii)there is no vertex in G' of weight $\frac{Y}{k'}$ together with $k' \leq x$. Here our main goal is to get a good upper bound on the number of nodes in the branching tree.

For any instance (G', w, k', W', x, Y) of Proc2, we call it one of Type-I, Type-II, Type-III, and Type-IV depending on values of k' and W'. In the case $E(G') = \emptyset$, Proc2 correctly returns a vertex cover of size at most k with weight at most W, so we assume that $E(G') \neq \emptyset$. Now, let us define above mentioned four different types of instances.

Type-I: (G_1, w, k_1, W_1, x, Y) is a Type-I instance if $k_1 \leq x$ and $W_1 \leq Y$.

Type-II: (G_2, w, k_2, W_2, x, Y) is a Type-II instance if $k_2 \leq x$, $W_2 > Y$, and there is no vertex of weight Y/k_2 in G_2.

Type-III: (G_3, w, k_3, W_3, x, Y) is a Type-III instance if $k_3 \leq x$, $W_3 > Y$, and there exists at least one vertex of weight at least Y/k_3 in G_3.

Type-IV: All other instances $((G_4, w, k_4, W_4, x, Y)$ such that $k_4 > x)$.

A node in the branching tree of Proc2 is said to be of Type-I ,Type-II, Type-III, or Type-IV if its corresponding instance is a Type-I, Type-II, Type-III, or Type-IV instance respectively.

If any instance is of Type-I or Type-II, then Proc2 immediately applies the Algorithm of Theorem 1 to those instances. So, we continue branching on an instance when the instance is of Type-III or Type-IV. Also, note that the Type-IV nodes can be in the first δk layers of the branching tree only. Otherwise, after δk layers, the size-budget (k) will decrease by at least δk and thus the node cannot be of Type-IV anymore. For this reason, the total number of Type-IV nodes is bounded by $2^{\delta k}$. Let us assign two values α and β to every node of the branching tree of Proc2 where α and β are increased exactly as in the previous section (depending on whether u or $N(u)$ is included in the solution). Now let us prove the following lemma.

Lemma 10. (∗) *If* (G', w, k', W', x, Y) *is an instance of Type-III then* α *corresponding to the instance* (G', w, k', W', x, Y) *is at most* $k'(\frac{\delta}{1-\delta})$.

Next we want to compute the number of Type-III nodes at each layer of the branching tree. It is clear that upto δk layers, there can be a total of $\mathcal{O}(2^{\delta k})$ Type-III or Type-IV nodes. Also, note that for any node at any layer j, we have $\alpha + \beta = j$ corresponding to that node. Now we prove the following lemmas which determine the number of Type-III nodes at layers beyond the first δk layers.

Lemma 11. (∗) *There are no Type-III nodes at layer $\frac{k}{2}$.*

We have already seen that there are no Type-IV nodes at layer $\frac{k}{2}$, and the number of Type-IV nodes is bounded by $2^{\delta k}$. Also, since no branching is done for Type-I or Type-II instances, the branching tree does not have any nodes beyond layer $\frac{k}{2}$. Observe that $\delta k + i = k/2$ for $i = \frac{\delta k(1-2\delta)}{2\delta}$, and hence we only need to bound the number of Type-III nodes for layers of the form $\delta k + i$ where $i \leq \frac{\delta k(1-2\delta)}{2\delta}$. This is what we do in the following lemma.

Lemma 12. (∗) *At layer $\delta k + i$, the number of Type-III nodes can be at most $2^{\delta k + i}$ when $0 \leq i \leq \frac{\delta k(1-2\delta)}{1+2\delta}$ and bounded by $\mathcal{O}^*\left(\binom{\delta k + i}{\delta k - \frac{2i\delta}{1-2\delta}}\right)$ when $\frac{\delta k(1-2\delta)}{1+2\delta} \leq i \leq \frac{\delta k(1-2\delta)}{2\delta}$.*

Now we are ready to prove the lemma about the running time of Algorithm 2.

Lemma 13. *Algorithm 2 runs in time $\mathcal{O}^*\left(\sum_{i=\frac{\delta k(1-2\delta)}{1+2\delta}}^{\frac{\delta k(1-2\delta)}{2\delta}} \binom{\delta k + i}{\delta k - \frac{2i\delta}{1-2\delta}}\right)$.*

Proof. We need to show that the total number of nodes in the branching tree is bounded by $\mathcal{O}^*(\Gamma)$, where $\Gamma := \left(\sum_{i=\frac{\delta k(1-2\delta)}{1+2\delta}}^{\frac{\delta k(1-2\delta)}{2\delta}} \binom{\delta k + i}{\delta k - \frac{2i\delta}{1-2\delta}}\right)$. For that, first observe that the number of Type-I and Type-II nodes can be at most twice the number of Type-III and Type-IV nodes in the branching tree, as the parent of every Type-I and Type-II node is either a Type-III node or a Type-IV node and the branching tree is a binary tree. So all we need to do is to bound the number of Type-III and Type-IV nodes by $\mathcal{O}^*(\Gamma)$.

The first term in the expression of Γ is $\binom{\delta k + i}{\delta k - \frac{2i\delta}{1-2\delta}}$ for $i = \frac{\delta k(1-2\delta)}{1+2\delta}$, which turns out to be $\binom{\frac{2\delta k}{1+2\delta}}{\frac{\delta k}{1+2\delta}}$. We know that the central binomial coefficient, $\binom{n}{n/2} \geq 2^n/(n+1)$. This gives us

$$\binom{\frac{2\delta k}{1+2\delta}}{\frac{\delta k}{1+2\delta}} \geq \frac{2^{\frac{2\delta k}{1+2\delta}}}{\frac{2\delta k}{1+2\delta} + 1}.$$

This implies that

$$2^{\frac{2\delta k}{1+2\delta}} = \mathcal{O}^*\left(\binom{\frac{2\delta k}{1+2\delta}}{\frac{\delta k}{1+2\delta}}\right) = \mathcal{O}^*(\Gamma).$$

Since the number of Type-III and Type-IV nodes in the fist δk layers is bounded by $2^{\delta k}$, which is $\mathcal{O}(2^{\frac{2\delta k}{1+2\delta}})$ for $\delta < 0.5$, we get that the number of Type-III and Type-IV nodes in the first δk layers is bounded by $\mathcal{O}^*(\Gamma)$. We have already seen that there are no Type-IV nodes beyond the first δk layers, so this bounds the total number of Type-IV nodes by $\mathcal{O}^*(\Gamma)$.

Number of Type-III nodes at layers $\delta k + i$, where $\frac{\delta k(1-2\delta)}{1+2\delta} \leq i \leq \frac{\delta k(1-2\delta)}{2\delta}$, is already bounded by $\mathcal{O}^*(\Gamma)$ due to Lemma 12. So all we are left to bound now is the number of Type-III nodes at layers $\delta k + i$, where $0 \leq i \leq \frac{\delta k(1-2\delta)}{1+2\delta}$. From Lemma 12, each such layer contains at most $2^{\delta k+i}$ nodes of Type-III. This attains the maximum value of $2^{\frac{2\delta k}{1+2\delta}}$ at $i = \frac{\delta k(1-2\delta)}{2\delta}$, which we have already seen to be bounded by $\mathcal{O}^*(\Gamma)$. Since there are at most $k/2$ such layers, the total number of Type-III nodes, and hence the total number of nodes in the branching tree is bounded by $\mathcal{O}^*(\Gamma)$.

It is also clear from the algorithm that we are either spending only polynomial time at every node of the branching tree or applying our $(2,2)$-approximation algorithm which also runs in polynomial time. Hence the total time taken by the algorithm will be at most $\mathcal{O}^*\left(\sum_{i=\frac{\delta k(1-2\delta)}{1+2\delta}}^{\frac{\delta k(1-2\delta)}{2\delta}} \binom{\delta k+i}{\delta k-\frac{2i\delta}{1-2\delta}}\right)$. This completes the proof of the Lemma 13. $\qquad\square$

The proof of the following theorem follows from Lemma 8, Lemma 9 and Lemma 13.

Theorem 6. *There exists a $(2-\delta, 2-\delta)$-approximation algorithm for (k,W)-* VERTEX COVER *with running time $\mathcal{O}^*\left(\sum_{i=\frac{\delta k(1-2\delta)}{1+2\delta}}^{\frac{\delta k(1-2\delta)}{2\delta}} \binom{\delta k+i}{\delta k-\frac{2i\delta}{1-2\delta}}\right)$.*

7.1   Computation of the Running Time

We write a computer program to compute the running time of Algorithm 2 for any given $\delta \in (0, 0.5)$. The program finds M_δ where the running time of Algorithm 2 is of the form $(M_\delta)^k$. Following table contains the value of M_δ and the running time for some values of δ. More details of how M_δ is computed are deferred to the full version.

Table of running times for different values of δ.

No.	δ	approximation factor	M_δ	Run Time
1	0.05	$(1.95, 1.95)$	1.0943	1.0943^k
2	0.10	$(1.90, 1.90)$	1.1510	1.1510^k
3	0.15	$(1.85, 1.85)$	1.1968	1.1968^k
4	0.20	$(1.80, 1.80)$	1.2365	1.2365^k
5	0.25	$(1.75, 1.75)$	1.2720	1.2720^k
6	0.30	$(1.70, 1.70)$	1.3045	1.3045^k
7	0.35	$(1.65, 1.65)$	1.3345	1.3345^k
8	0.40	$(1.60, 1.60)$	1.3626	1.3626^k
9	0.45	$(1.55, 1.55)$	1.3891	1.3891^k

8   Conclusion

In this paper, we considered the WEIGHTED VERTEX COVER problem parameterized by the solution size. We gave an exact algorithm running in time $\mathcal{O}^*(1.4656^k)$ and $(2 - \delta, 2 - \delta)$- approximation algorithms for $\delta \in [0, 0.5]$ which run faster than the exact algorithm. It will be interesting to obtain a parameterized approximation scheme for (k, W)-VERTEX COVER which works for every $\delta \in (0, 1)$ where the running time scales between the running time of $(2, 2)$-approximation algorithm and the running time of the exact algorithm.

Acknowledgment. The third author would like to thank Andreas Emil Feldmann and Cornelius Brand for some initial discussions on the problem.

References

1. Bourgeois, N., Escoffier, B., Paschos, V.T.: Approximation of max independent set, min vertex cover and related problems by moderately exponential algorithms. Discret. Appl. Math. **159**(17), 1954–1970 (2011)
2. Brankovic, L., Fernau, H.: A novel parameterised approximation algorithm for minimum vertex cover. Theor. Comput. Sci. **511**, 85–108 (2013)
3. Chen, J., Fomin, F.V., Liu, Y., Lu, S., Villanger, Y.: Improved algorithms for feedback vertex set problems. J. Comput. Syst. Sci. **74**(7), 1188–1198 (2008)
4. Chen, J., Kanj, I.A., Xia, G.: Improved upper bounds for vertex cover. Theoret. Comput. Sci. **411**(40–42), 3736–3756 (2010)
5. Dinur, I., Khot, S., Kindler, G., Minzer, D., Safra, M.: Towards a proof of the 2-to-1 games conjecture? In: Proceedings of the 50th Annual ACM SIGACT Symposium on Theory of Computing, STOC 2018, Los Angeles, CA, USA, June 25–29, 2018, pp. 376–389. ACM (2018)
6. Feldmann, A.E., Karthik, C.S., Lee, E., Manurangsi, P.: A survey on approximation in parameterized complexity: hardness and algorithms. Algorithms **13**(6), 146 (2020)

7. Fellows, M.R., Kulik, A., Rosamond, F.A., Shachnai, H.: Parameterized approximation via fidelity preserving transformations. J. Comput. Syst. Sci. **93**, 30–40 (2018)
8. Fomin, F.V., Gaspers, S., Saurabh, S.: Branching and treewidth based exact algorithms. In: Asano, T. (ed.) ISAAC 2006. LNCS, vol. 4288, pp. 16–25. Springer, Heidelberg (2006). https://doi.org/10.1007/11940128_4
9. Galby, E., Marx, D., Schepper, P., Sharma, R., Tale, P.: Parameterized complexity of weighted multicut in trees. In: Bekos, M.A., Kaufmann, M. (eds.) Graph-Theoretic Concepts in Computer Science: 48th International Workshop, WG 2022, Tübingen, Germany, June 22–24, 2022, Revised Selected Papers, pp. 257–270. Springer International Publishing, Cham (2022). https://doi.org/10.1007/978-3-031-15914-5_19
10. Karp, R.M.: Reducibility among combinatorial problems. In: Proceedings of a symposium on the Complexity of Computer Computations, held March 20–22, 1972, pp. 85–103. The IBM Research Symposia Series, Plenum Press, New York (1972)
11. Khot, S., Regev, O.: Vertex cover might be hard to approximate to within 2-epsilon. J. Comput. Syst. Sci. **74**(3), 335–349 (2008)
12. Kim, E.J., Kratsch, S., Pilipczuk, M., Wahlström, M.: Directed flow-augmentation. In: STOC '22: 54th Annual ACM SIGACT Symposium on Theory of Computing, Rome, Italy, June 20–24, 2022, pp. 938–947. ACM (2022)
13. Kim, E.J., Pilipczuk, M., Sharma, R., Wahlström, M.: On weighted graph separation problems and flow-augmentation. CoRR abs/2208.14841 (2022)
14. Kulik, A., Shachnai, H.: Analysis of two-variable recurrence relations with application to parameterized approximations, pp. 762–773. IEEE (2020)
15. Nemhauser, G.L., Trotter, L.E.: Vertex packings: structural properties and algorithms. Math. Program. **8**(1), 232–248 (1975)
16. Niedermeier, R., Rossmanith, P.: On efficient fixed-parameter algorithms for weighted vertex cover. J. Algorithms **47**(2), 63–77 (2003)
17. Saurabh, S.: What's next? future directions in parameterized complexity. Recent Advances in Parameterized Complexity school, Tel Aviv (December (2017)
18. Shachnai, H., Zehavi, M.: A multivariate framework for weighted FPT algorithms. J. Comput. Syst. Sci. **89**, 157–189 (2017)

Parameterized Algorithms for Minimum Sum Vertex Cover

Shubhada Aute[1]([✉])[iD] and Fahad Panolan[2][iD]

[1] Indian Institute of Technology Hyderabad, Sangareddy, India
cs21resch11001@iith.ac.in
[2] School of Computing, University of Leeds, Leeds, UK
f.panolan@leeds.ac.uk

Abstract. Minimum sum vertex cover of an n-vertex graph G is a bijection $\phi : V(G) \rightarrow [n]$ that minimizes the cost $\sum_{\{u,v\} \in E(G)} \min\{\phi(u), \phi(v)\}$. Finding a minimum sum vertex cover of a graph (the MSVC problem) is NP-hard. MSVC is studied well in the realm of approximation algorithms. The best-known approximation factor in polynomial time for the problem is 16/9 [Bansal, Batra, Farhadi, and Tetali, SODA 2021]. Recently, Stankovic [APPROX/RANDOM 2022] proved that achieving an approximation ratio better than 1.014 for MSVC is NP-hard, assuming the Unique Games Conjecture. We study the MSVC problem from the perspective of parameterized algorithms. The parameters we consider are the size of a minimum vertex cover and the size of a minimum clique modulator of the input graph. We obtain the following results.

- MSVC can be solved in $2^{2^{O(k)}} n^{O(1)}$ time, where k is the size of a minimum vertex cover.
- MSVC can be solved in $f(k) \cdot n^{O(1)}$ time for some computable function f, where k is the size of a minimum clique modulator.

Keywords: FPT · Vertex Cover · Integer Quadratic Programming

1 Introduction

A vertex cover in a graph is vertex subset such that each edge has at least one endpoint in it. Finding a vertex cover of minimum size is NP-complete, and it is among the renowned 21 NP-complete problems proved by Karp in 1972 [24]. Since then, it has been extensively studied in both the fields of approximation algorithms and parameterized algorithms. The approximation ratio 2 of minimum vertex cover is easily achievable using any maximal matching of the graph and it is optimal assuming Unique Games Conjecture [25]. The best-known FPT algorithm for vertex cover has running time $O(1.2738^k + kn)$ [8], where k is the minimum vertex cover size. In this paper, we study the well-known MINIMUM SUM VERTEX COVER, defined below. For $n \in \mathbb{N}$, $[n] = \{1, 2, \ldots, n\}$.

Definition 1. *Let G be an n-vertex graph, and let $\phi : V(G) \rightarrow [n]$ be a bijection. The weight (or cover time or cost) of an edge $e = \{u, v\}$ is $w(e, \phi) = \min\{\phi(u), \phi(v)\}$. The cost of the ordering ϕ for the graph G is defined as,*

© The Author(s), under exclusive license to Springer Nature Switzerland AG 2024
J. A. Soto and A. Wiese (Eds.): LATIN 2024, LNCS 14579, pp. 193–207, 2024.
https://doi.org/10.1007/978-3-031-55601-2_13

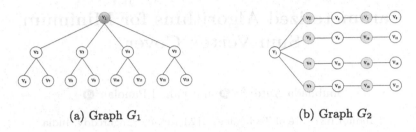

(a) Graph G_1 (b) Graph G_2

Fig. 1. Counter example

$$\mu_G(\phi) = \sum_{e \in E(G)} w(e, \phi).$$

The objective of MINIMUM SUM VERTEX COVER (MSVC) problem is to find an ordering ϕ that minimizes $\mu_G(\phi)$ over all possible orderings. MINIMUM SUM VERTEX COVER came up in the context of designing efficient algorithms for solving semidefinite programs [7]. It was first studied by Feige et al. [12], though their main focus was on MINIMUM SUM SET COVER (MSSC), which is a generalization of the MSVC on hypergraphs.

As MSVC requires the sum of the weights of the edges to be minimized, it is natural to think of prioritizing vertices based on their degree. But this isn't always an optimal approach. For instance, v_1 is the vertex of the maximum degree in the graph G_1 as shown in Fig. 1a. Consider an ordering ψ with v_1 in the first position. For position 2 onwards, any vertex from $G_1 - v_1$ can cover at most two edges. So, the cost of such an ordering is at least 32. However, if the vertices v_2, v_3, v_4, v_5 take the first four locations in an ordering, say ϕ, then cost of ϕ is 30. Though greedy isn't an optimal strategy, it is interesting to find the graphs for which preferring a vertex of highest degree at each location from the remaining graph yields an optimal solution for MSVC. Mohan et al. gave a sufficiency condition for graphs admitting greedy solution as optimal [28]. Interestingly, the greedy approach performs no worse than 4 times the optimum cost. However, certain bipartite graphs yield a solution from the greedy algorithm that is precisely four times the optimal solution for MSVC [12]. The greedy algorithm achieves factor 4 approximation for MSSC on hypergraphs as well [3].

Both the problems MSVC and MSSC are well-studied in the realm of approximation algorithms. An improvement over the greedy algorithm is a 2-factor approximation for MSVC using linear programming [12]. The approximation ratio for MSVC was further improved to 1.999946 [4]. The best approximation ratio for MSVC is 16/9, which was achieved using a generalized version of MSSC, called the generalized min-sum set cover (GMSSC) [2]. Input to GMSSC is a hypergraph $H = (V, E)$ in which every hyperedge has a covering requirement k_e where $k_e \leq |e|$. The first location where k_e number of vertices of an edge e has appeared in the ordering is the cover time of e. The GMSSC problem is the problem of finding an ordering of vertices that minimizes the sum of cover time of all the hyperedges. For all hyperedges of H, when $k_e = 1$, it boils down to the MSSC problem. Bansal et al. provide a 4.642 approximation algorithm for GMSSC, which is close to the approximation factor of 4 for MSSC [2]. For every $\epsilon > 0$, it is NP-hard to approximate MSSC within a ratio of $4 - \epsilon$ [12].

We now state some results from the literature regarding the hardness of approximability of MSVC and MSSC. It is NP-hard to approximate MSVC within a ratio better than $1 + \epsilon$ for some $\epsilon > 0$. For some $\rho < 4/3$ and every d, MSVC can be approximated within a ratio of ρ on d-regular graphs, but such a result doesn't hold in the case of MSSC. For every $\epsilon > 0$, there exists r, d such that it is NP-hard to approximate MSSC within a ratio better than $2 - \epsilon$ on r-uniform d-regular hypergraphs [12]. Recently, it was proved that under the assumption of Unique Games Conjecture, MSVC can not be approximated within 1.014 [32].

Although MSVC is NP-hard, there are some classes of graphs for which it is polynomial time solvable. Gera et al. provide the cost of MSVC in polynomial time for complete bipartite graphs, biregular bipartite graphs, multi-stars, hypercubes, prisms, etc. [19]. They also provide upper and lower bounds for the cost of MSVC in terms of independence number, the girth of a graph, and vertex cover number. MSVC is polynomial-time solvable for split graphs and caterpillars [31]. Interestingly, it is an open question whether MSVC for trees is polynomial-time solvable or NP-hard for a long time [31].

MSSC is studied in the realm of online algorithms, too. A constructive deterministic upper bound of $O(r^{3/2} \cdot n)$, where r is an upper bound for the cardinality of subsets, for online MSSC was given by Fotakis [15]. This bound was then improved to $O(r^4)$ by Bienkowski and Mucha [6]. Though this bound removed the dependency on n, this was existential. Basiak et al. gave a constructive and improved bound of $O(r^2)$ [5].

Our Methods and Results. We study MSVC from the perspective of parameterized complexity. A natural parameter to consider for the FPT algorithm is the solution size. But, for any connected graph G on n vertices, and an optimal ordering ϕ, $\mu_G(\phi) \geq n - 1$. Hence, MSVC parameterized by the solution size is trivially in FPT. Many NP-hard problems are generally tractable when parameterized by treewidth. However, vertex ordering problems are an exception to that. For MSVC, it is even harder to consider the treewidth as a parameter because we do not know if MSVC on trees is polynomial-time solvable or not.

The vertex cover number is a parameter used to prove the tractability of many vertex ordering problems. As MSVC aims to minimize the sum of the cover time of all the edges, it is natural to consider the vertex cover number as the parameter. For a vertex cover S of size k, we define a relation on $I = V \setminus S$ such that two vertices of I are related to each other if their neighborhood is the same. This is an equivalence relation that partitions I into equivalence classes. Interestingly, we prove that all vertices in the same equivalence class appear consecutively in an optimal ordering. So, we guess the relative ordering of vertices in S and then guess the locations of equivalence classes. This gives an FPT algorithm for MSVC parameterized by vertex cover.

MSVC is polynomial-time solvable for a few classes of graphs, with complete graphs being one among them. In fact, any ordering of vertices of K_n is an optimal ordering. A set $M \subset V(G)$ is called a *clique modulator* of the graph G, if the graph induced on $V(G) \setminus M$ is a clique. Finding an optimal solution for MSVC is non-trivial if G has a clique modulator of size k. Hence, we consider the size of a clique modulator as another parameter for MSVC. Vertex cover and clique modulator are complementary parameters. If a graph G has a vertex cover of size k, then the same set of vertices is a clique modulator for the complement of graph G. Hence, these two parameters fit well together in understanding the complete picture of tractability. Although clique modulator is complementary parameter of vertex cover, finding an optimal solution of MSVC with clique modulator is more challenging than that of vertex cover. We use similar approach by defining the same neighborhood relation on clique vertices. Unfortunately, all the vertices of an equivalence class need not be consecutive here in any optimal ordering. But, we prove that there is an optimal ordering σ with the following property. Let $\ell_1, \ldots, \ell_k$ be the location of the vertices from the modulator. Then, for any $i \in [k-1]$ and any equivalence class A, the vertices from A in the locations $\ell_i + 1, \ldots, \ell_{i+1} - 1$ are consecutive in σ. Also, for any equivalence class A, the vertices from A in the locations $1, \ldots, \ell_1 - 1$ as well as in the locations $\ell_k - 1, \ldots, n$ are consecutive. We formulate an Integer Quadratic Programming for finding the number of vertices from each equivalence class present between two modulator vertices and provide an FPT algorithm. We summarize our results as follows:

- MSVC can be solved in $[k!(k+1)^{2^k} + (1.2738)^k]n^{O(1)}$ time, where k is the size of a minimum vertex cover.
- MSVC can be solved in $f(k) \cdot n^{O(1)}$ time for some computable function f, where k is the size of a minimum clique modulator.

Other Related Works. Given a graph $G(V, E)$ on n vertices and a bijection $\phi : V(G) \to [n]$, let a $cost(G, \phi)$ be defined on the vertex ordering ϕ of G. Vertex ordering problems are to minimize the $cost(G)$ over all possible orderings ϕ. Based on the definition of the cost function, there are various vertex ordering problems studied in the literature.

BANDWIDTH of a graph G, a vertex ordering problem, has been extensively studied since 1960. It has applications in speeding up many matrix computations of symmetric matrices. It is NP-complete [30], even for many restricted classes of graphs [16,29]. Approximating BANDWIDTH with a constant factor is NP-hard even for trees [11], but it has an FPT approximation algorithm [10]. It is W[1]-hard parameterized by cluster vertex deletion number [20]. Though BANDWIDTH parameterized by its solution size is in XP [21], it is fixed-parameter tractable parameterized by vertex cover number [13], and neighborhood diversity [1].

CUTWIDTH of a graph G, often referred to as the Min-Cut Linear Arrangement problem in the literature, is NP-complete [18], even when restricted to graphs with maximum degree 3 [27]. However, it is polynomial-time solvable on trees [34]. For a given k, there is a linear time algorithm that outputs the cut if the CUTWIDTH of G is at most k, else outputs that the graph has CUTWIDTH more than k [33]. CUTWIDTH is FPT parameterized by vertex cover [13].

The OPTIMAL LINEAR ARRANGEMENTS (OLA) problem came up in the minimization of error-correcting codes and was first studied on n-dimensional cubes

[23]. It is NP-complete [17], but polynomial-time solvable for trees [9]. OLA parameterized above-guaranteed value (guaranteed value: $|E(G)|$) [14,22], vertex cover [26] is in FPT.

These vertex ordering problems are studied in the field of parameterized algorithms and approximation algorithms. MSVC is well-explored in the realm of approximation algorithms. But to the best of our knowledge, it is not studied in the realm of parameterized complexity.

2 Preliminaries and Notations

Let G be a graph with $V(G)$ as the set of vertices and $E(G)$ as the set of edges. Let σ be an ordering of $V(G)$ and $X \subseteq V(G)$. The ordering σ restricted on X is denoted by $\sigma|_X$. It represents the relative ordering between the vertices of X. A bijection $\sigma|_X : X \to \{1, 2, \ldots, |X|\}$ is defined such that for all $u, v \in X$,

$$\sigma(u) < \sigma(v) \text{ if and only if } \sigma|_X (u) < \sigma|_X (v).$$

For an ordering ϕ, we represent $\phi(u) < \phi(v)$ as $u \prec_\phi v$ as well. We omit the subscript ϕ when the ordering is clear. Let X and Y be disjoint subsets of $V(G)$, such that in an ordering ϕ, $u \prec v$, for all $u \in X, v \in Y$. If the vertices of X are consecutive in ϕ; likewise, the vertices of Y are also consecutive, though there could be vertices between X and Y, we represent this as $\phi(X) \prec \phi(Y)$.

Definition 2. *The right degree of a vertex v in an ordering ϕ is defined as* $rd_\phi(v) = |\{u \in N_G(v) : \phi(u) > \phi(v)\}|$.

Lemma 1 ($\star$). [1]*In an optimal ordering ϕ, the sequence of right degrees of vertices is non-increasing.*

Lemma 2 ($\star$). *In an ordering ϕ of a graph G, locations of two consecutive non-adjacent vertices of equal right degrees can be swapped to get a new ordering with the same cost.*

The input to an Integer Quadratic Programming (IQP) problem is an $n \times n$ integer matrix Q, an $m \times n$ integer matrix A, and an m-dimensional integer vector b. The task is to find a vector $x \in \mathbb{Z}^+$ such that $x^T Q x$ is minimized subject to the constraints $Ax \le b$. Let L be the total number of bits required to encode the input IQP.

Proposition 1 (Lokshtanov [26]). *There exists an algorithm that, given an instance of Integer Quadratic Programming, runs in time $f(n, \alpha)L^{O(1)}$, and determines whether the instance is infeasible, feasible and unbounded, or feasible and bounded. If the instance is feasible and bounded, the algorithm outputs an optimal solution. Here α is the largest absolute value of an entry of Q and A.*

---

[1] Proofs of results marked with $\star$ are omitted due to paucity of space.

3   MSVC Parameterized by Vertex Cover

Let G be a graph with the size of a minimum vertex cover of G at most k, where k is a positive integer. Placing the vertices of a minimum vertex cover first in the ordering appears to be an appealing approach to achieve MSVC. However, graph G_2 in Fig. 1b shows this isn't always optimal. The red-colored vertices form the minimum vertex cover of G_2. Consider an ordering ψ with the first eight locations having these vertices. The order among these vertices doesn't matter, as each vertex covers exactly two unique edges (each edge is covered by exactly one vertex of vertex cover). And the remaining vertices of the graph are located 9th location onwards, in any order. The cost of such an ordering is 72. Consider an ordering ϕ, such that $\phi(v_1) = 1$, and v_1 is followed by $v_6, v_9, v_{12}, v_{15}, v_7, v_{10}, v_{13}, v_{16}$ respectively. The remaining vertices are ordered in any way after v_{16}. The cost of ϕ is 62. So, positioning the vertices of minimum vertex cover of size k, in the first k locations, doesn't always yield an optimal solution. Hence, a careful analysis is required for designing an FPT algorithm.

Let $S = \{v_1, v_2, \ldots, v_k\}$ be a minimum vertex cover of graph G. Then, $I = V(G) \setminus S$ is an independent set. And $N_S(u) = \{v \in S | \{u, v\} \in E(G)\}$. There are total $k!$ relative orderings of vertices of S. Consider one such ordering, $v_{i_1} \prec v_{i_2} \prec \cdots \prec v_{i_k}$, where $i_1, i_2, \ldots i_k$ is a permutation of $[k]$. The set of vertices of I that appear before v_{i_1} is represented by $Block$ 1. The set of vertices between v_{i_j} and $v_{i_{j+1}}$ is represented as $Block$ $j + 1$, for $j \in [k-1]$. And the set of vertices after v_{i_k} is called $Block$ $k + 1$. This depiction is shown in Fig. 2. In an optimal ordering, a vertex from I can be in any one of the $k + 1$ blocks. Consider the relation R on I as follows: uRw if $N_S(u) = N_S(w)$. This equivalence relation R partitions I into at most 2^k equivalence classes.

Fig. 2. $k + 1$ blocks in an ordering of $V(G)$

Lemma 3 ($\star$). *There is an optimal ordering ϕ such that, in every block, all the vertices from the same equivalence class are consecutive.*

Lemma 4. *Let S be a minimum vertex cover of G. There is an optimal ordering such that for every equivalence class A, all the vertices from A are consecutive.*

Proof. To prove the lemma, it is enough to show that there is an optimal ordering σ such that the following conditions are satisfied:

(i) Each equivalence class is contained in exactly one block.
(ii) For every block, all the vertices from the same equivalence class are consecutive.

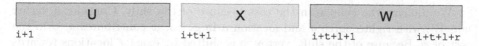

Fig. 3. Vertices of equivalence class A distributed in different blocks

(a) Ordering ψ_1 (b) Ordering ψ_2

Fig. 4. Cases of Lemma 4

Given an ordering ϕ, let f_ϕ be a function defined on the equivalence classes as

$f_\phi(A)$: the number of blocks that contain vertices from equivalence class A.

Notice that $f_\phi(A) \geq 1$, for all A. Let $Q(\phi) = \sum_A f_\phi(A)$.

Any optimal ordering can be modified to satisfy (ii) by Lemma 3, with the cost being unaffected. Let α be the number of equivalence classes formed by the relation R on $I = V(G) \setminus S$. We only need to prove there is an optimal ordering σ, satisfying (ii), with $Q(\sigma) = \alpha$. For contradiction, assume in every optimal solution satisfying (ii), there is at least one equivalence class whose vertices are not entirely contained in a single block. Consider one with the least value of Q, say ϕ, among all the optimal orderings, satisfying (ii). At least one equivalence class A exists whose vertices are not in one block.

Let $U = \{u_1, u_2, \ldots, u_t\}$ be a subset of A, with $\phi(u_1) = i + 1, \phi(u_2) = i + 2, \ldots, \phi(u_t) = i + t$, where $t \geq 1$ and $i \geq 0$. And $\phi^{-1}(i + t + 1) \notin A$. Let $X = \{\phi^{-1}(i + t + 1), \ldots, \phi^{-1}(i + t + \ell)\}$ be such that $X \cap S \neq \emptyset$. Let us denote these vertices in X as $x_1, x_2, \ldots, x_\ell$. Let $W = \{w_1, w_2, \ldots, w_r\}$ be the first appearance of vertices from A after U, with $\phi(w_1) = i + t + \ell + 1, \phi(w_2) = i + t + \ell + 2, \ldots, \phi(w_t) = i + t + \ell + r$, with $r \geq 1$, as shown in Fig. 3.

Let $v_1, v_2, \ldots, v_p$ be the vertices from S, adjacent to A, such that $\phi(v_1) = i_1, \phi(v_2) = i_2, \ldots, \phi(v_p) = i_p$, where $i + t < i_1 < i_2 < \cdots < i_p \leq i + t + \ell$. There can be more vertices from S in X that are not adjacent to A, but we do not care about their location. Let the vertices in W have the right degree d. Then the vertices in U have the right degree $d + p$.

We claim that $p \geq 1$. If $p = 0$, then the right degree of vertices in U and W is d. And by Lemma 1, all the vertices between U and W have the right degree d. Hence, by repeated application of Lemma 2, U can be shifted to W or W to U without affecting the cost and minimizing the value of $Q(\phi)$, a contradiction.

Case(i): Consider a new ordering ψ_1, where the entire set W is shifted next to U. The weight for all the edges incident on vertices up to U and after W, from ϕ will remain unchanged. The weight changes for edges incident on X and W. The set W shifts to the left by ℓ locations, and the set X shifts to the right by r locations, as shown in Fig. 4a.

We analyze the change in the cost because of the shift. Consider a vertex $w \in W$, with right degree d and location a in ϕ. So, w contributes $a \cdot d$ in the cost of ϕ. Because of the shift, vertex w is shifted by exact ℓ locations towards the left. It now contributes $(a - \ell) \cdot d$ in the cost of ψ. The change in the cost because of w is $-\ell \cdot d$. This holds true for each vertex in W. As there are r vertices, each with the right degree d, the total cost change is $-\ell d r$.

The degrees of all the vertices of X, which are not adjacent to W, remain unchanged by this shift. Let x_i be one such vertex at location b. It contributes $b \cdot rd_\phi(x_i)$ in ϕ. Because of the shift of the r locations, it now contributes $(b + r) \cdot rd_\phi(x_i)$ in ψ. Hence, the change in the cost is $r \cdot rd_\phi(x_i)$. The increased cost due to all such vertices is

$$\left[\sum_{i \in [\ell] \setminus \{i_1, i_2, \ldots i_p\}} rd_\phi(x_i) \right] \cdot r.$$

A modulator vertex v_p is adjacent to all the vertices of W. But $w \prec_{\psi_1} v_p$. So, the right degree of v_p decreases by $|W| = r$. And its location shifts by r locations. As it is the same for all the vertices $v_1, \ldots, v_p$, the change in the cost due to these vertices is

$$\left[\sum_{i \in \{i_1, \ldots i_p\}} (rd_\phi(x_i) - r) \right] \cdot r$$

The weight of an edge $e = \{v_1, w_1\}$ is i_1 in ϕ. But, $w_1 \prec_{\psi_1} v_1$, and e's weight now is the location of w_1 in ψ_1, i.e. $i + t + 1$. Hence, the change in the weight is $(i + t + 1 - i_1)$. Similarly, for the edge $\{v_1, w_j\}$, for $j \in [r]$, change in its weight is $(i + t + j - i_1)$. In the same manner, the change in the weight of the edges $\{v_2, w_j\}$ becomes $(i + t + j - i_2)$, for $j \in [r]$. And it follows a similar pattern for edges incident on $v_3, \ldots, v_p$ as well. The weight of the rest of the edges remains unaffected. Hence, the total change in the cost is: $\mu_G(\psi_1) - \mu_G(\phi) =$

$$- \ell d r + \left[\sum_{i \notin \{i_1, i_2, \ldots i_p\}, i=1}^{\ell} rd_\phi(x_i) \right] \cdot r + \left[\sum_{i \in \{i_1, \ldots i_p\}} (rd_\phi(x_i) - r) \right] \cdot r$$

$$+ \sum_{j=1}^{r} (i + t + j - i_1) + \sum_{j=1}^{r} (i + t + j - i_2) + \cdots + \sum_{j=1}^{r} (i + t + j - i_p)$$

$$= -\ell d r + r \cdot \sum_{i=1}^{\ell} rd_\phi(x_i) - pr^2 - r(i_1 + \ldots i_p) + p[(i + t + 1) + \ldots (i + t + r)]$$

$$= -\ell d r + r \cdot \sum_{i=1}^{\ell} rd_\phi(x_i) - pr^2 - r(i_1 + \cdots + i_p) + pir + ptr + pr(r+1)/2$$

$$= r \cdot \sum_{i=1}^{l} rd_\phi(x_i) - r \cdot y - pr^2/2 \quad (\text{where, } y = \ell d + (i_1 + \ldots i_p) - pi - pt - p/2)$$

So, we have, $\mu_G(\psi_1) - \mu_G(\phi) = r \cdot \left[\sum_{i=1}^{l} rd_\phi(x_i) - y - pr/2 \right]$.

As $r > 0$, if $\displaystyle\sum_{i=1}^{l} rd_\phi(x_i) < y + pr/2$, then $\mu_G(\psi_1) - \mu_G(\phi) < 0$.   (1)

Case(ii): Let ψ_2 be a new ordering obtained by swapping the positions of U and X. It is shown in Fig. 4b. Weights of the edges $\{u_j, v_1\}$, $j \in [t]$, change from $i + j$ to $i_1 - t$. Similarly, weights of the edges $\{u_j, v_q\}$, $j \in [t]$, $q \in [p]$, change from $i + j$ to $i_q - t$. For the remaining d edges incident on U, their weight increases by ℓ. As there are t vertices in U each of degree d, the total increase cost is ℓdt. And for the remaining edges of X, their weights are reduced by the factor t. The total change in the cost function is as follows: $\mu_G(\psi_2) - \mu_G(\phi) =$

$$\ell dt - \left[\sum_{i=1}^{\ell} rd_\phi(x_i) \right] \cdot t + \sum_{j=1}^{t} [i_1 - t - (i + j)]$$

$$+ \sum_{j=1}^{t} [i_2 - t - (i + j)] + \cdots + \sum_{j=1}^{t} [i_p - t - (i + j)]$$

$$= \ell dt - t \sum_{i=1}^{l} rd_\phi(x_i) - pt^2 + t(i_1 + \cdots + i_p) - p[(i + 1) + (i + 2) + \cdots + (i + t)]$$

$$= -t \cdot \sum_{i=1}^{l} rd_\phi(x_i) + ty - pt^2/2 \quad (\text{as, } y = \ell d + (i_1 + \cdots + i_p) - pi - pt - p/2)$$

We have $\mu_G(\psi_2) - \mu_G(\phi) = t \cdot \left[-\sum_{i=1}^{\ell} rd_\phi(x_i) + y - pt/2 \right]$.

As $t > 0$, we have, if $\displaystyle\sum_{i=1}^{\ell} rd_\phi(x_i) > y - pt/2$, then $\mu_G(\psi_2) - \mu_G(\phi) < 0$ (2)

Notice that $p, t, r > 0$, this implies $\sum_{i=1}^{l} rd_\phi(x_i) > y - pt/2$ or $\sum_{i=1}^{l} rd_\phi (x_i) < y + pr/2$ (or both). Hence, corresponding to the sum of the right degrees of vertices in X, shifting U towards W or vice versa gives an ordering with a cost less than the optimal value, a contradiction. Hence, our assumption that Q is greater than the number of equivalence classes in every optimal ordering is wrong.   □

Let $\mathcal{S} = \{\sigma_1, \sigma_2, \ldots \sigma_{k!}\}$ be the set of all permutations of vertices in S, and $A_1, A_2, \ldots, A_q$ be all the equivalence classes. By Lemma 4, each equivalence class is contained in exactly one block, and vertices from each equivalence class are consecutive. With $k + 1$ blocks and q equivalence classes, there are $(k + 1)^q$ possible choices. We denote a choice as c_j for $j \in \{1, 2, \ldots, (k + 1)^q\}$. Let $\mathcal{E}$ denote the set of all possible choices. A configuration (σ_i, c_j) corresponds to the permutation σ_i of $\mathcal{S}$, and choice c_j of $\mathcal{E}$ for the equivalence classes, where $i \in [k!], j \in [(k+1)^q]$. An equivalence class, once fixed in one of the $k+1$ blocks, in a configuration, (σ_i, c_j), its right degree can be easily calculated, as it is adjacent to only vertices from vertex cover. In a configuration (σ_i, c_j), the relative order between the equivalence classes in any block is decided by their right degrees by Lemma 1. The equivalence classes are arranged such that their right degrees are non-increasing. A configuration with minimum cost is an optimal ordering for MSVC.

It takes $(1.2738)^k n^{O(1)}$ time to find a vertex cover of size at most k. Hence, the FPT running time of the algorithm is $[k!(k+1)^{2^k} + (1.2738)^k]n^{O(1)}$. Correctness follows from Lemma 4.

4   MSVC Parameterized by Clique Modulator

In this section, we prove that the MSVC parameterized by the clique modulator is fixed-parameter tractable.

Definition 3. *A set $M \subset V(G)$ is called a clique modulator of the graph G, if the graph induced on $V(G) \setminus M$ is a clique.*

The input of the problem is a positive integer k, and graph G with the size of a minimum clique modulator at most k. The question is to find an optimal ordering for MSVC of G. We use Integer Quadratic Programming (IQP) to construct the solution for MSVC parameterized by the clique modulator. IQP can be solved in time $f(t, \alpha)n^{O(1)}$ time, where f is some computable function, and t is the number of variables in IQP, and α is an upper bound for the coefficients of variables in IQP [26].

Let $M = \{v_1, v_2, \ldots, v_k\}$ be a clique modulator of the graph G. Then, $Q = V(G) \setminus M$ is a clique. Consider the relation R on Q as follows: a vertex u is related to w if $N_M(u) = N_M(w)$. It is an equivalence relation, partitioning Q into equivalence classes $A_1, A_2, \ldots, A_\ell$, where $\ell \le 2^k$.

In any optimal ordering, each equivalence class doesn't need to be entirely contained within one block, unlike when the parameter is a vertex cover. But, in an optimal solution, vertices from each equivalence class can be arranged consecutively within each block, with the cost being unaffected (see Lemma 6). Once the relative ordering of modulator vertices is guessed, Lemma 1 enables us to determine the ordering of equivalence classes in each block. The distinguishing factor to get this ordering is their neighbors in M. For a given relative ordering σ_M of M, for each block B_i, $i \in [k+1]$, we define right modulator degree for a vertex $u \in B_i$ as

$$\mathrm{rm}_{\sigma_M}(u, i) = |\{v \in N_M(u) : \sigma_M(v) \ge i\}|.$$

For any two vertices u and v in an equivalence class A, $\mathrm{rm}_{\sigma_M}(u, i) = \mathrm{rm}_{\sigma_M}(v, i)$. We call this the right modulator degree of equivalence class A, $\mathrm{rm}_{\sigma_M}(A, i)$ in B_i.

First, we prove that for any ordering σ of $V(G)$, if we sort the equivalence classes based on their right modulator degree in each block, then the cost of this new ordering is at most the cost of σ. Let $\sigma_M = \sigma|_M$.

Lemma 5 ($\star$). *Let σ_M be a permutation of M. Let $g_1 : [\ell] \to \{A_1, A_2, \ldots A_\ell\}$ be a bijection such that $\mathrm{rm}_{\sigma_M}(g_1(1), 1) \geq \mathrm{rm}_{\sigma_M}(g_1(2), 1) \geq \cdots \geq \mathrm{rm}_{\sigma_M}(g_1(\ell), 1)$. For any ordering σ of $V(G)$, with $\sigma|_M = \sigma_M$, let $\widehat{\sigma}$ be an ordering of $V(G)$ such that*

$$\widehat{\sigma}(Y \cap g_1(1)) \prec \widehat{\sigma}(Y \cap g_1(2)) \prec \cdots \prec \widehat{\sigma}(Y \cap g_1(\ell)) \prec \sigma_{>|Y|}$$

where the set of vertices of the clique $(V(G) \setminus M)$ in block 1 of σ is denoted by Y, and $\sigma_{>|Y|}$ denotes the ordering of vertices after the first $|Y|$ locations. Then, $\mu_G(\widehat{\sigma}) \leq \mu_G(\sigma)$.

Now, we extend this analysis to the remaining blocks.

Lemma 6. *Let σ_M, a permutation of M, be given. Let $g_i : [\ell] \to \{A_1, A_2, \ldots A_\ell\}$ be a bijection such that $\mathrm{rm}_{\sigma_M}(g_i(1), i) \geq \mathrm{rm}_{\sigma_M}(g_i(2), i) \geq \cdots \geq \mathrm{rm}_{\sigma_M}(g_i(\ell), i)$, for all $i \in [k + 1]$. For any ordering σ of $V(G)$, with $\sigma|_M = \sigma_M$, let $\widehat{\sigma}$ be an ordering of $V(G)$ such that*

$$\widehat{\sigma} = \widehat{\sigma}_1(Y_1) \prec \sigma_M^{-1}(1) \prec \widehat{\sigma}_2(Y_2) \prec \cdots \prec \sigma_M^{-1}(k) \prec \widehat{\sigma}_{k+1}(Y_{k+1}).$$

where the set of vertices of the clique $(V(G) \setminus M)$ in block i of σ is denoted by Y_i and for each $i \in [k + 1]$, $\widehat{\sigma}_i(Y_i) = (Y_i \cap g_i(1)) \prec (Y_i \cap g_i(2)) \prec \cdots \prec (Y_i \cap g_i(\ell))$. Then, $\mu_G(\widehat{\sigma}) \leq \mu_G(\sigma)$.

Proof. Repeatedly apply the technique of Lemma 5 for each block $i \in [k+1]$, to get the ordering $\widehat{\sigma}$. The proof follows from Lemma 5. $\square$

We call a permutation of the form $\widehat{\sigma}$ in Lemma 6, a nice permutation.

4.1   Integer Quadratic Programming

Let G be the input graph and M be a clique modulator of size k. And $Q = V(G) \setminus M$ is a clique of size $n - k$. Let $A_1, \ldots, A_\ell$ be the number of equivalence classes that partition Q.

As a first step, we guess the ordering σ_M of M such that there is an optimum ordering σ with $\sigma|_M = \sigma_M$. Lemma 6 implies that we know the ordering of vertices within each block irrespective of the number of vertices from equivalence classes in each block. We use IQP to determine the number of vertices from each equivalence class that will be present in different blocks. Thus, for each $i \in [\ell]$ and $j \in [k+1]$, we use a variable x_{ij} that represents the number of vertices from A_i in the block j.

Recall that Q is a clique of size $n - k$. Consider an arbitrary ordering of vertices of Q. Its cost, denoted by $\mu(Q)$ is

$$\mu(Q) = 1 \cdot (n - k - 1) + 2 \cdot (n - k - 2) + \cdots + (n - k - 1)(1).$$

This cost remains consistent for any ordering of Q.

For any ordering ϕ of $V(G)$, $\mu_G(\phi) \geq \mu(Q)$. And $\mu(Q)$ can be found beforehand. We treat $\mu(Q)$ as the base cost for G. We need to minimize the cost after introducing the modulator vertices. Let the first modulator vertex v_1 be introduced in an ordering of Q. Assume, there are n_1 vertices of Q after v_1. Then, the weight of each edge whose both endpoints are after v_1 increases by 1. There are $\binom{n_1}{2}$ such clique edges. Hence, the total increase in the cost with respect to the clique edges is $\binom{n_1}{2}$. A modulator vertex v_i is introduced after v_{i-1} and $v_{i-1} \prec v_i$, for all $i \in \{2, \ldots, k\}$. Let n_i be the number of clique vertices after v_i in the ordering. Hence, inclusion of v_i causes the cost to increase by $\binom{n_i}{2}$. The total increase in the weights of clique edges after including all the modulator vertices is $\sum_{i=1}^{k} \binom{n_i}{2}$.

Thus, in the IQP, we also use variables $n_1, n_2, \ldots, n_k$. To minimize the cost, we need to find the ordering of modulator vertices to be introduced and their location. Since σ_M is fixed, we know the relative ordering of equivalence classes in each block using Lemma 6. Recall that we have a variable x_{ij} for each equivalence class A_i and block B_j. The variable n_i represents the number of vertices from Q after v_i in the hypothetical optimum solution σ. This implies that in IQP, we need to satisfy the constraints of the following form.

$$n_p = \sum_{j=p+1}^{k+1} \sum_{i=1}^{\ell} x_{ij} \qquad \text{for all } p \in [k]$$

$$|A_i| = \sum_{j=1}^{k+1} x_{ij} \qquad \text{for all } i \in [\ell]$$

We explained that after introducing the modulator vertices, the increase in cost due to edges in the clique is given by the expression $\sum_{i=1}^{k} \binom{n_i}{2}$. Next, we need an expression regarding the increase in cost due to edges incident on modulator vertices. Towards that, we introduce a variable y_p for each p to indicate the location of the modulator vertex v_p in the hypothetical optimal ordering σ. Clearly, we should satisfy the following constraints.

$$y_p = n - (n_p + k - p) \qquad \text{for all } p \in [k]$$

There are n_p vertices from Q after v_p and $k - p$ vertices from M after v_p in the hypothetical optimum ordering σ. Now we explain how to get the increase in the cost due to edges incident on the modulator vertex v_p. We know its position is y_p. We need to find its right degree. We use a variable d_p to denote the right degree of p. Let $\text{rm}_p = |\{v \in N_M(v_p) : \sigma_M(v) > \sigma_M(v_p)\}|$. That is, rm_p is the number of edges between the modulator vertices with one endpoint v_p and the other in $\{v_{p+1}, \ldots, v_k\}$. Let $I_p \subseteq [\ell]$ such that $i \in I_p$ if and only if v_p is adjacent to the vertices in A_i. Then, the number of neighbors of v_p in the clique Q that appear to the right of v_p in the hypothetical solution is $\sum_{j=p+1}^{k+1} \sum_{i \in I_p} x_{ij}$. This implies that we have the following constraints.

$$d_p = \text{rm}_p + \sum_{j=p+1}^{k+1} \sum_{i \in I_p} x_{ij} \qquad \text{for all } p \in [k]$$

The increase in cost due to edges incident on modulator vertices such that the endpoints of those edges in the modulator appear before the other endpoints is $\sum_{p=1}^{k} d_p \cdot y_p$. Now, we need to consider the cost of edges between the clique vertices and the modulator vertices such that clique vertices appear before the modulator vertex. We use r_{ij} to denote the right modulator degree of vertices in the equivalence class A_i from j^{th} block. It is defined as follows. Let $u \in A_i$ be a fixed vertex. Then, $r_{ij} = |\{v \in N_M(u) : \sigma_M(v) \geq j\}|$. Notice that r_{ij} is a constant, less than or equal to k. Let y_{ij} denote the location of the first vertex from the equivalence class A_i in block j in the hypothetical solution σ. Then, we should have the following constraint. Here, we use Lemma 6 and recall the bijection g_j. Let $J_{i,j} \subseteq [\ell]$ be such that $q \in J_{i,j}$ if and only if $g_j^{-1}(A_q) < g_j^{-1}(A_i)$.

$$y_{ij} = y_{j-1} + \sum_{q \in J_{i,j}} x_{qj}.$$

Here, we set $y_0 = 0$. The cost due to edges from A_i in block j to modulator vertices to its right in the ordering is

$$r_{ij}[y_{ij} + (y_{ij} + 1) + \ldots + (y_{ij} + x_{ij} - 1)] = r_{ij}\left(x_{ij} \cdot y_{ij} + \binom{x_{ij}}{2}\right).$$

Thus, we summarize our IQP as follows.

Minimize $\sum_{p=1}^{k} d_p \cdot y_p + \sum_{i=1}^{k} \binom{n_i}{2} + \sum_{i=1}^{\ell} \sum_{j=1}^{k+1} r_{ij}\left(x_{ij} \cdot y_{ij} + \binom{x_{ij}}{2}\right)$

Subject to

$$n_p = \sum_{j=p+1}^{k+1} \sum_{i=1}^{\ell} x_{ij} \qquad \text{for all } p \in [k]$$

$$|A_i| = \sum_{j=1}^{k+1} x_{ij} \qquad \text{for all } i \in [\ell]$$

$$y_p = n - (n_p + k - p) \qquad \text{for all } p \in [k]$$

$$d_p = \mathrm{rm}_p + \sum_{j=p+1}^{k+1} \sum_{i \in I_p} x_{ij} \qquad \text{for all } p \in [k]$$

$$y_{ij} = y_{j-1} + \sum_{q \in J_{i,j}} x_{qj} \qquad \text{for all } i \in [\ell] \text{ and } j \in [k+1]$$

Here, n, k, r_{ij} and rm_p are constants. Also, $r_{ij}, \mathrm{rm}_p \leq k$ for all i, j and p. This completes the construction of an IQP instance for a fixed permutation σ_M of M. Notice that, there are at most $2^k(k+1)$ variables x_{ij} and $2^k(k+1)$ variables y_{ij}. And there are k variables for y_p and n_p each. Hence, the total number of variables in the above IQP formulation is at most $2^{k+1} \cdot (k+1) + 2k$.

Next, we prove that the coefficients in the objective functions and the constraints are upper bounded by a function of k. The coefficient of $y_p \cdot d_p$ is at most 1 in the objective function. The coefficients of n_i and n^2 are also upper bounded by a constant. As $r_{ij} \leq k$, the coefficients of $(x_{ij})^2$, $x_{ij} \cdot y_{ij}$, and x_{ij} are all at most k. Similarly, coefficients are bounded above by $O(k)$ in the constraints. Thus, by Proposition 1, we can solve the above IQP instance in time $f(k)n^{O(1)}$ time for a computable function f.

Lemma 7 ($\star$). *Let σ_M be an ordering of M such that there is an optimum ordering σ with $\sigma|_M = \sigma_M$. An optimal solution to the IQP defined above is an optimal solution to MSVC, and vice versa.*

For each ordering σ_M of M, we construct an IQP instance as described above and solve. Finally, we consider the best solution among them and construct an ordering from it according to the explanation given for the IQP formulation. Correctness of the algorithm follows from Lemma 7.

References

1. Bakken, O.R.: Arrangement problems parameterized by neighbourhood diversity. Master's thesis, The University of Bergen (2018)
2. Bansal, N., Batra, J., Farhadi, M., Tetali, P.: Improved approximations for min sum vertex cover and generalized min sum set cover. In: Proceedings of the 2021 ACM-SIAM Symposium on Discrete Algorithms (SODA), pp. 998–1005 (2021)
3. Bar-Noy, A., Bellare, M., Halldórsson, M.M., Shachnai, H., Tamir, T.: On chromatic sums and distributed resource allocation. Inf. Comput. **140**(2), 183–202 (1998)
4. Barenholz, U., Feige, U., Peleg, D., et al.: Improved approximation for min-sum vertex cover. **81**, 06–07 (2006). http://wisdomarchive.wisdom.weizmann.ac.il
5. Basiak, M., Bienkowski, M., Tatarczuk, A.: An improved deterministic algorithm for the online min-sum set cover problem. arXiv preprint arXiv:2306.17755 (2023)
6. Bienkowski, M., Mucha, M.: An improved algorithm for online min-sum set cover. In: AAAI, pp. 6815–6822 (2023)
7. Burer, S., Monteiro, R.D.: A projected gradient algorithm for solving the maxcut sdp relaxation. Optim. Methods Softw. **15**(3–4), 175–200 (2001)
8. Chen, J., Kanj, I.A., Xia, G.: Improved upper bounds for vertex cover. Theoret. Comput. Sci. **411**(40–42), 3736–3756 (2010)
9. Chung, F.R.: On optimal linear arrangements of trees. Comput. Math. Appl. **10**(1), 43–60 (1984)
10. Dregi, M.S., Lokshtanov, D.: Parameterized complexity of bandwidth on trees. In: Esparza, J., Fraigniaud, P., Husfeldt, T., Koutsoupias, E. (eds.) Automata, Languages, and Programming, pp. 405–416. Springer Berlin Heidelberg, Berlin, Heidelberg (2014). https://doi.org/10.1007/978-3-662-43948-7_34
11. Dubey, C., Feige, U., Unger, W.: Hardness results for approximating the bandwidth. J. Comput. Syst. Sci. **77**(1), 62–90 (2011)
12. Feige, U., Lovász, L., Tetali, P.: Approximating min sum set cover. Algorithmica **40**, 219–234 (2004)
13. Fellows, M.R., Lokshtanov, D., Misra, N., Rosamond, F.A., Saurabh, S.: Graph layout problems parameterized by vertex cover. In: Hong, S.-H., Nagamochi, H., Fukunaga, T. (eds.) Algorithms and Computation, pp. 294–305. Springer Berlin Heidelberg, Berlin, Heidelberg (2008). https://doi.org/10.1007/978-3-540-92182-0_28

14. Fernau, H.: Parameterized algorithmics for linear arrangement problems. Discret. Appl. Math. **156**(17), 3166–3177 (2008)
15. Fotakis, D., Kavouras, L., Koumoutsos, G., Skoulakis, S., Vardas, M.: The online min-sum set cover problem. In: 47th International Colloquium on Automata, Languages, and Programming, ICALP, pp. 51:1–51:16 (2020)
16. Garey, M.R., Graham, R.L., Johnson, D.S., Knuth, D.E.: Complexity results for bandwidth minimization. SIAM J. Appl. Math. **34**(3), 477–495 (1978)
17. Garey, M.R., Johnson, D.S., Stockmeyer, L.: Some simplified np-complete problems. In: Symposium on the Theory of Computing (STOC), pp. 47–63 (1974)
18. Gavril, F.: Some np-complete problems on graphs. In: Proceedings of the Conference on Information Science and Systems, 1977, pp. 91–95 (1977)
19. Gera, R., Rasmussen, C., Stanica, P., Horton, S.: Results on the min-sum vertex cover problem. Tech. rep, NAVAL POSTGRADUATE SCHOOL MONTEREY CA DEPT OF APPLIED MATHEMATICS (2006)
20. Gima, T., Kim, E.J., Köhler, N., Melissinos, N., Vasilakis, M.: Bandwidth parameterized by cluster vertex deletion number. arXiv preprint arXiv:2309.17204 (2023)
21. Gurari, E.M., Sudborough, I.H.: Improved dynamic programming algorithms for bandwidth minimization and the mincut linear arrangement problem. J. Algorithms **5**(4), 531–546 (1984)
22. Gutin, G., Rafiey, A., Szeider, S., Yeo, A.: The linear arrangement problem parameterized above guaranteed value. Theor. Comput. Syst. **41**, 521–538 (2007)
23. Harper, L.H.: Optimal assignments of numbers to vertices. J. Soc. Ind. Appl. Math. **12**(1), 131–135 (1964)
24. Karp, R.M.: Reducibility among combinatorial problems. In: Complexity of computer computations Proc. Sympos., pp. 85–103 (1972)
25. Khot, S., Regev, O.: Vertex cover might be hard to approximate to within 2- ε. J. Comput. Syst. Sci. **74**(3), 335–349 (2008)
26. Lokshtanov, D.: Parameterized integer quadratic programming: Variables and coefficients. arXiv preprint arXiv:1511.00310 (2015)
27. Makedon, F.S., Papadimitriou, C.H., Sudborough, I.H.: Topological bandwidth. SIAM J. Algebraic Discr. Methods **6**(3), 418–444 (1985)
28. Mohan, S.R., Acharya, B., Acharya, M.: A sufficiency condition for graphs to admit greedy algorithm in solving the minimum sum vertex cover problem. In: International Conference on Process Automation, Control and Computing, pp. 1–5. IEEE (2011)
29. Monien, B.: The bandwidth minimization problem for caterpillars with hair length 3 is np-complete. SIAM J. Algebraic Discr. Methods **7**(4), 505–512 (1986)
30. Papadimitriou, C.H.: The np-completeness of the bandwidth minimization problem. Computing **16**(3), 263–270 (1976)
31. Rasmussen, C.W.: On efficient construction of minimum-sum vertex covers (2006)
32. Stanković, A.: Some results on approximability of minimum sum vertex cover. In: Approximation, Randomization, and Combinatorial Optimization. Algorithms and Techniques (APPROX/RANDOM 2022). vol. 245, pp. 50:1–50:16 (2022)
33. Thilikos, D.M., Serna, M., Bodlaender, H.L.: Cutwidth i: a linear time fixed parameter algorithm. J. Algorithms **56**(1), 1–24 (2005)
34. Yannakakis, M.: A polynomial algorithm for the min-cut linear arrangement of trees. J. ACM (JACM) **32**(4), 950–988 (1985)

A Polynomial Kernel for Proper Helly Circular-Arc Vertex Deletion

Akanksha Agrawal[1], Satyabrata Jana[2(✉)], and Abhishek Sahu[3]

[1] Indian Institute of Technology Madras, Chennai, India
akanksha@cse.iitm.ac.in
[2] The Institute of Mathematical Sciences, HBNI, Chennai, India
satyamtma@gmail.com
[3] National Institute of Science, Education and Research, An OCC of Homi Bhabha National Institute, Bhubaneswar 752050, Odisha, India

Abstract. A *proper Helly circular-arc graph* is an intersection graph of a set of arcs on a circle such that none of the arcs properly contains any other arc and every set of pairwise intersecting arcs has a common intersection. The PROPER HELLY CIRCULAR-ARC VERTEX DELETION problem takes as input a graph G and an integer k, and the goal is to check if we can remove at most k vertices from the graph to obtain a proper Helly circular-arc graph; the parameter is k. Recently, Cao et al. [MFCS 2023] obtained an FPT algorithm for this (and related) problem. In this work, we obtain a polynomial kernel for the problem.

1 Introduction

The development of parameterized complexity is much owes much to the study of graph modification problems, which have inspired the evolution of many important tools and techniques. One area of parameterized complexity is data reduction, also known as kernelization, which focuses on the family of graphs $\mathcal{F}$ and the $\mathcal{F}$-MODIFICATION problem. Given a graph G and an integer k, this problem asks whether it is possible to obtain a graph in $\mathcal{F}$ using at most k modifications in G, where the modifications are limited to vertex deletions, edge deletions, edge additions, and edge contractions. The problem has been extensively studied, even when only a few of these operations are allowed.

Here we deal on the parameterization of the $\mathcal{F}$-VERTEX DELETION problem, which is a special case of $\mathcal{F}$-MODIFICATION where the objective is to find the minimum number of vertex deletions required to obtain a graph in $\mathcal{F}$. This problem encompasses several well-known NP-complete problems, such as VERTEX COVER, FEEDBACK VERTEX SET, ODD CYCLE TRANSVERSAL, PLANAR VERTEX DELETION, CHORDAL VERTEX DELETION, and INTERVAL VERTEX DELETION, which correspond to $\mathcal{F}$ being the family of graphs that are edgeless, forests, bipartite, planar, chordal and interval, respectively. Unfortunately, most of these problems are known to be NP-complete, and therefore have been extensively studied in paradigms such as parameterized complexity designed to cope

J. A. Soto and A. Wiese (Eds.): LATIN 2024, LNCS 14579, pp. 208–222, 2024.
https://doi.org/10.1007/978-3-031-55601-2_14

with NP-hardness. There have been many studies on this topic, including those referenced in this paper, but this list is not exhaustive.

In this article, we focus on the $\mathcal{F}$-VERTEX DELETION problem, specifically when $\mathcal{F}$ refers to the family of proper Helly circular-arc graphs. We refer to this problem as PROPER HELLY CIRCULAR-ARC VERTEX DELETION (PHCAVD) for brevity. A circular-arc graph is a graph whose vertices can be assigned to arcs on a circle such that there is an edge between two vertices if and only if their corresponding arcs intersect. If none of the arcs properly contains one another, the graph is a proper circular-arc graph. These graphs have been extensively studied, and their structures and recognition are well understood [9,12,16]. These graphs also arise naturally when considering the clique graphs of a circular-arc graph. However, the lack of the Helly property, which dictates that every set of intersecting arcs has a common intersection, contributes to the complicated structures of circular-arc graphs. A Helly circular-arc graph is a graph that admits a Helly arc representation. All interval graphs are Helly circular-arc graphs since every interval representation is Helly. The class of proper Helly circular-arc graphs lies between proper circular-arc graphs and proper interval graphs. A graph is a proper Helly circular-arc graph if it has a proper and Helly arc representation. Circular-arc graphs are a well-studied graph class due to their intriguing combinatorial properties and modeling power [11]. Additionally, there exists a linear-time algorithm to determine if a given graph is a circular-arc graph and construct a corresponding arc representation if so [19], even for Helly circular-arc graphs, such algorithm exists [17].

For graph modification problems, the number of allowed modifications, k, is considered the *parameter*. With respect to k, such a problem is said to be *fixed-parameter tractable* (FPT) if it admits an algorithm running in time $f(k)n^{\mathcal{O}(1)}$ for some computable function f. Also, the problem is said to have a polynomial kernel if in polynomial time (with respect to the size of the instance) one can obtain an equivalent instance of polynomial size (with respect to the parameter), i.e., for any given instance (G, k) of the problem, it can be reduced in time $n^{\mathcal{O}(1)}$ to an equivalent instance (G', k') where $|V(G')|$ and k' are upper bounded by $k^{\mathcal{O}(1)}$. A kernel for a problem immediately implies that it admits an FPT algorithm, but kernels are also interesting in their own right. In particular, kernels allow us to model the performance of polynomial-time preprocessing algorithms. The field of kernelization has received considerable attention, especially after the introduction of methods to prove kernelization lower bounds [4]. We refer to the books [6,8], for a detailed treatment of the area of kernelization.

Designing polynomial kernels for problems such as CHORDAL VERTEX DELETION [1] and INTERVAL VERTEX DELETION [2] posed several challenges. In fact, kernels for these problems were obtained only recently, after their status being open for quite some time. PROPER HELLY CIRCULAR-ARC VERTEX DELETION has remained an interesting problem in this area. Recently, Cao et al. [5] studied this problem and showed that it admits a factor 6-approximation algorithm, as well as an FPT algorithm that runs in time $6^k \cdot n^{\mathcal{O}(1)}$.

A natural follow-up question to the prior work on this problem is to check whether PHCAVD admits a polynomial kernel. In this paper, we resolve this question in the affirmative way.

PROPER HELLY CIRCULAR-ARC VERTEX DELETION (PHCAVD)
Input: A graph G and an integer k.
Parameter: k
Output: Does there exist a subset $S \subseteq V(G)$ of size at most k such that $G - S$ is a proper Helly circular-arc graph?

Theorem 1. PROPER HELLY CIRCULAR-ARC VERTEX DELETION *admits a polynomial kernel.*

1.1 Methods

Our kernelization heavily uses the characterization of proper Helly circular-arc graphs in terms of their *forbidden induced subgraphs*, also called *obstructions*. Specifically, a graph H is an obstruction to the class of proper Helly circular-arc graphs if H is not proper Helly circular-arc graph but $H - \{v\}$ is proper Helly circular-arc graph for every vertex $v \in V(H)$. A graph G is a proper Helly circular-arc graph if and only if it does not contain any of the following obstructions as induced subgraphs, which are $\overline{C_3^*}$ (claw), S_3 (tent), $\overline{S_3}$ (net), W_4 (wheel of size 4) , W_5 (wheel of size 5), $\overline{C_6}$ as well as a family of graphs: C_ℓ^*, $\ell \geq 4$ referred to as a Monad of size ℓ (see Fig. 2) [5,15]. We call any obstruction of size less than 12 a *small obstruction*, and call all other obstructions *large obstructions*. Note that every *large obstruction* is a Monad of size at least 12.

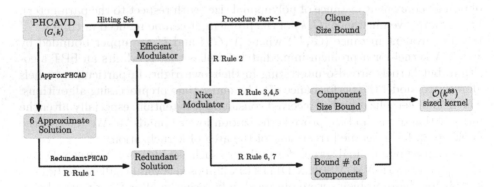

Fig. 1. Flowchart of the Kernelization algorithm for PHCAVD.

The first ingredient of our kernelization algorithm is the factor 6 polynomial-time approximation algorithm for PHCAVD given by Cao et al. [5]. We use this algorithm to obtain an approximate solution of size at most $6k$, or conclude that there is no solution of size at most k. We grow (extend) this approximate solution to a set T_1 of size $\mathcal{O}(k^{12})$, such that every set $Y \subseteq V(G)$ of size at most k is a minimal hitting set for all *small* obstructions in G if and only if Y is a minimal hitting set for all *small* obstructions in $G[T_1]$. Notice that $G - T_1$ is a proper Helly circular-arc graph (we call T_1 as an *efficient modulator*, description prescribed in Lemma 1), where for any minimal (or minimum) solution S of size at most k, the only purpose of vertices in $S \setminus T_1$ is to hit *large* obstructions. This T_1 is the first part of the *nice modulator* T that we want to construct. The other part is M, which is a 5-redundant solution (see Definition 3) of size $\mathcal{O}(k^6)$, which we obtain in polynomial time following the same construction procedure given by [2]. This gives us the additional property that any obstruction of size at least 5 contains at least 5 vertices from M and hence also from $T = T_1 \cup M$. We bound the size of such a *nice modulator* T by $\mathcal{O}(k^{12})$. Next, we analyze the graph $G - T$ and reduce its size by applying various reduction rules.

For the kernelization algorithm, we look at $G - T$, which is a proper Helly circular-arc graph and hence has a *"nice clique partition"* (defined in Sect. 2). Let $\mathcal{Q} = \{Q_1, Q_2, \ldots\}$ denote such a *nice clique partition* of $G-T$. This partition is similar to the clique partition used by Ke et al. [13] to design a polynomial kernel for vertex deletion to proper interval graphs.

In the first phase, we bound the size of a clique Q_i for each $Q_i \in \mathcal{Q}$. Our clique-reduction procedure is based on "irrelevant vertex rule" [18]. In particular, we find a vertex that is not necessary for a solution of size at most k, and delete it. And after this procedure, we reduce the size of each clique in $G - T$ to $k^{\mathcal{O}(1)}$.

In the second phase, we bound the size of each connected component in $G-T$. Towards this, we bound the number of cliques in $Q_1, Q_2, \ldots, Q_t$ that contain a neighbor of a vertex in T (say *good cliques*). We use *small* obstructions, and in particular, the claw, to bound the number of good cliques by $k^{\mathcal{O}(1)}$. This automatically divides the clique partition into *chunks*. A chunk is a maximal set of *non-good cliques* between a pair of *good cliques* where the *non-good cliques* along with the *good cliques* induce a connected component. We show that the number of chunks is upper bounded by $k^{\mathcal{O}(1)}$. Finally, we use a structural analysis to bound the size of each chunk, which includes the design of a reduction rule that computes a minimum cut between the two cliques of a certain distance from the border of the chunk. With this, we bound the number of cliques in each chunk and hence the size of each chunk as well as every connected component by $k^{\mathcal{O}(1)}$.

In the third and final phase of our kernelization algorithm using the claw obstruction, we bound the number of connected components in $G - T$ by $k^{\mathcal{O}(1)}$. Using this bound, together with the facts that $|T| \leq k^{\mathcal{O}(1)}$, and that each connected component is of size $k^{\mathcal{O}(1)}$, we are able to design a polynomial kernel for PHCAVD. The proofs of the results marked with $\star$ can be found in the full version [3] . We conclude this section by summarizing all the steps in our kernelization algorithm (Fig. 1).

2   Preliminaries

Sets and Graph Notations. We denote the set of natural numbers by $\mathbb{N}$. For $n \in \mathbb{N}$, by $[n]$ and $[n]_0$, we denote the sets $\{1, 2, \cdots, n\}$ and $\{0, 1, 2, \cdots, n\}$,

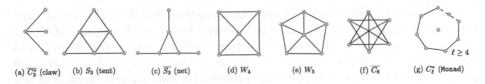

(a) $\overline{C_3^*}$ (claw)   (b) S_3 (tent)   (c) $\overline{S_3}$ (net)   (d) W_4 (e) W_5 (f) $\overline{C_6}$ (g) C_ℓ^* (Monad)

Fig. 2. Forbidden induced subgraphs of proper Helly circular-arc graphs.

respectively. For a graph G, $V(G)$ and $E(G)$ denote the set of vertices and edges, respectively. The neighborhood of a vertex v, denoted by $N_G(v)$, is the set of vertices adjacent to v. For $A, B \subseteq V(G)$ with $A \cap B = \emptyset$, $E(A, B)$ denotes the set of edges with one endpoint in A and the other in B. For a set $S \subseteq V(G)$, $G - S$ is the graph obtained by removing S from G and $G[S]$ denotes the subgraph of G induced on S. A *path* $P = v_1, \ldots, v_\ell$ is a sequence of distinct vertices where every consecutive pair of vertices is adjacent. We say that P *starts* at v_1 and *ends* at v_ℓ. The vertex set of P, denoted by $V(P)$, is the set $\{v_1, \ldots, v_\ell\}$. The *internal vertices* of P is the set $V(P) \setminus \{v_1, v_\ell\}$. The *length* of P is defined as $|V(P)| - 1$. A *cycle* is a sequence $v_1, \ldots, v_\ell$ of vertices such that $v_1, \ldots, v_\ell$ is a path and $v_\ell v_1$ is an edge. A cycle (or path) $v_1, \ldots, v_\ell$ is also represented as the ordered set $\{v_1, \ldots, v_\ell\}$. A set $Q \subseteq V(G)$ of pairwise adjacent vertices is called a *clique*. A *hole* is an induced cycle of length at least four. A vertex is *isolated* if it has degree zero. For a pair of sets $A, B \subseteq V(G)$, we say S is an A-B cut in G if there is no edge (u, v) where $u \in A \setminus S$, $v \in B \setminus S$. Such a S with minimum cardinality is called as *minimum A-B cut*. The *distance* between two vertices u and v denoted by $d_G(u, v)$ is the length of a shortest uv path in the graph G. The complement graph $\overline{G}$ of a graph G is defined in the same set of vertex $V(G)$ such that $(u, v) \in E(\overline{G})$ if and only if $(u, v) \notin E(G)$. For $\ell \geq 3$, we use C_ℓ to denote an induced cycle on ℓ vertices; if we add a new vertex to a C_ℓ and make it adjacent to none or each vertex in C_ℓ we end with C_ℓ^* or W_ℓ, respectively. A Monad is a C_ℓ^* with $\ell \geq 4$. We call the C_ℓ as M-Hole and the corresponding isolated vertex as centre of the Monad. For graph-theoretic terms and definitions not stated explicitly here, we refer to [7].

Proper Helly Circular-arc Graphs. A *proper Helly circular-arc graph* is an intersection graph of a set of arcs on a circle such that none of the arcs properly contains another (proper) and every set of pairwise intersecting arcs has a common intersection (Helly). The following is a characterization of proper Helly circular arc graphs.

Proposition 1 ([15])**.** *A graph is a proper Helly circular-arc graph if and only if it contains neither claw, net, tent, wheel of size 4, wheel of size 5, complement of cycle of length 6, nor* Monad *of length at least 4 as induced subgraphs.*

Proposition 2 (Theorem 1.3 [5]). PHCAVD *admits a polynomial-time 6-approximation algorithm, called* ApproxPHCAD.

Nice Clique Partition. For a connected graph G, a clique partition $\mathbb{Q} = (Q_1, Q_2, \ldots, Q_{|\mathbb{Q}|}(= Q_0))$ is called a *nice clique partition* of G if (i) $\bigcup_i V(Q_i) = V(G)$, (ii) $V(Q_i) \cap V(Q_j) = \emptyset$ if $i \neq j$, and (iii) $E(Q_i, Q_j) = \emptyset$ if $|i - j| > 1$ holds. In such a *nice clique partition* every edge of G is either inside a clique in $\mathbb{Q}$ or present between vertices from adjacent cliques. For a proper circular-arc graph such a partition always exists and can be obtained in $n^{\mathcal{O}(1)}$ time using a procedure similar to that for a proper interval graph [14].

Kernelization. An instance of a parameterized problem $\Pi \subseteq \Gamma^* \times \mathbb{N}$ consists of (X, k), where k is called the parameter. The notion of kernelization is formally defined as follows. A kernelization algorithm, or in short, a kernelization, for a parameterized problem $\Pi \subseteq \Gamma^* \times \mathbb{N}$ is an algorithm that, given $(X, k) \in \Gamma^* \times \mathbb{N}$, outputs in time polynomial in $|X| + k$ a pair $(X', k') \in \Gamma^* \times \mathbb{N}$ such that (i) $(X, k) \in \Pi$ if and only if $(X', k') \in \Pi$, (ii) $k' < k$, and (iii) $|X'| \leq g(k)$, where g is some computable function depending only on k. If $g(k) \in k^{\mathcal{O}(1)}$, then we say that Π admits a polynomial kernel.

3 Constructing an Efficient Modulator

We classify the set of obstructions for proper Helly circular-arc graphs as follows. Any obstruction of size less than 12 is known as a *small* obstruction, while other obstructions are said to be *large*. In this section we construct an *efficient modulator* T_1, of size $\mathcal{O}(k^{12})$ such that $G - T_1$ is a proper Helly circular-arc graph with some additional properties that are mentioned in later part.

Proposition 3 (Lemma 3.2 [10]). *Let $\mathcal{F}$ be a family of sets of cardinality at most d over a universe U and let k be a positive integer. Then there is an $\mathcal{O}(|\mathcal{F}|(k + |\mathcal{F}|))$ time algorithm that finds a non-empty family of sets $\mathcal{F}' \subseteq \mathcal{F}$ such that*

1. *For every $Z \subseteq U$ of size at most k, Z is a minimal hitting set of $\mathcal{F}$ if and only if Z is a minimal hitting set of $\mathcal{F}'$; and*
2. *$|\mathcal{F}'| \leq d!(k + 1)^d$.*

Using Proposition 3 we identify a vertex subset of $V(G)$, which allows us to forget about *small* obstructions in G and concentrate on *large* obstructions for the kernelization algorithm for PHCAVD.

Lemma 1 ($\star$). *Let (G, k) be an instance of PHCAVD. In polynomial-time, either we conclude that (G, k) is a No-instance, or we can construct a vertex subset T_1 such that*

1. *Every set $Y \subseteq V(G)$ of size at most k is a minimal hitting set for all small obstructions in G if and only if it is a minimal hitting set for all small obstructions in $G[T_1]$; and*
2. *$|T_1| \leq 12!(k + 1)^{12} + 6k$.*

Let S be a minimal (or minimum) solution of size at most k. Then, the only purpose of the vertices in $S \cap (V(G) \setminus T_1)$ is to hit *large* obstructions. We call the modulator constructed above an *efficient modulator*. We summarize these discussions in the next lemma.

Lemma 2. *Let* (G, k) *be an instance of* PHCAVD. *In polynomial time, we can either construct an* efficient *modulator* $T_1 \subseteq V(G)$ *of size* $\mathcal{O}(k^{12})$, *or conclude that* (G, k) *is a No-instance.*

4   Computing a Redundant Solution

In this section, our main purpose is to prove Lemma 7. Intuitively, this lemma asserts that in $n^{\mathcal{O}(1)}$ time we can compute an r-redundant solution M whose size is polynomial in k (for a fixed constant r). Such a set M plays a crucial role in many of the reduction rules that follow this section while designing our kernelization algorithm. We remark that in this section we use the letter ℓ rather than k to avoid confusion, as we will use this result with $\ell = k + 2$. Towards the definition of redundancy, we require the following notions and definitions.

Definition 1 (t-solution). *Let* (G, k) *be an instance of* PHCAVD. *A subset* $S \subseteq V(G)$ *of size at most* t *such that* $G - S$ *is a proper Helly circular-arc graph is called a* t-solution.

Definition 2 (t-necessary). *A family* $\mathcal{W} \subseteq 2^{V(G)}$ *is called* t-necessary *if and only if every* t-solution *is a hitting set for* $\mathcal{W}$.

Given a family $\mathcal{W} \subseteq 2^{V(G)}$, we say that an obstruction $\mathbb{O}$ is *covered by* $\mathcal{W}$ if there exists $W \in \mathcal{W}$, such that $W \subseteq V(\mathbb{O})$.

Definition 3 (t-redundant). Given a family $\mathcal{W} \subseteq 2^{V(G)}$ and $t \in \mathbb{N}$, a subset $M \subseteq V(G)$ is t *-redundant with respect to* $\mathcal{W}$ if for every obstruction $\mathbb{O}$ that is not covered by $\mathcal{W}$, it holds that $|M \cap V(\mathbb{O})| > t$.

Definition 4. Let G be a graph, $U \subseteq V(G)$, and $t \in \mathbb{N}$. Then, $\mathsf{copy}(G, U, t)$ is defined as the graph G' in the vertex set $V(G) \cup \{v^i \mid v \in U, i \in [t]\}$ and the edge set $E(G) \cup \{(u^i, v) \mid (u, v) \in E(G), u \in U, i \in [t]\} \cup \{(u^i, v^j) \mid (u, v) \in E(G), u, v \in U, i, j \in [t]\} \cup \{(v, v^i) \mid v \in U, i \in [t]\} \cup \{(v^i, v^j) \mid v \in U, i, j \in [t], i \neq j\}$.

Informally, $\mathsf{copy}(G, U, t)$ is simply the graph G where for every vertex $u \in U$, we add t twins that (together with u) form a clique. Intuitively, this operation allows us to make a vertex set "undeletable"; in particular, this enables us to test later whether a vertex set is "redundant" and therefore we can grow the redundancy of our solution, or whether it is "necessary" and hence we should update $\mathcal{W}$ accordingly. Before we turn to discuss computational issues, let us first assert that the operation in Definition 4 does not change the class of the graph, which means it remains a proper Helly circular-arc graph. We verify this in the following lemma.

Lemma 3 ($\star$). *Let G be a graph, $U \subseteq V(G)$, and $t \in \mathbb{N}$. If G is a proper Helly circular-arc graph, then $G' = \mathsf{copy}(G, U, t)$ is also a proper Helly circular-arc graph.*

Now, we present two simple claims that exhibit relations between the algorithm `ApproxPHCAD` and Definition 4. After presenting these two claims, we will be ready to give our algorithm for computing a redundant solution. Generally speaking, the first claim shows the meaning of a situation where `ApproxPHCAD` returns a "large" solution; intuitively, for the purpose of the design of our algorithm, we interpret this meaning as an indicator to extend $\mathcal{W}$.

Lemma 4 ($\star$). *Let G be a graph, $U \subseteq V(G)$, and $\ell \in \mathbb{N}$. If the algorithm `ApproxPHCAD` returns a set A of size larger than 6ℓ when called with $G' = \mathsf{copy}(G, U, 6\ell)$ as input, then $\{U\}$ is ℓ-necessary.*

Complementing our first claim, the second claim exhibits the meaning of a situation where `ApproxPHCAD` returns a "small" solution A; we interpret this meaning as an indicator of growing the redundancy of our current solution M by adding A — indeed, this lemma implies that every obstruction is hit one more time by adding A to a subset $U \subseteq M$ (to grow the redundancy of M, every subset $U \subseteq M$ will have to be considered).

Lemma 5 ($\star$). *Let G be a graph, $U \subseteq V(G)$, and $\ell \in \mathbb{N}$. If the algorithm `ApproxPHCAD` returns a set A of size at most 6ℓ when called with $G' = \mathsf{copy}(G, U, 6\ell)$ as input, then for every obstruction $\mathbb{O}$ of G, $|V(\mathbb{O}) \cap U| + 1 \leq |V(\mathbb{O}) \cap (U \cup (A \cap V(G)))|$.*

Now, we describe our algorithm, `RedundantPHCAD`, which computes a redundant solution. First, `RedundantPHCAD` initializes M_0 to be the 6-approximate solution to PHCAVD with (G, ℓ) as input, $\mathcal{W}_0 := \emptyset$ and $\mathcal{T}_0 := \{(v) \mid v \in M_0\}$. If $|M_0| > 6\ell$, then `RedundantPHCAD` concludes that (G, ℓ) is a No-instance. Otherwise, for $i = 1, 2, \ldots, r$ (in this order), the algorithm executes the following steps (Step 3 in the figure below) and eventually, it outputs the pair $(M_r, \mathcal{W}_r)$. In the `RedundantPHCAD` algorithm, by `ApproxPHCAD` (H) we mean the 6-approximate solution returned by the approximation algorithm to the input graph H.

Let us comment that in this algorithm we make use of the sets $\mathcal{T}_{i-1}$ rather than going over all subsets of size i of M_{i-1} in order to obtain a substantially better algorithm in terms of the size of the redundant solution produced.

The properties of the algorithm `RedundantPHCAD` that are relevant to us are summarized in the following lemma and observation, which are proved by induction and by making use of Lemmata Lemma 3, Lemma 4 and Lemma 5. Roughly speaking, we first assert that, unless (G, ℓ) is concluded to be a No-instance, we compute sets $\mathcal{W}_i$ that are ℓ-necessary as well as that the tuples in $\mathcal{T}_i$ "hit more vertices" of the obstructions in the input as i grows larger.

Algorithm 2: RedundantPHCAD (G, ℓ, r)

1. Initialization:
   $M_0 := \text{ApproxPHCAD}(G)$,
   $\mathcal{W}_0 := \emptyset$,
   $\mathcal{T}_0 := \{(v) \mid v \in M_0\}$.
2. If $|M_0| > 6\ell$, return "(G, ℓ) is a No-instance".
   Otherwise, $i = 1$ and go to Step 3.
3. While $i \leq r$, for every tuple $(v_0, v_1, \ldots, v_{i-1}) \in \mathcal{T}_{i-1}$:
   (a) $A := \text{ApproxPHCAD}(\text{copy}(G, \{v_0, v_1, \ldots, v_{i-1}\}, 6\ell))$.
   (b) If $|A| > 6\ell$, $\mathcal{W}_i := \mathcal{W}_{i-1} \cup \{\{v_0, v_1, \ldots, v_{i-1}\}\}$.
   (c) Otherwise,
       $M_i := M_{i-1} \cup \{u \mid u \in (A \cap V(G)) \setminus \{v_0, v_1, \ldots, v_{i-1}\}\}$,
       $\mathcal{T}_i \quad := \quad \mathcal{T}_{i-1} \cup \{(v_0, v_1, \ldots, v_{i-1}, u) \mid u \in (A \cap V(G)) \setminus \{v_0, v_1, \ldots, v_{i-1}\}\}$.
   (d) $i = i + 1$;
4. Return $(M_r, \mathcal{W}_r)$.

Lemma 6 ($\star$). *Consider a call to RedundantPHCAD with (G, ℓ, r) as input that did not conclude that (G, ℓ) is a No-instance. For all $i \in [r]_0$, the following conditions hold:*

1. *For any set $W \in \mathcal{W}_i$, every solution S of size at most ℓ satisfies $W \cap S \neq \emptyset$.*
2. *For any obstruction $\mathbb{O}$ of G that is not covered by $\mathcal{W}_i$, there exists $(v_0, v_1, \ldots, v_i) \in \mathcal{T}_i$ such that $\{v_0, v_1, \ldots, v_i\} \subseteq V(\mathbb{O})$.*

Towards showing that the output set M_r is "small", let us upper bound the sizes of the sets M_i and $\mathcal{T}_i$.

Observation 1 ($\star$). *Consider a call to RedundantPHCAD with (G, ℓ, r) as input that did not conclude that (G, ℓ) is a No-instance. For all $i \in [r]_0$, $|M_i| \leq \sum_{j=0}^{i} (6\ell)^{j+1}$, $|\mathcal{T}_i| \leq (6\ell)^{i+1}$ and every tuple in $\mathcal{T}_i$ consists of distinct vertices.*

By the specification of RedundantPHCAD, as a corollary to Lemma 6 and Observation 1, we directly obtain the following result.

Corollary 1. *Consider a call to RedundantPHCAD with (G, ℓ, r) as input that did not conclude that (G, ℓ) is a No-instance. For all $i \in [r]_0$, $\mathcal{W}_i$ is an ℓ-necessary family and M_i is a $\sum_{j=0}^{i} (6\ell)^{j+1}$-solution that is i-redundant with respect to $\mathcal{W}_i$.*

Lemma 7 ($\star$). *Let $r \in \mathbb{N}$ be a fixed constant, and (G, ℓ) be an instance of PHCAVD. In polynomial-time, it is possible to either conclude that (G, ℓ) is a No-instance, or compute an ℓ-necessary family $\mathcal{W} \subseteq 2^{V(G)}$ and a set $M \subseteq V(G)$, such that $\mathcal{W} \subseteq 2^M$ and M is a $(r+1)(6\ell)^{r+1}$-solution that is r-redundant with respect to $\mathcal{W}$.*

In light of Lemma 7, from now on, we suppose that we have an ℓ-necessary family $\mathcal{W} \subseteq 2^{V(G)}$ along with a $(r+1)(6\ell)^{r+1}$-solution M that is r-redundant with respect to $\mathcal{W}$ for $r = 5$. Let us note that, any obstruction in G that is not covered by $\mathcal{W}$ intersects M in at least six vertices. We have the following reduction rule that follows immediately from Lemma 6.

Reduction Rule 1. *Let v be a vertex such that $\{v\} \in \mathcal{W}$. Then, output the instance $(G - \{v\}, k - 1)$.*

From here onwards we assume that each set in $\mathcal{W}$ has a size at least 2.

Nice Modulator. Once we construct both the *efficient modulator* T_1 and *redundant solution* M, we take their union and consider that set of vertices as a modulator, we called it as *nice modulator*.

From here onwards, for the remaining sections, we assume that

---

We have a *nice modulator* $T \subseteq V(G)$ along with $(k+2)$-necessary family $\mathcal{W} \subseteq 2^T$ satisfying the following:

- $G - T$ is a proper Helly circular-arc graph.
- $|T| \leq \mathcal{O}(k^{12})$.
- For any *large* obstruction $\mathbb{O}$ containing no $W \in \mathcal{W}$, we have $|V(\mathbb{O}) \cap T| \geq 6$.

---

5 Bounding the Size of Each Clique

In this section, we consider a *nice modulator* T of G obtained in the previous section and we bound the size of each clique in a *nice clique partition* $\mathcal{Q}$ of $G - T$ in polynomial time. If there is a *large* clique in $\mathcal{Q}$ of size more than $\mathcal{O}(k^{12})$, we can safely find and remove an *irrelevant vertex* from the clique, thus reducing its size. Next, we prove a simple result that will later be used to bound the size of each clique in $G - T$.

Lemma 8 ($\star$). *Let H be an induced path in G. Consider a vertex $v \in V(G) \setminus V(H)$. If v has more than four neighbors in $V(H)$ then $G[V(H) \cup \{v\}]$ contains a small obstruction (claw).*

Marking Scheme. We start with the following marking procedure, which marks $k^{\mathcal{O}(1)}$ vertices in each clique $\mathcal{Q}_i \in \mathcal{Q}$.

We will now bound the size of the set $T(\mathcal{Q}_i)$.

Remark 1. Observe that the procedure Mark-1 can be executed in polynomial time. Also, note that $|T(\mathcal{Q}_i)| \leq 2(k+1)|T|^4$.

Reduction Rule 2. *If there exists a vertex $v \in \mathcal{Q}_i \setminus T(\mathcal{Q}_i)$ for some clique $\mathcal{Q}_i \in \mathcal{Q} \subseteq V(G) \setminus T$, then delete v.*

Lemma 9 ($\star$). *Reduction Rule 2 is safe.*

Procedure Mark-1. Let $\mathcal{Q}_i$ be a clique. For a pair of disjoint subsets $A, B \subseteq T$, where $|A| \leq 2$ and $|B| \leq 2$, let $\mathtt{Mark}_i[A, B]$ be the set defined by $\{v \in \mathcal{Q}_i \mid A \subseteq N(v),\ B \cap N(v) = \emptyset\}$. We initialize $T(\mathcal{Q}_i) = \emptyset$, and do as follows:

- If $|\mathtt{Mark}_i[A, B]| \leq 2(k+1)$, we add all vertices from the set $\mathtt{Mark}_i[A, B]$ to $T(\mathcal{Q}_i)$.
- Else, we add the left most $(k + 1)$ vertices (clockwise order of vertices according to their corresponding arc representation) and the right most $(k + 1)$ vertices (anticlockwise order) in $\mathtt{Mark}_i[A, B]$ to $T(\mathcal{Q}_i)$.

With the help of Reduction Rule 2, after deleting all unmarked vertices from each $\mathcal{Q}_i \in \mathcal{Q}$, size of each clique $\mathcal{Q}_i$ is reduced to $k^{\mathcal{O}(1)}$. Therefore, we have the following result. Notice that $\mathcal{Q}$ (with the reduced cliques) is also a *nice clique partition* of $G' - T$ in the reduced instance (G', k).

Lemma 10. *Given an instance* (G, k) *of* PHCAVD *and a nice modulator* $T \subseteq V(G)$ *of size* $k^{\mathcal{O}(1)}$, *in polynomial time, we can construct an equivalent instance* (G', k) *such that* $T \subseteq V(G')$ *and there exists a nice clique partition* $\mathcal{Q}$ *of* $G' - T$ *such that the size of each clique in* $\mathcal{Q}$ *is bounded by* $k^{\mathcal{O}(1)}$.

6   Bounding the Size of Each Connected Component

From Lemma 10, we can assume that the size of every clique in the *nice clique partition* $\mathbb{Q} = (Q_1, \ldots)$ of $G - T$ for a given instance (G, k) is bounded by $k^{\mathcal{O}(1)}$. In this section, we will bound the size of each connected component in $G - T$. For this purpose, it is sufficient to bound the number of cliques Q_i's from $\mathbb{Q}$ appearing in each connected component.

Let $\mathcal{C}$ be such a connected component. Without loss of generality, we assume that $\mathcal{C} = \bigcup_i (Q_i)$ i.e. in the *nice clique partition* $\mathbb{Q}$, in the connected component $\mathcal{C}$, the cliques appear in clockwise direction starting from Q_1 as $Q_1, Q_2, \ldots$ etc. We denote $(Q_1, Q_2, \ldots)$ from $\mathcal{C}$ by $\mathbb{Q}_{\mathcal{C}}$.

Reduction Rule 3. *Let v be a vertex in T. If v is contained in at least $k + 1$ distinct claws* (v, a_i, b_i, c_i) *intersecting exactly at* $\{v\}$, *where* $a_i, b_i, c_i \in V(G) \setminus T$ *then delete v from G, and reduce k by 1. The resultant instance is* $(G - v, k - 1)$.

The correctness of the above reduction rule is easy to see as every solution to (G, k) of PHCAVD must contain the vertex v. From here onward we assume that the Reduction Rule 3 is no longer applicable.

Reduction Rule 4. *Let v be a vertex in T. If v has neighbors in more than* $6(k + 1)$ *different* Q_i's *(a_i's being the corresponding neighbors), then remove v from G and reduce k by 1. The resultant instance is* $(G - v, k - 1)$.

Lemma 11 ($\star$). *Reduction Rule 4 is safe.*

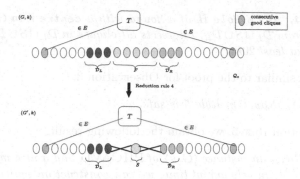

Fig. 3. Description of Reduction Rule 5.

From now on, we assume that the Reduction Rules 3 and 4 are no longer applicable i.e. every vertex $v \in T$ has neighbors in at most $6(k+1)$ different Q_i' from $\mathbb{Q}_{\mathcal{C}}$. And we have the following result.

Lemma 12. *Let $\mathcal{C}$ be a connected component in $G - T$. Then there are at most $6(k+1)|T|$ many distinct cliques Q_i's from $\mathbb{Q}_{\mathcal{C}}$ such that $N(T) \cap Q_i \neq \emptyset$.*

If $\mathbb{Q}_{\mathcal{C}}$ has more than $300|T|k(k+1)$ cliques, then by the pigeonhole principle and Lemma 12, there are at least $50k$ consecutive cliques that do not contain any vertex from $N(T)$. Let $Q_1, Q_2, \ldots, Q_{50k}$ be the set of $50k$ such consecutive cliques in $\mathbb{Q}_{\mathcal{C}}$ which are disjoint from $N(T)$. Let $\mathcal{D}_L = \{Q_i \mid i \in [15k, 20k]\}$, $\mathcal{D}_R = \{Q_i \mid i \in [30k, 35k]\}$, $F = \{Q_i \mid i \in [20k+1, 30k-1]\}$ and $Z = \mathcal{D}_L \cup \mathcal{D}_R \cup F$. Observe that, for a vertex $v \in Z$ and a vertex $u \in T$, $\text{dist}_G(u, v) \geq 15k$. And hence there can not be any *small* obstruction containing vertices from Z (Observation 2) which we will use to our advantage in many proofs throughout the current section. Let τ be the size of minimum $(Q_{20k} - Q_{30k})$ cut in $\mathbb{Q}_{\mathcal{C}}$.

Reduction Rule 5. *Let F be as defined above. Delete all the vertices of F from G. Introduce a new clique S of size τ. Also, add edges such that $G[V(Q_{20k}) \cup S]$ and $G[V(Q_{30k}) \cup S]$ are complete graphs. The cliques appear in the order Q_{20k}, S, Q_{30k}.*

Let G' be the reduced graph after application of the Reduction Rule 5. For an illustration, see Fig. 3. Notice that $G' - T$ is a proper Helly circular-arc graph by construction.

Observation 2. (⋆) *There are no* small *obstructions containing any vertices from $\mathcal{D}_L \cup F \cup \mathcal{D}_R$ in G. Similarly, there are no* small *obstructions containing vertices of $\mathcal{D}_L \cup S \cup \mathcal{D}_R$ in G'.*

Observation 3. (⋆) *Any M-Hole H of a Monad with a centre v in G which contains a vertex from $\mathcal{D}_L \cup F \cup \mathcal{D}_R$, intersects all cliques in $\mathcal{D}_L \cup F \cup \mathcal{D}_R$. And such an H has size at least $20k$.*

Observation 4. *Any M-Hole H of a Monad with a centre v in G' which contains a vertex from $\mathcal{D}_L \cup S \cup \mathcal{D}_R$, intersects all cliques in $\mathcal{D}_L \cup S \cup \mathcal{D}_R$. And such an H has size at least $20k$.*

Proof. Proof is similar to the proof for Observation 3.   □

Lemma 13 (⋆). *Reduction Rule 5 is safe.*

With Reduction Rule 5, we obtain the following result.

Lemma 14. *Given an instance (G, k) of PHCAVD and a nice modulator $T \subseteq V(G)$ of size $k^{\mathcal{O}(1)}$, in polynomial time, we can construct an equivalent instance (G', k) such that, $T \subseteq V(G')$ is a nice modulator for G' and for each connected component $\mathcal{C}$ of $G' - T$, the number of cliques in $\mathbb{Q}_\mathcal{C}$ is at most $300 \cdot |T| \cdot k(k+1) = \mathcal{O}(k^2 \cdot |T|)$.*

7   Bounding the Number of Connected Components

Until now we have assumed that $G - T$ is connected. Further, in Sect. 6, we showed that the size of any connected component is upper bounded by $k^{\mathcal{O}(1)}$. In this section, we show that the number of connected components in $G - T$ can be upper bounded by $k^{\mathcal{O}(1)}$. This together with the fact that $|T| \leq k^{\mathcal{O}(1)}$, results in a polynomial kernel for PHCAVD.

Here we bound the number of connected components with an argument similar to the one using which we bounded the neighborhood of the modulator. We make use of the claw obstruction to get the desired bound. Notice that if any vertex v in T has neighbors in three different components in $G - T$, then we get a claw.

Reduction Rule 6. *Let v be a vertex in T such that v has neighbors in at least $3(k + 1)$ different components in $G - T$ then delete v from G, and reduce k by 1. The resultant instance is $(G - v, k - 1)$.*

The correctness of the above reduction rule is easy to see as every solution to (G, k) of PHCAVD must contain v. From now onwards we assume that Reduction Rule 6 is not applicable. And this leads to the following lemma.

Lemma 15. *T can have neighbors in at most $3(k + 1)|T|$ many different components.*

Now we bound the number of connected components that have no neighbor in T. Towards that, we classify all such connected components into two classes: interval connected components (which admit an interval representation) and non-interval connected components. Here non-interval connected components cover the entire circle whereas others partially cover the underlying circle.

Claim 1 (⋆). *The number of non-interval connected components in $G - T$ is at most one.*

Reduction Rule 7. *If there are more than $(k+1)$ interval connected components in $G - T$ that have no neighbor in T, delete all but $(k+1)$ components.*

Lemma 16. ($\star$). *Reduction Rule 7 is safe.*

From now onwards we assume that Reduction Rules 6 and 7 are not applicable. Now these two reduction rules and Lemma 15 implies the following result:

Lemma 17. *Given an instance (G, k) and a nice modulator $T \subseteq V(G)$ of size $\mathcal{O}(k^{12})$, in polynomial-time, we can construct an equivalent instance (G', k') such that the number of connected component in $G' - T$ is $\mathcal{O}(k \cdot |T|^2)$.*

8 Kernel Size Analysis

Now we are ready to prove the main result of our paper, that is, Theorem 1. Before proceeding with the proof, let us state all the bounds that contributes to the kernel size.

Size of *nice modulator* T: $\mathcal{O}(k^{12})$ (from Sect. 4).
Number of connected components in $G - T$: $\mathcal{O}(k \cdot |T|^2)$ (by Lemma 17).
Number of cliques in any connected component in $G - T$: $\mathcal{O}(k^2 \cdot |T|)$ (by Lemma 14).
Size of any clique $\mathcal{Q}_i$ in $G - T$: $2(k+1)|T|^4$ (by Remark 1).

Proof (Proof of theorem 1). From Lemma 2 and Lemma 7, in polynomial-time, we can obtain a *nice modulator* $T \subseteq V(G)$ of size $\mathcal{O}(k^{12})$ or conclude that (G, k) is a No-instance. Note that, $G - T$ is a proper Helly circular-arc graph. Next, we take a *nice clique partition* of $G - T$. Now by Lemma 17, in polynomial-time we return a graph G such that $G - T$ has $\mathcal{O}(k \cdot |T|^2)$ components. By Lemma 14, in polynomial-time, we can reduce the graph G such that any connected component in $G - T$ has at most $\mathcal{O}(k^2 \cdot |T|)$ cliques. Next, we bound the size of each clique in $G - T$ by $2(k+1)|T|^4$ from Lemma 10. Hence the graph $G - T$ has at most $\mathcal{O}(k \cdot |T|^2) \cdot \mathcal{O}(k^2 \cdot |T|) \cdot 2(k+1)|T|^4$, that is, $\mathcal{O}(k^4 \cdot |T|^7)$ many vertices. Recall that $|T| = \mathcal{O}(k^{12})$. Therefore, the size of the obtained kernel is $\mathcal{O}(k^4 \cdot |T|^7)$, that is, $\mathcal{O}(k^{88})$. □

9 Conclusion

In this paper, we studied PHCAVD from the perspective of kernelization complexity, and designed a polynomial kernel of size $\mathcal{O}(k^{88})$. We remark that the size of a kernel can be further optimized with more careful case analysis. However, getting a kernel of a significantly smaller size might require an altogether different approach.

Acknowledgements. The first author acknowledges Science and Engineering Research Board for supporting this research via Startup Research Grant (SRG/2022/000962).

References

1. Agrawal, A., Lokshtanov, D., Misra, P., Saurabh, S., Zehavi, M.: Feedback vertex set inspired kernel for chordal vertex deletion. In: (SODA 2017), pp. 1383–1398 (2017)
2. Agrawal, A., Misra, P., Saurabh, S., Zehavi, M.: Interval vertex deletion admits a polynomial kernel. In: Chan, T.M., (ed.) SODA 2019, San Diego, California, USA, pp. 1711–1730. SIAM (2019)
3. Jana, S., Agrawal, A., Sahu, A.: A polynomial kernel for proper Helly circular-arc vertex deletion (2024). arXiv:2401.03415
4. Bodlaender, H.L., Downey, R.G., Fellows, M.R., Hermelin, D.: On problems without polynomial kernels. J. Comput. Syst. Sci. 75(8), 423–434 (2009)
5. Cao, Y., Yuan, H., Wang, J.: Modification problems toward proper (Helly) circular-arc graphs. In: Leroux, J., Lombardy, S., Peleg, D., (eds.) MFCS 2023, vol. 272, pp. 31:1–31:14. LIPIcs (2023)
6. Cygan, M., et al.: Parameterized Algorithms. Springer, Cham (2015). https://doi.org/10.1007/978-3-319-21275-3
7. Reinhard Diestel. Graph Theory, 4th Edition, volume 173 of Graduate texts in mathematics. Springer, 2012
8. Downey, R.G., Fellows, M.R.: Fundamentals of Parameterized Complexity. TCS, Springer, London (2013). https://doi.org/10.1007/978-1-4471-5559-1
9. Durán, G., Grippo, L.N., Safe, M.D.: Structural results on circular-arc graphs and circle graphs: a survey and the main open problems. Discrete Appl. Math. 164, 427–443 (2014)
10. Fomin, F.V., Saurabh, S., Villanger, Y.: A polynomial kernel for proper interval vertex deletion. SIAM J. Discret. Math. 27(4), 1964–1976 (2013)
11. Golumbic, M.C.: Algorithmic Graph Theory and Perfect Graphs. Academic Press (1980)
12. Kaplan, H., Nussbaum, Y.: Certifying algorithms for recognizing proper circular-arc graphs and unit circular-arc graphs. Discret. Appl. Math. 157(15), 3216–3230 (2009)
13. Ke, Y., Cao, Y., Ouyang, X., Li, W., Wang, J.: Unit interval vertex deletion: fewer vertices are relevant. J. Comput. Syst. Sci. 95, 109–121 (2018)
14. Krithika, R., Sahu, A., Saurabh, S., Zehavi, M.: The parameterized complexity of cycle packing: indifference is not an issue. Algorithmica 81(9), 3803–3841 (2019)
15. Lin, M.C., Soulignac, F.J., Szwarcfiter, J.L.: Normal Helly circular-arc graphs and its subclasses. Discret. Appl. Math. 161, 7–8 (2013)
16. Lin, M.C., Szwarcfiter, J.L.: Characterizations and recognition of circular-arc graphs and subclasses: a survey. Discrete Math. 309(18), 5618–5635 (2009)
17. Lin, M.C., Szwarcfiter, J.L.: Characterizations and linear time recognition of Helly circular-arc graphs. In: Chen, D.Z., Lee, D.T. (eds.) COCOON 2006. LNCS, vol. 4112, pp. 73–82. Springer, Heidelberg (2006). https://doi.org/10.1007/11809678_10
18. Marx, D.: Chordal deletion is fixed-parameter tractable. Algorithmica 57(4), 747–768 (2010)
19. McConnell, R.M.: Linear-time recognition of circular-arc graphs. Algorithmica 37(2), 93–147 (2003)

Max-SAT with Cardinality Constraint Parameterized by the Number of Clauses

Pallavi Jain[1,5], Lawqueen Kanesh[1,5], Fahad Panolan[2,5], Souvik Saha[3,5(✉)] (iD),
Abhishek Sahu[4,5], Saket Saurabh[3,5], and Anannya Upasana[3,5]

[1] Indian Institute of Technology Jodhpur, Jodhpur, India
{pallavi,lawqueen}@iitj.ac.in
[2] School of Computing, University of Leeds, Leeds, UK
f.panolan@leeds.ac.uk
[3] The Institute of Mathematical Sciences, HBNI, Chennai, India
{souviks,saket,anannyaupas}@imsc.res.in
[4] National Institute of Science Education and Research, HBNI, Mumbai, India
abhisheksahu@niser.ac.in
[5] University of Bergen, Bergen, Norway

Abstract. MAX-SAT with cardinality constraint (CC-MAX-SAT) is one of the classical NP-complete problems. In this problem, given a CNF-formula Φ on n variables, positive integers k, t, the goal is to find an assignment β with at most k variables set to true (also called a weight k-assignment) such that the number of clauses satisfied by β is at least t. The problem is known to be W[2]-hard with respect to the parameter k. In this paper, we study the problem with respect to the parameter t. The special case of CC-MAX-SAT, when all the clauses contain only positive literals (known as MAXIMUM COVERAGE), is known to admit a $2^{\mathcal{O}(t)} n^{\mathcal{O}(1)}$ algorithm. We present a $2^{\mathcal{O}(t)} n^{\mathcal{O}(1)}$ algorithm for the general case, CC-MAX-SAT. We further study the problem through the lens of kernelization. Since MAXIMUM COVERAGE does not admit polynomial kernel with respect to the parameter t, we focus our study on $K_{d,d}$-free formulas (that is, the clause-variable incidence bipartite graph of the formula that excludes $K_{d,d}$ as a subgraph). Recently, in [Jain et al., SODA 2023], an $\mathcal{O}(dt^{d+1})$ kernel has been designed for the MAXIMUM COVERAGE problem on $K_{d,d}$-free incidence graphs. We extend this result to MAX-SAT on $K_{d,d}$-free formulas and design a $\mathcal{O}(d4^{d^2} t^{d+1})$ kernel.

Keywords: FPT · Kernel · Max-SAT

1 Introduction

SAT, the first problem that was shown to be NP-complete, is one of the most fundamental, important, and well-studied problems in computer science. In this problem, given a CNF-formula Φ, the goal is to decide if there exists an assignment β that sets the variables of the given formula to true or false such that the formula is satisfied. The optimisation version of the problem is MAX-SAT, in which the goal is to find an assignment that satisfies the maximum number of clauses. There are several applications of MAX-SAT e.g., cancer therapy design, resource allocation, formal verification, and many more [10,15]. In this paper, we study the MAX-SAT problem under cardinality constraint, known as CC-MAX-SAT, which is a generalisation of MAX-SAT. In this problem, given a CNF-formula Φ, positive integers k and t; the goal is to find an assignment β with at most k variables set to true (also called a weight k-assignment) such that the number of clauses satisfied by β is at least t.

CC-MAX-SAT has been studied from the approximation viewpoint. It admits a $(1 - \frac{1}{e})$-factor approximation algorithm [13]. Here, e is the base of the natural logarithm. Feige [5] showed that this approximation algorithm is optimal even for a special case of CC-MAX-SAT, where the clauses contain only positive literals. This special case of CC-MAX-SAT is known as MAXIMUM COVERAGE, in which we are given a family of subsets, $\mathcal{F}$, over a universe U, and the goal is to find a subfamily $\mathcal{F}' \subseteq \mathcal{F}$ of size at most k such that the number of elements of U that belong to a set in $\mathcal{F}'$ is at least t.

In this paper, we study the problem in the realm of parameterized complexity. The MAXIMUM COVERAGE problem is well-studied from the parameterized viewpoint and is known to be W[2]-hard with respect to the parameter k [3]. In fact, it is known that assuming GAP-ETH, we cannot hope for an approximation algorithm with factor $(1 - \frac{1}{e} + \epsilon)$ in running time $f(k, \epsilon)n^{\mathcal{O}(1)}$ for MAXIMUM COVERAGE [9,12]. Thus, the next natural parameter to study is t (the minimum number of clauses to be satisfied).

As early as 2003, Bläser [2] showed that MAXIMUM COVERAGE admits a $2^{\mathcal{O}(t)}n^{\mathcal{O}(1)}$ algorithm. We first generalise this result to CC-MAX-SAT.

Theorem 1. *There exists a deterministic algorithm that solves the* CC-MAX-SAT *problem in time* $2^{\mathcal{O}(t)}n \log n$.

The next natural question is, "Does the problem admit a polynomial kernel with respect to the parameter t?". That is, does there exist a polynomial time algorithm that, given an instance (Φ, k, t) of CC-MAX-SAT returns an equivalent instance (Φ', k', t') of CC-MAX-SAT whose size is bounded by a polynomial in t. Unfortunately, MAXIMUM COVERAGE does not admit a polynomial kernel with respect to the parameter t unless PH = Σ^3_p [4]. Thus, we cannot hope for a polynomial kernel with respect to the parameter t for CC-MAX-SAT. So a natural question is for which families of input does the problem admit a polynomial kernel.

In 2018, Agrawal et al. [1] designed a kernel for a special case of MAXIMUM COVERAGE when every element of the universe appears in at most d sets. In this kernel, the universe is bounded by $\mathcal{O}(dt^2)$, and the size of the family is bounded by $\mathcal{O}(dt)$. Recently, in 2023, Jain et al. [7] designed an $\mathcal{O}(dt^{d+1})$ kernel for the MAXIMUM COVERAGE problem on $K_{d,d}$-free incident graphs. Here, by an incident graph, we mean a bipartite graph $G = (A \cup B, E)$ where A contains a vertex for each element of the set family and B contains a vertex for each

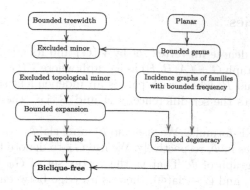

Fig. 1. Inclusion relation between the class of biclique-free graphs considered in this paper and well-studied graph classes in the literature (figure based on [14]). If there is an arrow of the form $A \to B$, then class A is a subclass of B.

element of the ground set and there is an edge between $u \in A$ and $v \in B$ if the set corresponding to u contains the element corresponding to v. $K_{d,d}$ denotes a complete bipartite graph with bipartitions of size d each. $K_{d,d}$-free (also called biclique-free) graphs are very extensive classes of graphs and include many well-studied graph classes such as bounded treewidth graphs, graphs that exclude a fixed minor, graphs of bounded expansion, nowhere-dense graphs, and graphs of bounded degeneracy. That is, for any of the classes - bounded treewidth graphs, graphs that exclude a fixed minor, graphs of bounded expansion, nowhere-dense graphs, and graphs of bounded degeneracy, there is a p such that the class is contained in the class of $K_{p,p}$-free graphs (see Fig. 1 for an illustration of the inclusion relation between these classes). For a CNF-formula Φ, let G_Φ denote the clause-variable incident bipartite graph of Φ. That is, the vertex set of G_Φ is $\text{var}(\Phi) \uplus \text{cla}(\Phi)$. For each $v \in \text{var}(\Phi)$ and $C \in \text{cla}(\Phi)$, there is an edge between v and C in G_Φ if and only if v or $\overline{v}$ belongs to the clause C. We say that Φ is a $K_{d,d}$-free formula, if G_Φ excludes $K_{d,d}$ as an induced subgraph [8].

Theorem 2. CC-MAX-SAT *in* $K_{d,d}$-free *formulae admits a kernel of size* $\mathcal{O}(d4^{d^2} t^{d+1})$.

One may ask why not design a polynomial kernel for CC-MAX-SAT in $K_{d,d}$-free formulae with parameter k or $k + d$. We would like to point out that CC-MAX-SAT is W[1]-hard when parameterized by $k + d$, as explained in the below reduction (also the problem generalizes the MAX COVERAGE problem). A simple reduction from a W[1]-hard problem, named PARTIAL VERTEX COVER is as follows [6]. Here, we want to cover the maximum number of edges in the input graph using k vertices. The construction of a formula for CC-MAX-SAT is as follows. We will have a variable x_v for each vertex v in the input graph G. For an edge $e = \{u, v\}$, we will construct a clause $C_e = (x_v \vee x_u)$. The resulting formula will be $K_{2,2}$-free for a simple graph G. This implies that the CC-MAX-SAT problem is W[1]-hard when parameterized by $k + d$.

2   Preliminaries

For a graph G, we denote its vertex set by $V(G)$ and its edge set by $E(G)$. We define a bipartite graph $G = (A, B, E)$ as a graph where $V(G)$ can be partitioned into two parts A and B, where each part is an independent set. For two sets A and B, $A \backslash B$ denotes the set difference of A and B, i.e., the set of elements in A, but not in B.

Let Φ be a CNF-formula. We use $\mathsf{var}(\Phi)$ and $\mathsf{cla}(\Phi)$ to denote the set of variables and clauses in Φ, respectively. We use G_Φ to denote the clause-variable incident bipartite graph of Φ. That is, the vertex set of G_Φ is $\mathsf{var}(\Phi) \uplus \mathsf{cla}(\Phi)$. For each $v \in \mathsf{var}(\Phi)$ and $C \in \mathsf{cla}(\Phi)$, there is an edge between v and C in G_Φ if and only if v or $\overline{v}$ belongs to the clause C.

An *all zero* assignment $\beta_\emptyset$ assigns false value to every variable of the input formula. With a slight abuse of terminology, for a clause $C \in \mathsf{cla}(\Phi)$, by $\mathsf{var}(C)$, we denote variables contained in C. For a subset Y of variables, $N_\Phi^+(Y)$ is the set of clauses in Φ that contains at least one variable from Y positively. We denote $N_\Phi^-(Y)$ as the set of clauses in Φ that contains at least one variable from Y negatively. For a subset Y of variables $N_\Phi(Y)$ is the set of clauses in Φ that contains at least one variable from Y either positively or negatively, that is the set $N_\Phi^+(Y) \cup N_\Phi^-(Y)$. For a clause C and variable $u \in \mathsf{var}(\Phi)$, by $C \backslash \{u\}$, we denote the clause C after removing the literal corresponding to u from C. For example let $C = u \vee v \vee w$, then $C \backslash \{u\} = v \vee w$. If $u \notin \mathsf{var}(C)$, then $C \backslash \{u\}$ is simply the clause C.

$K_{d,d}$ denotes the complete bipartite graph with each bipartition containing d vertices. A bipartite graph $G = (A, B, E)$ is $K_{d,d}$-free when G excludes $K_{d,d}$ as a subgraph (no set of d vertices in A together have d common neighbors in B).

3   FPT Algorithm for CC-Max-SAT parameterized by t

In this section, we design an FPT algorithm for CC-Max-SAT problem parameterized by t, the number of satisfied clauses.

We use the algorithm for Partial Hitting Set given in the following proposition.

Proposition 1 (Bläser [2]). *There is an algorithm that given a universe U, a family of sets $\mathcal{F}$ over U, and positive integers k and t, runs in time $2^{\mathcal{O}(t)} n^{\mathcal{O}(1)}$, and checks whether there exists a k-sized subset of U, that hits at least t sets from $\mathcal{F}$. If such a subset exists, then the algorithm will output such a subset.*

We use the above proposition to get the following lemma.

Lemma 1. *There exists a randomized algorithm that solves the CC-Max-SAT problem and outputs a satisfying assignment in time $2^{\mathcal{O}(t)} n^{\mathcal{O}(1)}$.*

Algorithm 1. Procedure: FPT Algorithm for CC-MAX-SAT

1: Randomly assign values to all the variables in $\mathsf{var}(\phi)$ from $\{0,1\}$. Let T be the set of variables assigned 1 and F be the set of variables assigned 0.
2: Delete all clauses C in $\mathsf{cla}(\phi)$ that contains a negative literal of a variable $v \in F$. Let t_1 be the number of deleted clauses and ϕ' be the resultant formula.
3: Construct a family $\mathcal{F}$ as follows.
4: **for** each clause $C \in \mathsf{cla}(\phi')$, **do**
5:      Add $\mathsf{var}(C) \setminus F$ to $\mathcal{F}$.
6: **end for**
7: Run the algorithm from Proposition 1 by converting to hitting set instance $(\mathcal{U}, \mathcal{F}, k, t - t_1)$ where $\mathcal{U} = \mathsf{var}(\phi') \setminus F$.
8: **if** the algorithm returns a NO. **then return** NO
9: **else** let S be the solution returned by the algorithm. We construct γ as follows.
10:      **if** $v \in S$ **then**
11:          let $\gamma(v) = 1$.
12:      **else** $\gamma(v) = 0$.
13:      **end if**
14:      **return** γ.
15: **end if**

Proof. We apply the procedure in Algorithm 1, 2^t times and if in any of the iteration it returns an assignment γ, then we output γ, otherwise we return NO. Next we prove the correctness of the algorithm.

Let (ϕ, k, t) be a YES instance for the problem. Let $C_1, C_2, \cdots C_t$ be t clauses that are satisfied by a particular assignment, say α. For each clause C_i, let x_i be the variable that is "responsible" for satisfying it by a feasible assignment α, that is assignment where at most k variables are assigned 1. By "responsible" we mean that the clause C_i is satisfied even if we give any other assignment to all the variables except x_i. Thus, $x_1, x_2, \cdots x_t$ are the "responsible" variables for the clauses $C_1, C_2, \cdots C_t$ by the assignment α. Let α_t be the assignment α restricted to these t variables. Now consider any random assignment α' : $\mathsf{var}(\phi) \to \{0,1\}$ of $\mathsf{var}(\phi)$. Then the assignments α and α' agree on the variables $x_1, x_2, \cdots x_t$ with probability $\frac{1}{2^t}$. Let α' be the assignment obtained in step 1 of our procedure. Therefore with probability $\frac{1}{2^t}$ the assignment α' satisfies the clauses $C_1, C_2, \cdots C_t$.

W.l.o.g. let $C_1, C_2, \cdots C_{t_1}$ be the set of clauses deleted in step 2 of our algorithm. Every clause C_i that is not deleted in step 2 must contain one of the variables from $x_1, x_2, \cdots x_t$ which is assigned 1. There are at most k such variables and hence the corresponding PARTIAL HITTING SET instance in step 7 will return a solution. Since we repeat the algorithm 2^t times, we get a success probability of $1 - \frac{1}{e}$.

Running Time: Observe that the running time of Procedure FPT Algorithm for CC-MAX-SAT depends on the algorithm used in step 7 which runs in time $2^{\mathcal{O}(t)} n^{\mathcal{O}(1)}$. As we repeat the Procedure 2^t times, our algorithm runs in time $2^{\mathcal{O}(t)} n^{\mathcal{O}(1)}$. $\qquad\square$

We now derandomize the algorithm using Universal Sets. We deterministically construct a family $\mathcal{F}$ of functions $f : [n] \to [2]$ instead of randomly assigning values such that it is assured that one of the assignments when restricted to the t variables $x_1, x_2, \cdots x_t$ matches with the assignment α_t. For this we state the following definitions.

Definition 1 ((n, ℓ)-universal set). *An (n, ℓ)-universal set is a family $\mathcal{U}$ of subsets of $[n]$ such that for every subset $S \subseteq [n]$ of size at most ℓ, the family $\{U \cap S : U \in \mathcal{U}\}$ contains all $2^{|S|}$ subsets of S.*

Proposition 2 (Naor et al. [11]). *There is an algorithm that given integers $n, \ell \in \mathbb{N}$, runs in time $2^\ell \ell^{\mathcal{O}(\log \ell)} n \log n$, and outputs an (n, ℓ)-universal set of cardinality at most $2^\ell \ell^{\mathcal{O}(\log \ell)} \log n$.*

We construct an (n, t)-universal set $\mathcal{U}$ and then for every element of $\mathcal{U}$ we construct an equivalent assignment by assigning *true* to the variables represented by the elements in the subset and *false* to the elements outside. By the definition of universal sets we will have an assignment which when restricted to the t responsible variables will be equal to α_t which corresponds to the good event in our random experiment. Thus, we have the following theorem.

Theorem 1. *There exists a deterministic algorithm that solves the CC-MAX-SAT problem in time $2^{\mathcal{O}(t)} n \log n$.*

4   Polynomial Kernel for CC-MAX-SAT in $K_{d,d}$-free Formulas

In this section, we design a polynomial time kernelization algorithm for CC-MAX-SAT where the input is a $K_{d,d}$-free formula, parameterized by the number of clauses to be satisfied by a solution – that is, the parameter is t. We begin by defining and recalling some notations.

An *all zero* assignment $\beta_\emptyset$ assigns false value to every variable of the input formula. With a slight abuse of terminology, for a clause $C \in \mathsf{cla}(\Phi)$, by $\mathsf{var}(C)$, we denote variables appearing in C. Recall that for a subset Y of variables $N_\Phi^+(Y)$ is the set of clauses in Φ that contains at least one variable from Y positively. We denote $N_\Phi^-(Y)$ as the set of clauses in Φ that contains at least one variable from Y negatively. We denote $d_\Phi^+(Y) = |N_\Phi^+(Y)|$ and $d_\Phi^-(Y) = |N_\Phi^-(Y)|$. For a subset Y of variables $N_\Phi(Y)$ is the set of clauses in Φ that contains at least one variable from Y either positively or negatively, that is the set $N_\Phi^+(Y) \cup N_\Phi^-(Y)$. For a clause C and variable $u \in \mathsf{var}(\Phi)$, by $C \setminus \{u\}$, we denote the clause C after removing the literal corresponding to u from C. For example let $C = u \vee v \vee w$, then $C \setminus \{u\} = v \vee w$. If $u \notin \mathsf{var}(C)$, then $C \setminus \{u\}$ is simply the clause C. Next, we give an outline of our kernel.

Outline of the Kernel: Consider an instance (Φ, k, t) of CC-MAX-SAT, where Φ is a $K_{d,d}$-free formula. Our kernelization algorithm works in three phases. In the first phase, we apply some simple sanity check reduction rules to eliminate trivial YES/NO instances of CC-MAX-SAT. Reduction rules in this phase (1) upper bounds the frequency of variables in Φ by t, and (2) leads to an observation that any minimum weight assignment can satisfy at most $2t$ clauses. The above facts are useful to establish proofs in the next two phases.

Suppose (Φ, k, t) is a YES instance and let β be its minimum weight assignment and let $\mathcal{C}_\beta$ be the set of clauses satisfied by β. In the second phase, we bound the size of clauses in Φ by a function of t and d, say $f(t, d)$. Here, we use $K_{d,d}$-free property crucially. By the definition of $K_{d,d}$-free formula, a set of d variables cannot appear simultaneously in a set of d clauses. We generalize this idea together with the frequency bound on variables (obtained in phase one) to bound the size of a set of variables that appear together in $p \in [d-1]$ clauses by identifying and deleting some "redundant" variables.

For a set $Y \subseteq \mathrm{var}(\Phi)$, let $\mathsf{claInt}(Y)$ denote the set of all the clauses that contains all the variables in Y, that is the set $\{C \mid C \in \mathsf{cla}(\Phi), Y \subseteq \mathrm{var}(C)\}$.

For an intuition, suppose we have already managed to bound the size of every subset of $\mathrm{var}(\Phi)$, that appear together in at least two clauses, by $f(t, d)$. Now consider a set $Y \subseteq \mathrm{var}(\Phi)$ such that variables in Y appear together in at least one clause say C, that is $|\mathsf{claInt}(Y)| \geq 1$ and $C \in \mathsf{claInt}(Y)$. Now suppose that there is a clause $C^* \in \mathcal{C}_\beta \setminus \mathsf{claInt}(Y)$ such that $\mathrm{var}(C^*) \cap Y \neq \emptyset$. Then observe that for the variable set $\mathrm{var}(C^*) \cap Y$, C and C^* are common clauses. Now by considering the bound on sizes of sets of variables that have at least two common clauses (clauses in which they appear simultaneously), we obtain that $|\mathrm{var}(C^*) \cap Y|$ is also bounded by $f(t, d)$. We will now mark variables of all the clauses in $\mathcal{C}_\beta \setminus \mathsf{claInt}(Y)$ in Y and will conclude that if Y is sufficiently "large", then there exists a *redundant* variable in Y, that can be removed from Φ. Employing the above discussion, by repeatedly applying a careful deletion procedure, we manage to bound the size of sets of variables that have at least one common clause. By repeating these arguments we can show that to bound the size of sets of variables with at least 2 common clauses, all we require is to bound size of sets of variables with at least 3 common clauses. Thus, inductively, we bound size of sets of variables with $d-1$ common clauses, by using the fact that the input formula is $K_{d,d}$-free. Thus the algorithm starts for $d-1$ and applies a reduction rule to bound size of sets of variables with $d-1$ common clauses. Once we apply reduction rule for $d-1$ exhaustively, we apply for $d-2$ and by inductive application we reach the one common clause case. To apply reduction rule for $p \in [d-1]$ we assume that reduction rules for $d-1$ common clauses case have already been applied exhaustively. The bound on size of the sets of variables with one common clause also gives bound on the size of clause by $f(k, d)$.

Finally in the third phase, the algorithm applies a reduction rule to remove all the variables which are not among first $g(t, d)$ high positive degree variables and not among first $g(t, d)$ high negative degree variables when sorted in non decreasing ordering of their positive (negative) degrees, which together with the upper bound obtained on size frequency of variables and size of clauses obtained in phase one gives the desired kernel size.

We now formally introduce our reduction rules. Our algorithm applies each reduction rule exhaustively in the order in which they are stated. We begin by stating some simple sanity check reduction rules.

Reduction Rule 1. *If $k < 0$, or $k = 0$, $t \geq 1$ and $\beta_\emptyset$ satisfies less than t clauses in $\mathsf{cla}(\Phi)$, then return that (Φ, k, t) is a NO instance of CC-MAX-SAT.*

The safeness of Reduction Rule 1 follows from the fact that the cardinality of number of variables assigned 1 cannot be negative and since $k = 0$, all the variables must be assigned 0 values and an all zero assignment must satisfy at least t clauses for (Φ, k, t) to be a YES instance of CC-MAX-SAT.

Reduction Rule 2. *If at least one of the following holds, then return that (Φ, k, t) is a YES instance of CC-MAX-SAT.*

1. *$\beta_\emptyset$ satisfies at least t clauses in $\mathsf{cla}(\Phi)$.*
2. *$k \geq 0$ and there exists a variable $v \in \mathsf{var}(\Phi)$ such that $d_\Phi^+(v) \geq t$.*

The safeness of Reduction Rule 2 follows from the following facts: For the first condition, if an all zero assignment satisfies at least t clauses, then trivially (Φ, k, t) is a YES instance of CC-MAX-SAT. For the second condition, there exists an assignment which assigns v to true in case $d_\Phi^+(v) \geq t$, then the variable v alone can satisfy at least t clauses in $\mathsf{cla}(\Phi)$.

When none of the Reduction Rules 1 and 2 are applicable, we obtain the following observation.

Observation 1. *Consider a minimum weight assignment β of $\mathsf{var}(\Phi)$ that satisfies at least t clauses in $\mathsf{cla}(\Phi)$. Then the number of clauses satisfied by β is at most $2t$.*

Proof. As Reduction Rule 2 is no longer applicable β is not an all zero assignment. Let u be a variable that has been assigned true value by β. Now consider another assignment β', where for every $u' \in \mathsf{var}(\Phi)$, $u' \neq u$, $\beta'(u') = \beta(u')$, and $\beta'(u)$ is false. As Reduction Rule 2 is no longer applicable, β' is also not an all zero assignment. Also, β' satisfies strictly less than t clauses, as otherwise it contradicts that β is a minimum weight assignment. Moreover, as Reduction Rule 2 is no longer applicable, $d_\Phi^+(u) < t$ and $d_\Phi^-(u) < t$. Notice that the difference between clauses satisfied by β and β' is exactly the clauses where u appears. The assignment β satisfies clauses where u appears positively, while β' satisfies clauses where u appears negatively. Above observations implies that β satisfies at most $t - d_\Phi^-(u) + d_\Phi^+(u) \leq 2t$ clauses. $\square$

Lemma 2. *Consider a set $Y \subseteq \mathsf{var}(\Phi)$. Let $|\mathsf{clalnt}(Y)| = \ell$. If $|Y| \geq 2^\ell \cdot \tau + 1$, for some positive integer τ, then in polynomial time we can find sets $\widehat{Y}_{\mathsf{pos}}, \widehat{Y}_{\mathsf{neg}} \subseteq Y$ and a variable $\widehat{v} \in Y \setminus (\widehat{Y}_{\mathsf{pos}} \cup \widehat{Y}_{\mathsf{neg}})$ such that the following holds:*

1. $|\widehat{Y}_{\text{pos}}| = |\widehat{Y}_{\text{neg}}| = \frac{\tau}{2}$.
2. Let $\widehat{Y} = \widehat{Y}_{\text{pos}} \cup \widehat{Y}_{\text{neg}} \cup \{\widehat{v}\}$. For every pair of variables $u, u' \in \widehat{Y}$ and every clause $C \in \text{claInt}(Y)$, u appears in C positively (negatively) if and only if u' appears in C positively (negatively).
3. For every variable $u \in \widehat{Y}_{\text{pos}}$, $d_{\Phi}^{+}(\widehat{v}) \le d_{\Phi}^{+}(u)$.
4. For every variable $u \in \widehat{Y}_{\text{neg}}$, $d_{\Phi}^{-}(\widehat{v}) \le d_{\Phi}^{-}(u)$.

Proof. Let $\text{claInt}(Y) = \{C_1, \cdots, C_\ell\}$. For each $u \in Y$, we define a string Γ_u on $\{0, 1\}$ of length ℓ by setting i-th bit of Γ_u as 1 (0) if u appears as positively (negatively) in C_i. By simple combinatorial arguments, we have that the number of different Γ strings that we can obtain are at most 2^ℓ. Since $|Y| \ge 2^\ell \cdot \tau + 1$, by pigeonhole principle there exists a set $Y' \subseteq Y$ of size at least $\tau + 1$ such that for every pair u, u' of variables in Y', $\Gamma_u = \Gamma_{u'}$. Also for every clause $C \in \text{claInt}(Y)$, u appears in C positively (negatively) if and only if u' appears in C positively (negatively). Now we obtain $\widehat{Y}_{\text{pos}}$ and $\widehat{Y}_{\text{neg}}$ from Y'.

We let $\widehat{Y}_{\text{pos}}$ be a subset of Y' of size $\frac{\tau}{2}$ such that for every $u \in \widehat{Y}_{\text{pos}}$ and every $u' \in Y' \setminus \widehat{Y}_{\text{pos}}$, $d_{\Phi}^{+}(u') \le d_{\Phi}^{+}(u)$. We let $\widehat{Y}_{\text{neg}}$ be a subset of Y' of size $\frac{\tau}{2}$ such that for every $u \in \widehat{Y}_{\text{neg}}$ and every $u' \in Y' \setminus \widehat{Y}_{\text{neg}}$, $d_{\Phi}^{-}(u') \le d_{\Phi}^{-}(u)$, that is, $\widehat{Y}_{\text{pos}}$ is the set of first $\frac{\tau}{2}$ variables in Y' when sorted by their positive degrees. Similarly $\widehat{Y}_{\text{neg}}$ is the set of first $\frac{\tau}{2}$ variables in Y' when sorted by their negative degrees.

Observe that since $|\widehat{Y}_{\text{pos}}| + |\widehat{Y}_{\text{neg}}| < |Y'|$, therefore $Y' \setminus (\widehat{Y}_{\text{pos}} \cup \widehat{Y}_{\text{neg}}) \ne \emptyset$. We let $\widehat{v}$ be an arbitrary variable in $Y' \setminus (\widehat{Y}_{\text{pos}} \cup \widehat{Y}_{\text{neg}})$. By the above description, clearly $\widehat{v}, \widehat{Y}_{\text{pos}}$, and $\widehat{Y}_{\text{neg}}$ satisfies the properties stated in lemma and are computed in polynomial time. $\square$

Next, we will describe reduction rules that help us bound the size of clauses in $\text{cla}(\Phi)$. For each $p \in [d-1]$, we introduce Reduction Rule 3.p. We apply Reduction Rule 3.p in the increasing order of p. That is, first apply Reduction Rule 3.1 exhaustively, and for each $p \in [d-1] \setminus \{1\}$, apply Reduction Rule 3.p only if Reduction Rule 3.$(p-1)$ has been applied exhaustively. We apply our reduction rule on a "large" subset of $\text{var}(\Phi)$. To quantify "large" we introduce the following definition (Fig. 2).

Definition 2. For each $p \in [d-1]$, we define an integer z_p as follows:

- If $p = 1$, then $z_1 = 2^{d+1} \cdot (t(d-1) + 1)$, and
- $z_p = 2^{d-p+2} \cdot (t \cdot z_{p-1} + 1)$, otherwise.

The following observation will be helpful to bound the size of our kernel.

Observation 2 $z_{d-1} \le 4^{d^2}(d \cdot t^d + 1)$.

Reduction Rule 3. For each $p \in [d-1]$ we introduce Reduction Rule 3.p as follows. If there exists $Y \subseteq \text{var}(\Phi)$ such that $|\text{claInt}(Y)| = d - p$ and $|Y| = z_p + 1$. Use Lemma 2 to find sets $\widehat{Y}_{\text{pos}}, \widehat{Y}_{\text{neg}} \subseteq Y$ and a variable $\widehat{v} \in Y \setminus (\widehat{Y}_{\text{pos}} \cup \widehat{Y}_{\text{neg}})$ which satisfies properties stated in Lemma 2. Remove $\widehat{v}$ from $\text{var}(\Phi)$ and return the instance (Φ', k, t). Here, Φ' is the formula with variable set $\text{var}(\Phi) \setminus \{\widehat{v}\}$ and clause set $\bigcup_{C \in \text{cla}(\Phi)} C \setminus \{\widehat{v}\}$.

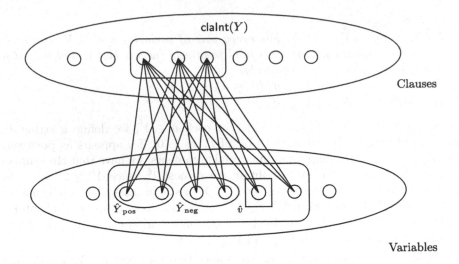

Fig. 2. A visual representation of Reduction Rule 3

Lemma 3. *Reduction Rule 3 is safe.*

Proof. To show that the lemma holds, we will show that (Φ, k, t) is a YES instance of CC-MAX-SAT if and only if (Φ', k, t) is a YES instance of CC-MAX-SAT, for each $p \in [d-1]$. We prove the lemma by induction on p.

Base Case: $p = 1$. We have $z_1 = 2^{d+1} \cdot (t(d-1)+1)$. By Lemma 2 we have $|\widehat{Y}_{\text{pos}}| = |\widehat{Y}_{\text{neg}}| = 2t(d-1)+2$. In the forward direction suppose that (Φ, k, t) is a YES instance of CC-MAX-SAT and let β be its minimum weight assignment. Let $\mathcal{C}_\beta \subseteq \text{cla}(\Phi)$ be the set of clauses satisfied by β and $\widetilde{\mathcal{C}_\beta} = \mathcal{C}_\beta \setminus \text{claInt}(Y)$ be the set of clauses satisfied by β but not in $\text{claInt}(Y)$. Observe the following:

1. By Observation 1, the number of clauses satisfied by β is at most $2t$, that is $|\mathcal{C}_\beta| \le 2t$.
2. Suppose that $C \in \widetilde{\mathcal{C}_\beta}$. Then, since Φ is a $K_{d,d}$-free formula and $|\text{claInt}(Y) \cup \{C\}| = d$, we have $|\text{var}(C) \cap Y| \le d-1$. As otherwise the set of variables $\text{var}(C) \cap Y$ and set of clauses $\text{claInt}(Y) \cup \{C\}$ will contradict $K_{d,d}$-free property.

By (1) and (2) the set of variables in Y, that appear in clauses in the set $\widetilde{\mathcal{C}_\beta}$ is bounded by $2t(d-1)$. Hence the set of variables in $\widehat{Y}_{\text{pos}}$ and in $\widehat{Y}_{\text{neg}}$, that appear in clauses in the set $\widetilde{\mathcal{C}_\beta}$ is bounded by $2t(d-1)$. That is $|\bigcup_{C \in \widetilde{\mathcal{C}_\beta}} \text{var}(C) \cap \widehat{Y}_{\text{pos}}| \le 2t(d-1)$ and $|\bigcup_{C \in \widetilde{\mathcal{C}_\beta}} \text{var}(C) \cap \widehat{Y}_{\text{neg}}| \le 2t(d-1)$.

Recall that $|\widehat{Y}_{\text{pos}}| = |\widehat{Y}_{\text{neg}}| = 2t(d-1)+2$. Therefore, there exists two variables say $w_1, w_2 \in \widehat{Y}_{\text{pos}}$ such that w_1, w_2 do not appear in any clause in $\widetilde{\mathcal{C}_\beta}$. Similarly, there exists two variables say $u_1, u_2 \in \widehat{Y}_{\text{neg}}$ such that u_1, u_2 do not appear in any clause in $\widetilde{\mathcal{C}_\beta}$. That is every clause that is satisfied by β and contains any of w_1, w_2, u_1, u_2 is contained in $\text{claInt}(Y)$. This also implies that every clause that is satisfied by any of the variables w_1, w_2, u_1, u_2 in assignment β, is contained in claInt(Y). Now consider the following cases:

Case 1: $\beta(\widehat{v}) = 1$. First we claim that w_1 and w_2 are both set to false by β. Suppose for a contradiction that one of them say w_1 is set to true, that is $\beta(w_1) = 1$. By the properties of $\widehat{Y}_{\text{pos}}$, and $\widehat{v}$ (See Lemma 2), in every clause $C \in$ claInt(Y), either both $\widehat{v}, w_1$ appear positively in C or both $\widehat{v}, w_1$ appear negatively in C. Further, since w_1 can only satisfy clauses in claInt(Y) by assignment β, we have that w_1 satisfies same set of clauses as $\widehat{v}$ in claInt(Y). That is $N_\Phi(w_1) \cap C_\beta = N_\Phi(\widehat{v}) \cap C_\beta$. In this case, we can obtain another assignment of smaller weight by setting w_1 to false, which contradicts that β is a minimum weight assignment. By similar arguments we can show that w_2 is set to false by β.

Now we will construct an assignment β' for variables var(Φ') = var(Φ) \ $\{\widehat{v}\}$, by setting the value of w_1 to true and we will show that β' satisfies as many clauses in Φ' as satisfied by β in Φ. We define β' formally as follows: $\beta'(w_1) = 1$, and for every $u \in$ var(Φ) \ $\{\widehat{v}, w_1\}$, $\beta'(u) = \beta(u)$. Notice that C_β comprises of the following set of clauses satisfied by β (i) clauses that do not contain variables $\widehat{v}, w_1$, (ii) clauses that are in claInt(Y) and that contains variable $\widehat{v}$ positively, (iii) clauses that are in claInt(Y) and that contains variable w_1 negatively, and lastly (iv) clauses that are not in claInt(Y) and that contains variable $\widehat{v}$ positively.

Clearly cla(Φ') contains every clause in the set (i) and they are also satisfied by β'. For every clause $C \in$ cla(Φ) that contain $\widehat{v}$, we have a clause $C \setminus \{\widehat{v}\}$ in cla(Φ'). If C is a clause in the set (ii), then C contains $\widehat{v}$ positively and thus also contains w_1 positively. Therefore, $C \setminus \{\widehat{v}\}$ is satisfied by β'. Hence for every clause in the set (ii), we have a clause in cla(Φ') satisfied by w_1 in β'. Next, consider a clause C in the set (iii), then C contains w_1 negatively and thus also contains w_2 negatively. As argued before $\beta(w_2) = \beta'(w_2) = 0$. Therefore, $C \setminus \{\widehat{v}\}$ is satisfied by w_2 in β'. Hence for every clause in set (iii), we have a clause in cla(Φ') satisfied by β'.

Now we are only remaining to show that β' compensate for the clauses in the set (iv) for Φ'. For that purpose recall that we have $d_\Phi^+(\widehat{v}) \leq d_\Phi^+(w_1)$, by the properties of $\widehat{Y}_{\text{pos}}$, and $\widehat{v}$ (See Lemma 2). Therefore $|N_\Phi^+(w_1)| \geq |N_\Phi^+(\widehat{v})|$, and hence $|N_{\Phi'}^+(w_1)| \geq |N_\Phi^+(\widehat{v})|$. Also $(N_\Phi^+(w_1) \cap C_\beta) \setminus$ claInt(Y) $= \emptyset$. That is, the clauses which contains w_1 positively and are not in claInt(Y) were not satisfied by β. As β' sets w_1 to true, now clauses in $N_{\Phi'}^+(w_1)$ are satisfied by β'. We obtain the following:

$$|(N_\Phi^+(w_1) \cap C_\beta)| - |N_\Phi^+(w_1) \cap \text{claInt}(Y)| \geq |(N_\Phi^+(\widehat{v}) \cap C_\beta)| - |N_\Phi^+(\widehat{v}) \cap \text{claInt}(Y)|.$$

$$|(N_{\Phi'}^+(w_1) \cap C_{\beta'})| - |N_\Phi^+(w_1) \cap \text{claInt}(Y)| \geq |(N_\Phi^+(\widehat{v}) \cap C_\beta)| - |N_\Phi^+(\widehat{v}) \cap \text{claInt}(Y)|.$$

All the above discussion concludes that the number of clauses satisfied by β' in Φ' are at least the number of clauses satisfied by β in Φ. Hence, (Φ', k, t) is a YES instance of CC-MAX-SAT.

Case 2: $\beta(\widehat{v}) = 0$. We can show that at least one of u_1, u_2 is set to false by β, by using analogous arguments that were used in Case 1 to show that when $\widehat{v}$ is set to true then both w_1 and w_2 are set to false. Without loss of generality suppose that u_1 is set to false by β. We will show that assignment β restricted to $\mathsf{var}(\Phi') = \mathsf{var}(\Phi) \setminus \{\widehat{v}\}$ satisfies as many clauses in Φ' as satisfied by β in Φ. We define β' as β restricted to $\mathsf{var}(\Phi')$ formally as follows: for every $u \in \mathsf{var}(\Phi) \setminus \{\widehat{v}\}$, $\beta'(u) = \beta(u)$. Notice that $\mathcal{C}_\beta$ comprises of the following set of clauses satisfied by β (i) clauses that do contain variable $\widehat{v}$, (ii) clauses that are in $\mathsf{clalnt}(Y)$ and that contains variable $\widehat{v}$ negatively, and lastly (iii) clauses that are not in $\mathsf{clalnt}(Y)$ and that contains variable $\widehat{v}$ negatively.

Clearly $\mathsf{cla}(\Phi')$ contains every clause in the set (i) and they are also satisfied by β'. For every clause $C \in \mathsf{cla}(\Phi)$ that contain $\widehat{v}$, we have a clause $C \setminus \{\widehat{v}\}$ in $\mathsf{cla}(\Phi')$. If C is a clause in the set (ii), then C contains $\widehat{v}$ negatively and thus also contains u_1 negatively. Therefore, $C \setminus \{\widehat{v}\}$ is satisfied by β'. Hence for every clause in the set (ii), we have a clause in $\mathsf{cla}(\Phi')$ satisfied by u_1 in β'. Now we are only remaining to consider the clauses in the set (iii). That is the set of clauses that are not in $\mathsf{clalnt}(Y)$ and contains $\widehat{v}$ negatively. We claim that there is no clause in set (iii), that is $N_\Phi^-(\widehat{v}) \setminus \mathsf{clalnt}(Y) = \emptyset$.

To prove the claim first recall that we have $d_\Phi^-(\widehat{v}) \leq d_\Phi^-(u_1)$, by the properties of $\widehat{Y}_{\mathsf{neg}}$, and $\widehat{v}$ (See Lemma 2). Therefore $|N_\Phi^-(u_1)| \geq |N_\Phi^-(v)|$. Also $N_\Phi^-(u_1) \cap \mathsf{clalnt}(Y) = N_\Phi^-(\widehat{v}) \cap \mathsf{clalnt}(Y)$. Further, $(N_\Phi^-(u_1) \cap \mathcal{C}_\beta) \setminus \mathsf{clalnt}(Y) = \emptyset$. Therefore, $N_\Phi^-(u_1) \setminus \mathsf{clalnt}(Y) = \emptyset$, as u_1 is set to false by β. If $(N_\Phi^-(\widehat{v}) \cap \mathcal{C}_\beta) \setminus \mathsf{clalnt}(Y) \neq \emptyset$, then $N_\Phi^-(\widehat{v}) \setminus \mathsf{clalnt}(Y) \neq \emptyset$. Therefore $|N_\Phi^-(u_1) \setminus \mathsf{clalnt}(Y)| < |N_\Phi^-(\widehat{v}) \setminus \mathsf{clalnt}(Y)|$. Hence, the following holds:

$$d_\Phi^-(u_1) = |(N_\Phi^-(u_1) \cap \mathsf{clalnt}(Y))| + |(N_\Phi^-(u_1) \cap \mathsf{clalnt}(Y))|,$$

$$d_\Phi^-(u_1) = |(N_\Phi^-(u_1) \cap \mathsf{clalnt}(Y))| + |(N_\Phi^-(\widehat{v}) \cap \mathsf{clalnt}(Y))|,$$

$$d_\Phi^-(u_1) < |(N_\Phi^-(\widehat{v}) \cap \mathsf{clalnt}(Y))| + |(N_\Phi^-(\widehat{v}) \cap \mathsf{clalnt}(Y))|.$$

Thus $d_\Phi^-(u_1) < d_\Phi^-(\widehat{v})$, a contradiction. All the above discussion concludes that the number of clauses satisfied by β' in Φ' are at least the number of clauses satisfied by β in Φ. Hence, (Φ', k, t) is a YES instance of CC-MAX-SAT.

It is easy to see that in the backward direction if (Φ', k, t) is a YES instance of CC-MAX-SAT then (Φ, k, t) is a YES instance of CC-MAX-SAT. As any assignment β' of $\mathsf{var}(\Phi')$ can be extended to an assignment β of Φ by setting $\widehat{v}$ to false and assigning every variable $u \neq \widehat{v}$ as $\beta(u)$. By the definition of Φ', β satisfies as many clauses as β' and weight of β is equal to weight of β'.

Induction Hypothesis: Assume that Reduction Rule 3. p is safe for all $p < q$, $q \in [d-2]$.

Inductive Case: $p = q$. We have $z_p = 2^{d-p+2} \cdot (t \cdot z_{p-1} + 1) + 1$. By Lemma 2 we have $|\widehat{Y}_{\mathsf{pos}}| = |\widehat{Y}_{\mathsf{neg}}| = 2t \cdot z_{p-1} + 2$. In the forward direction suppose that (Φ, k, t) is a YES instance of CC-MAX-SAT and let β be its minimum weight assignment. Let $\mathcal{C}_\beta \subseteq \mathsf{cla}(\Phi)$ be the set of clauses satisfied by β and $\widetilde{\mathcal{C}_\beta} = \mathcal{C}_\beta \setminus \mathsf{clalnt}(Y)$ be the set of clauses satisfied by β but not in $\mathsf{clalnt}(Y)$. Observe the following:

1. By Observation 1, the number of clauses satisfied by β is at most $2t$, that is $|\mathcal{C}_\beta| \leq 2t$.

2. Suppose that $C \in \widetilde{\mathcal{C}_\beta}$. Then, since Reduction Rule 3.$(p-1)$ is not applicable and $|\mathsf{claInt}(Y) \cup \{C\}| = d - p + 1 = d - (p-1)$, we have $|\mathsf{var}(C) \cap Y| \leq z_{p-1}$. As otherwise the set of variables $\mathsf{var}(C) \cap Y$ and set of clause $\mathsf{claInt}(Y) \cup \{C\}$ will contradict that Reduction Rule 3.$(p-1)$ is not applicable.

By (1) and (2) the number of variables in Y, that appear in clauses in the set $\widetilde{\mathcal{C}_\beta}$ is bounded by $2t \cdot z_{p-1}$. Hence the number of variables in $\widehat{Y}_{\mathsf{pos}}$ and in $\widehat{Y}_{\mathsf{neg}}$, that appear in clauses in the set $\widetilde{\mathcal{C}_\beta}$ is bounded by $2t \cdot z_{p-1}$. That is $|\bigcup_{C \in \widetilde{\mathcal{C}_\beta}} \mathsf{var}(C) \cap \widehat{Y}_{\mathsf{pos}}| \leq 2t \cdot z_{p-1}$ and $|\bigcup_{C \in \widetilde{\mathcal{C}_\beta}} \mathsf{var}(C) \cap \widehat{Y}_{\mathsf{neg}}| \leq 2t \cdot z_{p-1}$.

Recall that $|\widehat{Y}_{\mathsf{pos}}| = |\widehat{Y}_{\mathsf{neg}}| = 2t \cdot z_{p-1} + 2$. Therefore, there exists two variables say $w_1, w_2 \in \widehat{Y}_{\mathsf{pos}}$ such that w_1, w_2 do not appear in any clause in $\widetilde{\mathcal{C}_\beta}$. Similarly, there exists two variables say $u_1, u_2 \in \widehat{Y}_{\mathsf{neg}}$ such that u_1, u_2 do not appear in any clause in $\widetilde{\mathcal{C}_\beta}$. Now by analogous arguments as in Case 1 and Case 2 of the base case, it follows that (Φ', k, t) is a YES instance of CC-Max-SAT. Backward direction also follows similar to the base case. This completes the proof. □

When Reduction Rule 3 is no longer applicable, we have that every set $Y \subseteq \mathsf{var}(\Phi)$ such that $|\mathsf{claInt}(Y)| \geq d - (d-1) = 1$, satisfies $|Y| \leq z_{d-1}$. In other words for every clause $C \in \mathsf{cla}(\Phi)$, $|\mathsf{var}(C)| \leq z_{d-1}$. We record this observation in the following.

Observation 3. *When Reduction Rules 1–3 are not applicable, then for every clause $C \in \mathsf{cla}(\Phi)$, $|\mathsf{var}(C)| \leq z_{d-1}$.*

For stating our next reduction rule, we define two sets $\widehat{V}_{\mathsf{pos}}, \widehat{V}_{\mathsf{neg}} \subseteq \mathsf{var}(\Phi)$ of size $\min\{n, z_{d-1} + 1\}$ with the following properties: (1) For every variable $u \in \widehat{V}_{\mathsf{pos}}$ and every variable $u' \in \mathsf{var}(\Phi) \setminus \widehat{V}_{\mathsf{pos}}$, $d_\Phi^+(u) \geq d_\Phi^+(u')$. (2) For every variable $u \in \widehat{V}_{\mathsf{neg}}$ and every variable $u' \in \mathsf{var}(\Phi) \setminus \widehat{V}_{\mathsf{neg}}$, $d_\Phi^-(u) \geq d_\Phi^-(u')$. Clearly, if $\mathsf{var}(\Phi) \geq 2t \cdot z_{d-1} + 3$, then by pigeonhole principle $\mathsf{var}(\Phi) \setminus (\widehat{V}_{\mathsf{pos}} \cup \widehat{V}_{\mathsf{neg}}) \neq \emptyset$.

Reduction Rule 4. *If $|\mathsf{var}(\Phi)| \geq 2t \cdot z_{d-1} + 3$, then let $u \in \mathsf{var}(\Phi) \setminus (\widehat{V}_{\mathsf{pos}} \cup \widehat{V}_{\mathsf{neg}})$. Remove u from $\mathsf{var}(\Phi)$ and return the instance (Φ', k, t). Here Φ' is the formula with variable set $\mathsf{var}(\Phi) \setminus \{u\}$ and clause set $\bigcup_{C \in \mathsf{cla}(\Phi)} C \setminus \{u\}$.*

Lemma 4. *Reduction Rule 4 is safe.*

Proof. In the forward direction suppose that (Φ, k, t) is a YES instance of CC-Max-SAT. Let β be its minimum weight assignment and let X be the set of clauses satisfied by β. Since Reduction Rule 2 is not applicable β is not an all zero assignment.

(1) By Observation 1, $|X| \leq 2t$.

(2) By Observation 3, for every $C \in \mathsf{cla}(\Phi)$, $|\mathsf{var}(C)| \leq z_{d-1}$.

Let $Y = \bigcup_{C \in X} \mathsf{var}(C)$, then by (1) and (2), $|Y| \leq 2t \cdot z_{d-1}$. Therefore, there exists a variable say $w_1 \in \widehat{V}_{\mathsf{pos}}$ and there exists a variable say $w_2 \in \widehat{V}_{\mathsf{neg}}$ such that $w_1, w_2 \notin Y$ and hence, $N_\Phi(w_1) \cap X = N_\Phi(w_2) \cap X = \emptyset$. Since β is a minimum weight assignment, $\beta(w_1) = \beta(w_2) = 0$.

Case 1: $\beta(u) = 1$. We obtain another assignment by setting the value of w_1 to true. That is we construct a new assignment β' of $\mathsf{var}(\Phi') = \mathsf{var}(\Phi) \setminus \{u\}$ as follows: $\beta'(w_1) = 1$, and for every $v \in \mathsf{var}(\Phi) \setminus \{u, w_1\}$, $\beta'(v) = \beta(v)$. We have $d_\Phi^+(u) \leq d_\Phi^+(w_1)$ (by the definition of $\widehat{V}_{\mathsf{pos}}$). Let $X' = (X \setminus N_\Phi^+(u)) \cup N_\Phi^+(w_1)$ and $X'' = \bigcup_{C \in X'} C \setminus \{u\}$. Then observe that X'' is the set of clauses satisfied by β'. Also $|X''| = |X'| \geq |X| \geq t$. This implies that β' is a solution to (Φ', k, t) of CC-MAX-SAT.

Case 2: $\beta(u) = 0$. We have $d_\Phi^-(u) \leq d_\Phi^-(w_2)$ (by the definition of $\widehat{V}_{\mathsf{neg}}$) and $N_\Phi^-(w_2) \cap X = \emptyset$. As both u, w_2 are set to false by β, $N_\Phi^-(w_2) \setminus X = N_\Phi^-(u) \setminus X = \emptyset$. Therefore $N_\Phi^-(u) \cap X = \emptyset$, as otherwise $|N_\Phi^-(u) \cap X| > |N_\Phi^-(w_2) \cap X|$ which contradicts $d_\Phi^-(u) \leq d_\Phi^-(w_2)$. Then observe that X'' is the set of clauses satisfied by β restricted to $\mathsf{var}(\Phi) \setminus \{u\} = \mathsf{var}(\Phi')$. Also $|X''| = |X'| = |X| \geq t$. This implies that assignment β restricted to $\mathsf{var}(\Phi')$ is a solution to (Φ', k, t) of CC-MAX-SAT.

It is easy to see that in the backward direction if (Φ', k, t) is a YES instance of CC-MAX-SAT then (Φ, k, t) is a YES instance of CC-MAX-SAT. As any assignment β' of $\mathsf{var}(\Phi)$ can be extended to an assignment β of Φ by setting u to false and assigning every variable $v \neq u$ as $\beta(v)$. Observe that β satisfies as many clauses as β' and weight of β is equal to weight of β'. This completes the proof. □

Observe that Reduction Rules 1,2 and 4 can be applied in polynomial time. Reduction Rule 3 can be applied in $n^{\mathcal{O}(4^{d^2})}$ time. Each of our reduction rule is applicable only polynomial many times. Hence, the kernelization algorithm runs in polynomial time. Each of our reduction rule is safe. When Reduction Rules 1-4 are no longer applicable, the size of the set $\mathsf{var}(\Phi)$ is bounded by $2t(z_{d-1})+2$, and number of clauses a variable appear in is bounded by $2t$, and there are no variables which do not appear in any clause. Therefore by using Observation 1, the number of variables and clauses in Φ is bounded by $\mathcal{O}(d4^{d^2+1}t^{d+1})$.

Theorem 2. CC-MAX-SAT *in* $K_{d,d}$-free *formulae admits a kernel of size* $\mathcal{O}(d4^{d^2}t^{d+1})$.

References

1. Agrawal, A., Choudhary, P., Jain, P., Kanesh, L., Sahlot, V., Saurabh, S.: Hitting and covering partially. In: COCOON, pp. 751–763 (2018)
2. Bläser, M.: Computing small partial coverings. Inf. Process. Lett. **85**(6), 327–331 (2003). https://doi.org/10.1016/S0020-0190(02)00434-9
3. Cygan, M., et al.: Parameterized Algorithms. Springer, Cham (2015). https://doi.org/10.1007/978-3-319-21275-3
4. Dom, M., Lokshtanov, D., Saurabh, S.: Kernelization lower bounds through colors and ids. ACM Trans. Algorithms **11**(2), 13:1–13:20 (2014)
5. Feige, U.: A threshold of ln n for approximating set cover. J. ACM **45**(4), 634–652 (1998). https://doi.org/10.1145/285055.285059

6. Guo, J., Niedermeier, R., Wernicke, S.: Parameterized complexity of vertex cover variants. Theory Comput. Syst. **41**(3), 501–520 (2007). https://doi.org/10.1007/s00224-007-1309-3
7. Jain, P., Kanesh, L., Panolan, F., Saha, S., Sahu, A., Saurabh, S., Upasana, A.: Parameterized approximation scheme for biclique-free max k-weight SAT and max coverage. In: Bansal, N., Nagarajan, V. (eds.) Proceedings of the 2023 ACM-SIAM Symposium on Discrete Algorithms, SODA 2023, Florence, Italy, 22–25 January 2023, pp. 3713–3733. SIAM (2023)
8. Lokshtanov, D., Panolan, F., Ramanujan, M.S.: Backdoor sets on nowhere dense SAT. In: Bojanczyk, M., Merelli, E., Woodruff, D.P. (eds.) 49th International Colloquium on Automata, Languages, and Programming, ICALP 2022, July 4-8, 2022, Paris, France. LIPIcs, vol. 229, pp. 91:1–91:20. Schloss Dagstuhl - Leibniz-Zentrum für Informatik (2022). https://doi.org/10.4230/LIPIcs.ICALP.2022.91
9. Manurangsi, P.: Tight running time lower bounds for strong inapproximability of maximum k-coverage, unique set cover and related problems (via t-wise agreement testing theorem). In: Chawla, S. (ed.) Proceedings of the 2020 ACM-SIAM Symposium on Discrete Algorithms, SODA 2020, Salt Lake City, UT, USA, 5-8 January 2020, pp. 62–81. SIAM (2020). https://doi.org/10.1137/1.9781611975994.5
10. Muise, C.J., Beck, J.C., McIlraith, S.A.: Optimal partial-order plan relaxation via maxsat. J. Artif. Intell. Res. **57**, 113–149 (2016). https://doi.org/10.1613/jair.5128
11. Naor, M., Schulman, L.J., Srinivasan, A.: Splitters and near-optimal derandomization. In: 36th Annual Symposium on Foundations of Computer Science, Milwaukee, Wisconsin, USA, 23-25 October 1995, pp. 182–191. IEEE Computer Society (1995). https://doi.org/10.1109/SFCS.1995.492475
12. Skowron, P., Faliszewski, P.: Chamberlin-courant rule with approval ballots: approximating the maxcover problem with bounded frequencies in FPT time. J. Artif. Intell. Res. **60**, 687–716 (2017). https://doi.org/10.1613/jair.5628
13. Sviridenko, M.: Best possible approximation algorithm for MAX SAT with cardinality constraint. Algorithmica **30**(3), 398–405 (2001)
14. Telle, J.A., Villanger, Y.: FPT algorithms for domination in sparse graphs and beyond. Theor. Comput. Sci. **770**, 62–68 (2019). https://doi.org/10.1016/j.tcs.2018.10.030
15. Zhang, L., Bacchus, F.: MAXSAT heuristics for cost optimal planning. In: Hoffmann, J., Selman, B. (eds.) Proceedings of the Twenty-Sixth AAAI Conference on Artificial Intelligence, 22-26 July 2012, Toronto, Ontario, Canada, pp. 1846–1852. AAAI Press (2012). https://doi.org/10.1609/aaai.v26i1.8373

Automata Theory and Formal Languages

Counting Fixed Points and Pure 2-Cycles of Tree Cellular Automata

Volker Turau(✉) (iD)

Institute of Telematics, Hamburg University of Technology, Hamburg, Germany
turau@tuhh.de

Abstract. Cellular automata are synchronous discrete dynamical systems used to describe complex dynamic behaviors. The dynamic is based on local interactions between the components, these are defined by a finite graph with an initial node coloring with two colors. In each step, all nodes change their current color synchronously to the least/most frequent color in their neighborhood and in case of a tie, keep their current color. After a finite number of rounds these systems either reach a fixed point or enter a 2-cycle. The problem of counting the number of fixed points for cellular automata is #P-complete. In this paper we consider cellular automata defined by a tree. We propose an algorithm with runtime $O(n\Delta)$ to count the number of fixed points, here Δ is the maximal degree of the tree. We also prove upper and lower bounds for the number of fixed points. Furthermore, we obtain corresponding results for pure cycles, i.e., instances where each node changes its color in every round. We provide examples demonstrating that the bounds are sharp.

Keywords: Tree cellular automata · Fixed points · Counting problems

1 Introduction

A widely used abstraction of classical distributed systems such as multi-agent systems are *graph automata*. They evolve over time according to some simple local behaviors of its components. They belong to the class of synchronous discrete-time dynamical systems. A common model is as follows: Let G be a graph, where each node is initially either black or white. In discrete-time rounds, all nodes simultaneously update their color based on a predefined local rule. Locality means that the color associated with a node in round t is determined by the colors of the neighboring nodes in round $t-1$. As a local rule we consider the minority and the majority rule that arises in various applications and as such have received wide attention in recent years, in particular within the context of information spreading. Such systems are also known as graph cellular automata. It is well-known [9,19] that they always converge to configurations that correspond to cycles either of length 1 – a.k.a. fixed points – or of length 2, i.e., such systems eventually reach a stable configuration or toggle between two configurations.

© The Author(s), under exclusive license to Springer Nature Switzerland AG 2024
J. A. Soto and A. Wiese (Eds.): LATIN 2024, LNCS 14579, pp. 241–256, 2024.
https://doi.org/10.1007/978-3-031-55601-2_16

One branch of research so far uses the assumption that the initial configuration is random. Questions of interest are on the expected stabilization time of this process [25] and the dominance problem [18].

In this paper we focus on counting problems related to cellular automata, in particular counting the number of fixed points and pure 2-cycles, i.e., instances where each node changes its color in every round. This research is motivated by applications of so-called Boolean networks (BN) [11], i.e., discrete-time dynamical systems, where each node (e.g., gene) takes either 0 (passive) or 1 (active) and the states of nodes change synchronously according to regulation rules given as Boolean functions. Since the problem of counting the fixed points of a BN is in general #P-complete [3,8,21], it is interesting to find graph classes, for which the number of fixed points can be efficiently determined. These counting problems have attracted a lot of research in recent years [4,6,13].

We consider tree cellular automata, i.e., the defining graphs are finite trees. The results are based on a characterization of fixed points and pure 2-cycles for tree cellular automata [22]. The authors of [22] describe algorithms to enumerate all fixed points and all pure cycles. Since the number of fixed points and pure 2-cycles can grow exponentially with the tree size, these algorithms are unsuitable to efficiently compute these numbers. We prove the following theorem.

Theorem 1. *The number of fixed points and the number of pure 2-cycles of a tree with n nodes and maximal node degree Δ can be computed in time $O(n\Delta)$.*

We also prove the following theorem with upper and lower bounds for the number of fixed points of a tree improving results of [22] (parameter r is explained in Sect. 4.3). In the following, the ith Fibonacci number is denoted by $\mathbb{F}_i$.

Theorem 2. *A tree with n nodes, diameter D and maximal node degree Δ has at least $\max\left(2^{r/2+1}, 2\mathbb{F}_D\right)$ and at most $\min\left(2^{n-\Delta}, 2\mathbb{F}_{n-\lceil\Delta/2\rceil}\right)$ fixed points.*

For the number of pure cycles we prove the following result, which considerably improves the bound of [22].

Theorem 3. *A tree with maximal degree Δ has at most $\min\left(2^{n-\Delta}, 2\mathbb{F}_{\lfloor n/2\rfloor}\right)$ pure 2-cycles.*

We provide examples demonstrating ranges where these bounds are sharp. All results hold for the minority and the majority rule. We also formulate several conjectures about counting problems and propose future research directions. A long version of the paper including all proofs and more results is available [23].

2   State of the Art

The analysis of fixed points of minority/majority rule cellular automata received limited attention so far. Královič determined the number of fixed points of a complete binary tree for the majority process [12]. For the majority rule he showed that this number asymptotically tends to $4n(2\alpha)^n$, where n is the number of nodes and $\alpha \approx 0.7685$. Agur et al. did the same for ring topologies [2], the number of fixed point is in the order of Φ^n, where $\Phi = (1 + \sqrt{5})/2$. In both cases the number of fixed points is an exponentially small fraction of all configurations.

A related concept are Boolean networks (BN). They have been extensively used as mathematical models of genetic regulatory networks. The number of fixed points of a BN is a key feature of its dynamical behavior. A gene is modeled by binary values, indicating two transcriptional states, active or inactive. Each network node operates by the same nonlinear majority rule, i.e., majority processes are a particular type of BN [24]. The number of fixed points is an important feature of the dynamical behavior of a BN [5]. It is a measure for the general memory storage capacity. A high number implies that a system can store a large amount of information, or, in biological terms, has a large phenotypic repertoire [1]. However, the problem of counting the fixed points of a BN is in general #P-complete [3]. There are only a few theoretical results to efficiently determine this set [10]. Aracena determined the maximum number of fixed points regulatory Boolean networks, a particular class of BN [5].

Recently, Nakar and Ron studied the dynamics of a class of synchronous one-dimensional cellular automata for the majority rule [16]. They proved that fixed points and 2-cycles have a particular spatially periodic structure and give a characterization of this structure. Concepts related to fixed points of the minority/majority process are global defensive 0-alliance or monopolies [14].

Most research on discrete-time dynamical systems on graphs is focused on bounding the stabilization time. Good overviews for the majority (resp. minority) process can be found in [25] (resp. [17]). Rouquier et al. studied the minority process in the asynchronous model, i.e., not all nodes update their color concurrently [20]. They showed that the stabilization time strongly depends on the topology and observe that the case of trees is non-trivial.

2.1   Notation

Let $T = (V, E)$ be a finite, undirected tree with $n = |V|$. The maximum degree of T is denoted by $\Delta(T)$, the diameter by $D(T)$. The parameter T is omitted in case no ambiguity arises. A *star graph* is a tree with $n - 1$ leaves. A *l-generalized star graph* is obtained from a star graph by inserting $l - 1$ nodes into each edge, i.e., $n = l\Delta + 1$. For $F \subseteq E$ and $v \in V$ denote by $deg_F(v)$ the number of edges in F incident to v. Note that $deg_F(v) \leq deg(v)$. For $i \geq 2$ denote by $E^i(T)$ the set of edges of T, where each end node has degree at least i. For $v \in V$ denote the set of v's neighbors by $N(v)$. For $e = (v_1, v_2) \in E^2(T)$ let T_i be the subtree of T consisting of e and the connected component of $T \setminus e$ that contains v_i. We call T_i the *constituents* of T for e. T_1 and T_2 together have $n + 2$ nodes. We denote the ith Fibonacci number by $\mathbb{F}_i$, i.e., $\mathbb{F}_0 = 0, \mathbb{F}_1 = 1$, and $\mathbb{F}_i = \mathbb{F}_{i-1} + \mathbb{F}_{i-2}$.

3   Synchronous Discrete-Time Dynamical Systems

Let $G = (V, E)$ be a finite, undirected graph. A coloring c assigns to each node of G a value in $\{0, 1\}$ with no further constraints on c. Denote by $\mathcal{C}(G)$ the set of all colorings of G, i.e., $|\mathcal{C}(G)| = 2^{|V|}$. A transition process $\mathcal{M}$ is a mapping $\mathcal{M} : \mathcal{C}(G) \longrightarrow \mathcal{C}(G)$. Given an initial coloring c, a transition process produces a sequence of colorings $c, \mathcal{M}(c), \mathcal{M}(\mathcal{M}(c)), \ldots$. We consider two transition processes: *Minority* and *Majority* and denote the corresponding mappings by $\mathcal{MIN}$ and $\mathcal{MAJ}$. They are local mappings in the sense that the new color of a node is based on the current colors of its neighbors. To determine $\mathcal{M}(c)$ the local mapping is executed in every *round* concurrently by all nodes. In the minority (resp. majority) process each node adopts the minority (resp. majority) color among all neighbors. In case of a tie the color remains unchanged (see Fig. 1). Formally, the minority process is defined for a node v as follows:

$$\mathcal{MIN}(c)(v) = \begin{cases} c(v) & \text{if } |N^{c(v)}(v)| \leq |N^{1-c(v)}(v)| \\ 1 - c(v) & \text{if } |N^{c(v)}(v)| > |N^{1-c(v)}(v)| \end{cases}$$

$N^i(v)$ denotes the set of v's neighbors with color i ($i = 0, 1$). The definition of $\mathcal{MAJ}$ is similar, only the binary operators $\leq$ and $>$ are reversed. Some results hold for both the minority and the majority process. To simplify notation we use the symbol $\mathcal{M}$ as a placeholder for $\mathcal{MIN}$ and $\mathcal{MAJ}$.

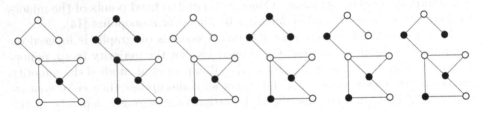

Fig. 1. For the coloring on the left $\mathcal{MIN}$ reaches after 5 rounds a fixed point. $\mathcal{MAJ}$ reaches for the same initial coloring after one round a monochromatic coloring.

Let $c \in \mathcal{C}(G)$. If $\mathcal{M}(c) = c$ then c is called a *fixed point*. It is called a *2-cycle* if $\mathcal{M}(c) \neq c$ and $\mathcal{M}(\mathcal{M}(c)) = c$. A 2-cycle is called *pure* if $\mathcal{M}(c)(v) \neq c(v)$ for each node v of G, see Fig. 2. Denote by $\mathcal{F}_{\mathcal{M}}(G)$ (resp. $\mathcal{P}_{\mathcal{M}}(G)$) the set of all $c \in \mathcal{C}(G)$ that constitute a fixed point (resp. a pure 2-cycle) for $\mathcal{M}$.

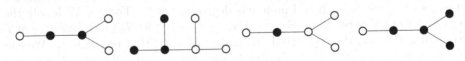

Fig. 2. Examples for the $\mathcal{MIN}$ rule. The coloring of the first (resp. second) tree is a fixed point (resp. a pure 2-cycle). The right two colorings are a non-pure 2-cycle.

Let T be a tree. The following results are based on a characterization of $\mathcal{F}_{\mathcal{M}}(T)$ and $\mathcal{P}_{\mathcal{M}}(T)$ by means of subsets of $E(T)$ [22]. Let $E_{fix}(T)$ be the set of all $\mathcal{F}$-*legal* subsets of $E(T)$, where $F \subseteq E(T)$ is $\mathcal{F}$-*legal* if $2deg_F(v) \leq deg(v)$ for each $v \in V$. Each $\mathcal{F}$-legal set is contained in $E^2(T)$, hence $|E_{fix}(T)| \leq 2^{|E^2(T)|}$. Theorem 1 of [22] proves that $|\mathcal{F}_\mathcal{M}(T)| = 2|E_{fix}(T)|$, see Fig. 3. Let $E_{pure}(T)$ be the set of all $\mathcal{P}$-*legal* subsets of $E(T)$, where $F \subseteq E(T)$ is $\mathcal{P}$-*legal* if $2deg_F(v) < deg(v)$ for each $v \in V$. Thus, $\mathcal{P}$-*legal* subsets are contained in $E^3(T)$ and therefore $|E_{pure}(T)| \leq 2^{|E^3(T)|}$. Theorem 4 of [22] proves that $|\mathcal{P}_\mathcal{M}(T)| = 2|E_{pure}(T)|$. For the tree in Fig. 3 we have $E_{pure}(T) = \{\emptyset\}$, thus $|\mathcal{P}_\mathcal{M}(T)| = 2$. The pure colorings are the two monochromatic colorings. Given these results it is unnecessary to treat $\mathcal{MIN}$ and $\mathcal{MAJ}$ separately. To determine the number of fixed points (resp. pure 2-cycles) it suffices to compute $|E_{fix}(T)|$ (resp. $|E_{pure}(T)|$).

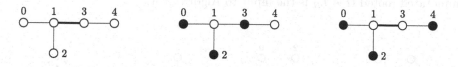

Fig. 3. A tree T with $E_{fix}(T) = \{\emptyset, \{(1,3)\}\}$ and the corresponding fixed points for $\mathcal{MIN}$, the other two can be obtained by inverting colors.

4   Fixed Points

In this section we propose an efficient algorithm to determine $|\mathcal{F}_\mathcal{M}(T)|$, we provide upper and lower bounds for $|\mathcal{F}_\mathcal{M}(T)|$ in terms of n, Δ, and D, and discuss the quality of these bounds. As stated above, it suffices to consider $E_{fix}(T)$ and there is no need to distinguish the minority and the majority model. The following lemma is crucial for our results. It allows to recursively compute $|E_{fix}(T)|$. For a node v define $E_{fix}(T, v) = \{F \in E_{fix}(T) \mid 2(deg_F(v) + 1) \leq deg(v)\}$.

Lemma 1. *Let T be a tree, $e = (v_1, v_2) \in E^2(T)$, and T_i the constituents of T for e. Then $|E_{fix}(T)| = |E_{fix}(T_1, v_1)||E_{fix}(T_2, v_2)| + |E_{fix}(T_1)||E_{fix}(T_2)|$.*

Proof. Let $A = \{F \in E_{fix}(T) \mid e \in F\}$ and $B = \{F \in E_{fix}(T) \mid e \notin F\}$. Then $A \cap B = \emptyset$ and $A \cup B = E_{fix}(T)$, i.e., $|E_{fix}(T)| = |A| + |B|$. If $F \in A$ then $F \setminus e \cap T_i \in E_{fix}(T_i, v_i)$ since $T_1 \cap T_2 = \{e\}$. Hence, $|A| \leq |E_{fix}(T_1, v_1)||E_{fix}(T_2, v_2)|$. If $F \in B$ then $F \cap T_i \in E_{fix}(T_i)$, i.e., $|B| \leq |E_{fix}(T_1)||E_{fix}(T_2)|$. This yields,

$$|E_{fix}(T)| \leq |E_{fix}(T_1, v_1)||E_{fix}(T_2, v_2)| + |E_{fix}(T_1)||E_{fix}(T_2)|.$$

If $F_i \in E_{fix}(T_i, v_i)$, then $F_1 \cup F_2 \cup \{e\} \in A$. If $F_i \in E_{fix}(T_i)$, then $F_1 \cup F_2 \in B$. Hence, $|E_{fix}(T_1, v_1)||E_{fix}(T_2, v_2)| + |E_{fix}(T_1)||E_{fix}(T_2)| \leq |E_{fix}(T)|$. $\square$

If $deg_T(v_i) \equiv 0(2)$, then $E_{fix}(T_i, v_i) = E_{fix}(T_i \setminus \{v_i\})$. This yields a corollary.

Corollary 1. *Let P_n be a path with n nodes, then $|E_{fix}(P_n)| = F_{n-1}$.*

4.1 Computing $|\mathcal{F}_\mathcal{M}(T)|$

Algorithm 1 of [22] enumerates all elements of $E_{fix}(T)$ for a tree T. Since $|E_{fix}(T)|$ can grow exponentially with the size of T, it is unsuitable to efficiently determine $|E_{fix}(T)|$. In this section we propose an efficient novel algorithm to compute $|E_{fix}(T)|$ in time $O(n\Delta)$ based on Lemma 1. The algorithm operates in several steps. Let us define the input for the algorithm. First, each node v_i is annotated with $b_i = \lfloor deg(v_i)/2 \rfloor$. Let T_R be the tree obtained form T by removing all leaves of T; denote by t the number of nodes of T_R. Select a node of T_R as a root and assign numbers $1, \ldots, t$ to nodes in T_R using a postorder depth-first search. Direct all edges towards higher numbers, i.e., the numbers of all predecessors of a node i are smaller than i, see Fig. 4 for an example. The annotated rooted tree T_R is the input to Algorithm 1.

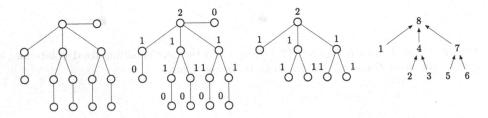

Fig. 4. From left to right: Tree T, annotation of T, T_R, a postorder numbering of T_R.

Algorithm 1 recursively operates on two types of subtrees of T_R which are defined next. For $k = 1, \ldots, t - 1$ denote by T_k the subtree of T_R consisting of k's parent together with all nodes connected to k's parent by paths using only nodes with numbers at most k (see Fig. 5). Note that $T_R = T_{t-1}$. For $k = 1, \ldots, t$ denote by S_k the subtree of T_R consisting of all nodes from which node k can be reached. In particular $S_t = T_R$, and if k is a leaf then S_k consist of node k only.

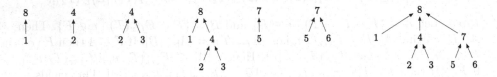

Fig. 5. The subtrees $T_1, \ldots T_7$ of the tree from Fig. 4. Note that $S_4 = T_3$ and $S_3 = \{3\}$.

For a subtree S of T_R with largest node s and $b \geq 0$ denote by $w(S, b)$ the number of subsets F of $E(S)$ with $deg_F(i) \leq b_i$ for all nodes i of S (recall that b_i is defined above) and $deg_F(s) \leq b$. Let $w(S, -1) = 0$. Clearly if $b \geq b_s$ then $w(S, b) = w(S, b_s)$. If S consists of a single node s then s is a leaf and $E(S) = \emptyset$;

therefore $w(S, b) = 1$ for all $b \geq 0$. Note that $w(S, b_s - 1) = |E_{fix}(S, s)|$. The following observation shows the relation between $|E_{fix}(T)|$ and $w(T_k, b)$.

Lemma 2. $|E_{fix}(T)| = w(T_{t-1}, b_t)$ *for any tree* T.

The next lemma shows how to recursively compute $w(T_k, b)$ using Lemma 1.

Lemma 3. *Let i be an inner node of T_R, k a child of i, and $b \geq 0$. Let $\delta_{b,0} = 0$ if $b = 0$ and 1 otherwise. If k is the smallest child of i, then*

$$w(T_k, b) = w(S_k, b_k) + w(S_k, b_k - 1)\delta_{b,0}.$$

Otherwise let $j \neq k$ be the largest child of i such that $j < k$. Then

$$w(T_k, b) = w(T_j, b)w(S_k, b_k) + w(T_j, b - 1)w(S_k, b_k - 1).$$

Proof. The proof for both cases is by induction on k. Consider the first case. If k is a leaf then $w(S_k, b) = 1$ for all $b \geq 0$ and $w(S_k, -1) = 0$. This is the base case. Assume k is not a leaf. T_k consists of node i, S_k and the edge $e = (i, k)$. If $b = 0$ then $\delta_{b,0} = 0$ and $w(T_k, 0) = w(S_k, b_k)$ by definition. Let $b > 0$, i.e., $\delta_{b,0} = 1$. Let f_i (resp. f_o) be the number of $F \subseteq E(T_k)$ with $deg_F(l) \leq b_l$ for all nodes l of T_k, $deg_F(i) \leq b$ and $e \in F$ (resp. $e \notin F$). Let $F \subseteq E(T_k)$. If $e \in F$ let $\hat{F} = F \setminus e$. Then $\hat{F} \in E(S_k)$, $deg_{\hat{F}}(l) \leq b_l$ for all nodes $l \neq k$ in S_k and $deg_{\hat{F}}(k) \leq b_k - 1$, hence $f_i \leq w(S_k, b_k - 1)$. If $e \notin F$ then $F \in E(S_k)$ and $deg_F(l) \leq b_l$ for all nodes l in S_k, hence $f_o \leq w(S_k, b_k)$. Thus, $w(T_k, b) \leq w(S_k, b_k) + w(S_k, b_k - 1)$. On the other hand, let $F \in E(S_k)$ such that $deg_F(l) \leq b_l$ for all nodes $l \neq k$ in S_k. If $deg_F(k) \leq b_k - 1$ then $F \cup \{e\}$ contributes to $w(T_k, b)$ and if $deg_F(k) \leq b_k$ then F contributes to $w(T_k, b)$. Thus $w(T_k, b) \geq w(S_k, b_k) + w(S_k, b_k - 1)$.

Consider the second statement. The case that k is a leaf follows immediately from Lemma 1. Assume that k is not a leaf. We apply Lemma 1 to T_k and edge (i, k). Then $T_j \cup (i, k)$ and $S_k \cup (i, k)$ are the constituents of T_k. Note that $E_{fix}(S_k \cup (i, k), i) = w(S_k, b_k - 1)$ and $E_{fix}(T_j \cup (i, k), k) = w(T_j, b - 1)$. $\qquad\square$

Algorithm 1 makes use of Lemma 2 and 3 to determine $w(T_{t-1}, b_t)$, which is equal to $|E_{fix}(T)|$. Let $B = \max\{b_i \mid i = 1, \ldots, t\}$, clearly $B \leq \Delta/2$. Algorithm 1 uses an array W of size $[0, t-1] \times [0, B]$ to store the values of $w(\cdot, \cdot)$. The first index is used to identify the tree T_k. To simplify notation this index can also have the value 0. To store the values of $w(S_k, b)$ in the same array we define for each inner node k an index $l(k)$ as follows $l(k) = k - 1$ if k is not a leaf and $l(k) = 0$ otherwise. Then clearly $S_k = T_{l(k)}$ if k is not a leaf. More importantly, the value of $w(S_k, b)$ is stored in $W(l(k), b)$ for all k and b.

The algorithm computes the values of W for increasing values of $k < t$ beginning with $k = 0$. If $W(j, b)$ is known for all $j < k$ and all $b \in [0, B]$ we can compute $W(k, b)$ for all values of b in $[0, B]$ using Lemma 3. Finally we have $w(t - 1, b_t)$ which is equal to $|E_{fix}(T)|$. Theorem 1 follows from Lemma 2 and 3.

Algorithm 1: Computation of $W(k,b)$ for all k and b using T_R.

> **for** $b = 0, \ldots, B$ **do**
> > $W(0,b) = 1$
>
> **for** $k = 1, \ldots, t - 1$ **do**
> > **if** k is the smallest child of its parent in T_R **then**
> > > $W(k,0) := W(l(k), b_k)$
> > > **for** $b = 1, \ldots, B$ **do**
> > > > $W(k,b) := W(l(k), b_k) + W(l(k), b_k - 1)$
> >
> > **else**
> > > let j be the largest sibling of k with $j < k$ in T_R
> > > $W(k,0) := W(j,0)W(l(k), b_k)$
> > > **for** $b = 1, \ldots, B$ **do**
> > > > $W(k,b) := W(j,b)W(l(k), b_k) + W(j, b-1)W(l(k), b_k - 1)$

4.2   Upper Bounds for $|\mathcal{F}_\mathcal{M}(T)|$

The definition of $E_{fix}(T)$ immediately leads to a first upper bound for $|\mathcal{F}_\mathcal{M}(T)|$.

Lemma 4. $|E_{fix}(T)| \leq 2^{n-\Delta-1}$.

Proof. Theorem 1 of [22] implies $|E_{fix}(T)| \leq 2^{|E^2(T)|}$. Note that $|E^2(T)| = n - 1 - l$, where l is the number of leaves of T. It is well known that $l = 2 + \sum_{j=3}^{\Delta}(j-2)D_j$, where D_j denotes the number of nodes with degree j. Thus,

$$|E^2(T)| = n - 3 - \sum_{j=3}^{\Delta}(j-2)D_j \leq n - 3 - (\Delta - 2) = n - \Delta - 1.$$

$\square$

For $\Delta < n - \lceil n/3 \rceil$ the bound of Lemma 4 is not attained. Consider the case $n = 8, \Delta = 4$. The tree from Fig. 6 with $x = 1$ has 14 fixed points, this is the maximal attainable value.

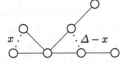

Fig. 6. If $x \geq \Delta/2$ then $|\mathcal{F}_\mathcal{M}(T)| = 2^{n-\Delta}$.

For $\Delta < n/2$ we will prove a much better bound than that of Lemma 4. For this we need the following technical result.

Lemma 5. *Let $T = (V, E)$ be a tree with a single node v that has degree larger than 2. Let $\mathcal{D} = [dist(w, v) \mid w \in V, w \neq v]$ be the multi-set with the distances of all nodes to v. Then*

$$|E_{fix}(T)| = \sum_{S \subset \mathcal{D}, |S| \leq \Delta/2} \prod_{s \in S} \mathbb{F}_{s-1} \prod_{s \in \mathcal{D} \setminus S} \mathbb{F}_s.$$

Proof. Let $\mathcal{P}$ be the set of all Δ paths from v to a leaf of T. Let $F \in E_{fix}(T)$. Then $\mathcal{P}_F = \{P \in \mathcal{P} \mid F \cap P \notin E_{fix}(P)\}$. Let $P \in \mathcal{P}_F$ and v_P the node of P adjacent to v. Then $(v, v_P) \in F$. This yields $|\mathcal{P}_F| \leq \Delta/2$. For $P \in \mathcal{P}$ let $\hat{P}$ be an extension of P by an edge (v, x) with a new node x. Then $F \cap P \in E_{fix}(\hat{P})$ for each $P \in \mathcal{P}_F$. By Cor. 1 there are $\mathbb{F}_{|P|-2}$ possibilities for $F \cap P$. Let $\bar{\mathcal{P}}_F = \mathcal{P} \setminus \mathcal{P}_F$. For $P \in \bar{\mathcal{P}}_F$ we have $(v, v_P) \notin F$. Hence, $F \cap P \in E_{fix}(P)$. By Cor. 1 there are $\mathbb{F}_{|P|-1}$ possibilities for $F \cap P$. Also $|\bar{\mathcal{P}}_F| = \Delta - |\mathcal{P}_F|$.

Let $\mathcal{P}_1 \subset \mathcal{P}$ with $|\mathcal{P}_1| \leq \Delta/2$ and $\hat{F}_P \in E_{fix}(\hat{P})$ for all $P \in \mathcal{P}_1$ and $F_P \in E_{fix}(P)$ for all $P \in \mathcal{P} \setminus \mathcal{P}_1$. Then the union of all $\hat{F}_P$ and all F_P is a member of $E_{fix}(T)$. This yields the result. □

Corollary 2. *Let T be a l-generalized star graph. Then*

$$|E_{fix}(T)| = \sum_{i=0}^{\lfloor \Delta/2 \rfloor} \binom{\Delta}{i} \mathbb{F}_{l-1}^i \mathbb{F}_l^{\Delta-i}.$$

The corollary yields that a star graph (i.e. $l = 1$) has two fixed points. For $l = 2$ we have the following result.

Lemma 6. *Let T be a 2-generalized star graph. Then $|E_{fix}(T)| \leq \mathbb{F}_{n-\lceil \Delta/2 \rceil}$.*

Proof. We use Corollary 2. If $\Delta \equiv 0(2)$ then

$$|E_{fix}(T)| = \sum_{i=0}^{\lfloor \Delta/2 \rfloor} \binom{\Delta}{i} = \frac{1}{2} \left(2^{\Delta} + \binom{\Delta}{\Delta/2} \right) \leq \mathbb{F}_{3\Delta/2+1} = \mathbb{F}_{n-\Delta/2},$$

otherwise $|E_{fix}(T)| = 2^{\Delta-1} \leq \mathbb{F}_{n-\lceil \Delta/2 \rceil}$. □

In Theorem 2 we prove that the upper bound of Lemma 6 holds for all trees. First, we prove two technical results.

Lemma 7. *Let T be a tree and v a leaf of T with neighbor w. Let n_l (resp. n_i) be the number of neighbors of w that are leaves (resp. inner nodes). If $n_l > n_i$ then $E_{fix}(T) = E_{fix}(T \setminus v)$.*

Proof. Clearly, $E_{fix}(T \setminus v) \subseteq E_{fix}(T)$. Let $F \in E_{fix}(T)$. Then $deg_F(w) \leq n_i$. Thus, $2deg_F(w) \leq 2n_i \leq n_l - 1 + n_i = deg_T(w) - 1$. Hence, $F \in E_{fix}(T \setminus v)$, i.e., $E_{fix}(T) \subseteq E_{fix}(T \setminus v)$. □

Lemma 8. *Let T be a tree and v, w, u a path with $deg(v) = 1$ and $deg(w) = 2$. Then $|E_{fix}(T)| \leq |E_{fix}(T_v)| + |E_{fix}(T_w)|$ with $T_v = T \setminus v$ and $T_w = T_v \setminus w$.*

Proof. Let $F \in E_{fix}(T)$ and $e = (u, w)$. If $e \in F$ then $F \setminus e \in E_{fix}(T_w)$ otherwise $F \in E_{fix}(T_v)$. This proves the lemma. □

Lemma 9. $|E_{fix}(T)| \leq \mathbb{F}_{n - \lceil \Delta/2 \rceil}$ *for a tree T with n nodes.*

Proof. The proof is by induction on n. If $\Delta = 2$ the result holds by By Cor. 1. If T is a star graph then $|\mathcal{F}_\mathcal{M}(T)| = 2$, again the result is true. Let $\Delta > 2$ and T not a star graph. Thus, $n > 4$. There exists an edge (v, w) of T where v is a leaf and all neighbors of w but one are leaves. If $deg(w) > 2$ then there exists a neighbor $u \neq v$ of w that is a leaf. Let $T_u = T \setminus u$. Then $|E_{fix}(T)| = |E_{fix}(T_u)|$ by Lemma 7. Since $\Delta(T_u) \geq \Delta(T) - 1$ we have by induction $|E_{fix}(T)| = |E_{fix}(T_u)| \leq \mathbb{F}_{n - 1 - \lceil \Delta(T_u)/2 \rceil} \leq \mathbb{F}_{n - \lceil \Delta(T)/2 \rceil}$. Hence, we can assume that $deg(w) = 2$.

Let $u \neq v$ be the second neighbor of w. Denote by T_v (resp. T_w) the tree $T \setminus v$ (resp. $T \setminus \{v, w\}$). By Lemma 8 we have $|E_{fix}(T)| \leq |E_{fix}(T_v)| + |E_{fix}(T_w)|$. If there exists a node different from u with degree Δ then $|E_{fix}(T)| \leq \mathbb{F}_{n - 1 - \lceil \Delta/2 \rceil} + \mathbb{F}_{n - 2 - \lceil \Delta/2 \rceil} = \mathbb{F}_{n - \lceil \Delta/2 \rceil}$ by induction. Hence we can assume that u is the only node with degree Δ. Repeating the above argument shows that T is 2-generalized star graph with center node u. Hence, $|E_{fix}(T)| \leq \mathbb{F}_{n - \lceil \Delta/2 \rceil}$ by Lemma 6. □

Lemmas 4 and 9 prove the upper bound of Theorem 2. The bound of Lemma 9 is sharp for paths. Note that for a fixed value of n the monotone functions $\mathbb{F}_{n - \lceil \Delta/2 \rceil}$ and $2^{n - \Delta}$ intersect in $\Delta \in (\lceil n/2 \rceil, \lceil n/2 \rceil + 1)$.

4.3   Lower Bounds for $|\mathcal{F}_\mathcal{M}(T)|$

A trivial lower bound for $|\mathcal{F}_\mathcal{M}(T)|$ for all trees is $1 + |E_{fix}(T)|$. It is sharp for star graphs. For a better bound other graph parameters besides n are required.

Lemma 10. *Let T be a tree and T_L the tree obtained from T by removing all leaves. Then $|E_{fix}(T)| \geq 2^{r/2}$, where r is the number of inner nodes of T_L.*

Proof. By induction we prove that T_L has a matching M with $r/2$ edges. Then $M \subseteq E_{fix}(T)$ and each subset of M is $\mathcal{F}$-legal. □

Applying Lemma 10 to a 2-generalized star graph yields a lower bound of 1, which is far from the real value. Another lower bound for $|E_{fix}(T)|$ uses D, the diameter of T. Any tree T with diameter D contains a path of length $D + 1$. Thus, $|E_{fix}(T)| \geq F_D$ by Cor. 1. This completes the proof of Theorem 2. We show that there are trees for which $|E_{fix}(T)|$ is much larger than F_D. Let $n, D, h \in \mathbb{N}$ with $n - 2 > D \geq 2(n - 1)/3$ and $n - D - 1 \leq h \leq 2D - n + 1$. Let $T_{n,D,h}$ be a tree with n nodes that consists of a path $v_0, \ldots, v_D$ and another path of length $n - D - 2$ attached to v_h. Clearly, $T_{n,D,h}$ has diameter D. Also $deg(v_h) = 3$, all other nodes have degree 1 or 2. Figure 7 shows $T_{n,D,h}$.

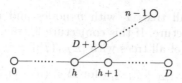

Fig. 7. A tree $T_{n,D,h}$.

Lemma 11. $|E_{fix}(T_{n,D,h})| = \mathbb{F}_D \mathbb{F}_{n-D-1} + \mathbb{F}_h \mathbb{F}_{D-h} \mathbb{F}_{n-D-2}.$

Proof. Let $e = (v_h, v_{D+1})$ and $T_{v_h}, T_{v_{D+1}}$ the constituents of T for e. Then
$|E_{fix}(T)| = |E_{fix}(T_{v_h}, v_h)||E_{fix}(T_{v_{D+1}}, v_{D+1})| + |E_{fix}(T_{v_h})||E_{fix}(T_{v_{D+1}})|$ by
Lemma 1. Clearly, $|E_{fix}(T_{v_h}, v_h)| = \mathbb{F}_h \mathbb{F}_{D-h}$ and $|E_{fix}(T_{v_{D+1}}, v_{D+1})| = \mathbb{F}_{n-D-2}$
by Cor. 1. $\qquad\square$

Lemma 17 of [23] determines the value h_0 for which $|E_{fix}(T_{n,D,h})|$ is maximal
and show that $|E_{fix}(T_{n,D,h_0})| = \mathbb{F}_D \mathbb{F}_{n-D-1} + \mathbb{F}_{h_0} \mathbb{F}_{D-h_0} \mathbb{F}_{n-D-2}$. Let $T_{n,D}$ be
a tree that maximizes the number of fixed points among all trees with n nodes
and diameter D. An interesting question is about the structure of $T_{n,D}$. By an
exhaustive search among all trees with $n \leq 34$ there was just a single case where
this was not a star-like tree (all nodes but one have degree 1 or 2) that maximizes
the number of fixed points (see [23]). We have the following conjecture.

Conjecture 1. Except for a finite number of cases for each combination of n and
D there exists a star-like graph that maximizes the number of fixed points.

4.4 Special Cases

Lemma 12. *Let T be a tree with $n \geq 4$ and $\Delta \geq n - \lceil n/3 \rceil$ then $|E_{fix}(T)| \leq 2^{n-\Delta-1}$. This bound is sharp.*

Proof. First we construct a tree T_m realizing this bound. Let T_m be a tree with
a single node v with degree Δ, $x = 2\Delta - n + 1$ neighbors of v are leaves and the
remaining $\Delta - x = n - \Delta - 1$ neighbors have degree 2 (see Fig. 6). Assumption
$\Delta \geq n - \lceil n/3 \rceil$ implies $\Delta - x \leq \Delta/2$. Hence $E_{fix}(T_m) = \sum_{i=0}^{\Delta-x} \binom{\Delta-x}{i} = 2^{\Delta-x} = 2^{n-\Delta-1}$ (see also Lemma 5).

Let v be a node of T with degree Δ. Assume that at most two neighbors of v
are leaves. Then $n \geq \Delta + 1 + \Delta - 2 = 2\Delta - 1$. This yields $\Delta \leq (n+1)/2$, which
contradicts the assumption $\Delta \geq n - \lceil n/3 \rceil$. Hence, at least three neighbors of
v are leaves. Let w be a non-leaf neighbor of v and $e = (v, w)$. Without loss of
generality we can assume that $deg(w) = 2$ and that the neighbor $v' \neq v$ is a
leaf. Next we apply Lemma 1. Let x be a neighbor of v that is a leaf. Note that
$E_{fix}(T_v, u) = E_{fix}(T_v \setminus \{w, x\})$. By induction we have $|E_{fix}(T_v, u)| \leq 2^{n-\Delta-2}$.
We also have $|E_{fix}(T_v)| \leq 2^{n-\Delta-2}$. This yields the upper bound. $\qquad\square$

Let T be a tree with $n - \lceil n/3 \rceil > \Delta > (n-1)/2$. In [23] it is proved that
$|E_{fix}(T)| \leq \sum_{i=0}^{\lfloor \Delta/2 \rfloor} \binom{n-\Delta-1}{i}$. This bound is sharp. Let $\tau_{n,\Delta}$ be the maximal

value of $|E_{fix}(T)|$ for all trees T with n nodes and maximal degree Δ. We have the following conjecture. If this conjecture is true, it would be possible to determine the structure of all trees with $|E_{fix}(T)| = \tau_{n,\Delta}$.

Conjecture 2. $\tau_{n,\Delta} = \tau_{n-1,\Delta} + \tau_{n-2,\Delta}$ for $\Delta < (n-1)/2$.

5 Pure 2-Cycles

In this section we prove an upper bound for $|\mathcal{P}_{\mathcal{M}}(T)|$. We use the fact $|\mathcal{P}_{\mathcal{M}}(T)| = 2|E_{pure}(T)|$ from [22]. Algorithm 1 is easily adopted to compute $|E_{pure}(T)|$. The difference between $\mathcal{F}$-legal and $\mathcal{P}$-legal is that instead of $2deg_F(v) \leq deg(v)$ condition $2deg_F(v) < deg(v)$ is required. If $deg(v)$ is odd then the conditions are equivalent. Thus, it suffices to define for each node v_i with even degree $b_i = \lfloor deg(v_i)/2 \rfloor - 1$. Hence, $|E_{pure}(T)|$ can be computed in time $O(n\Delta)$.

Note that $E_{pure}(T) \subseteq E_{fix}(T)$ with $E_{pure}(T) = E_{fix}(T)$ if all degrees of T are odd. Thus, $|E_{pure}(T)| \leq F_{n-\lceil \Delta/2 \rceil}$. In this section we prove the much better general upper bound stated in Theorem 3. We start with an example. Let $n \equiv 0(2)$ and H_n the tree with n nodes consisting of a path P_n of length $n/2$ and a single node attached to each inner node of P_n (see Fig. 8). Since all non-leaves have degree 3 we have $E_{pure}(H_n) = E_{fix}(P_n)$, thus, $|E_{pure}(H_n)| \leq F_{n/2}$ by Corollary 1. We prove that $F_{\lfloor n/2 \rfloor}$ is an upper bound for $|E_{pure}(T)|$ in general.

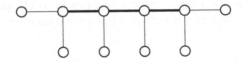

Fig. 8. The graph H_{10}, edges of $E^3(H_{10})$ are depicted as solid lines.

We first state a few technical lemmas and then prove Theorem 3. The proof of the first Lemma is similar to Lemma 7.

Lemma 13. *Let T be a tree, v a leaf with neighbor w, and $T' = T \setminus v$. Let n_l (resp. n_i) be the number of neighbors of w that are leaves (resp. inner nodes). If $n_l > n_i + 1$ then $|E_{pure}(T)| = |E_{pure}(T')|$.*

For a tree T let T^2 be the tree obtained from T by recursively removing each node v with degree 2 and connecting the two neighbors of v by a new edge. Note that T^2 is uniquely defined and $deg_{T^2}(v) = deg_T(v)$ for each node v of T^2.

Lemma 14. $E_{pure}(T) \subseteq E_{pure}(T^2)$ *for each tree T.*

Proof. Let w be a node v of T with degree 2. Let T' be the tree obtained from T by removing w and connecting the two neighbors of w by a new edge. Let $F \in E_{pure}(T)$. Since $deg(w) = 2$ no edge of F is incident to w. Thus, for all $v \neq w$ we have $2deg_F(v) < deg_T(v) = deg_{T'}(v)$, i.e., $F \in E_{pure}(T')$. Hence, $E_{pure}(T) \subseteq E_{pure}(T')$ and the statement follows by induction. $\qquad\square$

Proof (Proof of Theorem 3). Proof by induction on n. The statement is true for $n \leq 5$ as can be seen by a simple inspection of all cases. Let $n \geq 6$. By Lemma 14 we can assume that no node of T has degree 2. Let $\hat{T}$ be the tree induced by the edges in $E^3(T)$. $\hat{T}$ includes all inner nodes of T. Let u be a node of $\hat{T}$ such that all neighbors of u in $\hat{T}$ except one are leaves. Denote the neighbors of u in $\hat{T}$ that are leaves by $v_1, \ldots, v_d$ with $d \geq 1$, i.e., $deg_{\hat{T}}(u) = d+1$. Let $e_i = (u, v_i)$. By Lemma 13 we can assume $deg_T(v_i) = 3$ for $i = 1, \ldots, d$. Denote the two neighbors of v_i in $T \setminus \hat{T}$ by v_i^a and v_i^b. Let $H_1 = \{F \in E_{pure}(T) \mid e_1 \notin F\}$ and $T' = T \setminus \{v_1^a, v_2^b\}$. Clearly $H_1 = E_{pure}(T')$. Thus, $|H_1| \leq \mathbb{F}_{\lfloor n/2 \rfloor - 1}$ by induction. Assume that u has a neighbor u' in T that is a leaf in T. Let $F \in E_{pure}(T) \setminus H_1$. Then $F \setminus \{e_1\} \in E_{pure}(T \setminus \{u', v_1, v_1^a, v_1^b\})$. Thus, by induction $|E_{pure}(T) \setminus H_1| \leq \mathbb{F}_{\lfloor n/2 \rfloor - 2}$ and hence, $|E_{pure}(T)| \leq \mathbb{F}_{\lfloor n/2 \rfloor}$. Therefore we can assume that $deg_T(u) = d+1$, i.e., $d \geq 2$.

Next we expand the definition of H_1 as follows. For $i = 1, \ldots, d$ let $H_i = \{F \in E_{pure}(T) \mid e_i \notin F \text{ and } e_1, \ldots, e_{i-1} \in F\}$. Thus, $H_i = \emptyset$ for $i \geq (d+3)/2$ since $2deg_F(u) < d+1$. Let $d_0 = \lfloor (d+1)/2 \rfloor$. Then $E_{pure}(T) = H_1 \cup \ldots \cup H_{d_0}$. We claim that $|H_i| \leq \mathbb{F}_{\lfloor n/2 \rfloor - (2i-1)}$. The case $i = 1$ was already proved above. Let $i \geq 2$ and define $T_i = T \setminus \{v_j, v_j^a, v_j^b \mid j = i-1, i\}$. Then $|E_{pure}(T_i)| \leq \mathbb{F}_{\lfloor n/2 \rfloor - 3}$ by induction, hence $|H_2| = |E_{pure}(T_2)| \leq \mathbb{F}_{\lfloor n/2 \rfloor - 3}$. Let $i \geq 3$ and $F \in H_i$. Then $F \setminus \{e_{i-1}\} \in E_{pure}(T_i)$. Clearly, $E_{pure}(T_i) \setminus H_i$ consists of all $F \in E_{pure}(T_i)$ for which $\{e_1, \ldots, e_{i-2}\} \cap F \neq \emptyset$ holds. Hence,

$$|H_i| \leq |E_{pure}(T_i)| - \left| \bigcup_{j=1}^{i-2} E_j^i \right| \leq \mathbb{F}_{\lfloor n/2 \rfloor - 3} - \left| \bigcup_{j=1}^{i-2} E_j^i \right|$$

where $E_j^i = \{F \in E_{pure}(T_i) \mid e_j \notin F\}$. We use the following well known identity for $I = \{1, \ldots, i-2\}$.

$$\left| \bigcup_{i \in I} E_i \right| = \sum_{\emptyset \neq J \subseteq I} (-1)^{|J|+1} \left| \bigcap_{j \in J} E_i \right|$$

Note that $|\bigcap_{j \in J} E_j^i| \leq \mathbb{F}_{\lfloor n/2 \rfloor - (|J|+3)}$ by induction. Let $k = i - 2$. Then

$$|H_i| \leq \mathbb{F}_{\lfloor n/2 \rfloor - 3} - \sum_{j=1}^{k} (-1)^{j+1} \binom{k}{j} \mathbb{F}_{\lfloor n/2 \rfloor - (j+3)} = \sum_{j=0}^{k} (-1)^j \binom{k}{j} \mathbb{F}_{\lfloor n/2 \rfloor - (j+3)}$$

Note that $\lfloor n/2 \rfloor \geq 2k+3$. Lemma 25 of [23] implies $|H_i| \leq \mathbb{F}_{\lfloor n/2 \rfloor - (2i-1)}$. Then Lemma 15 yields

$$|E_{pure}(T)| \leq \sum_{i=1}^{d_0} |H_i| \leq \sum_{i=1}^{d_0} \mathbb{F}_{\lfloor n/2 \rfloor - (2i-1)} \leq \mathbb{F}_{\lfloor n/2 \rfloor}.$$

$\square$

Lemma 15. *For each $c \geq 1$ and $d \leq \frac{c+1}{2}$ we have $\sum_{i=1}^{d} \mathbb{F}_{c-(2i-1)} \leq \mathbb{F}_c$.*

Proof. The proof is by induction on d. The cases $d \leq 2$ clearly hold. Let $d > 2$. By induction we get

$$\sum_{i=1}^{d} \mathbb{F}_{c-(2i-1)} = \mathbb{F}_{c-1} + \sum_{i=2}^{d} \mathbb{F}_{c-(2i-1)} = \mathbb{F}_{c-1} + \sum_{i=1}^{d-1} \mathbb{F}_{c-2-(2i-1)} \overset{\text{Ind.}}{\leq} \mathbb{F}_{c-1} + \mathbb{F}_{c-2} = \mathbb{F}_c.$$

6   Conclusion and Open Problems

The problem of counting the fixed points for general cellular automata is #P-complete. In this paper we considered counting problems associated with the minority/majority rule of tree cellular automata. In particular we examined fixed points and pure 2-cycles. The first contribution is a novel algorithm that counts the fixed points and the pure 2-cycles of such an automata. The algorithms run in time $O(\Delta n)$. It utilizes a characterization of colorings for these automata in terms of subsets of the tree edges. This relieved us from separately treating the minority and the majority rule. The second contribution are upper and lower bounds for the number of fixed points and pure 2-cycles based on different graph parameters. We also provided examples to demonstrate the cases when these bounds are sharp. The bounds show that the number of fixed points (resp. 2-cycles) is a tiny fraction of all colorings.

There are several open questions that are worth pursuing. Firstly, we believe that it is possible to sharpen the provided bounds and to construct examples for these bounds. In particular for the case $\Delta \leq n/2$, we believe that our bounds can be improved. Another line of research is to ext minority/majority rules. One option is to consider the minority/majority not just in the immediate neighborhood of a node but in the r-hop neighborhood for $r > 1$ [15]. We believe that this complicates the counting problems considerably. Finally the predecessor existence problem and the corresponding counting problem [7] have not been considered for tree cellular automata. The challenge is to find for a given $c \in \mathcal{C}(T)$ a coloring $c' \in \mathcal{C}(T)$ with $\mathcal{M}(c') = c$. A coloring $c \in \mathcal{C}(T)$ is called a *garden of Eden* coloring if there doesn't exist a $c' \in \mathcal{C}(T)$ with $\mathcal{M}(c') = c$. The corresponding counting problem for tree cellular automata is yet unsolved.

References

1. Agur, Z.: Fixed points of majority rule cellular automata with application to plasticity and precision of the immune system. Complex Syst. **5**(3), 351–357 (1991)
2. Agur, Z., Fraenkel, A., Klein, S.: The number of fixed points of the majority rule. Discret. Math. **70**(3), 295–302 (1988)
3. Akutsu, T., Kuhara, S., Maruyama, O., Miyano, S.: A system for identifying genetic networks from gene expression patterns produced by gene disruptions and overexpressions. Genome Inform. **9**, 151–160 (1998)
4. Aledo, J.A., Diaz, L.G., Martinez, S., Valverde, J.C.: Enumerating periodic orbits in sequential dynamical systems over graphs. J. Comput. Appl. Math. **405**, 113084 (2022)

5. Aracena, J.: Maximum number of fixed points in regulatory Boolean networks. Bull. Math. Biol. **70**(5), 1398 (2008)

6. Aracena, J., Richard, A., Salinas, L.: Maximum number of fixed points in and-or-not networks. J. Comput. Syst. Sci. **80**(7), 1175–1190 (2014)

7. Barrett, C., et al.: Predecessor existence problems for finite discrete dynamical systems. Theoret. Comput. Sci. **386**(1–2), 3–37 (2007)

8. Bridoux, F., Durbec, A., Perrot, K., Richard, A.: Complexity of fixed point counting problems in Boolean networks. J. Comput. Syst. Sci. **126**, 138–164 (2022)

9. Goles, E., Olivos, J.: Periodic behaviour of generalized threshold functions. Discret. Math. **30**(2), 187–189 (1980)

10. Irons, D.: Improving the efficiency of attractor cycle identification in Boolean networks. Physica D **217**(1), 7–21 (2006)

11. Kauffman, S., et al.: The origins of order: Self-organization and selection in evolution. Oxford University Press, USA (1993)

12. Královič, R.: On majority voting games in trees. In: Pacholski, L., Ružička, P. (eds.) SOFSEM 2001. LNCS, vol. 2234, pp. 282–291. Springer, Heidelberg (2001). https://doi.org/10.1007/3-540-45627-9_25

13. Mezzini, M., Pelayo, F.L.: An algorithm for counting the fixed point orbits of an and-or dynamical system with symmetric positive dependency graph. Mathematics **8**(9), 1611 (2020)

14. Mishra, S., Rao, S.: Minimum monopoly in regular and tree graphs. Discret. Math. **306**(14), 1586–1594 (2006). https://doi.org/10.1016/j.disc.2005.06.036

15. Moran, G.: The r-majority vote action on 0–1 sequences. Discr. Math. **132**(1–3), 145–174 (1994)

16. Nakar, Y., Ron, D.: The structure of configurations in one-dimensional majority cellular automata: from cell stability to configuration periodicity. In: Chopard, B., Bandini, S., Dennunzio, A., Arabi Haddad, M. (eds.) 15th International Conference on Cellular Automata. LNCS, vol. 13402, pp. 63–72. Springer, Cham (2022). https://doi.org/10.1007/978-3-031-14926-9_6

17. Papp, P., Wattenhofer, R.: Stabilization time in minority processes. In: 30th International Symposium on Algorithms & Computation. LIPIcs, vol. 149, pp. 43:1–43:19 (2019)

18. Peleg, D.: Local majorities, coalitions and monopolies in graphs: a review. Theoret. Comput. Sci. **282**(2), 231–257 (2002)

19. Poljak, S., Sura, M.: On periodical behaviour in societies with symmetric influences. Combinatorica **3**(1), 119–121 (1983)

20. Rouquier, J., Regnault, D., Thierry, E.: Stochastic minority on graphs. Theoret. Comput. Sci. **412**(30), 3947–3963 (2011)

21. Tošić, P.T., Agha, G.A.: On computational complexity of counting fixed points in symmetric boolean graph automata. In: Calude, C.S., Dinneen, M.J., Păun, G., Pérez-Jímenez, M.J., Rozenberg, G. (eds.) UC 2005. LNCS, vol. 3699, pp. 191–205. Springer, Heidelberg (2005). https://doi.org/10.1007/11560319_18

22. Turau, V.: Fixed points and 2-cycles of synchronous dynamic coloring processes on trees. In: Parter, M. (eds.) 29th International Colloquium on Structural Information and Communication Complexity - Sirocco. pp. 265–282. Springer, Cham (2022). https://doi.org/10.1007/978-3-031-09993-9_15

23. Turau, V.: Counting Problems in Trees, with Applications to Fixed Points of Cellular Automata. arXiv preprint arXiv:2312.13769 (2023)
24. Veliz-Cuba, A., Laubenbacher, R.: On the computation of fixed points in Boolean networks. J. Appl. Math. Comput. **39**, 145–153 (2012)
25. Zehmakan, A.: On the Spread of Information Through Graphs. Ph.D. thesis, ETH Zürich (2019)

Semantics of Attack-Defense Trees for Dynamic Countermeasures and a New Hierarchy of Star-Free Languages

Thomas Brihaye[1]([✉]), Sophie Pinchinat[2], and Alexandre Terefenko[1,2]

[1] University of Mons, Mons, Belgium
thomas.brihaye@umons.ac.be
[2] University of Rennes, Rennes, France

Abstract. We present a mathematical setting for attack-defense trees (adts), a classic graphical model to specify attacks and countermeasures. We equip adts with (trace) language semantics allowing to have an original dynamic interpretation of countermeasures. Interestingly, the expressiveness of adts coincides with star-free languages, and the nested countermeasures impact the expressiveness of adts. With an adequate notion of countermeasure-depth, we exhibit a strict hierarchy of the star-free languages that does not coincide with the classic one. Additionally, driven by the use of adts in practice, we address the decision problems of trace membership, non-emptiness, and equivalence, and study their computational complexities parameterized by the countermeasure-depth.

1 Introduction

Security is nowadays a subject of increasing attention as means to protect critical information resources from disclosure, theft or damage. The informal model of *attack trees* is due to Bruce Schneier[1] to graphically represent and reason about possible threats one may use to attack a system. Attack trees have then been widespread in the industry and are advocated since the 2008 NATO report to govern the evaluation of the threat in risk analysis. The attack tree model has attracted the interest of the academic community in order to develop their mathematical theory together with formal methods (see the survey [25]).

Originally in [21], the model of attack tree aimed at describing how an attack goal refines into subgoals, by using two *operators OR* and *AND* to coordinate those refinements. The subgoals are understood in a "static" manner in the sense that there is no notion of temporal precedence between them. Still, with this limited view, many analysis can be conducted (see for example [6,8]). Then, the academic community considered two extensions of attack trees. The first one, called *adt* (adt, for short), is obtained by augmenting attack trees with

---

This work has been partly supported by the F.R.S.- FNRS under grant n°T.0027.21 and a PHC Tournesol project.

[1] https://www.schneier.com/academic/archives/1999/12/attack_trees.html.

J. A. Soto and A. Wiese (Eds.): LATIN 2024, LNCS 14579, pp. 257–271, 2024.
https://doi.org/10.1007/978-3-031-55601-2_17

nodes representing countermeasures [9,11]. The second one, initiated by [7,17], concerns a "dynamic" view of attacks with the ability to specify that the sub-goals must be achieved in a given order. This way to coordinate the subgoals is commonly specified by using operator $SAND$ (for Sequential AND). In [1], the authors proposed a *path semantics* for attack trees with respect to a given a tran-sition system (a model of the real system). However, a unifying formal semantics amenable to the coexistence of both extensions of attack trees – namely with the defense and the dynamics – has not been investigated yet.

In this paper, we propose a formal language semantics of adts, in the spirit of the trace semantics by [2] (for defenseless attack trees), which allows coun-termeasure features via the new operator CO (for "countermeasure"). Interest-ingly, because in adts, countermeasures of countermeasures exist, we define the *countermeasure-depth* (maximum number of nested CO operators) and analyze its role in terms of expressiveness of the model.

First, we establish the Small Model Property for adts with countermeasure-depth bounded by one (Theorem 1), which ensure the existence of small traces in a non-empty semantics. This not so trivial result is a stepping stone to prove further results.

Second, since our model of adts is very close to *star-free extended regular expressions* (SEREs for short), which are star-free regular expressions extended with intersection and complementation, we provide a two-way translation from the former to the latter (Theorem 2). It is known that the class of languages denoted by SEREs coincides with the class of *star-free languages* [19], which can also be characterized as the class of languages definable in first-order logic over strings (FO[<]). We make explicit a translation from adts into FO[<] (Lemma 1) to shed light on the role played by the countermeasure-depth. Our translation is reminiscent of the constructions in [14] for an alternative proof of the result in [23] that relates the classic dot-depth hierarchy of star-free languages and the FO[<] quantifier alternation hierarchy. In particular, we show (Lemma 2) that any language definable by an adt with countermeasure-depth less than equal to k is definable in Σ_{k+1}, the $(k+1)$-th level of the first-order quantifier alternation hierarchy.

Starting from the proof used in [23] to show the strictness of the dot-depth hierarchy, we demonstrate that there exists an infinite family of languages whose definability by an adt requires arbitrarily large countermeasure-depths. It should be noticed that our notion of countermeasure-depth slightly differs from the complementation-depth considered in [22] for extended regular expressions[2], because the new operator AND is rather a relative complementation. As a result, the countermeasure-depth of adts induces a new hierarchy of all star-free lan-guages, that we call the *ADT-hierarchy*, that coincides (at least on the very first levels) neither with the dot-depth hierarchy, nor with the first-order logic quantifier alternation hierarchy.

Third, we study three natural decision problems for adts, namely the *mem-bership problem* (ADT-MEMB), the *non-emptiness problem* (ADT-NE) and the

---

[2] Arbitrary regular expressions extended with intersection and complementation.

equivalence problem (*ADT*-EQUIV). The problem *ADT*-MEMB is to determine if a trace is in the semantics of an adt. From a practical security point of view, *ADT*-MEMB addresses the ability to recognize an attack, say, in a log file. The problem *ADT*-NE consists in, given an adt, deciding if its semantics is non-empty. Otherwise said, whether the information system can be attacked or not. Finally, the problem *ADT*-EQUIV consists of deciding whether two adts describe the same attacks or not. Our results are summarized in Table 1.

The paper is organized as follows. Section 2 proposes an introductory example. Next, we define our model of adts in Sect. 3, their trace semantics and countermeasure-depth, and present the Small Model Property for adts with countermeasure-depth bounded by one. We then show in Sect. 4 that adts coincide with star-free languages. We next study the novel hierarchy induced by the countermeasure-depth (Sect. 5) and study decision problems on adts (Sect. 6). An extended version of the present paper, including proofs, is available in [3].

2 Introductory Example

Consider a thief (the proponent) who wants to steal two documents inside respective safes Safe 1 and Safe 2, without being seen. Safe 1 and Safe 2 are located in adjacent rooms of a building, Room 1 and Room 2 respectively, separated by a door; the entrance/exit door of the building leads to Room 1. Each room has a window. Initially, the thief is outside of the building. A strategy for the proponent to steal the documents is to attempt to open Safe 1 until she succeeds, then open Safe 2 until she succeeds, and finally to exit the building. However, such a strategy may easily be countered by the company, say by hiring a security guard visiting the rooms on a regular basis.

Security experts would commonly use an adt to describe how the proponent may achieve her goal and, at the same time, the ways her opponent (the company) may prevent the proponent from reaching her goal. An informal adt expressing the situation is depicted in Fig. 1a, where, traditionally, the proponent's goals are represented in red circles, while the opponent countermeasures are represented in green squares. An arrow from a left sibling to a right sibling specifies that the former goal must be achieved before starting the latter. A countermeasure targeting a proponent's goal is represented with dashed lines.

As said, the graphical model of Fig. 1a is informal and cannot be exploited by any automated tool for reasoning. With the setting proposed in this contribution, we make a formal model, and in particular we work out a new binary operator CO for "countermeasure", graphically reflected with a curved dashed line between two siblings: the left sibling is a proponent's goal while the right sibling is the opponent's countermeasure – with this convention, we can unambiguously retrieve the player's type of an adt node. The more formal version of the adt in Fig. 1a is drawn in Fig. 1b (details for its construction can be found in Example 4).

The proposed semantics for adts also allow us to consider nested CO operators to express countermeasures of countermeasures. For example, a proponent

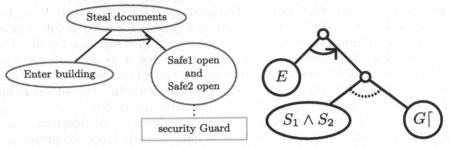

(a) A countermeasure from the Company.   (b) Formal representation of the adt

Fig. 1. Adts for the thief and company problem.

countermeasure against the company countermeasure could be to dress up as an employee working in the building; we formalise this situation in Example 4. It should be observed that nested countermeasures is a core aspect of our contribution and the main subject of Sect. 5.

3   Attack-Defense Trees and Countermeasure-Depth

Preliminary Notations. For the rest of this paper, we fix *Prop* a finite set of propositions and we assume that the reader is familiar with propositional logic. We use the typical symbol γ for propositional formulas over *Prop* and write a valuation of the propositional variables a as an element of $\Sigma := 2^{Prop}$, which will be viewed as an alphabet. A *trace* t over *Prop*, is finite word over Σ, that is a finite sequence of valuations. We denote the empty trace by ε, and we define $\Sigma^+ := \Sigma^* \backslash \{\varepsilon\}$. For a trace $t = a_1...a_n$, we define $|t| := n$, its *length*, as the number of valuations appearing in t, and we let $t[i] := a_i$, for each $i \in \{1,...,n\}$. We define the classic *concatenation* of traces: given two traces $t = a_1...a_n$ and $t' = a'_1...a'_m$, we define $t \cdot t' := a_1...a_n a'_1...a'_m$. We also lift this operator to sets of traces in the usual way: given two sets of traces L and $L' \subseteq \Sigma^*$, we let $L \cdot L' := \{t \cdot t' \ : \ t \in L \text{ and } t' \in L'\}$. For a trace $t = a_1...a_n \in \Sigma^*$ and $1 \leq i \leq n$, the trace $t' = a_1...a_i$ is a *prefix* of t, written $t' \preceq t$.

We define attack-defense trees (adts) over *Prop*, as well as their *trace semantics* and their *countermeasure-depth*, and develop enlightening examples. Adts are standard labeled finite trees with a dedicated set of labels based on the special ϵ label and propositional formulas for leaves and on the set $\{\text{OR}, \text{SAND}, \text{AND}, \text{CO}\}$ for internal nodes.

Definition 1. *The set ADT of adts over Prop is inductively defined by:*

– *the empty-word leaf ϵ and every propositional formula γ over Prop are in ADT;*

- if trees $\tau_1, ..., \tau_n$ are in ADT, so are $\text{OR}(\tau_1, ..., \tau_n)$, $\text{SAND}(\tau_1, ..., \tau_n)$, and $\text{AND}(\tau_1, ..., \tau_n)$;
- if trees τ and τ' are in ADT, so is $\text{CO}(\tau, \tau')$.

The *size of an adt* τ, written $|\tau|$ is defined as the sum of the sizes of its leaves, provided the size of ϵ is 1, while the size of γ is its size when seen as a propositional formula.

Regarding the semantics, adts describe a set of traces over alphabet $\Sigma :=$ 2^{Prop}, hence their *trace semantics*. Formally, for an adt τ, we define the language $Traces(\tau) \subseteq \Sigma^*$. First, we set $Traces(\epsilon) = \{\epsilon\}$. Now, a leaf adt hosting formula γ denotes the reachability goal γ, that is the set of traces ending in a valuation satisfying γ (we use the classic notations $\top = p \vee \neg p$ and $\bot = \neg \top$ with $p \in Prop$). We make our trace semantics compositional by providing the semantics of the four operators OR, SAND, CO, and AND in terms of how the subgoals described by their arguments interact. Operator OR tells that at least one of the subgoals has to be achieved. Operator SAND requires that all the subgoals need being achieved in the left-to-right order. The binary operator CO requires to achieve the first subgoal without achieving the second one. Finally, operator AND tells that all subgoals need to be achieved, regardless of the order. Without any countermeasure, AND can be seen as a relaxation of the SAND, but it is not true in general (see example 3).

At the level of the property described by an adt, i.e. a trace language, the operators correspond to specific language operations: OR corresponds to union, SAND to concatenation, CO to a relativized complementation. Only AND corresponds to a less classic operation: a trace t belongs to the language of $\text{AND}(\tau_1, \tau_2)$ if t belongs to the language of τ_1 and has a prefix in the language of τ_2 or vice-versa. Formally:

Definition 2. *Let L_1, L_2 be two languages over alphabet Σ. We let the language $L_1 \sqcap L_2$ be defined by $L_1 \sqcap L_2 := (L_1 \cdot \Sigma^* \cap L_2) \cup (L_2 \cdot \Sigma^* \cap L_1)$.*

Although tedious, one can show that operator $\sqcap$ is associative, so that language $L_1 \sqcap L_2 \sqcap \ldots \sqcap L_n$ is well-defined and actually equal to $\bigcup_{i \in \{1, ..., n\}} (L_i \cap \bigcap_{j \neq i} L_j \cdot \Sigma^*)$.

Example 1. A word $w = a_1 a_2 \ldots a_m$ belongs to $L_1 \sqcap L_2 \sqcap L_3$ whenever there are three (possibly equal) positions $i_1, i_2, i_3 = m$ such that, for each $j \in \{1, 2, 3\}$, the word prefix $a_1 \ldots a_{i_j} \in L_{\pi(j)}$, for some permutation π of $(1, 2, 3)$.

We can now formally define the adt semantics.

Definition 3.

- $Traces(\epsilon) := \{\varepsilon\}$ and $Traces(\gamma) := \{a_1 ... a_n \in \Sigma^* : a_n \models \gamma\}$;
  In particular, $Traces(\top) = \Sigma^+$ and $Traces(\bot) = \emptyset$;
- $Traces(\text{OR}(\tau_1, ..., \tau_n)) := Traces(\tau_1) \cup ... \cup Traces(\tau_n)$;
- $Traces(\text{SAND}(\tau_1, ..., \tau_n)) := Traces(\tau_1) \cdot ... \cdot Traces(\tau_n)$;

- $Traces(\text{CO}(\tau_1, \tau_2)) := Traces(\tau_1) \setminus Traces(\tau_2);$
- $Traces(\text{AND}(\tau_1, ..., \tau_n)) = Traces(\tau_1) \sqcap ... \sqcap Traces(\tau_n).$

In the rest of the paper, we say for short that *an adt is non-empty*, written $\tau \neq \emptyset$, whenever $Traces(\tau) \neq \emptyset$. We say that two adts τ and τ' are *equivalent*, whenever $Traces(\tau) = Traces(\tau')$.

Remark 1. Since all operators $\cup$, $\cdot$ and $\sqcap$ over trace languages are associative, the trees of the form $\text{OP}(\tau_1, \text{OP}(\tau_2, \tau_3))$, $\text{OP}(\text{OP}(\tau_1, \tau_2), \tau_3)$, and $\text{OP}(\tau_1, \tau_2, \tau_3)$ are all equivalent, when OP ranges over $\{\text{OR}, \text{SAND}, \text{AND}\}$. As a consequence, we may sometimes assume that nodes with such operators are binary.

We now introduce some notations for particular adts to ease our exposition and provide some examples of adts with their corresponding trace property.

We define a family of adts of the form $|{\geq}\ell|, |{<}\ell|, |{=}\ell|$, where ℓ is a non-zero natural. We let $|{\geq}\ell| := \text{SAND}(\top, ..., \top)$ where $\top$ occurs ℓ times; $|{<}\ell| := \text{CO}(\top, |{\geq}\ell|)$; and $|{=}\ell| := \text{CO}(|{\geq}\ell|, |{\geq}\ell + 1|)$. It is easy to establish that adt $|{\geq}\ell|$ (resp. $|{<}\ell|$, $|{=}\ell|$) denotes the set of traces of length at least (resp. at most, exactly) ℓ.

We also consider particular adts and constructs for them.

- $\text{ALL} := \text{OR}(\epsilon, \top)$,
- $\text{NOT}(\tau) := \text{CO}(\text{ALL}, \tau)$, and $\text{INTER}(\tau_1, \tau_2) := \text{NOT}(\text{OR}(\text{NOT}(\tau_1), \text{NOT}(\tau_2)))$,
- $]\tau[:= \text{SAND}(\text{ALL}, \tau, \text{ALL})$, $]\tau := \text{SAND}(\text{ALL}, \tau)$ and $\tau[:= \text{SAND}(\tau, \text{ALL})$.
- Given a formula γ over *Prop*, we let $\underline{\gamma} := \text{CO}(\gamma, |{\geq}2|)$.

Based on these notations, we develop further examples.

Example 2. – $Traces(\text{ALL}) = \Sigma^*$;
- $Traces(\text{NOT}(\tau)) = \Sigma^* \setminus Traces(\tau)$;
- $Traces(\text{INTER}(\tau_1, \tau_2)) = Traces(\tau_1) \cap Traces(\tau_2)$;
- $Traces(\underline{\gamma})$ is set of one-length traces whose unique valuation satisfies γ; in particular, when a valuation a is understood as a formula, namely formula $\bigwedge_{p \in a} p \wedge \bigwedge_{p \notin a} \neg p$, the adt $\underline{a} := \text{CO}(a, |{\geq}2|)$ is such that $Traces(\underline{a}) = \{a\}$;
- For a valuation a,
  - $Traces(]\underline{a}[) = Traces(]\underline{a}[) = Traces(\underline{a}[) = \Sigma^* a \Sigma^*$;
  - $Traces(]\underline{a}) = Traces(]\underline{a}) = Traces(\underline{a}) = \Sigma^* a$;
  - $Traces(\underline{a}[) = a\Sigma^*$.

Example 3. Note that AND cannot be seen as a kind of relaxation of SAND. For the set of propositions $\{p\}$, if we consider the formula p as a leaf, $Traces(p) = \{p\}$ and $Traces(\neg p) = \{\emptyset\}$. Thus $\text{SAND}(\neg p, p) = \{t\}$ with trace $t = \emptyset p$. However $\text{AND}(\neg p, p) = \emptyset$. Let us notice that the construction of this example uses the CO operator (hidden in $\underline{p}$ and $\neg p$).

Example 4. We come back to the situation of our introductory example (Sect. 2). First, we discuss the formal semantics of the informal tree in Fig. 1a. To do so, we propose the following set of propositions: $Prop = \{E, S_1, S_2, G\}$ where E

holds when the thief is entering the building, S_1 (resp. S_2) holds when the first (resp. second) safe is open, and G is true if a guard is in the building. The situation can be described by the following adt: $\tau_{ex_1} = \text{SAND}(E, \text{CO}(S_1 \wedge S_2, G\lceil))$, represented in Fig. 1b, where we distinguish SAND with a curved line and CO with a dashed line. We have $Traces(\tau_{ex_1}) = \{a_1...a_n \in \Sigma^* : a_n \models E\} \cdot (\{a_1...a_n \in \Sigma^* : a_n \models S_1 \wedge S_2\} \setminus \{a_1...a_n \in \Sigma^* : \exists i \text{ such that } a_i \models G\})$. If we write $\gamma_\varphi = \{a \in 2^{Prop} : a \models \varphi\}$, we have $Traces(\tau_{ex_1}) = \Sigma^* \gamma_E \cdot (\Sigma^* \gamma_{S_1 \wedge S_2} \setminus \Sigma^* \gamma_G \Sigma^*)$. In other words, we want all traces where E holds at some point and, after it, G cannot be true and finish by a valuation where $S_1 \wedge S_2$ holds. Notice that we did not use $\text{AND}(S_1, S_2)$ as we want to reach a state where both safes are open.

In order to illustrate the nesting of countermeasures, we now allow the thief to disguise himself as an employee (assuming that when disguised, the guard does not identify him as a thief). To do so, we extended the set of propositions: $Prop' = \{E, S_1, S_2, G, D\}$, where D holds when the thief is disguised. The situation is now described by the following adt: $\tau_{ex_2} = \text{SAND}(E, \text{CO}(S_1 \wedge S_2, \rceil\text{CO}(G, D)\lceil))$. The semantics for τ_{ex_2} is all traces where E holds at some point and, after it, G cannot be true, except if D holds at the same time, and finish by a valuation where $S_1 \wedge S_2$ holds.

We now stratify the set ADT of adts according to their *countermeasure-depth* which denotes the maximum number of nested countermeasures.

Definition 4. *The* countermeasure-depth *of an adt τ, written $\delta(\tau)$, is inductively defined by:*

- $\delta(\epsilon) := \delta(\gamma) = 0$;
- $\delta(OP(\tau_1, ..., \tau_n)) := \max\{\delta(\tau_1), ..., \delta(\tau_n)\}$ *for every* $OP \in \{\text{OR}, \text{SAND}, \text{AND}\}$;
- $\delta(\text{CO}(\tau_1, \tau_2)) := \max\{\delta(\tau_1), \delta(\tau_2) + 1\}$

We let $ADT_k := \{\tau \in ADT : \delta(\tau) \leq k\}$ be the set of adts with countermeasure-depth at most k. Clearly $ADT_0 \subseteq ADT_1 \subseteq ... ADT_k \subseteq ADT_{k+1} \subseteq ...$, and $ADT = \bigcup_{k \in \mathbb{N}} ADT_k$.

Example 5. We list a couple of examples. $\delta(\text{NOT}(\tau)) = 1 + \delta(\tau)$; $\delta(\text{INTER}(\tau_1, \tau_2)) = 2 + \max\{\delta(\tau_1), \delta(\tau_2)\}$; $\delta(\text{CO}(\text{CO}(\gamma_1, \gamma_2), \gamma_3)) = 1$ while $\delta(\text{CO}(\gamma_3, \text{CO}(\gamma_2, \gamma_3))) = 2$; $\delta(|<\ell|) = \delta(|=\ell|) = 1$ while $\delta(|\geq\ell|) = 0$; $\delta(\gamma) = 1$; In Example 4, $\delta(\tau_{ex_1}) = 1$ and $\delta(\tau_{ex_2}) = 2$.

Also, $\delta(|\geq\ell|) = 0$, $\delta(|<\ell|) = \delta(|=\ell|) = 1$, so that $|\geq\ell| \in ADT_0$; $|<\ell|$ and $|=\ell| \in ADT_1$. Moreover, $\tau_{ex_1} \in ADT_1$ and $\tau_{ex_2} \in ADT_2$.

We say that a language L is ADT_k-*definable* (resp. *ADT-definable*), written $L \in ADT_k$ (resp. ADT), whenever $Traces(\tau) = L$, for some $\tau \in ADT_k$ (resp. for some k).

It can be established that non-empty adts in ADT_1 enjoy small traces, i.e. smaller than the size of the tree.

Theorem 1 (Small model property for ADT_1). *An adt $\tau \in ADT_1$ is non-empty if, and only if, there is a trace $t \in Traces(\tau)$ with $|t| \leq |\tau|$.*

Notice that Theorem 1 also holds for ADT_0, since $ADT_0 \subseteq ADT_1$.

4   Adts, Star-Free Languages, and First-Order Logic

We prove that adts coincide with star-free languages and first-order formulas.

4.1   Reminders on Star-Free Languages and First-Order Logic

The class of *star-free languages* introduced by [5,12,16] (over alphabet Σ) is obtained from the finite languages (or alternatively languages consisting of a single one-length word in Σ) by finitely many applications of Boolean operations ($\cup$, $\cap$ and $\sim$ for the complement) and the concatenation product (see [19, Chapter 7]). Alternatively, one characterizes star-free languages by first considering extended regular expressions – that are regular expressions augmented with intersection and complementation, and second by restricting to *star-free extended regular expressions* (SEREs, for short) that are extended regular expressions with no Kleene-star operator. Regarding computational complexity aspects, we recall the subclass of SEREs of extended regular expressions. The word membership problem (i.e., whether a given word belongs to the language denoted by a SERE) is in PTIME [10, Theorem 2], while the non-emptiness problem (i.e., is the denoted language empty?) and the equivalence problem (i.e., do two SEREs denote the same language?) are hard, both non-elementary [22, p. 162].

We now recall classical results on the first-order logic on finite words FO[<] (see details in [15, Chapter 29]). The signature of FO[<], say for words over an alphabet Σ, is composed of a unary predicate $a(x)$ for each $a \in \Sigma$, whose meaning is the "letter at position x of the word is a", and the binary predicate $x < y$ that states "position x is strictly before position y in the word". For a FO[<]-formula, we define its *size* $|\psi|$ as the size of the expression considered as a word. A language L is FO[<]-*definable* whenever there exists a FO[<]-formula ψ such that a word $w \in L$ if, and only if, w is a model of ψ. Similarly, we say that an adt τ is FO[<]-definable if $Traces(\tau)$ is FO[<]-definable. It is well-known that FO[<]-definable languages coincide with star-free languages [12,14,23].

Also, for a fine-grained inspection of FO[<], let us denote by Σ_ℓ (resp. Π_ℓ) the fragments of FO[<] consisting of formulas with at most ℓ alternation of $\exists$ and $\forall$ quantifier blocks, starting with $\exists$ (resp. $\forall$). The folklore results regarding satisfiability of FO[<]-formulas [13,22] are: (a) The satisfiability problem for FO[<] is non-elementary; (b) The satisfiability for Σ_ℓ is in $(\ell - 1)$-EXPSPACE.[3]

We are not aware of any result that establishes a tight lower bound complexity for the satisfiability problem on the Σ_ℓ fragments of FO[<].

We lastly recall the definition of the *dot-depth hierarchy* of star-free languages: level 0 of this hierarchy is $B_0 := \{L \subseteq 2^\Sigma : L \text{ is finite or co-finite}\}$, and level ℓ is $B_\ell := \{L \subseteq 2^\Sigma : L \text{ is a Boolean combination of languages of the form } L_1 \cdot ... \cdot L_n \text{ where } L_1, ..., L_n \in B_{\ell-1}\}$. The dot-depth hierarchy has a tight connection with FO[<] fragments [23]: for every $\ell > 0$, $\Sigma_\ell \subseteq B_\ell \subseteq \Sigma_{\ell+1}$.

---

[3] With the convention that 0-EXPSPACE =PSPACE.

4.2   Expressiveness of Adts

The first result of this section consists in showing that adts and star-free extended regular expressions share the same expressiveness.

Theorem 2. *A language L is star-free if, and only if, L is ADT-definable.*

For the "only if" direction of Theorem 2, we reason by induction on the class of star-free languages. For a language of the form $\{a\}$ where $a \in \Sigma$ one can take the adt $\underline{a}$ (that is $\text{CO}(a, |{\geq}2|)$). Now we can inductively build adequate adts for compound star-free languages by noticing that language operations of union and concatenation are captured by adts operators OR and SAND respectively, while complementation and intersection are obtained from the NOT(.) and INTER(.,.) as formalized in Example 2. One easily verifies that the size of the adt corresponding to a SERE E is in $O(|E|)$, where $|E|$ denotes the size (number of characters) of E.

For the "if" direction of Theorem 2, it is easy to translate an adt into an SERE: the leaf ϵ translates into ϵ, a leaf adt γ translates into $\bigcup_{a\models\gamma} a$ – notice that this translation is exponential. For non-leaf adts, since every operator occurring in the adt has its language-theoretic counterpart the translation goes smoothly. However, the translation is exponential because of the adt operator AND, see Definition 2.

We now dig into the ADT-hierarchy induced by the countermeasure-depth and compare it with the FO[<] fragments Σ_ℓ and Π_ℓ.

We first design a translation from ADT into FO[<], inductively over adts. The translation of an adt τ into a formula is denoted by ψ_τ. For the base cases of adts ϵ and γ, and we let: $\psi_\epsilon := \forall x \bot$ and $\psi_\gamma := \exists x (\forall y \neg(x < y) \wedge \bigvee_{a\models\gamma} a(x))$.

Now, regarding compound adts, operator OR is reflected by the logical disjunction: $\psi_{\text{OR}(\tau_1,\tau_2)} := \psi_{\tau_1} \vee \psi_{\tau_2}$, while operator CO is reflected by the logical conjunction with the negated second argument: $\psi_{\text{CO}(\tau_1,\tau_2)} := \psi_{\tau_1} \wedge \neg\psi_{\tau_2}$. On the contrary, the two remaining operators SAND and AND require to split the trace into pieces, which can be captured by the folklore operation of *left (resp. right) position relativizations* of FO[<]-formulas w.r.t. a position [14, Proposition 2.1] (see also formulas of the form $\phi^{[x,y]}$ in [20]). Formally, given a position x in the trace t and an FO[<]-formula ψ, we define formula $\psi^{\leq x}$ (resp. $\psi^{>x}$) that holds for t if the prefix (resp. suffix) of t up to (resp. from) position x satisfies ψ, as follows. For $\bowtie \in \{\leq, >\}$, we let:

$$
\begin{aligned}
a(y)^{\bowtie x} &= a(y) \\
(y < z)^{\bowtie x} &= (y < z) \\
p(y)^{\bowtie x} &= p(y)
\end{aligned}
\qquad
\begin{aligned}
(\psi \vee \psi')^{\bowtie x} &= \psi^{\bowtie x} \vee \psi'^{\bowtie x} \\
(\neg\psi)^{\bowtie x} &= \neg\psi^{\bowtie x} \\
(\exists y\, \psi)^{\bowtie x} &= \exists y\, (y \bowtie x \wedge \psi^{\bowtie x})
\end{aligned}
$$

Additionally, we write $\psi^{\leq 0}$ and $\psi^{>0}$ as the formulas obtained from $\psi^{\leq x}$ and $\psi^{>x}$ by replacing every occurrence of expressions $y \leq x$ and $y > x$ by $\bot$ and $\top$ respectively.

Remark 2. For every formula $\psi \in \Sigma_\ell$, we also have $\psi^{\leq x}, \psi^{>x} \in \Sigma_\ell$.

We can now complete the translation from ADT into FO[<] by letting (w.l.o.g., by Remark 1, we can consider binary SAND and AND):
$$\psi_{\text{SAND}(\tau_1,\tau_2)} := \exists x [\psi_{\tau_1}{}^{\leq x} \wedge \psi_{\tau_2}{}^{>x}] \vee (\psi_{\tau_1}{}^{\leq 0} \wedge \psi_{\tau_2})$$
$$\psi_{\text{AND}(\tau_1,\tau_2)} := \exists x [(\psi_{\tau_1}{}^{\leq x} \wedge \psi_{\tau_2}) \vee (\psi_{\tau_2}{}^{\leq x} \wedge \psi_{\tau_1})] \vee (\psi_{\tau_1}{}^{\leq 0} \wedge \psi_{\tau_2}) \vee (\psi_{\tau_2}{}^{\leq 0} \wedge \psi_{\tau_1}).$$

Lemma 1. *1. For any $t \in \Sigma^*$, $t \models \psi_\tau$ iff $t \in Traces(\tau)$.*
2. For any adt τ, formula ψ_τ is of size exponential in $|\tau|$.

In the rest of the paper, we use mere inclusion symbol $\subseteq$ between subclasses of ADT and subclasses of FO[<], with the canonical meaning regarding the denoted trace languages.

An accurate inspection of the translation $\tau \mapsto \psi_\tau$ entails that every ADT_k-definable adt can be equivalently represented by a Σ_{k+2}-formula, namely $ADT_k \subseteq \Sigma_{k+2}$. However, we significantly refine this expressiveness upperbound for ADT_k.

Lemma 2. *1. $ADT_0 \subseteq \Sigma_2 \cap \Pi_2$ – with an effective translation.*
2. For every $k > 0$, $ADT_k \subseteq \Sigma_{k+1}$ – with an effective translation.

Regarding Item 1 of Lemma 2, it can be observed that, whenever $\tau \in ADT_0$, the quantifiers $\forall$ and $\exists$ commute in ψ_τ. Now, for Item 2, the proof is conducted by induction over k. We sketch here the case $k > 1$. First, remark that if $\tau_1, ..., \tau_n \in ADT_{k-1}$ are Σ_k-definable, then $\text{OR}(\tau_1, ..., \tau_n), \text{SAND}(\tau_1, ..., \tau_n)$ and $\text{AND}(\tau_1, ..., \tau_n)$ remain Σ_k-definable, as formulas are obtained from conjunctions or disjunctions of Σ_k-definable formulas. Moreover, if τ_1 and τ_2 are Σ_k-definable, then $\tau = \text{CO}(\tau_1, \tau_2)$ is Σ_{k+1}-definable as a formula can be obtained from a boolean combination of two Σ_k-definable formulas. Finally, with k still fixed, it can be shown by induction over the size of an adt τ that, if $\tau \in ADT_k$, since all its countermeasures operators are of the form $\text{CO}(\tau_1, \tau_2)$ where $\tau_1 \in ADT_k$ and $\tau_2 \in ADT_{k-1}$, we have that adt τ is also Σ_{k+1}-definable, which concludes the proof of Item 2 of Lemma 2.

We can also establish lowerbounds in the ADT-hierarchy.

Lemma 3. *1. $B_\ell \subseteq ADT_{2\ell+2}$, and therefore $\Sigma_\ell \subseteq ADT_{2\ell+2}$*
2. $\Sigma_1 \subseteq ADT_0$ – with an effective translation.

Item 1 of Lemma 3 is obtained by an induction of ℓ. Regarding Item 2, the translation consists in putting the main quantifier-free subformula of a Σ_1-formula in disjunctive normal form, and to focus for each conjunct on the set of "ordering" literals of the form $x < y$ or $\neg(x < y)$ (leaving aside the other literals of the form $a(x)$ or $\neg a(x)$ for a while). Each ordering literal naturally induces a partial order between the variables. We expend this partial order constraint over the variables as a disjunction of all its possible linearizations. For example, the conjunct $x < y \wedge \neg(y < z)$ is expended as the equivalent formula $(x < y \wedge y = z) \vee (x < z \wedge z < y) \vee (z = x \wedge x < y) \vee (z < x \wedge x < y)$. Now, each disjunct of this new formula, together with the constraints $a(x)$ (or $\neg a(x)$), can easily be specified by a SAND-rooted adt (i.e. the root is a SAND). The initial Σ_1-formula then is associated with the OR-rooted tree that gathers all the aforementioned SAND-rooted subtrees. Notice that the translation may induce at least an exponential blow-up.

The reciprocal of Item 2 in Lemma 3 is an open question. Still, we have little hope that it holds because ψ_γ seems to require a property expressible in $\Sigma_2 \setminus \Sigma_1$.

5   Strictness of the ADT-Hierarchy

One can notice that the adt $|<2|\ \in ADT_1$ defines the finite language of traces of length at most 1, while it can be established that languages arising from adts in ADT_0 are necessarily infinite – if inhabited by a non-empty word. Thus, we can easily deduce $ADT_0 \subsetneq ADT_1$. This section aims at showing that the entire ADT-hierarchy is strict:

Proposition 1. *For every* $k \in \mathbb{N}$, $ADT_k \subsetneq ADT_{k+1}$, *even if* $Prop = \{p\}$.

To show that $ADT_k \subsetneq ADT_{k+1}$, we use a family of languages, originally introduced in [24], which we denote $\{W_k\}_{k\in\mathbb{N}}$, over the two-letter alphabet Σ obtained from $Prop = \{p\}$. For readability, we use symbol a (resp. b) for the valuation $\{p\}$ (resp. $\emptyset$) of Σ. Formally, we define $W_k \subseteq \Sigma^*$ as follows – where $\|w\|$ denotes the number of occurrences of a minus the number of occurrences of b in the word w:

We let $w \in W_k$ whenever all the following holds.

- $\|w\| = 0$;
- for every $w' \preceq w$, $0 \le \|w'\| \le k$;
- there exists $w'' \preceq w$ s.t. $\|w''\| = k$.

In [24, Theorem 2.1], it is shown that $W_k \in B_k \setminus B_{k-1}$, for all $k \ge 1$.

We now determine the position of W_k languages in the ADT-hierarchy. We show Proposition 2.

Proposition 2. *For each* $k > 0$, $W_k \in ADT_{k+1} \setminus ADT_{k-2}$.

First, $W_k \notin ADT_{k-2}$ because W_k is not Σ_{k-1}-definable ([24]). By an inductive argument over k, we can build an adt $\mu_k \in ADT_{k+1}$ that captures W_k. We only sketch here the case $k = 1$. For W_1, we set $\mu_1 :=$ CO(b, OR($\underline{b}\lceil\rceil$, $\rceil$CO($|\ge 2|$, AND($a\lceil$, $b\lceil$))$\lceil$)), depicted in Fig. 2a. Note that $\delta(\mu_1) = 2$ and that by a basic use of semantics, we have $Traces(\mu_1) = (ab)^+ = W_1$.

Now we have all the material to prove Proposition 1. Indeed, assuming the hierarchy collapses at some level will contradict Proposition 2. Notice that this argument is not constructive as we have no witness of $ADT_k \subsetneq ADT_{k+1}$ but for $k = 1$ where it can be shown that $(ab)^+ \in ADT_2 \setminus ADT_1$. Our results about the ADT-hierarchy are depicted on Fig. 2b.

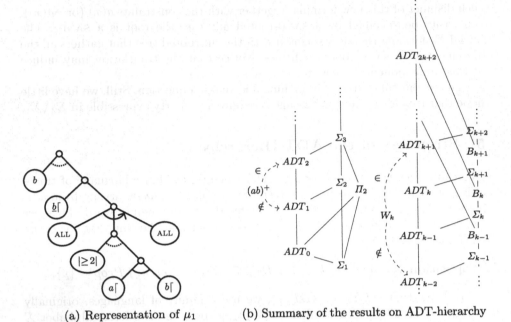

(a) Representation of μ_1 (b) Summary of the results on ADT-hierarchy

Fig. 2. An example of tree in ADT_2 and the ADT-hierarchy.

6   Decision Problems on Attack-Defense Trees

We study classical decision problems on languages, through the lens of adts, with a focus on the role played by the countermeasure-depth in their complexities. The problems are the following.

– The *membership* problem, written ADT-MEMB, is defined by:
   Input: τ an attack-defense tree and t a trace.
   Output: "YES" if $t \in Traces(\tau)$, "NO" otherwise.
– The *non-emptiness* problem, written ADT-NE, is defined by:
   Input: τ an attack-defense tree.
   Output: "YES" if $Traces(\tau) \neq \emptyset$, "NO" otherwise.
– The *equivalence* problem, written ADT-EQUIV, is is defined by:
   Input: τ_1 and τ_2, two attack-defense trees.
   Output: "YES" if $Traces(\tau_1) = Traces(\tau_2)$, "NO" otherwise.

We use notations ADT_k-MEMB and ADT_k-NE whenever the input adts of the respective decision problems are in ADT_k, with a fixed k. Our results are summarized in Table 1, and we below comment on them, row by row.

Regarding ADT-MEMB (first row of Table 1), we recall that adts and SEREs are expressively equivalent, but with a translation (Theorem 2) from the former to the latter that is not polynomial. We therefore cannot exploit [10, Theorem

Table 1. Computational complexities of decision problems on adts.

	ADT_0	ADT_1	ADT_k $(k \geq 2)$	ADT
ADT_{MEMB}	PTIME	PTIME	PTIME	PTIME
ADT-NE	NP-comp	NP-comp	$(k+1)$-EXPSPACE $^a \geq \text{NSPACE}(g(k-5, c\sqrt{\frac{n-1}{3}}))$	non-elem
ADT-EQUIV	coNP-comp	4-EXPSPACE	$(k+2)$-EXPSPACE $^b \geq \text{NSPACE}(g(k-4, c\sqrt{\frac{n-1}{3}}))$	non-elem

^{a}If $k \geq 5$.
^{b}If $k \geq 4$.

2] for a PTIME complexity of the word membership problem for SEREs, and have instead developed a dedicated alternating logarithmic-space algorithm.

Regarding ADT-NE (second row of Table 1), and because SEREs can be translated as adts (see "if" direction in the proof of Theorem 2), the problem ADT-NE inherits from the hardness of the non-emptiness of SEREs [22, p. 162]. In its full generality, ADT-NE is therefore non-elementary (last column). Moreover, by our exponential translation of ADT_k into the FO[<]-fragment Σ_{k+1} (Lemma 2), we obtain the $(k+1)$-EXPSPACE upper-bound complexity for ADT_k-NE (recall satisfiability problem for Σ_ℓ is $(\ell-1)$-EXPSPACE). Additionally, a lower-bound for ADT_k-NE follows from [22, Theorem 4.29]: for SEREs that linearly translate as adts with countermeasure-depth k, their non-emptiness is at least $\text{NSPACE}(g(k-5, c\sqrt{\frac{n-1}{3}}))$. Interestingly, the Small Model Property (Theorem 1) yields an NP upper-bound complexity for ADT_1-NE, also applicable for ADT_0-NE, which is optimal since one can reduce the NP-complete satisfiability of propositional formulas to the non-emptiness of leaf adts.

Finally regarding ADT-EQUIV (last row of Table 1), one can observe that ADT-NE and ADT-EQUIV are very close. First, ADT-NE is a particular case of ADT-EQUIV with the second input adt $\tau_2 := \emptyset$. As a consequence, ADT-EQUIV inherits from the hardness of ADT-NE, and is therefore non-elementary (last column). Also, because deciding the equivalence between τ_1 and τ_2 amounts to deciding whether $\text{OR}(\text{CO}(\tau_1, \tau_2), \text{CO}(\tau_2, \tau_1)) = \emptyset$, we get reduction from ADT_k-EQUIV into ADT_{k+1}-NE, which provides the results announced in Columns 1–3.

7 Discussion

First, we discuss our model of adts with regard to the literature. However, we do not compare with settings where adts leaves are actions [11], as they yield only finite languages, and address other issues [25].

Our adts have particular features, but remain somehow standard. Regarding the syntax, firstly, even though we did not type our nodes as proponent/opponent, the countermeasure operator fully determines the alternation between attack and defense. Also, we introduced the non-standard leaf ϵ for the singleton empty-trace language, not considered in the literature. Still, it is a very natural object in the formal language landscape, and anyhow does not impact our overall computational complexity analysis. Remark that in the setting where adt leaves are interpreted as reachability goals, the semantics of AND operator proposed in [1,2] and ours coincide.

Second, we discuss our results. We showed the strictness of the ADT-hierarchy in a non-constructive manner. However, exhibiting an element of $ADT_{k+1} \setminus ADT_k$ ($k \geq 2$) is still an open question. Since $W_1 \in ADT_2 \setminus ADT_1$, languages W_k are natural candidates. For future work, we intend to explore Erenfeucht-Fraïssé (EF)-like games for adts (in the spirit of [24] for SEREs) as a mean to precisely locate languages W_k in our hierarchy. Moreover, EF-like games for adts may also help to better compare the ADT-hierarchy and the FO[<] alternation hierarchy, in particular, whether the hierarchies eventually coincide. For now, finding a tighter inclusion of Σ_ℓ in some ADT_k for each ℓ seems difficult; recall that we established $\Sigma_\ell \subseteq ADT_{2\ell+2}$. Any progress in this direction would be of great help to obtain tight complexity bounds for ADT_k-NE and ADT_k-EQUIV.

Finally, determining the level of a language in the ADT-hierarchy seems as hard as determining its level in the dot-depth hierarchy, recognised as a difficult question [4,18].

References

1. Audinot, M., Pinchinat, S., Kordy, B.: Is my attack tree correct? In: Foley, S.N., Gollmann, D., Snekkenes, E. (eds.) ESORICS 2017. LNCS, vol. 10492, pp. 83–102. Springer, Cham (2017). https://doi.org/10.1007/978-3-319-66402-6_7

2. Brihaye, T., Pinchinat, S., Terefenko, A.: Adversarial formal semantics of attack trees and related problems. In: Ganty, P., Monica, D.D. (eds.) Proceedings of the 13th International Symposium on Games, Automata, Logics and Formal Verification, GandALF 2022, Madrid, 21–23 September 2022, vol. 370 of EPTCS, pp. 162–177 (2022)

3. Brihaye, T., Pinchinat, S., Terefenko, A.: Semantics of attack-defense trees for dynamic countermeasures and a new hierarchy of star-free languages. arXiv preprint arXiv:2312.00458 (2023)

4. Diekert, V., Gastin, P.: First-order definable languages. In: Flum, J., Grädel, E., Wilke, T. (eds.) Logic and Automata: History and Perspectives [in Honor of Wolfgang Thomas], vol. 2 of Texts in Logic and Games, pp. 261–306. Amsterdam University Press (2008)

5. Eilenberg, S.: Automata, Languages, and Machines. Academic Press (1974)

6. Gadyatskaya, O., Hansen, R.R., Larsen, K.G., Legay, A., Olesen, M.C., Poulsen, D.B.: Modelling attack-defense trees using timed automata. In: Fränzle, M., Markey, N. (eds.) Formal Modeling and Analysis of Timed Systems. LNCS, vol. 9884, pp. 35–50. Springer, Cham (2016). https://doi.org/10.1007/978-3-319-44878-7_3

7. Jhawar, R., Kordy, B., Mauw, S., Radomirović, S., Trujillo-Rasua, R.: Attack trees with sequential conjunction. In: Federrath, H., Gollmann, D. (eds.) SEC 2015, pp. 339–353. Springer, Cham (2015). https://doi.org/10.1007/978-3-319-18467-8_23

8. Kordy, B., Mauw, S., Radomirović, S., Schweitzer, P.: Attack-defense trees. J. Log. Comput. 24(1), 55–87 (2014)

9. Kordy, B., Pouly, M., Schweitzer, P.: Computational aspects of attack–defense trees. In: Bouvry, P., et al. (eds.) Security and Intelligent Information Systems, pp. 103–116. Springer, Heidelberg (2012). https://doi.org/10.1007/978-3-642-25261-7_8

10. Kupferman, O., Zuhovitzky, S.: An improved algorithm for the membership problem for extended regular expressions. In: Diks, K., Rytter, W. (eds.) Mathematical Foundations of Computer Science 2002: 27th International Symposium, MFCS 2002 Warsaw, 26–30 August 2002 Proceedings, pp. 446–458. Springer, Heidelberg (2002). https://doi.org/10.1007/3-540-45687-2_37

11. Mauw, S., Oostdijk, M.: Foundations of attack trees. In: Won, D.H., Kim, S. (eds.) Information Security and Cryptology - ICISC 2005. LNCS, vol. 3935, pp. 186–198. Springer, Heidelberg (2006). https://doi.org/10.1007/11734727_17

12. McNaughton, R., Papert, S.A.: Counter-Free Automata. MIT Research Monograph, vol. 65. The MIT Press (1971)

13. Meyer, A.R.: Weak monadic second order theory of succesor is not elementary-recursive. In: Parikh, R. (ed.) Logic Colloquium. LNM, vol. 453. Springer, Heidelberg (1975). https://doi.org/10.1007/BFb0064872

14. Perrin, D., Pin, J.-E.: First-order logic and star-free sets. J. Comput. Syst. Sci. **32**(3), 393–406 (1986)

15. Pin, J. (ed.): Handbook of Automata Theory. European Mathematical Society Publishing House, Zürich (2021)

16. Pin, J.E., Schützenberger, M.P.: Variétés de Langages Formels, vol. 17. Masson, Paris (1984)

17. Pinchinat, S., Acher, M., Vojtisek, D.: Towards synthesis of attack trees for supporting computer-aided risk analysis. In: Canal, C., Idani, A. (eds.) Software Engineering and Formal Methods: SEFM 2014. LNCS, vol. 8938, pp. 363–375. Springer, Cham (2015). https://doi.org/10.1007/978-3-319-15201-1_24

18. Place, T., Zeitoun, M.: The tale of the quantifier alternation hierarchy of first-order logic over words. ACM SIGLOG News **2**(3), 4–17 (2015)

19. Rozenberg, G., Salomaa, A.: Handbook of Formal Languages, vol. 3 Beyond Words. Springer (2012)

20. Schiering, I., Thomas, W.: Counter-free automata, first-order logic, and star-free expressions extended by prefix oracles. Developments in Language Theory, II (Magdeburg, 1995), pp. 166–175. World Science Publishing, River Edge (1996)

21. Schneier, B.: Attack trees. Dr. Dobb's J. **24**(12), 21–29 (1999)

22. Stockmeyer, L.J.: The complexity of decision problems in automata theory and logic. Ph.D. thesis, Massachusetts Institute of Technology (1974)

23. Thomas, W.: Classifying regular events in symbolic logic. J. Comput. Syst. Sci. **25**(3), 360–376 (1982)

24. Thomas, W.: An application of the ehrenfeucht-fraïssé game in formal language theory. Bull. Soc. Math. France **16**(1), 1–21 (1984)

25. Wideł, W., Audinot, M., Fila, B., Pinchinat, S.: Beyond 2014: formal methods for attack tree-based security modeling. ACM Comput. Surv. **52**(4), 1–36 (2019)

Asymptotic (a)Synchronism Sensitivity and Complexity of Elementary Cellular Automata

Isabel Donoso Leiva[1,3], Eric Goles[1], Martín Ríos-Wilson[1(✉)], and Sylvain Sené[2,3]

[1] Facultad de Ingeniería y Ciencias, Universidad Adolfo Ibáñez, Peñalolén, Chile
martin.rios@uai.cl
[2] Université Publique, Marseille, France
[3] Aix Marseille Univ, CNRS, LIS, Marseille, France

Abstract. Among the fundamental questions in computer science is that of the impact of synchronism/asynchronism on computations, which has been addressed in various fields of the discipline: in programming, in networking, in concurrence theory, in artificial learning, etc. In this paper, we tackle this question from a standpoint which mixes discrete dynamical system theory and computational complexity, by highlighting that the chosen way of making local computations can have a drastic influence on the performed global computation itself. To do so, we study how distinct update schedules may fundamentally change the asymptotic behaviors of finite dynamical systems, by analyzing in particular their limit cycle maximal period. For the message itself to be general and impacting enough, we choose to focus on a "simple" computational model which prevents underlying systems from having too many intrinsic degrees of freedom, namely elementary cellular automata. More precisely, for elementary cellular automata rules which are neither too simple nor too complex (the problem should be meaningless for both), we show that update schedule changes can lead to significant computational complexity jumps (from constant to superpolynomial ones) in terms of their temporal asymptotes.

1 Introduction

In the domain of discrete dynamical systems at the interface with computer science, the generic model of automata networks, initially introduced in the 1940s through the seminal works of McCulloch and Pitts on formal neural networks [20] and of Ulam and von Neumann on cellular automata [30], has paved the way for numerous fundamental developments and results; for instance with the introduction of finite automata [17], the retroaction cycle theorem [26], the undecidability of all nontrivial properties of limit sets of cellular automata [15], the Turing universality of the model itself [11,28], ... and also with the generalized use of Boolean networks as a representational model of biological regulation networks since the works of Kauffman [16] and Thomas [29]. Informally speaking, automata networks

J. A. Soto and A. Wiese (Eds.): LATIN 2024, LNCS 14579, pp. 272–286, 2024.
https://doi.org/10.1007/978-3-031-55601-2_18

are collections of discrete-state entities (the automata) interacting locally with each other over discrete time which are simple to define at the static level but whose global dynamical behaviors offer very interesting intricacies.

Despite major theoretical contributions having provided since the 1980s a better comprehension of these objects [11,27] from computational and behavioral standpoints, understanding their sensitivity to (a)synchronism remains an open question on which any advance could have deep implications in computer science (around the thematics of synchronous versus asynchronous computation and processing [2,3]) and in systems biology (around the temporal organization of genetic expression [10,13]). In this context, numerous studies have been published by considering distinct settings of the concept of synchronism/asynchronism, i.e. by defining update modes which govern the way automata update their state over time. For instance, (a)synchronism sensitivity has been studied *per se* according to deterministic and non-deterministic semantics in [1,12,21] for Boolean automata networks and in [7–9,14] for cellular automata subject to stochastic semantics.

In these lines, the aim of this paper is to increase the knowledge on Boolean automata networks and cellular automata (a)synchronism sensitivity. To do so, we choose to focus on the impact of different kinds of periodic update modes on the dynamics of elementary cellular automata: from the most classical parallel one to the more general local clocks one [22]. Because we want to exhibit the very power of update modes on dynamical systems and concentrate on it, the choice of elementary cellular automata is quite natural: they constitute a restricted and "simple" cellular automata family which is well known to have more or less complex representatives in terms of dynamical behaviors [4,18,31] without being too much permissive. For our study, we use an approach derived from [25] and we pay attention to the influence of update modes on the asymptotic dynamical behaviors they can lead to compute, in particular in terms of limit cycles maximal periods.

In this paper, we highlight formally that the choice of the update mode can have a deep influence on the dynamics of systems. In particular, two specific elementary cellular automata rules, namely rules 156 and 178 (a variation of the well-known majority function tie case) as defined by the Wolfram's codification, are studied here. They have been chosen from the experimental classification presented in [31], as a result of numerical simulations which have given the insight that they are perfect representatives for highlighting (a)synchronism sensitivity. Notably, they all belong to the Wolfram's class II, which means that, according to computational observations, these cellular automata evolved asymptotically towards a "set of separated simple stable or periodic structures". Since our (a)synchronism sensitivity measure consists in limit cycles maximal periods, this Wolfram's class II is naturally the most pertinent one in our context. Indeed, class I cellular automata converge to homogeneous fixed points, class III cellular automata leads to aperiodic or chaotic patterns, and class IV cellular automata, which are deeply interesting from the computational standpoint, are not relevant for our concern because of their global high expressiveness which would prevent from showing asymptotic complexity jumps depending on update modes. On this basis, for these two rules, we show between which kinds of update modes asymptotic complexity changes appear. What stands out is that each of these rules admits it own (a)synchronism sensitivity scheme (which one could call its own asymptotic complexity scheme with respect to synchronism), which supports interestingly the existence of a periodic update modes expressiveness hierarchy.

In Sect. 2, the main definitions and notations are formalized. The emphasizing of elementary cellular automata (a)synchronism sensitivity is presented in Sect. 3 through upper-bounds for the limit-cycle periods of rules 156 and 178 depending on distinct families of periodic update modes. The paper ends with Sect. 4 in which we discuss some perspectives of this work. Full proofs can be found in the long version of this paper available here.

2   Definitions and Notations

General notations Let $[\![n]\!] = \{0, \ldots, n-1\}$, let $\mathbb{B} = \{0, 1\}$, and let x_i denote the i-th component of vector $x \in \mathbb{B}^n$. Given a vector $x \in \mathbb{B}^n$, we can denote it classically as $(x_0, \ldots, x_{n-1})$ or as the word $x_0 \ldots x_{n-1}$ if it eases the reading.

2.1   Boolean Automata Networks and Elementary Cellular Automata

Roughly speaking, a Boolean automata network (BAN) applied over a grid of size n is a collection of n automata represented by the set $[\![n]\!]$, each having a state within $\mathbb{B}$, which interact with each other over discrete time. A *configuration* x is an element of $\mathbb{B}^n$, i.e. a Boolean vector of dimension n. Formally, a *BAN* is a function $f : \mathbb{B}^n \to \mathbb{B}^n$ defined by means of n local functions $f_i : \mathbb{B}^n \to \mathbb{B}$, with $i \in [\![n]\!]$, such that f_i is the ith component of f. Given an automaton $i \in [\![n]\!]$ and a configuration $x \in \mathbb{B}^n$, $f_i(x)$ defines the way that i updates its states depending on the state of automata *effectively* acting on it; automaton j "effectively" acts on i if and only if there exists a configuration x in which the state of i changes with respect to the change of the state of j; j is then called a *neighbor* of i.

An *elementary cellular automaton (ECA)* is a particular BAN dived into the cellular space $\mathbb{Z}$ so that *(i)* the evolution of state x_i of automaton i (rather called cell i in this context) over time only depends on that of cells $i-1$, i itself, and $i+1$, and *(ii)* all cells share the same and unique local function. As a consequence, it is easy to derive that there exist $2^{2^3} = 256$ distinct ECA, and it is well known that these ECA can be grouped into 88 equivalence classes up to symmetry.

In absolute terms, BANs as well as ECA can be studied as infinite models of computation, as it is classically done in particular with ECA. In this paper, we choose to focus on finite ECA, which are ECA whose underlying structure can be viewed as a torus of dimension 1 which leads naturally to work on $\mathbb{Z}/n\mathbb{Z}$, the ring of integers modulo n so that the neighborhood of cell 0 is $\{n-1, 0, 1\}$ and that of cell $n-1$ is $\{n-2, n-1, 0\}$.

Now the mathematical objects at stake in this paper are statically defined, let us specify how they evolve over time, which requires defining when the cells state update, by executing the local functions.

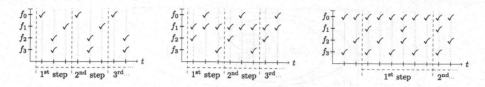

Fig. 1. Illustration of the execution over time of local transition functions of any BAN f of size 4 according to (left) $\mu_{\mathrm{BS}} = (\{0\}, \{2, 3\}, \{1\})$, (center) $\mu_{\mathrm{BP}} = \{(1), (2, 0, 3)\}$, and (right) $\mu_{\mathrm{LC}} = ((1, 3, 2, 2), (0, 2, 1, 0))$. The ✓ symbols indicate the moments at which the automata update their states; the vertical dashed lines separate periodical time steps from each other.

2.2 Update Modes

To choose an organization of when cells update their state over time leads to define what is classically called an update mode (aka update schedule or scheme). In order to increase our knowledge on (a)synchronism sensitivity, as evoked in the introduction, we pay attention in this article to deterministic and periodic update modes. Generally speaking, given a BAN f applied over a grid of size n, a *deterministic* (resp. *periodic*) *update mode* of f is an infinite (resp. a finite) sequence $\mu = (B_k)_{k \in \mathbb{N}}$ (resp. $\mu = (B_0, \ldots, B_{p-1})$), where B_i is a subset of $[\![n]\!]$ for all $i \in \mathbb{N}$ (resp. for all $i \in [\![p]\!]$). Another way of seeing the update mode μ is to consider it as a function $\mu^\star : \mathbb{N} \to \wp([\![n]\!])$ which associates each time step with a subset of $[\![n]\!]$ so that $\mu^\star(t)$ gives the automata which update their state at step t; furthermore, when μ is periodic, there exists $p \in \mathbb{N}$ such that for all $t \in \mathbb{N}$, $\mu^\star(t + p) = \mu^\star(t)$.

Three known update mode families are considered: the block-sequential [27], the block-parallel [6,23] and the local clocks [24] ones. Updates induced by each of them over time are depicted in Fig. 1.

A *block-sequential update mode* $\mu_{\mathrm{BS}} = (B_0, \ldots, B_{p-1})$ is an ordered partition of $[\![n]\!]$, with B_i a subset of $[\![n]\!]$ for all i in $[\![p]\!]$. Informally, μ_{BS} defines an update mode of period p separating $[\![n]\!]$ into p disjoint blocks so that all automata of a same block update their state in parallel while the blocks are iterated in series. The other way of considering μ_{BS} is: $\forall t \in \mathbb{N}, \mu_{\mathrm{BS}}^\star(t) = B_{t \bmod p}$.

A *block-parallel update mode* $\mu_{\mathrm{BP}} = \{S_0, \ldots, S_{s-1}\}$ is a partitioned order of $[\![n]\!]$, with $S_j = (i_{j,k})_{0 \le k \le |S_j| - 1}$ a sequence of $[\![n]\!]$ for all j in $[\![s]\!]$. Informally, μ_{BP} separates $[\![n]\!]$ into s disjoint subsequences so that all automata of a same subsequence update their state in series while the subsequences are iterated in parallel. Note that there exists a natural way to convert μ_{BP} into a sequence of blocks of period $p = lcm(|S_0|, \ldots, |S_{s-1}|)$. It suffices to define function φ as: $\varphi(\mu_{\mathrm{BP}}) = (B_\ell)_{\ell \in [\![p]\!]}$ with $B_\ell = \{i_{j,\ell \bmod |S_j|} \mid j \in [\![s]\!]\}$. The other way of considering μ_{BP} is: $\forall j \in [\![s]\!], \forall k \in [\![|S_j|]\!], i_{j,k} \in \mu_{\mathrm{BP}}^\star(t) \iff k = t \bmod |S_j|$.

A *local clocks update mode* $\mu_{\mathrm{LC}} = (P, \Delta)$, with $P = (p_0, \ldots, p_{n-1})$ and $\Delta = (\delta_0, \ldots, \delta_{n-1})$, is an update mode such that each automaton i of $[\![n]\!]$ is associated with a period $p_i \in \mathbb{N}^*$ and an initial shift $\delta_i \in [\![p_i]\!]$ such that $i \in \mu_{\mathrm{LC}}^\star(t) \iff t = \delta_i \bmod p_i$, with $t \in \mathbb{N}$.

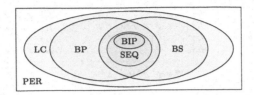

Fig. 2. Order of inclusion of the defined families of periodic update modes, where PER stands for "periodic".

Let us now introduce three particular cases or subfamilies of these three latter update mode families. The *parallel update mode* $\mu_{\text{PAR}} = (\llbracket n \rrbracket)$ makes every automaton update its state at each time step, such that $\forall t \in \mathbb{N}, \mu_{\text{PAR}}^{\star}(t) = \llbracket n \rrbracket$. A *bipartite update mode* $\mu_{\text{BIP}} = (B_0, B_1)$ is a block-sequential update mode composed of two blocks such that the automata in a same block do not act on each other (in our ECA framework, this definition induces that such update modes are associated to 1D tori of even size and that there are two such update modes). A *sequential update mode* $\mu_{\text{SEQ}} = (\phi(\llbracket n \rrbracket))$, where $\phi(\llbracket n \rrbracket) = \{i_0\}, \ldots, \{i_{n-1}\}$ is a permutation of $\llbracket n \rrbracket$, makes one and only one automaton update its state at each time step so that all automata have updated their state after n time steps depending on the order induced by ϕ. All these update modes follow the order of inclusion pictured in Fig. 2.

As we focus on periodic update modes, let us differentiate two kinds of time steps. A *substep* is a time step at which a subset of automata change their states. A *step* is the composition of substeps having occurred over a period.

2.3 Dynamical Systems

An ECA $f \in \llbracket 256 \rrbracket$ together with an update mode μ define a *discrete dynamical system* denoted by the pair (f, μ). (f, M) denotes by extension any dynamical system related to f under the considered update mode families, with $M \in \{\text{PAR}, \text{BIP}, \text{SEQ}, \text{BP}, \text{BS}, \text{LC}, \text{PER}\}$.

Let f be an ECA of size n and let μ be a periodic update mode represented as a periodical sequence of subsets of $\llbracket n \rrbracket$ such that $\mu = (B_0, \ldots, B_{p-1})$. Let $F = (f, \mu)$ be the global function from $\mathbb{B}^n$ to itself which defines the dynamical system related to ECA f and update mode μ. Let $x \in \mathbb{B}^n$ a configuration of F. The *trajectory* of x is the infinite path $\mathscr{T}(x) \overset{\Delta}{=} x^0 = x \to x^1 = F(x) \to \cdots \to x^t = F^t(x) \to \cdots$, where $t \in \mathbb{N}$ and

$$F(x) = f_{B_{p-1}} \circ \cdots \circ f_{B_0},$$

$$\text{where } \forall k \in \llbracket p \rrbracket, \forall i \in \llbracket n \rrbracket, f_{B_k}(x)_i = \begin{cases} f_i(x) & \text{if } i \in B_k, \\ x_i & \text{otherwise,} \end{cases}$$

$$\text{and } F^t(x) = \underbrace{F \circ \cdots \circ F}_{t \text{ times}}(x).$$

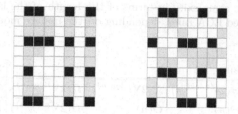

Fig. 3. Space-time diagrams (time going downward) representing the 3 first (periodical) steps of the evolution of configuration $x = (0, 1, 1, 0, 0, 1, 0, 1)$ of dynamical systems (left) $(156, \mu_{\text{BS}})$, and (right) $(178, \mu_{\text{BP}})$, where $\mu_{\text{BS}} = (\{1, 3, 4\}, \{0, 2, 6\}, \{5, 7\})$, and $\mu_{\text{BP}} = \{(1, 3, 4), (0, 2, 6), (5), (7)\}$, The configurations obtained at each step are depicted by lines with cells at state 1 in black. Lines with cells at state 1 in light gray represent the configurations obtained at substeps. Remark that x belongs to a limit cycle of length 3 (resp. 2) in $(156, \mu_{\text{BS}})$ (resp. $(178, \mu_{\text{BP}})$).

In the context of ECA, it is convenient to represent trajectories by *space-time diagrams* which give a visual aspect of the latter, as illustrated in Fig. 3. The *orbit* of x is the set $\mathcal{O}(x)$ composed of all the configurations which belongs to $\mathcal{T}(x)$. Since f is defined over a grid of finite size and the boundary condition is periodic, the temporal evolution of x governed by the successive applications of F leads it undoubtedly to enter into a *limit phase*, i.e.a cyclic subpath $\mathcal{C}(x)$ of $\mathcal{T}(x)$ such that $\forall y = F^k(x) \in \mathcal{C}(x), \exists t \in \mathbb{N}, F^t(y) = y$, with $k \in \mathbb{N}$. $\mathcal{T}(x)$ is this separated into two phases, the limit phase and the *transient phase* which corresponds to the finite subpath $x \to \ldots \to x^\ell$ of length ℓ such that $\forall i \in [\![\ell + 1]\!], \nexists t \in \mathbb{N}, x^{i+t} = x^i$. The *limit set* of x is the set of configurations belonging to $\mathcal{C}(x)$.

From these definitions, we derive that F can be represented as a graph $\mathcal{G}_F = (\mathbb{B}^n, T)$, where $(x, y) \in T \subseteq \mathbb{B}^n \times \mathbb{B}^n \iff y = F(x)$. In this graph, which is classically called a *transition graph*, the non-cyclic (resp. cyclic) paths represent the transient (resp. limit) phases of F. More precisely, the cycles of $\mathcal{G}_F$ are the *limit cycles* of F. When a limit cycle is of length 1, we call it a *fixed point*. Furthermore, if the fixed point is such that all the cells of the configuration has the same state, then we call it an *homogeneous fixed point*.

Eventually, we make use of the following specific notations. Let $x \in \mathbb{B}^n$ be a configuration and $[i, j] \subseteq [\![n]\!]$ be a subset of cells. We denote by $x_{[i,j]}$ the projection of x on $[i, j]$. Since we work on ECA over tori, such a projection defines a sub-configurations and can be of three kinds: either $i < j$ and $x_{[i,j]} = (x_i, x_{i+1}, \ldots, x_{j-1}, x_j)$, or $i = j$ and $x_{[i,j]} = (x_i)$, or $i > j$ and $x_{[i,j]} = (x_i, x_{i+1}, \ldots, x_n, x_0, \ldots, x_{j-1}, x_j)$. Thus, given $x \in \mathbb{B}^n$ and $i \in [\![n]\!]$, an ECA f can be rewritten as

$$f(x) = (f(x_{[n-1,1]}), f(x_{[0,2]}), \ldots, f(x_{[i,i+2]}), \ldots, f(x_{[n-3,n-1]}), f(x_{[n-2,0]})).$$

Abusing notations, the word $u \in \mathbb{B}^k$ is called a *wall* for a dynamical system if for all $a, b \in \mathbb{B}$, $f(aub) = u$, and we assume in this work that walls are of size 2, i.e. $k = 2$, unless otherwise stated. Such a word u is an *absolute wall* (resp. a *relative wall*) for an ECA rule if it is a wall for any update mode (resp. strict subset of

Table 1. Asymptotic complexity in terms of the length of the largest limit cycles of each of the two studied ECA rules, depending on the update modes.

ECA	M				
	PAR	BIP	BS	BP	LC
156	$\Theta(1)$	$\Theta(2^{\sqrt{n\log(n)}})$	$\Omega(2^{\sqrt{n\log(n)}})$	$\Omega(2^{\sqrt{n\log(n)}})$	$\Omega(2^{\sqrt{n\log(n)}})$
178	$\Theta(1)$	$\Theta(n)$	$O(n)$	$\Omega(2^{\sqrt{n\log(n)}})$	$\Omega(2^{\sqrt{n\log(n)}})$

update modes). We say that a rule F can *dynamically create new walls* if there is a time $t \in \mathbb{N}$ and an initial configuration $x^0 \in \mathbb{B}^n$ such that $x^t(=F^t(x^0))$ has a higher number of walls than x^0. Finally, we say that a configuration x is an *isle* of 1s (resp. an isle of 0s) if there exists an interval $I = [a, b] \subseteq [\![n]\!]$ such that $x_i = 1$ (resp. $x_i = 0$) for all $i \in I$ and $x_i = 0$ (resp. $x_i = 1$) otherwise.

3 Results

In this section, we will present the main results of our investigation related the asymptotic complexity of ECA rules 156 and 178, in terms of the lengths of their largest limit cycles depending on update modes considered. These results are summarized in Table 1 which highlights (a)synchronism sensitivity of ECA. As a reminder, these two rules have been chosen because they illustrate perfectly the very impact of the choice of update modes on their dynamics.

By presenting the results rule by rule, we clearly compare the complexity changes brought up by the different update modes. Starting with ECA rule 156, we prove that it has limit cycles of length at most 2 in parallel, but that in any other update mode can lead to reach limit cycles of superpolynomial length. Then, we show that for ECA rule 178, complexity increases less abruptly but still can reach very long cycles with carefully chosen update modes which fall into the category of update modes instantiating local function repetitions over a period.

3.1 ECA Rule 156

ECA rule 156 is defined locally by a transition table which associates any local neighborhood configuration $(x_{i-1}^t, x_i^t, x_{i+1}^t)$ at step $t \in \mathbb{N}$ with a new state x_i^{t+1}, where $i \in \mathbb{Z}/n\mathbb{Z}$, as follows:

$(x_{i-1}^t, x_i^t, x_{i+1}^t)$	000	001	010	011	100	101	110	111
x_i^{t+1}	0	0	1	1	1	0	0	1

Space-time diagrams depending on different update modes of a specific configuration under ECA rule 156 are given in Fig. 4. Each of them depicts a trajectory which gives insights about the role of walls together with the update modes in order to reach long limit cycles.

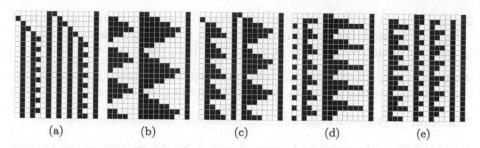

(a)   (b)   (c)   (d)   (e)

Fig. 4. Space-time diagrams (time going downward) of configuration 0000001100000 001 following rule 156 depending on: (a) the parallel update mode $\mu_{\text{PAR}} = (\llbracket 16 \rrbracket)$, (b) the bipartite update mode $\mu_{\text{BIP}} = (\{i \in \llbracket 16 \rrbracket \mid i \equiv 0 \mod 2\}, \{i \in \llbracket 16 \rrbracket \mid i \equiv 1 \mod 2\})$, (c) the block-sequential update mode $\mu_{\text{BS}} = (\{10, 15\}, \{0, 1, 5, 7, 8, 12\}, \{4, 6, 9, 11, 14\}, \{3, 13\}, \{2\})$, (d) the block-parallel update mode $\mu_{\text{BP}} = \{(0, 1), (2, 3, 4), (5), (6, 8, 7), (11, 10, 9), (14, 13, 12), (15)\}$, (e) the local clocks update mode $\mu_{\text{LC}} = (P = (2, 2, 2, 2, 4, 4, 4, 4, 3, 3, 3, 3, 1, 4, 1, 1), \Delta = (1, 1, 1, 0, 3, 3, 3, 2, 1, 1, 1, 0, 0, 3, 0, 0))$.

Lemma 1. *ECA rule 156 admits only one wall, namely the word $w = 01$.*

Proof. For all $a, b \in \mathbb{B}$, $f_{156}(a01) = 0$ and $f_{156}(01b) = 1$. By definition of the rule, no other word of length 2 gets this property. Thus w is the unique wall for ECA rule 156. □

Lemma 2. *ECA rule 156 can dynamically create new walls only if the underlying update mode makes two consecutive cells update their state simultaneously.*

Proof. Remark that to create walls in the trajectory of a configuration $x \in \mathbb{B}^n$, x need to have at least one wall. Indeed, the only configurations with no walls are 0^n and 1^n, and they are fixed points. Thus, with no loss of generality, let us focus on a subconfiguration $y \in \mathbb{B}^k$, with $k \le n - 2$, composed of free state cells surrounded by two walls. Notice that since the configurations are toric, the wall "at the left" and "at the right" of y can be represented by the same two cells.

Configuration y is necessarily of the form $y = (1)^\ell (0)^r$, with $k = \ell + r$. Furthermore, since $f_{156}(000) = 0$ and $f_{156}(100) = 1$, the only cells whose states can change are those where 1 meets 0, i.e. $y_{\ell-1}$ and y_ℓ. Such state changes depends on the schedule of updates between these two cells. Let us proceed with case disjunction:

1. Case of $\ell \ge 1$ and $r \ge 1$:
   - if y_ℓ is updated strictly before $y_{\ell-1}$, then $y^1 = (1)^{\ell+1}(0)^{r-1}$, and the number of 1s (resp. 0s) increases (resp. decreases);
   - if y_ℓ is updated strictly after $y_{\ell-1}$, then $y^1 = (1)^{\ell-1}(0)^{r+1}$, and the number of 1s (resp. 0s) decreases (resp. increases);
   - if y_ℓ and $y_{\ell-1}$ are updated simultaneously, then $y^1 = (1)^{\ell-1}(01)(0)^{r-1}$, and a wall is created.
2. Case of $\ell = 0$ (or of $y = 0^k$): nothing happens until y_0 is updated. Let us admit that this first state change has been done for the sake of clarity and focus on $y^1 = (1)^{\ell=1}(0)^{r=k-1}$, which falls into Case 1.

3. Case of $r = 0$ (or of $y = 1^k$): symmetrically to Case 2, nothing happens until y_{k-1} is updated. Let us admit that this first state change has been done for the sake of clarity and focus on $y^1 = (1)^{\ell=k-1}(0)^1$, which falls into Case 1.

As a consequence, updating two consecutive cells simultaneously is a necessary condition for creating new walls in the dynamics of ECA rule 156.   □

Theorem 1. $(156, \mathrm{PAR})$ *has only fixed points and limit cycles of length two.*

Proof. We base the proof on Lemmas 1 and 2. So, let us analyze the possible behaviors between two walls, since by definition, what happens between two walls is independent of what happens between two other walls.

Let us prove the results by considering the three possible distinct cases for a configuration between two walls $y \in \mathbb{B}^{k+4}$:

- Consider $y = (01)(0)^k(01)$ the configuration with only 0s between the two walls. Applying the rule twice, we obtain $y^2 = (01)^2(0)^{k-2}(01)$. So, a new wall appears every two iterations so that, for all $t < \frac{k}{2}$, $y^{2t} = (01)^{t+1}(0)^{k-2t}(01)$, until a step is reached where there is no room for more walls. This step is reached after k (resp $k-1$) iterations when k is even (resp. odd) and is such that there is only walls if k is even (which implies that the dynamics has converged to a fixed point), and only walls except one cell otherwise. In this case, considering that $t = \lfloor \frac{k}{2} \rfloor$, because $f_{156}(100) = 1$ and $f_{156}(110) = 0$, we have that $y^{2t} = (01)^{t+1}(0)(01) \to y^{2t+1} = (01)^{t+1}(1)(01) \to y^{2t} = (01)^{t+1}(0)(01)$, which leads to a limit cycle of length 2.
- Consider now that $y = (01)(1)^k(01)$. Symmetrically, for all $t < \frac{k}{2}$, $y^{2t} = (01)(0)^{k-2t}(01)^{t+1}$. The same reasoning applies to conclude that the length of the largest limit cycle is 2.
- Consider finally configuration $y = (01)(1)^\ell(0)^r(01)$ for which $k = \ell + r$. Applying f_{156} on it leads to $y^1 = (01)(1)^{\ell-1}(01)(0)^{r-1}(01)$, which falls into the two previous cases.   □

Theorem 2. $(156, \mathrm{BIP})$ *of size n has largest limit cycles of length $\Theta(2^{\sqrt{n \log(n)}})$.*

Proof. First, by definition of a bipartite update mode and by Lemmas 1 and 2, the only walls appearing in the dynamics are the ones present in the initial configuration. Let us prove that, given two walls u and v, distanced by k cells, the largest limit cycles of the dynamics between u and v are of length $k + 1$. We proceed by case disjunction depending on the nature of the subconfiguration $y \in \mathbb{B}^{k+4}$, with the bipartite update mode $\mu = (\{i \in [\![k+4]\!] \mid i \equiv 0 \mod 2\}, \{i \in [\![k+4]\!] \mid i \equiv 0 \mod 2\})$:

1. Case of $y = (01)(0)^k(01)$: we prove that configuration $(01)(1)^k(01)$ is reached after $\lceil \frac{k}{2} + 1 \rceil$ steps:

    – If k is odd, we have:

    $$y = \quad (01)(0)^k(01)$$
    $$y^1 = (01)(1)(0)^{k-1}(01)$$

    – If k is even, we have:

    $$y = \quad (01)(0)^k(01)$$
    $$y^1 = (01)(1)(0)^{k-1}(01)$$

$$y^2 = (01)111(0)^{k-3}(01) \qquad\qquad y^2 = (01)(111)(0)^{k-2}(01)$$

$$\vdots \qquad\qquad\qquad\qquad\qquad \vdots$$

$$y^{\lfloor \frac{k}{2} \rfloor} = (01)(1)^{k-2}(00)(01) \qquad y^{\frac{k}{2}} = (01)(1)^{k-1}(0)(01)$$
$$y^{\lceil \frac{k}{2} \rceil} = \quad (01)(1)^{k}(01) \qquad\qquad y^{\frac{k}{2}+1} = \quad (01)(1)^{k}(01)$$

2. Case of $y = (01)(1)^{k}(01)$: we prove that configuration $(01)(0)^{k}(01)$ is reached after $\lceil \frac{k}{2} + 1 \rceil$ steps:

    – If k is odd, we have: – If k is even, we have:

$$y = \quad (01)(1)^{k}(01) \qquad\qquad y = \quad (01)(1)^{k}(01)$$
$$y^1 = (01)(1)^{k-1}(0)(01) \qquad y^1 = (01)(1)^{k-2}(00)(01)$$
$$y^2 = (01)(1)^{k-3}(000)(01) \qquad y^2 = (01)(1)^{k-4}(0000)(01)$$

$$\vdots \qquad\qquad\qquad\qquad\qquad \vdots$$

$$y^{\lfloor \frac{k}{2} \rfloor} = (01)(11)(0)^{k-2}(01) \qquad y^{\frac{k}{2}} = (01)(11)(0)^{k-2}(01)$$
$$y^{\lceil \frac{k}{2} \rceil} = \quad (01)(0)^{k}(01) \qquad\qquad y^{\frac{k}{2}+1} = \quad (01)(0)^{k}(01)$$

3. Case of $y = (01)(1)^{\ell}(0)^{r}(01)$: this case is included in Cases 1 and 2.

As a consequence, the dynamics of any y leads indeed to a limit cycle of length $k + 1$. Remark that if we had chosen the other bipartite update mode as reference, the dynamics of any y would have been symmetric and led to the same limit cycle.

Finally, since the dynamics between two pairs of distinct walls of independent of each other, the asymptotic dynamics of a global configuration x such that $x = (01)(0)^{k_1}(01)(0)^{k_2}(01)\ldots(01)(0)^{k_m}$ is a limit cycle whose length equals to the least common multiple of the lengths of all limit cycles of the subconfigurations $(01)(0)^{k_1}(01), (01)(0)^{k_2}(01), \ldots, (01)(0)^{k_m}(01)$. We derive that the largest limit cycles are obtained when the $(k_i + 1)$s are distinct primes whose sum is equal to $n - 2m$, with m is constant. As a consequence, the length of the largest limit cycle is lower- and upper- bounded by the primorial of n (i.e. the maximal product of distinct primes whose sum is $\leq n$.), denoted by function $h(n)$. In [5], it is shown in Theorem 18 that when n tends to infinity, $\log h(n) \sim \sqrt{n \log n}$. Hence, we deduce that the length of the largest limit cycles of $(156, \text{BIP})$ of size n is $\Theta(2^{\sqrt{n \log(n)}})$. $\square$

Corollary 1. *The families* $(156, \text{BS})$, $(156, \text{BP})$ *and* $(156, \text{LC})$ *of size n have largest limit cycles of length* $\Omega(2^{\sqrt{n \log(n)}})$.

Proof. Since the bipartite update modes are specific block-sequential and block-parallel update modes, and since both block-sequential and block-parallel update modes are parts of local-clocks update modes, all of them inherit the property stating that the lengths of the largest limit cycles are lower-bounded by $2^{\sqrt{n \log(n)}}$. $\square$

3.2 ECA Rule 178

ECA rule 178 is defined locally by the following transition table:

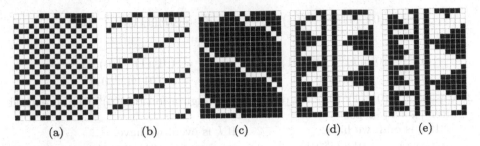

    (a)   (b)   (c)   (d)   (e)

Fig. 5. Space-time diagrams (time going downward) of configuration 0000011010111 110 following rule 178 depending on: (a) the parallel update mode $\mu_{\mathrm{PAR}} = (\llbracket 16 \rrbracket)$, (b) the bipartite update mode $\mu_{\mathrm{BIP}} = (\{i \in \llbracket 16 \rrbracket \mid i \equiv 0 \mod 2\}, \{i \in \llbracket 16 \rrbracket \mid i \equiv 1 \mod 2\})$, (c) the block-sequential update mode $\mu_{\mathrm{BS}} = (\{3, 9, 15\}, \{2, 4, 8, 10, 14\}, \{11, 5, 7, 11, 13\}, \{0, 6, 12\})$, (d) the block-parallel update mode $\mu_{\mathrm{BP}} = \{(0), (1), (2, 3), (4, 5), (6), (7), (8), (9), (10, 11), (12, 13), (14, 15)\}$, (e) the local clocks update mode $\mu_{\mathrm{LC}} = (P = (1, 1, 2, 2, 2, 2, 1, 1, 1, 4, 4, 4, 4, 4, 4, 4), \Delta = (0, 0, 1, 0, 1, 0, 0, 0, 0, 0, 1, 0, 1, 0, 1, 0))$.

$(x_{i-1}^t, x_i^t, x_{i+1}^t)$	000	001	010	011	100	101	110	111
x_i^{t+1}	0	1	0	0	1	1	0	1

Space-time diagrams depending on different update modes of a specific configuration under ECA rule 178 are given in Fig. 5. Each of them depict trajectories giving ideas of how to reach a limit cycle of high complexity.

Lemma 3. *ECA rule* 178 *admits two walls,* 01 *and* 10, *which are relative walls.*

Proof. Let $a, b \in \mathbb{B}$. Notice that by definition of the rule: $f(a00) \neq f(b00)$ and $f(00a) \neq f(00b)$, and $f(a11) \neq f(b11)$ and $f(11a) \neq f(11b)$ if $a \neq b$; $f(a01) = 1$ and $f(01a) = 0$; and $f(a10) = 0$ and $((10a) = 1$. Thus, neither 00 nor 01 nor 10 nor 11 are absolute walls. From what precedes, notice that the properties of 00 and 11 prevent them to be relative walls. Consider now the two words $u = 01$ and $v = 10$ and let us show that they constitute relative walls. Regardless the states of the cells surrounding u, every time both cells of u are updated simultaneously, u changes to v and similarly, v will change to u independently of the states of the cells that surround it, as long as both its cells are updated together. Thus, u and v are relative walls. □

We will show certain BP and LC update modes that are able to produce these relative walls.

Theorem 3. *Each representative of* (178, BP) *of size* n *has largest limit cycles of length* $\Theta(n)$.

Proof (Sketch of proof). From Lemma 3 we know that the representatives of (178, BP) do not have walls, because by definition of BP two consecutive cells cannot be updated simultaneously. Thus we can start by first considering configurations x composed by *one* isle of 1s and proving that if the number of 1s is odd then the dynamics lead to homogeneous fixed points. Meanwhile, if the number of 1s is even, then such configurations lead to limit cycles of length $\frac{n}{2}$. Then, we consider configurations with *several* isles of 1s, which can be analyzed by studying the behaviour with just two isles of 1s. These kinds of configurations are divided into 10 cases, depending on the parity of the size of the isles, and the local bipartite update mode that each of the isles follows. We prove that each case converges either to a homogeneous fixed point or to limit cycles of length $\frac{n}{2}$. □

Theorem 4. *The family* $(178, \text{BP})$ *of size* n *has largest limit cycles of length* $\Omega\left(2^{\sqrt{n \log n}}\right)$

Proof (Sketch of proof). First, we notice that since 1^n and 0^n are fixed points, the configurations of interest have at least one relative wall. By Lemma 3, we know that w_1 and w_2 are relative walls for the family $(178, \text{BP})$. Thus, we consider an initial configuration with at least one wall. Similar to the proof of Theorem 2, the idea is to focus on what can happen between two walls: $y = w_\ell y_1 y_2 \ldots y_k y_{k+1} w_r$, given that the dynamics of subconfigurations delimited by two pairs of walls are independent. Then, we analyse different cases depending on if $w_\ell = w_r = 01$ (similarly, $w_\ell = w_r = 10$), or $w_\ell = 01$ and $w_r = 10$ (similarly, $w_\ell = 10$ and $w_r = 01$) and we prove that each of the cases converges either to a fixed point or to limit cycles of length $k + 1$. And because of the independence of the dynamics between two pairs of walls, the asymptotic dynamics of a global configuration x is a limit cycle whose length equals the least common multiple of the lengths of all limit cycles of the subconfigurations, and with the same argument as was used in the proof of Theorem 2, we conclude that the largest limit cycles of the family $(178, \text{BP})$ applied over a grid of size n is $\Omega\left(2^{\sqrt{n \log n}}\right)$. □

Theorem 5. $(178, \text{SEQ})$ *of size* n *has largest limit cycles of length* $O(n)$.

Proof (Sketch of proof). We analyse the dynamics that start from an isle of 1s surrounded by 0s, dividing it on four cases, depending on if the first and last cells of the isle are updated before or after their neighboring 0. The first two cases result in fixed points, a third in which the isle of 1s shifts to the right, and a fourth in which the isle of 1s shifts to the left. Because of the condition necessary for the shift to occur, we know that an isle shifting left-to-right (or right-to-left) will complete a cycle in less than n iterations.

Thus, we consider an initial configuration x written as isles of 1s separated by 0s. Each of the isles must fall into one of the four cases analysed on the previous point. Isles that correspond to the first two cases either disappear without ever interacting with another isle or they fuse with the nearest isle, which will result in a new isle which case will be of the same type as the isle the original one fused with. We prove that an isle traveling left-to-right cannot interact with another one going in the same direction, but will fuse with one shifting right-to- left into a new isle which will correspond to the case where it will eventually disappearing with no possibility of ever interacting with another isle. We conclude that if in the initial configuration there is an equal number of isles that correspond to the cases where the limit cycle will be a fixed point, but if there is more of one of those cases than the other, the limit cycle will be of length less than n.   □

Theorem 6. $(178, \text{BS})$ *applied over a grid of size n has largest limit cycles of length $O(n)$.*

Proof (Sketch of proof). First, we use the fact that if two neighboring cells cannot be updated simultaneously, then the update mode is equivalent to a sequential update mode, whose limit cycles we have already proven to be strictly less than n. Then, we prove that if there are two consecutive cells that update simultaneously then limit cycle is of length at most 2. In order to prove this, we start by analysing the case with just one pair of cells that update on the same sub-step $\{s, s+1\}$, and just one isle of ones that stars on one of the neighboring cells of interest. This analysis is then divided in four cases, similar to the previous proof.

We determine that every case produces two isles of 1s, which travel in opposite directions. We know that they will have to meet on the other side, once their combined movement circumnavigates the ring, whereupon they will fuse, become one isle that will have to disappear. Then, we note that the previous analysis holds with an isle that starts in a different place of the configuration and eventually arrives to the cells s and $s+1$, and finally, it holds when there are multiple isles and/or multiple pairs of neighboring cells that update simultaneously.   □

Theorem 7. $(178, PAR)$ *has largest limit cycles of length 2.*

Proof. Direct from the proof of Theorem 6.   □

4   Discussion

In this paper, we have focused in the study of two ECA rules under different update modes. These rules are the rule 156 and the rule 178. These rule selections were made subsequent to conducting numerical simulations encompassing a set of 88 non-equivalent ECA rules, each subjected to diverse update modes. Rule 156 and Rule 178 emerged as pertinent subjects for further investigation due to their pronounced sensitivity to asynchronism, as substantiated by our simulation results. Following this insight we have analytically shown two different behaviors illustrated in Table 1:

- Rule 156: the maximum period of the attractors changes from constant to superpolynomial when a bipartite (block sequential with two blocks) update modes are considered.
- Rule 178: the same phenomenon is observed but"gradually increasing", from constant, to linear and then to superpolynomial.

The obtained results suggest that it might not be a unified classification according to our measure of complexity (the maximum period of the attractors). This observation presents an open question regarding what other complexity measures can be proposed to classify ECA rules under different update modes.

By analyzing our simulations we have found interesting observations about the dynamical complexity of ECA rules. An example that we identified in the simulations but is not studied in the paper is the one of the rule 184 (known also as the traffic rule). In this case we have shown that under the bipartite update mode there are only fixed points. This is interesting considering the fact that it is known that the maximum period can be linear in the size of the network for the parallel update mode [19]. Thus, in this case, the rule seems to exhibit a different kind of dynamical behaviour compared to rule 156 and 178 (asynchronism tend to produce simpler dynamics instead of increasing the complexity) and might be interesting to study from theoretical standpoint. In addition, it could be interesting to study if other rules present this particular behaviour.

Finally, from what we have been able to observe on our simulations, there exists a fourth class of rules where the length of the longest limit cycles remains constant regardless of the update mode (for example rule 4, 8, 12, 72, and 76 exhibit only fixed points when tested exhaustively for each configuration of size at most $n = 8$). Thus, this evidence might suggest the existence of rules that are robust with respect to asynchronism. In this sense, it could be interesting to analytically study some of these rules and try to determine which dynamical property makes them robust under asynchronism.

Acknowledgments. The authors are thankful to projects ANR-18-CE40-0002 "FANs" (MRW, SS), ANID-FONDECYT 1200006 (EG), ANID-FONDECYT Post-doctorado 3220205 (MRW), MSCA-SE-101131549 "ACANCOS" (EG, IDL, MRW, SS), STIC AmSud 22-STIC-02 (EG, IDL, MRW, SS) for their funding.

References

1. Aracena, J., Fanchon, É., Montalva, M., Noual, M.: Combinatorics on update digraphs in Boolean networks. Dicr. Appl. Math. **159**, 401–409 (2011)
2. Chapiro, D.M.: Globally-asynchronous locally-synchronous systems. PhD thesis, Stanford University (1984)
3. Charron-Bost, B., Mattern, F., Tel, G.: Synchronous, asynchronous, and causally ordered communication. Distrib. Comput. **9**, 173–191 (1996)
4. Culik, K., II., Yu, S.: Undecidability of CA classification schemes. Complex Syst. **2**, 177–190 (1988)
5. Deléglise, M., Nicolas, J.-L.: On the largest product of primes with bounded sum. J. Integer Sequences **18** 15.2.8 (2015)
6. Demongeot, J., Sené, S.: About block-parallel Boolean networks: a position paper. Nat. Comput. **19**, 5–13 (2020)
7. Dennunzio, A., Formenti, E., Manzoni, L., Mauri, G.: m-Asynchronous cellular automata: from fairness to quasi-fairness. Nat. Comput. **12**, 561–572 (2013)
8. Dennunzio, A., Formenti, E., Manzoni, L., Mauri, G., Porreca, A.E.: Computational complexity of finite asynchronous cellular automata. Theoret. Comput. Sci. **664**, 131–143 (2017)
9. Fatès, N., Morvan, M.: An experimental study of robustness to asynchronism for elementary cellular automata. Complex Syst. **16**, 1–27 (2005)

10. Fierz, B., Poirier, M.G.: Biophysics of chromatin dynamics. Annu. Rev. Biophys. **48**, 321–345 (2019)
11. Goles, E., Martinez, S.: Neural and automata networks: dynamical behavior and applications, volume 58 of Mathematics and Its Applications. Kluwer Academic Publishers (1990)
12. Goles, E., Salinas, L.: Comparison between parallel and serial dynamics of Boolean networks. Theor. Comput. Sci. **296**, 247-253 (2008)
13. Hübner, M.R., Spector, D.L.: Chromatin dynamics. Annu. Rev. Biophys. **39**, 471–489 (2010)
14. Ingerson, T.E., Buvel, R.L.: Structure in asynchronous cellular automata. Physica D **10**, 59–68 (1984)
15. Kari, J.: Rice's theorem for the limit sets of cellular automata. Theoret. Comput. Sci. **127**, 229–254 (1994)
16. Kauffman, S.A.: Metabolic stability and epigenesis in randomly constructed genetic nets. J. Theor. Biol. **22**, 437–467 (1969)
17. Kleene, S.C.: Automata studies, volume 34 of Annals of Mathematics Studies, chapter Representation of events in nerve nets and finite automata, pp. 3–41. Princeton Universtity Press (1956)
18. Kůrka, P.: Languages, equicontinuity and attractors in cellular automata. Ergodic Theory Dynam. Syst. **17**, 417–433 (1997)
19. Li, W.: Phenomenology of nonlocal cellular automata. J. Stat. Phys. **68**, 829–882 (1992)
20. McCulloch, W.S., Pitts, W.: A logical calculus of the ideas immanent in nervous activity. J. Math. Biophys. **5**, 115–133 (1943)
21. Noual, M., Sené, S.: Synchronism versus asynchronism in monotonic Boolean automata networks. Nat. Comput. **17**, 393–402 (2018)
22. Paulevé, L., Sené, S.: Systems biology modelling and analysis: formal bioinformatics methods and tools, chapter Boolean networks and their dynamics: the impact of updates. Wiley (2022)
23. Perrot, K., Sené, S., Tapin, L.: On countings and enumerations of block-parallel automata networks. arXiv:2304.09664 (2023)
24. Ríos-Wilson, M.: On automata networks dynamics: an approach based on computational complexity theory. PhD thesis, Universidad de Chile & Aix-Marseille Université (2021)
25. Ríos-Wilson, M., Theyssier, G.: On symmetry versus asynchronism: at the edge of universality in automata networks. arXiv:2105.08356 (2021)
26. Robert, F.: Itérations sur des ensembles finis et automates cellulaires contractants. Linear Algebra Appl. **29**, 393–412 (1980)
27. Robert, F.: Discrete Iterations: A Metric Study. Springer Berlin Heidelberg, Berlin, Heidelberg (1986)
28. Smith, A.R., III.: Simple computation-universal cellular spaces. J. ACM **18**, 339–353 (1971)
29. Thomas, R.: Boolean formalization of genetic control circuits. J. Theor. Biol. **42**, 563–585 (1973)
30. von Neumann, J.: Theory of self-reproducing automata. University of Illinois Press, 1966. Edited and completed by A. W. Burks (1966)
31. Wolfram, S.: Universality and complexity in cellular automata. Physica D **10**, 1–35 (1984)

Game Theory and Fairness

On Binary Networked Public Goods Game with Altruism

Arnab Maiti[1]([✉]) [iD] and Palash Dey[2] [iD]

[1] University of Washington, Seattle, USA
arnabm2@uw.edu
[2] Indian Institute of Technology Kharagpur, Kharagpur, India
palash.dey@cse.iitkgp.ac.in

Abstract. In the classical Binary Networked Public Goods (BNPG) game, a player can either invest in a public project or decide not to invest. Based on the decisions of all the players, each player receives a reward as per his/her utility function. However, classical models of BNPG game do not consider altruism which players often exhibit and can significantly affect equilibrium behavior. Yu et al. [25] extended the classical BNPG game to capture the altruistic aspect of the players. We, in this paper, first study the problem of deciding the existence of a Pure Strategy Nash Equilibrium (PSNE) in a BNPG game with altruism. This problem is already known to be NP-complete. We complement this hardness result by showing that the problem admits efficient algorithms when the input network is either a tree or a complete graph. We further study the Altruistic Network Modification problem, where the task is to compute if a target strategy profile can be made a PSNE by adding or deleting a few edges. This problem is also known to be NP-complete. We strengthen this hardness result by exhibiting intractability results even for trees. A perhaps surprising finding of our work is that the above problem remains NP-hard even for bounded degree graphs when the altruism network is undirected but becomes polynomial-time solvable when the altruism network is directed. We also show some results on computing an MSNE and some parameterized complexity results. In summary, our results show that it is much easier to predict how the players in a BNPG game will behave compared to how the players in a BNPG game can be made to behave in a desirable way.

1 Introduction

In a binary networked public goods (in short BNPG) game, a player can either decide to invest in a public project or decide not to invest in it. Every player however incurs a cost for investing. Based on the decision of all the players, each player receives a reward as per his/her externality function. The net utility is decided based on the reward a player receives and the cost a player incurs. Usually, the externality function and cost of investing differ for every player, making the BNPG game heterogeneous. In some scenarios, the externality function and cost of investing can be the same for every player, making the BNPG

© The Author(s), under exclusive license to Springer Nature Switzerland AG 2024
J. A. Soto and A. Wiese (Eds.): LATIN 2024, LNCS 14579, pp. 289–303, 2024.
https://doi.org/10.1007/978-3-031-55601-2_19

game fully homogeneous. Many applications of public goods, for example, wearing a mask [12], getting vaccinated [3], practicing social distancing [5], reporting crimes etc., involve binary decisions. Such domains can be captured using BNPG game. A BNPG game is typically modeled using a network of players which is an undirected graph [11].

We also observe that there are some societies where few people wear masks and/or get themselves vaccinated, and there are some other societies where most people wear masks and get themselves vaccinated [4,24]. This can be attributed to differences in altruistic behavior among various societies [2,6]. In an altruistic society, people consider their as well as their neighbors' benefit to take a decision. For example, young adults may wear a mask not only to protect themselves but also to protect their elderly parents and young children at home. Altruism can be modeled using an altruistic network which can be either an undirected graph or a directed graph [25]. Symmetric altruism (respectively asymmetric altruism) occurs when the altruistic network is undirected (respectively directed). The utility that a player receives depends on both the input network and the (incoming) incident edges in the altruistic network.

We study the BNPG game with altruism for two different problem settings. First, we look at the problem of deciding the existence of Pure Strategy Nash Equilibrium (PSNE) in the BNPG game with altruism. In any game, determining a PSNE is an important problem as it allows a social planner to predict the behaviour of players in a strategic setting and make appropriate decisions. It is known that deciding the existence of PSNE in a BNPG game (even without altruism) is NP-Complete [26]. This paper mainly focuses on deciding the existence of PSNE in special networks like trees, complete graphs, and graphs with bounded circuit rank. The circuit rank of an undirected graph is the minimum number of edges that must be removed from the graph to make it acyclic.

In the second problem setting, also known as Altruistic Network Modification (in short ANM), we can add or delete an edge from the altruistic network, and each such operation has a non-negative cost associated with it. The aim here is to decide if a target strategy profile can be made a PSNE by adding or deleting edges with certain budget constraints. This problem was first studied by [25] where they showed that ANM is an NP-Complete problem. This problem enables policymakers to strategically run campaigns to make a society more altruistic and achieve desirable outcomes like everyone wearing a mask and getting vaccinated. This paper mainly focuses on ANM in sparse input networks like trees and graphs with bounded degree.

1.1 Contribution

We show that the problem of deciding the existence of PSNE in BNPG game with asymmetric altruism is polynomial-time solvable if the input network is either a tree [Theorem 1], complete graph [Theorem 3] or graph with bounded circuit rank [Theorem 2]. Moreover, in Theorem 1, we formulated a non-trivial ILP (not the ILP that follows immediately from the problem definition) and depicted a greedy polynomial time algorithm [Algorithm 1] to solve it. This strengthens

Table 1. List of results (our results are in bold). PSNE existence results hold for both symmetric and asymmetric altruism.

Input graph type	PSNE existence	ANM symmetric altruism	ANM asymmetric altruism
Tree	**P**	**NP-hard**	**NP-hard**
Clique	**P**	NP-hard ($\star$)	NP-hard ($\star$)
Bounded degree	NP-hard ($\star\star$)	**NP-hard**	**P**
Bounded circuit rank	**P**	**NP-hard**	**NP-hard**

the tractable results for tree, complete graph and graph with bounded circuit rank in [18, 26] as the previous results were depicted for BNPG games without altruism. Hence, existence of a PSNE can be efficiently decided in an intimately connected society where everyone knows others and thus the underlying graph is connected, and for sparsely connected society where the circuit rank could be low. However, the problem is open for graphs with bounded treewidth, and it is known that the problem even without altruism is W[1]-hard for the parameter treewidth [18]. The problem of deciding the existence of PSNE in BNPG game even without altruism is known to be NP-complete [26]. A natural but often under-explored question here is if an MSNE can be computed efficiently. We show that computing an MSNE in BNPG game with symmetric altruism is PPAD-hard [Theorem 4].

ANM with either asymmetric or symmetric altruism is known to be NP-complete when the input network is a clique [25]. We complement this by showing that ANM with either asymmetric or symmetric altruism is NP-complete even when the input network is a tree (the circuit rank of which is zero) and the BNPG game is fully homogeneous [Theorems 5 and 6]. We also show that ANM with symmetric altruism is known to be para-NP-hard for the parameter maximum degree of the input network even when the BNPG game is fully homogeneous and the available budget is infinite [Theorem 9]. However, with asymmetric altruism, the problem is FPT for the parameter maximum degree of the input network [Theorem 8]. To show this result, we designed an $O(2^{n/2})$ time binary search based algorithm [Algorithm 2] for Minimum Knapsack problem. We are the first to provide an algorithm better than $O(2^n)$ time for Minimum Knapsack problem to the best of our knowledge.

In summary, our paper provides a more fine-grained complexity theoretic landscape for deciding if a PSNE exists in a BNPG game with altruism and the ANM problem which could be of theoretical as well as practical interest. We summarize all the main results (including that of prior work) in Table 1. There (⋆) denotes the results from [25] and (⋆⋆) denotes the results from [18]. All the hardness results in the table except for complete graph hold even for fully homogeneous BNPG game. We observe that the PSNE existence problem admits efficient algorithm for many settings compared to ANM. This seems to indicate that enforcing a PSNE is computationally more difficult than finding a PSNE.

1.2 Related Work

Our work is related to [26] who initiated the study of computing a PSNE in BNPG games. Their results were strengthened by [18] who studied the parameterized complexity of deciding the existence of PSNE in BNPG game. Recently, [23] studied about public goods games in directed networks and showed intractibility for deciding the existence of PSNE and for finding MSNE. Our work is also related to [25] who intiated the study of Altruistic Network Modification in BNPG game. [16,22] also discussed different ways to capture altruism. In the non-altruistic setting, [14] worked on modifying networks to induce certain classes of equilibria in BNPG game. Our work is part of graphical games where the fundamental question is to determine the complexity of finding an equilibrium [8,9,13]. Our model is also related to the best-shot games [7] as it is a special case of BNPG game. [11,15,17,21] also discussed some important variations of graphical games.

2 Preliminaries

Let $[n]$ denote the set $\{1, \ldots, n\}$. Let $\mathcal{G} = (\mathcal{V}, \mathcal{E})$ be an input network with n vertices (each denoting a player). The input network is always an undirected graph. Let $\mathcal{H} = (\mathcal{V}, \mathcal{E}')$ be an altruistic network on the same set of n vertices. The altruistic network can be directed or undirected graph. An undirected edge between $u, v \in \mathcal{V}$ is represented by $\{u, v\}$. Similarly a directed edge from u to v is represented by (u, v). N_v is the set of all neighbours (resp. out-neighbours) of the vertex v in an undirected (resp. directed) altrusitic network $\mathcal{H}$. Note that N_v is a subset of neightbours of v in $\mathcal{G}$. A Binary Networked Public Goods (BNPG) game with asymmetric (resp. symmetric) altruism can be defined on the input graph $\mathcal{G}$ and the directed (resp. undirected) altruistic network $\mathcal{H}$ as follows. We are given a set of players $\mathcal{V}$, and the strategy set of every player in $\mathcal{V}$ is $\{0, 1\}$. For a strategy profile $\mathbf{x} = (x_v)_{v \in \mathcal{V}} \in \{0,1\}^{|\mathcal{V}|}$, let $n_v = |\{u \in \mathcal{V} : \{u,v\} \in \mathcal{E}, x_u = 1\}|$. In this paper, we will be using playing 1 (resp. 0), investing (resp. not investing) and strategy $x_v = 1$ (resp. $x_v = 0$) interchangeably. Now the utility $U_v(\mathbf{x})$ of player $v \in \mathcal{V}$ is defined as follows.

$$U_v(\mathbf{x}) = g_v(x_v + n_v) + a \sum_{u \in N_v} g_u(x_u + n_u) - c_v \cdot x_v$$

where $g_v : \mathbb{N} \cup \{0\} \to \mathbb{R}^+ \cup \{0\}$ is a non-decreasing externality function in x and $a, c_v \in \mathbb{R}^+ \cup \{0\}$ are constants. c_v can also interpreted as the cost of investing for player v. We denote a BNPG game with altruism by $(\mathcal{G} = (\mathcal{V}, \mathcal{E}), \mathcal{H} = (\mathcal{V}, \mathcal{E}'), (g_v)_{v \in \mathcal{V}}, (c_v)_{v \in \mathcal{V}}, a)$. We also define $\Delta g(x) = g(x+1) - g(x)$ where $x \in \mathbb{N} \cup \{0\}$. In this paper, we study a general case of BNPG game called *heterogeneous* BNPG game where every player $v \in \mathcal{V}$ need not have the same externality function $g_v(.)$ and constant c_v. If nothing is mentioned, by BNPG game, we are referring to a heterogeneous BNPG game. In this paper, we also study a special case of BNPG game called *fully homogeneous* BNPG game where $g_v = g$ for all $v \in \mathcal{V}$ and $c_v = c$ for all $v \in \mathcal{V}$.

In this paper, we mainly focus on *pure-strategy Nash Equilibrium (PSNE)*. A strategy profile $\mathbf{x} = (x_v)_{v \in \mathcal{V}}$ is said to be a PSNE of a BNPG game with altruism if the following holds true for all $v \in \mathcal{V}$ and for all $x'_v \in \{0,1\}$

$$U_v(x_v, x_{-v}) \geqslant U_v(x'_v, x_{-v})$$

where $x_{-v} = (x_u)_{u \in \mathcal{V} \setminus \{v\}}$.

In this paper, we also look at ε-*Nash Equilibrium*. Let Δ_v be a distribution over that strategy set $\{0,1\}$. We define $\mathrm{Supp}(\Delta_v)$ to be the support of the distribution Δ_v, that is, $\mathrm{Supp}(\Delta_v) = \{x_v : x_v \in \{0,1\}, \Delta_v(x_v) > 0\}$ where $\Delta_v(x_v)$ denotes the probability of choosing the strategy x_v by player v. Now $(\Delta_v)_{v \in \mathcal{V}}$ is an ε-*Nash Equilibrium* if the following holds true for all $x'_v \in \{0,1\}$, for all $x_v \in \mathrm{Supp}(\Delta_v)$, for all $v \in \mathcal{V}$:

$$\mathbb{E}_{x_{-v} \sim \Delta_{-v}}[U_v(x_v, x_{-v})] \geqslant \mathbb{E}_{x_{-v} \sim \Delta_{-v}}[U_v(x'_v, x_{-v})] - \varepsilon$$

where $\Delta_{-v} = (\Delta_u)_{u \in \mathcal{V} \setminus \{v\}}$.

2.1 Altruistic Network Modification

In this paper we study a special case of Altruistic Network Modification (ANM) which was also studied by [25]. If nothing is mentioned, by ANM, we are referring to the special case which we will now discuss. We are given a target profile $\mathbf{x}^*$, BNPG game on an input graph $\mathcal{G}$, an initial altruistic network $\mathcal{H}$, a cost function $C(.)$ and budget B. In this setting, we can add or delete an edge e from $\mathcal{H}$ and each such operation has a non-negative cost $C(e)$ associated with it. We denote an instance of ANM with altruism by $(\mathcal{G} = (\mathcal{V}, \mathcal{E}), \mathcal{H} = (\mathcal{V}, \mathcal{E}'), (g_v)_{v \in \mathcal{V}}, (c_v)_{v \in \mathcal{V}}, a, C(.), B, \mathbf{x}^*)$. The aim of ANM with altruism is to add and delete edges in $\mathcal{H}$ such that $\mathbf{x}^*$ becomes a PSNE and the total cost for adding and deleting these edges is atmost B. Note that if the altruism is asymmetric (resp. symmetric), then we can add or delete directed (resp. undirected) edges only. We are not allowed to add any edge between two nodes u, v if $\{u, v\} \notin \mathcal{E}$.

2.2   Standard Definitions

Definition 1 (Circuit Rank). *[18] Let the number of edges and number of vertices in a graph $\mathcal{G}$ be m and n respectively. Then circuit rank is defined to be $m - n + c$ (c is the number of connected components in the graph). Note that circuit rank is not the same as feedback arc set.*

Definition 2 (FPT). *[19] A tuple (x, k), where k is the parameter, is an instance of a parameterized problem. Fixed parameter tractability (FPT) refers to solvability in time $f(k) \cdot p(|x|)$ for a given instance (x, k), where p is a polynomial in the input size $|x|$ and f is an arbitrary computable function of k.*

Definition 3 (Para-NP-Hard). *[19] We say a parameterized problem is para-NP-hard if it is NP-hard even for some constant values of the parameter.*

3   Results for Computing Equilibrium

In this section, we present the results for deciding the existence of PSNE and finding MSNE in BNPG game with altruism. Due to space constraints, we have omitted few proofs. They are marked by $(\star)$ and they are available in the full version [20].

[26] showed that the problem of checking the existence of PSNE in BNPG game without altruism is polynomial time solvable when the input network is a tree. We now provide a non-trivial algorithm to show that the problem of checking the existence of PSNE in BNPG game with asymmetric altruism is polynomial time solvable when the input network is a tree.

Theorem 1. *The problem of checking the existence of PSNE in BNPG game with asymmetric altruism is polynomial time solvable when the input network is a tree.*

Proof. For each player $v \in \mathcal{V}$, let d_v denote the degree of v. At each node v with parent u, we maintain a table of tuples (x_u, n_u, x_v, n_v) of valid configurations. A tuple (x_u, n_u, x_v, n_v) is said to be a valid configuration if there exists a strategy profile $\mathbf{x}' = (x'_v)_{v \in \mathcal{V}}$ such that the following holds true:

- $x'_u = x_u, x'_v = x_v$
- The number of neighbours of u and v playing 1 in $\mathbf{x}'$ is n_u and n_v respectively
- None of the players in the sub-tree rooted at v deviate from their strategy in $\mathbf{x}'$

Note that the root node r doesn't have a parent. Hence, we consider an imaginary parent p with $x_p = 0$ and $g_p(x) = 0$ for all $x \geqslant 0$. Hence if there is a tuple in the table of root node r then we can conclude that there is a PSNE otherwise we can conclude that there is no PSNE.

Leaf Nodes: We add a tuple (x_u, n_u, x_v, n_v) to the table if $n_v = x_u$, v does not deviate if it plays x_v and $x_v \leqslant n_u \leqslant d_u + x_v - 1$. Table for the leaf node can be clearly constructed in polynomial time.

Non-leaf Nodes: For each tuple (x_u, n_u, x_v, n_v) we do the following. If there is no child u' of v having a tuple of type $(x_v, n_v, x_{u'}, n_{u'})$ in its table, then we don't add (x_u, n_u, x_v, n_v) to the table of v. Similarly if $n_u > d_u + x_v - 1$ or $n_u < x_v$, then we don't add (x_u, n_u, x_v, n_v) to the table of v. Otherwise we do the following. Let U_1 be the set of children u' of v which have tuples of the type $(x_v, n_v, 1, n_{u'})$ in their table but don't have tuples of the type $(x_v, n_v, 0, n_{u'})$. Let U_0 be the set of children u' of v which have tuples of the type $(x_v, n_v, 0, n_{u'})$ in their table but don't have tuples of the type $(x_v, n_v, 1, n_{u'})$. Let U be the set of children u' of v which have tuples of the type $(x_v, n_v, 1, n_{u'})$ and $(x_v, n_v, 0, n_{u'})$ in their table. First let us consider the case when $x_v = 1$. Now for each $u' \in (U_1 \cup U) \cap N_v$, we find the tuple $(1, n_v, 1, n_{u'})$ in its table so that $a \cdot \Delta g_{u'}(n_{u'})$ is maximized and let this value be $y_{u'}$. Similarly for each $u' \in (U_0 \cup U) \cap N_v$, we find the tuple $(1, n_v, 0, n_{u'})$ in the table so that $a \cdot \Delta g_{u'}(n_{u'} - 1)$ is maximized and let this value be $z_{u'}$. Also $\forall u' \in (U_1 \cup U_0 \cup U) \setminus N_v$, $y_{u'} = z_{u'} = 0$. If $u \in N_v$ then $y_u = a \cdot \Delta g_u(x_u + n_u - 1)$ otherwise $y_u = 0$. Now we include the tuple $(x_u, n_u, x_v = 1, n_v)$ in the table if the optimal value of the following ILP is at least $c_v - \Delta g_v(n_v) - y_u - \sum_{u' \in U_1} y_{u'} - \sum_{u' \in U_0} z_{u'}$.

$$\max \quad \sum_{u' \in U} (x_1^{u'} y_{u'} + x_0^{u'} z_{u'})$$

$$\text{s.t.} \quad x_1^{u'} + x_0^{u'} = 1 \quad \forall u' \in U$$

$$\sum_{u' \in U} x_1^{u'} = n_v - x_u - |U_1|$$

$$x_1^{u'}, x_0^{u'} \in \{0, 1\} \quad \forall u' \in U$$

The above ILP can be solved in polynomial time as follows. First sort the values $a_{u'} = |y_{u'} - z_{u'}|$ in non-increasing order breaking ties arbitrarily and order the vertices in U as $\{u_1, \ldots, u_{|U|}\}$ as per this order, that is, $a_{u_i} \geqslant a_{u_j}$ if $i \leqslant j$. Then we traverse the list of values in non-increasing order and for the u' corresponding to the value, we choose $x_1^{u'} = 1$ if $y_{u'} > z_{u'}$ otherwise we choose $x_0^{u'} = 1$. We do this until $\sum_{u' \in U} x_1^{u'} = n_v'$ or $\sum_{u' \in U} x_0^{u'} = |U| - n_v'$ where $n_v' = n_v - x_u - |U_1|$. Remaining values are chosen in a way such that $\sum_{u' \in U} x_1^{u'} = n_v$ is satisfied. For a more detailed description, please see the Algorithm 1.

Now we show the correctness. Consider an optimal solution x^*. Let i be the smallest number such that $x_1^{u_i} = 1$ as per our algorithm and in optimal solution it is 0. Similarly let i' be the smallest number such that $x_1^{u_{i'}} = 0$ as per our algorithm and in optimal solution it is 1. We now swap the values of the variables $x_1^{u_i}$ and $x_1^{u_{i'}}$ (resp. $x_0^{u_i}$ and $x_0^{u_{i'}}$) in the optimal solution without decreasing the value of the objective function. Let us assume that $i < i'$. Then it must be the case that $y_{u_i} > z_{u_i}$ otherwise $\forall j > i$, we will have $x_1^{u_j} = 1$ as per our algorithm. Hence by swapping in the optimal solution, the value of the objective function increases by at least $a_{u_i} - a_{u_{i'}}$, which is a non-negative quantity. Similarly when $i' < i$, it must be the case that $y_{u_i} \leqslant z_{u_i}$ otherwise $\forall j > i$, we will have $x_1^{u_j} = 0$ as per our algorithm. Hence by swapping in the optimal solution, the value of the objective function increases by at least $a_{u_{i'}} - a_{u_i}$ which is a non-negative quantity. By repeatedtly finding such indices i, i' and then swaping the value of $x_1^{u_i}$ and $x_1^{u_{i'}}$ (resp. $x_0^{u_i}$ and $x_0^{u_{i'}}$) in the optimal solution leads to our solution.

An analogous procedure exists for the case where $x_v = 0$. For each $u' \in (U_1 \cup U) \cap N_v$, we find the tuple $(0, n_v, 1, n_{u'})$ in the table so that $a \cdot \Delta g_{u'}(n_{u'} + 1)$ is minimized and let this value be $y_{u'}$. Similarly for each $u' \in (U_0 \cup U) \cap N_v$, we find the tuple $(1, n_v, 0, n_{u'})$ in the table so that $a \cdot \Delta g_{u'}(n_{u'})$ is minimized and let this value be $z_{u'}$. Also $\forall u' \in (U_1 \cup U_0 \cup U) \setminus N_v$, $y_{u'} = z_{u'} = 0$. If $u \in N_v$ then $y_u = a \cdot \Delta g_u(x_u + n_u)$ otherwise $y_u = 0$. Now we include the tuple $(x_u, n_u, x_v = 0, n_v)$ in the table if the optimal value of the following ILP is at most $c_v - \Delta g_v(n_v) - y_u - \sum_{u' \in U_1} y_{u'} - \sum_{u' \in U_0} z_{u'}$.

$$\min \quad \sum_{u' \in U} (x_1^{u'} y_{u'} + x_0^{u'} z_{u'})$$

$$\text{s.t.} \quad x_1^{u'} + x_0^{u'} = 1 \quad \forall u' \in U$$

$$\sum_{u' \in U} x_1^{u'} = n_v - x_u - |U_1|$$

$$x_1^{u'}, x_0^{u'} \in \{0, 1\} \quad \forall u' \in U$$

The above ILP can be solved in polynomial time as follows. First sort the values $a_{u'} = |y_{u'} - z_{u'}|$ in non-increasing order breaking ties arbitrarily and order the vertices in U as $\{u_1, \ldots, u_{|U|}\}$ as per this order, that is, $a_{u_i} \geqslant a_{u_j}$ if $i \leqslant j$. Then we traverse the list in non-increasing order and for the u' corresponding to the value, we choose $x_0^{u'} = 1$ if $y_{u'} > z_{u'}$ otherwise we choose $x_1^{u'} = 1$. We do this until $\sum_{u' \in U} x_1^{u'} = n_v'$ or $\sum_{u' \in U} x_0^{u'} = |U| - n_v'$ where $n_v' = n_v - x_u - |U_1|$. Remaining values are chosen in a way such that $\sum_{u' \in U} x_1^{u'} = n_v$ is satisfied.

Now we show the correctness. Consider an optimal solution x^*. Let i be the smallest number such that $x_1^{u_i} = 1$ as per our algorithm and in optimal solution it is 0. Similarly let i' be the smallest number such that $x_1^{u_{i'}} = 0$ as per our algorithm and in optimal solution it is 1. We now swap the values of the variables $x_1^{u_i}$ and $x_1^{u_{i'}}$ (resp. $x_0^{u_i}$ and $x_0^{u_{i'}}$) in the optimal solution without increasing the value of the objective function. Let us assume that $i < i'$. Then it must be the case that $y_{u_i} \leqslant z_{u_i}$ otherwise $\forall j > i$, we will have $x_1^{u_j} = 1$ as per our algorithm. Hence by swapping in the optimal solution, the value of the objective function decreases by at least $a_{u_i} - a_{u_{i'}}$, which is a non-negative quantity. Similarly when $i' < i$, it must be the case that $y_{u_i} > z_{u_i}$ otherwise $\forall j > i$, we will have $x_1^{u_j} = 0$ as per our algorithm. Hence by swapping in the optimal solution, the value of the objective function increases by at least $a_{u_{i'}} - a_{u_i}$ which is a non-negative quantity. By repeatedtly finding such indices i, i' and then swapping the value of $x_1^{u_i}$ and $x_1^{u_{i'}}$ (resp. $x_0^{u_i}$ and $x_0^{u_{i'}}$) in the optimal solution leads to our solution.

As mentioned earlier if there is any tuple in the table of the root r, then we conclude that there is a PSNE otherwise we conclude that there is no such PSNE.

Algorithm 1. ILP Solver

1: $\forall u' \in U$, $a_{u'} \leftarrow |y_{u'} - z_{u'}|$.
2: Order the vertices in U as $\{u_1, \ldots, u_{|U|}\}$ such that $\forall i, j \in [n]$, we have $a_{u_i} \geqslant a_{u_j}$ if $i \leqslant j$.
3: $\forall u' \in U$, $x_0^{u'} \leftarrow 0$ and $x_1^{u'} \leftarrow 0$
4: $n_v' \leftarrow n_v - x_u - |U_1|$
5: **for** $i = 1$ to $|U|$ **do**
6:    **if** $\sum_{u' \in U} x_1^{u'} = n_v'$ **then**
7:       $x_0^{u_i} \leftarrow 1$ and $x_1^{u_i} \leftarrow 0$
8:    **else if** $\sum_{u' \in U} x_0^{u'} = |U| - n_v'$ **then**
9:       $x_0^{u_i} \leftarrow 0$ and $x_1^{u_i} \leftarrow 1$
10:    **else**
11:       **if** $y_{u_i} > z_{u_i}$ **then**
12:          $x_0^{u_i} \leftarrow 0$ and $x_1^{u_i} \leftarrow 1$
13:       **else**
14:          $x_0^{u_i} \leftarrow 1$ and $x_1^{u_i} \leftarrow 0$
15:       **end if**
16:    **end if**
17: **end for**

Corollary 1. *Given a BNPG game with asymmetric altruism on a tree $\mathcal{G} = (\mathcal{V}, \mathcal{E})$, a set $S \subseteq \mathcal{V}$ and a pair of tuples $(x_v')_{v \in S} \in \{0,1\}^{|S|}$ and $(n_v')_{v \in \mathcal{V}'} \in \{0, 1, \ldots, n\}^{|S|}$, we can decide in polynomial time if there exists a PSNE $\mathbf{x}^* = (x_v)_{v \in \mathcal{V}} \in \{0,1\}^n$ for the BNPG game with asymmetric altruism such that $x_v = x_v'$ for every $v \in S$ and the number of neighbors of v playing 1 in $\mathbf{x}^*$ is n_v' for every $v \in S$.*

Proof. In the proof of Theorem 1, just discard those entries from the table of $u \in S$ which don't have $x_u = x_u'$ and $n_u = n_u'$.

[18] showed that the problem of checking the existence of PSNE in BNPG game without altruism is polynomial time solvable when the input network is a graph with bounded circuit rank. By using the algorithm in Theorem 1 as a subroutine and extending the ideas of [18] to our setting, we show the following.

Theorem 2 ($\star$). *The problem of checking the existence of PSNE in BNPG game with asymmetric altruism is polynomial time solvable when the input network is a graph with bounded circuit rank.*

[18,26] showed that the problem of checking the existence of PSNE in BNPG game without altruism is polynomial time solvable when the input network is a complete graph. By extending their ideas to our setting, we show the following.

Theorem 3 ($\star$). *The problem of checking the existence of PSNE in BNPG game with asymmetric altruism is polynomial time solvable when the input network is a complete graph.*

The problem of deciding the existence of BNPG game with altruism where the atruistic network is empty is known to be NP-Complete [26]. Therefore, we look at the deciding the complexity of finding an ε-Nash equilibrium in BNPG game with symmetric altruism. We show that it is PPAD-Hard. Towards that, we reduce from an instance of Directed public goods game which is known to be PPAD-hard [23]. In directed public goods game, we are given a directed network of players and the utility $U'_u(x_v, x_{-v})$ of a player v is $Y(x_v + n_v^{in}) - p \cdot x_v$. Here $x_v \in \{0,1\}$, n_v^{in} is the number of in-neighbours of v playing 1 and $Y(x) = 0$ if $x = 0$ and $Y(x) = 1$ if $x > 0$.

Theorem 4. *Finding an ε-Nash equilibrium of the BNPG game with symmetric altruism is PPAD-hard, for some constant $\varepsilon > 0$.*

Proof. Let $(\mathcal{G}(\mathcal{V}, \mathcal{E}), p)$ be an input instance of directed public goods game. Now we create an instance $(\mathcal{G}' = (\mathcal{V}', \mathcal{E}'), \mathcal{H} = (\mathcal{V}', \mathcal{E}''), (g_v)_{v \in \mathcal{V}'}, (c_v)_{v \in \mathcal{V}'}, a)$ of BNPG game with symmetric altruism.

$$\mathcal{V}' = \{u_{in} : u \in \mathcal{V}\} \cup \{u_{out} : u \in \mathcal{V}\}$$
$$\mathcal{E}' = \{\{u_{in}, u_{out}\} : u \in \mathcal{V}'\} \cup \{\{u_{out}, v_{in}\} : (u, v) \in \mathcal{E}\}$$
$$\mathcal{E}'' = \{\{u_{in}, u_{out}\} : u \in \mathcal{V}'\}$$

Let the constant a be 1. $\forall u \in \mathcal{V}$, $c_{u_{in}} = 1 + 2\varepsilon$ and $c_{u_{out}} = p$. Now define the functions $g_w(.)$ as follows:

$$\forall u \in \mathcal{V}, g_{u_{in}}(x) = \begin{cases} 1 & x > 0 \\ 0 & \text{otherwise} \end{cases}$$

$$\forall u \in \mathcal{V}, g_{u_{out}}(x) = 0 \ \forall x \geqslant 0$$

Now we show that given any ε-Nash equilibrium of the BNPG game with altruism, we can find an ε-Nash equilibrium of the directed public goods game in polynomial time. Let $(\Delta_u)_{u \in \mathcal{V}'}$ be an ε-Nash equilibrium of the BNPG game with symmetric altruism.

For all $v \in \{u_{in} : u \in \mathcal{V}\}$, we have the following:

$$\mathbb{E}_{x_{-v} \sim \Delta_{-v}}[U_v(0, x_{-v})] - \varepsilon \geqslant -\varepsilon > -2\varepsilon = 1 - c_{u_{in}} = \mathbb{E}_{x_{-v} \sim \Delta_{-v}}[U_v(1, x_{-v})]$$

Hence 1 can't be in the support of Δ_v. Therefore $\forall u \in \mathcal{V}$, $\Delta_{u_{in}}(0) = 1$.

Now we show that $(\Delta_u)_{u \in \mathcal{V}}$ is an ε-Nash equilibrium of the directed public goods game where $\Delta_u = \Delta_{u_{out}} \ \forall u \in \mathcal{V}$. Now consider a strategy profile $(x_v)_{v \in \mathcal{V}'}$

such that $\forall u \in \mathcal{V}$, we have $x_{u_{in}} = 0$. Now let $(x_v)_{v \in \mathcal{V}}$ be a strategy profile such that $\forall u \in \mathcal{V}$ we have $x_u = x_{u_{out}}$. Then we have the following:

$$U_{v_{out}}(x_{v_{out}}, x_{-v_{out}}) = g_{v_{in}}(n_{v_{in}}) - p \cdot x_{v_{out}} = Y(x_v + n_v^{in}) - p \cdot x_v = U'_u(x_u, x_{-u})$$

Using the above equality and the fact that $\forall u \in \mathcal{V}$, $\Delta_{u_{in}}(0) = 1$, we have $\mathbb{E}_{x_{-u} \sim \Delta_{-u}}[U'_u(x_u, x_{-u})] = \mathbb{E}_{x_{-u_{out}} \sim \Delta_{-u_{out}}}[U_{u_{out}}(x_{u_{out}}, x_{-u_{out}})]$ where $x_u = x_{u_{out}}$. For all $u \in \mathcal{V}$, for all $x'_u \in \{0, 1\}$, for all $x_u \in \mathrm{Supp}(\Delta_u)$, we have the following:

$$\begin{aligned} \mathbb{E}_{x_{-u} \sim \Delta_{-u}}[U'_u(x_u, x_{-u})] &= \mathbb{E}_{x_{-u_{out}} \sim \Delta_{-u_{out}}}[U_{u_{out}}(x_{u_{out}} = x_u, x_{-u_{out}})] \\ &\geqslant \mathbb{E}_{x_{-u_{out}} \sim \Delta_{-u_{out}}}[U_{u_{out}}(x'_{u_{out}} = x'_u, x_{-u_{out}})] - \varepsilon \\ &\geqslant \mathbb{E}_{x_{-u} \sim \Delta_{-u}}[U_u(x'_u, x_{-u})] - \varepsilon \end{aligned}$$

Hence, given any ε-Nash equilibrium of the BNPG game with symmetric altruism , we can find an ε-Nash equilibrium of the directed public goods game in polynomial time. This concludes the proof of this theorem.

4 Results for Altruistic Network Modification

In this section, we present the results for Altruistic Network Modification. First let us call ANM with altruism as heterogeneous ANM with altruism whenever the BNPG game is heterogeneous. Similarly let us call ANM with altruism as fully homogeneous ANM whenever the BNPG game is fully homogeneous. [18] depicted a way to reduce heterogeneous BNPG game to fully homogeneous BNPG game. By extending their ideas to our setting, we show Lemmata 1 to 3 which will be helpful to prove the theorems on hardness in this section.

Lemma 1 ($\star$). *Given an instance of heterogeneous ANM with asymmetric altruism such that cost c_v of investing is same for all players v in the heterogeneous BNPG game, we can reduce the instance heterogeneous ANM with asymmetric altruism to an instance of fully homogeneous ANM with asymmetric altruism.*

Lemma 2 ($\star$). *Given an instance of heterogeneous ANM with symmetric altruism such that cost c_v of investing is same for all players v in the heterogeneous BNPG game, we can reduce the instance heterogeneous ANM with symmetric altruism to an instance of fully homogeneous ANM with symmetric altruism.*

Lemma 3 $(\star)$. *Given an instance of heterogeneous ANM with symmetric altruism such that input network has maximum degree 3, cost c_v of investing is same for all players v in the heterogeneous BNPG game and there are three types of externality functions, we can reduce the instance heterogeneous ANM with symmetric altruism to an instance of fully homogeneous ANM with symmetric altruism such that the input network has maximum degree 13.*

ANM with asymmetric altruism is known to be NP-complete when the input network is a clique [25]. We show a similar result for trees by reducing from Knapsack problem.

Theorem 5 $(\star)$. *For the target profile where all players invest, ANM with asymmetric altruism is NP-complete when the input network is a tree and the BNPG game is fully homogeneous.*

ANM with symmetric altruism is known to be NP-complete when the input network is a clique [25]. We show a similar result for trees by reducing from Knapsack problem.

Theorem 6 $(\star)$. *For the target profile where all players invest, ANM with symmetric altruism is NP-complete when the input network is a tree and the BNPG game is fully homogeneous.*

We now show that ANM with symmetric altruism is known to be para-NP-hard for the parameter maximum degree of the input network even when the BNPG game is fully homogeneous. Towards that, we reduce from an instance of $(3, B2)$-SAT which is known to be NP-complete [1]. $(3, B2)$-SAT is the special case of 3-SAT where each variable x_i occurs exactly twice as negative literal $\bar{x}_i$ and twice as positive literal x_i.

Theorem 7 $(\star)$. *For the target profile where all players invest, ANM with symmetric altruism is known to be para-NP-hard for the parameter maximum degree of the input network even when the BNPG game is fully homogeneous.*

We complement the previous result by showing that ANM with asymmetric altruism is FPT for the parameter maximum degree of the input graph.

Theorem 8. *For any target profile, ANM with asymmetric altruism can be solved in time $2^{\Delta/2} \cdot n^{O(1)}$ where Δ is the maximum degree of the input graph.*

Proof. [25] showed that solving an instance of asymmetric altruistic design is equivalent to solving n different instances of Minimum Knapsack problem and each of these instances have at most $\Delta + 1$ items. In Minimum Knapsack problem, we are give a set of items $1, \ldots, k$ with costs $p_1, \ldots p_k$ and weights $w_1, \ldots w_k$. The aim is to find a subset S of items minimizing $\sum_{i \in S} w_i$ subject to the constraint that $\sum_{i \in S} p_i \geqslant P$. We assume that $\sum_{i \in [k]} p_i \geqslant P$ otherwise we don't have any feasible solution. Let us denote the optimal value by OPT. Let $W := \sum_{i \in [k]} w_i$. Now consider the following integer linear program which we denote by ILP_w:

$$\max \quad \sum_{i \in [k]} x_i p_i$$

$$\text{s.t.} \quad \sum_{i \in [k]} x_i w_i \leqslant w, \ x_i \in \{0, 1\} \quad \forall i \in [k]$$

The above integer linear program can be solved in time $2^{k/2} \cdot k^{O(1)}$ [10]. Now observe that for all $w \geqslant OPT$, the optimal value OPT_w of ILP_w is at least P. Similarly for all $w < OPT$, the optimal value OPT_w of the ILP_w is less than P. Now by performing a binary search for w on the range $[0, W]$ and then solving the above ILP repeatedly, we can compute OPT in time $2^{k/2} \cdot \log W \cdot k^{O(1)}$. See Algorithm 2 for more details.

As discussed earlier, an Instance of asymmetric altruistic design is equivalent to solving n different instances of Minimum Knapsack problem and each of these instances have at most $\Delta + 1$ items. Hence ANM with asymmetric altruism can be solved in time $2^{\Delta/2} \cdot |x|^{O(1)}$ where x is the input instance of ANM with asymmetric altruism.

Algorithm 2. Minimum Knapsack Solver

1: $\ell \leftarrow 0, r \leftarrow W, w \leftarrow \lfloor \frac{\ell + r}{2} \rfloor$
2: **while true do**
3:    Solve ILP_w and ILP_{w+1}.
4:    **if** $OPT_w < P$ and $OPT_{w+1} \geqslant P$ **then**
5:       **return** $w + 1$
6:    **else if** $OPT_w < P$ and $OPT_{w+1} < P$ **then**
7:       $\ell \leftarrow w + 1$
8:       $w \leftarrow \lfloor \frac{\ell + r}{2} \rfloor$
9:    **else**
10:     $r \leftarrow w$
11:     $w \leftarrow \lfloor \frac{\ell + r}{2} \rfloor$
12:    **end if**
13: **end while**

We conclude our work by discussing about the approxibimility of ANM with symmetric altruism. [25] showed a $2 + \varepsilon$ approximation algorithm for ANM with symmetric altruism when the target profile has all players investing. However, for arbitrary target profile they showed that ANM with symmetric altruism is NP-complete when the input network is a complete graph and the budget is infinite. We show a similar result for graphs with bounded degree by reducing from (3, B2)-SAT.

Theorem 9 (⋆). *For an arbitrary target profile, ANM with symmetric altruism is known to be para-NP-hard for the parameter maximum degree of the input network even when the BNPG game is fully homogeneous and the budget is infinite.*

5 Conclusion and Future Work

In this paper, we first studied the problem of deciding the existence of PSNE in BNPG game with altruism. We depicted polynomial time algorithms to decide the existence of PSNE in trees, complete graphs and graphs with bounded We also that the problem of finding MSNE in BNPG game with altruism is PPAD-Hard. Next we studied Altruistic Network modification. We showed that ANM with either symmetric or asymmetric altruism is NP-complete for trees. We also showed that ANM with symmetric altruism is para-NP-hard for the parameter maximum degree whereas ANM with asymetric altruism is FPT for the parameter maximum degree. One important research direction in ANM is to maximize the social welfare while ensuring that the target profile remains a PSNE. Another research direction is to improve the approximation algorithms of [25] for ANM with asymmetric altruism for trees and graphs with bounded degree. Another interesting future work is to look at other graphical games by considering altruism.

References

1. Berman, P., Karpinski, M., Scott, A.: Approximation Hardness of Short Symmetric Instances of Max-3sat. Tech. Rep. (2004)
2. Bir, C., Widmar, N.O.: Social pressure, altruism, free-riding, and non-compliance in mask wearing by us residents in response to covid-19 pandemic. Soc. Sci. Human. Open 4(1), 100229 (2021)
3. Brito, D.L., Sheshinski, E., Intriligator, M.D.: Externalities and compulsary vaccinations. J. Public Econ. 45(1), 69–90 (1991)
4. Buchwald, E.: Why do so many Americans refuse to wear face masks? Politics is part of it - but only part (2020). https://www.marketwatch.com/story/why-do-so-many-americans-refuse-to-wear-face-masks-it-may-have-nothing-to-do-with-politics-2020-06-16
5. Cato, S., Iida, T., Ishida, K., Ito, A., McElwain, K.M., Shoji, M.: Social distancing as a public good under the covid-19 pandemic. Public Health 188, 51–53 (2020)
6. Cucciniello, M., Pin, P., Imre, B., Porumbescu, G.A., Melegaro, A.: Altruism and vaccination intentions: evidence from behavioral experiments. Soc. Sci. Med. 292, 114195 (2022). https://doi.org/10.1016/j.socscimed.2021.114195
7. Dall'Asta, L., Pin, P., Ramezanpour, A.: Optimal equilibria of the best shot game. J. Publ. Econ. Theory 13(6), 885–901 (2011)
8. Daskalakis, C., Goldberg, P.W., Papadimitriou, C.H.: The complexity of computing a Nash equilibrium. SIAM J. Comput. 39(1), 195–259 (2009)
9. Elkind, E., Goldberg, L.A., Goldberg, P.: Nash equilibria in graphical games on trees revisited. In: Proceedings of the 7th ACM Conference on Electronic Commerce, pp. 100–109 (2006)

10. Fomin, F.V., Kratsch, D.: Exact Exponential Algorithms. Springer, Heidelberg (2010)
11. Galeotti, A., Goyal, S., Jackson, M.O., Vega-Redondo, F., Yariv, L.: Network games. Rev. Econ. Stud. **77**(1), 218–244 (2010)
12. Ghosh, S.K.: Concept of public goods can be experimentally put to use in pandemic for a social cause (2020). https://indianexpress.com/article/opinion/columns/rbi-nobel-prize-in-economics-coronavirus-new-auction-/format-6779782/
13. Gottlob, G., Greco, G., Scarcello, F.: Pure nash equilibria: hard and easy games. J. Artif. Intell. Res. **24**, 357–406 (2005)
14. Kempe, D., Yu, S., Vorobeychik, Y.: Inducing equilibria in networked public goods games through network structure modification. International Foundation for Autonomous Agents and Multiagent Systems (AAMAS 2020), Richland, pp. 611–619 (2020)
15. Komarovsky, Z., Levit, V., Grinshpoun, T., Meisels, A.: Efficient equilibria in a public goods game. In: 2015 IEEE/WIC/ACM International Conference on Web Intelligence and Intelligent Agent Technology (WI-IAT), vol. 2, pp. 214–219. IEEE (2015)
16. Ledyard, J.O.: 2. Public Goods: A Survey of Experimental Research. Princeton University Press (2020)
17. Levit, V., Komarovsky, Z., Grinshpoun, T., Meisels, A.: Incentive-based search for efficient equilibria of the public goods game. Artif. Intell. **262**, 142–162 (2018)
18. Maiti, A., Dey, P.: On parameterized complexity of binary networked public goods game. arXiv preprint arXiv:2012.01880 (2020)
19. Maiti, A., Dey, P.: Parameterized algorithms for kidney exchange. arXiv preprint arXiv:2112.10250 (2021)
20. Maiti, A., Dey, P.: On binary networked public goods game with altruism. arXiv preprint arXiv:2205.00442 (2022)
21. Manshadi, V.H., Johari, R.: Supermodular network games. In: 2009 47th Annual Allerton Conference on Communication, Control, and Computing (Allerton), pp. 1369–1376. IEEE (2009)
22. Meier, D., Oswald, Y.A., Schmid, S., Wattenhofer, R.: On the windfall of friendship: inoculation strategies on social networks. In: Proceedings of the 9th ACM conference on Electronic Commerce, pp. 294–301 (2008)
23. Papadimitriou, C., Peng, B.: Public Goods Games in Directed Networks, pp. 745–762. Association for Computing Machinery, New York (2021). https://doi.org/10.1145/3465456.3467616
24. Wong, T.: Coronavirus: why some countries wear face masks and others don't (2020). https://www.bbc.com/news/world-52015486
25. Yu, S., Kempe, D., Vorobeychik, Y.: Altruism design in networked public goods games. In: Zhou, Z.H. (ed.) Proceedings of the Thirtieth International Joint Conference on Artificial Intelligence, IJCAI-21, pp. 493–499. International Joint Conferences on Artificial Intelligence Organization (2021). https://doi.org/10.24963/ijcai.2021/69
26. Yu, S., Zhou, K., Brantingham, P.J., Vorobeychik, Y.: Computing equilibria in binary networked public goods games. In: AAAI, pp. 2310–2317 (2020)

Proportional Fairness for Combinatorial Optimization

Minh Hieu Nguyen, Mourad Baiou, Viet Hung Nguyen$^{(\boxtimes)}$,
and Thi Quynh Trang Vo

INP Clermont Auvergne, Univ Clermont Auvergne, Mines Saint-Etienne, CNRS,
UMR 6158 LIMOS, 1 Rue de la Chebarde, Aubiere Cedex, France
viet_hung.nguyen@uca.fr

Abstract. Proportional fairness (PF) is a widely studied concept in the
literature, particularly in telecommunications, network design, resource
allocation, and social choice. It aims to distribute utilities to ensure
fairness and equity among users while optimizing system performance.
Under convexity, PF is equivalent to the *Nash bargaining solution*, a well-
known notion from cooperative game theory, and it can be obtained by
maximizing the product of the utilities. In contrast, when dealing with
non-convex optimization, PF is not guaranteed to exist; when it exists,
it is also the Nash bargaining solution. Consequently, finding PF under
non-convexity remains challenging since it is not equivalent to any known
optimization problem.

This paper deals with PF in the context of combinatorial optimization,
where the feasible set is discrete, finite, and non-convex. For this purpose,
we consider a general *Max-Max Bi-Objective Combinatorial Optimization*
(Max-Max BOCO) problem where its two objectives to be simultaneously maximized take only positive values. Then, we seek to find the
solution to achieving PF between two objectives, referred to as *proportional fair solution* (PF solution).

We first show that the PF solution, when it exists, can be obtained
by maximizing a suitable linear combination of two objectives. Subsequently, our main contribution lies in presenting an exact algorithm that
converges within a logarithmic number of iterations to determine the PF
solution. Finally, we provide computational results on the bi-objective
spanning tree problem, a specific example of Max-Max BOCO.

Keywords: Combinatorial Optimization · Bi-Objective Combinatorial
Optimization · Nash Bargaining Solution · Proportional Fairness

1 Introduction

Proportional fairness (PF) is a widely studied concept in the literature, particularly in telecommunications, network design, resource allocation, and social
choice. The goal of PF is to provide a compromise between the *utilitarian rule*
- which emphasizes overall system efficiency, and the *egalitarian rule* - which

J. A. Soto and A. Wiese (Eds.): LATIN 2024, LNCS 14579, pp. 304–319, 2024.
https://doi.org/10.1007/978-3-031-55601-2_20

emphasizes individual fairness. For example, in wireless communication systems, PF is often used to allocate transmission power, bandwidth, and data rates among mobile users to maximize overall system throughput while ensuring fair access for all users [1]. In network scheduling and traffic management, PF plays a significant role in packet scheduling algorithms. It helps ensure that packets from different flows or users are treated fairly, preventing any single user from dominating network resources [2]. In resource allocation, PF is applied to prevent resource starvation and improve system efficiency by dynamically distributing resources based on the individual demands of users, thereby enhancing overall system performance and ensuring a fair utilization of available resources [3].

The PF concept is proposed for multi-player problems where each player is represented by a utility function [4]. It has been demonstrated that if the feasible set is convex, PF is equivalent to the *Nash bargaining solution*, which always exists and can be obtained by maximizing the product of the utilities (or equivalently, by maximizing a logarithmic sum problem) [1,4,5]. In this case, finding the optimal solution is computationally expensive since the objective remains nonlinear. Thus, in practice, heuristic algorithms are often employed to find approximate solutions that achieve near-PF in some special scenarios (see, e.g., [2,6]). In contrast, when dealing with non-convex optimization, the existence of PF is not guaranteed, and if it exists, it is also the Nash bargaining solution [1,7]. To address such a case, popular approaches involve considering certain non-convex sets, which are convex after a logarithmic transformation [7]. An alternative approach is to introduce the concept of *local proportional fairness*, which is always achievable, and then analyze its properties [8].

This paper deals with PF in the context of combinatorial optimization, where the feasible set is discrete, finite, and non-convex. For this purpose, we consider a general *Max-Max Bi-Objective Combinatorial Optimization* (Max-Max BOCO) problem with two objectives to be simultaneously maximized. Notice that a general Max-Max BOCO can be considered a two-player problem with two utility functions. Then, we seek to find the solution achieving PF between two objectives, referred to as *proportional fair solution* (PF solution).

Some approaches using PF as a criterion for solving the bi-objective minimization problems have been introduced in [9–11]. In these scenarios, the solution achieved PF, referred to as the Nash Fairness (NF) solution, always exists and can be obtained by an iterative algorithm based on the application of the Newton-Raphson method. The NF solution is also generalized in bi-objective discrete optimization, where the objectives can be either maximized or minimized. As a result, the PF solution described in this paper represents a specific instance of the NF solution for Max-Max BOCO, as mentioned in [12]. However, it is important to note that in [12], the authors primarily introduced the concept of NF solution and focused on calculating the NF solution set for the cases when it always exists, and there may be many NF solutions (i.e., for Max-Min BOCO and Min-Min BOCO). Generally, determining the PF solution in combinatorial optimization remains a challenging question, especially when it is not guaranteed to exist and cannot be directly obtained by applying the Newton-Raphson method.

In this paper, we first show that the PF solution, when it exists, can be found by maximizing a suitable linear combination of two objectives. Then, our main contribution lies in presenting an exact algorithm that converges within a logarithmic (of fixed parameters depending on the data) number of iterations to determine efficiently the PF solution for Max-Max BOCO. Finally, we provide computational results on the Bi-Objective Spanning Tree Problem (BOSTP), a specific example of Max-Max BOCO. Notice that when the linear combination of two objectives can be solved in polynomial time, by the result in this paper, the PF solution, if it exists, can be determined in weakly polynomial time. To the best of our knowledge, this is the first approach in combinatorial optimization where an efficient algorithm has been developed for identifying PF. Furthermore, it can also be used in convex optimization.

The paper is organized as follows. In Sect. 2, we discuss the characterization of the PF solution for Max-Max BOCO. In Sect. 3, we propose an exact algorithm for finding the PF solution. Computational results on some instances of the BOSTP will be presented in Sect. 4. Finally, in Sect. 5, we give some conclusions and future works.

2   Characterization of the PF Solution

Since the objectives of Max-Max BOCO is to be simultaneously maximized, it can be formulated as

$$\max_{x \in \mathcal{X}} (P(x), Q(x)),$$

where $\mathcal{X}$ denotes the set of all feasible decision vectors x. For Max-Max BOCO, we suppose that $\mathcal{X}$ is a finite set and $P(x), Q(x) > 0, \forall x \in \mathcal{X}$.

Let $(P, Q) = (P(x), Q(x))$ denote the objective values corresponding to a decision vector $x \in \mathcal{X}$. Let $\mathcal{S}$ represent the set of pairs (P, Q) corresponding to all feasible decision vector solutions. This paper will characterize the feasible solutions for Max-Max BOCO using pairs (P, Q) instead of explicitly listing the decision vector solutions. Thus, two feasible solutions having the same values of (P, Q) will be considered equivalent. Throughout this paper, we use the notation "$\equiv$" to denote equivalent solutions. Since $\mathcal{X}$ is finite, the number of feasible solutions is also finite, implying that $\mathcal{S}$ is a finite set. For Max-Max BOCO, the PF solution (P^{PF}, Q^{PF}) should be such that, if compared to any other solution (P, Q), the aggregate proportional change is non-positive (see, e.g., [1,4]). Mathematically, we have

$$\frac{P - P^{PF}}{P^{PF}} + \frac{Q - Q^{PF}}{Q^{PF}} \leq 0 \iff \frac{P}{P^{PF}} + \frac{Q}{Q^{PF}} \leq 2, \ \forall (P, Q) \in \mathcal{S},$$

We recall that the PF solution does not always exist. For example, if $\mathcal{S}$ contains two feasible solutions $(P_1, Q_1) = (2, 2)$ and $(P_2, Q_2) = (1, 4)$ then none of them is PF solution since $P_1/P_2 + Q_1/Q_2 > 2$ and $P_2/P_1 + Q_2/Q_1 > 2$.

Proposition 1. *If* (P^{PF}, Q^{PF}) *is a PF solution for Max-Max BOCO, then it is the unique solution that maximizes the product PQ.*

Notice that the generalized version of Proposition 1 for non-convex optimization has been presented in [7]. In the following, we show that the PF solution, when it exists, can be obtained by maximizing a suitable linear combination of P and Q by considering the optimization problem:

$$\mathcal{F}(\alpha) = \max_{(P,Q) \in \mathcal{S}} f_\alpha(P, Q),$$

where $f_\alpha(P, Q) = P + \alpha Q$ and $\alpha \geq 0$ is a coefficient to be determined.

Notice that we assume the existence of the algorithms for maximizing the linear combinations of P and Q, including for maximizing P and Q.

Theorem 1. *[12]* $(P^{PF}, Q^{PF}) \in \mathcal{S}$ *is the PF solution if and only if* (P^{PF}, Q^{PF}) *is a solution of* $\mathcal{F}(\alpha^{PF})$ *with* $\alpha^{PF} = P^{PF}/Q^{PF}$.

As Theorem 1 provides a necessary and sufficient condition for the PF solution, the main question is how to propose an exact algorithm for determining the PF solution based on Theorem 1. We answer this question in the next section.

3 Algorithm for Determining the PF Solution

3.1 Algorithm Construction

In this section, we outline the idea of constructing an exact algorithm to determine the PF solution for Max-Max BOCO.

For a given $\alpha_k \geq 0$, let $T(\alpha_k) := P_k - \alpha_k Q_k$ where (P_k, Q_k) is a solution of $\mathcal{F}(\alpha_k)$. Throughout this paper, let (P^{PF}, Q^{PF}) denote the PF solution and α^{PF} denote the PF coefficient, a coefficient such that (P^{PF}, Q^{PF}) is a solution of $\mathcal{F}(\alpha^{PF})$ and $P^{PF} = \alpha^{PF} Q^{PF}$. We have $T(\alpha^{PF}) = P^{PF} - \alpha^{PF} Q^{PF} = 0$. According to Theorem 1 and the uniqueness of the PF solution when it exists, determining the PF solution is equivalent to determining the PF coefficient α^{PF}.

We first show the monotonic relationship between $\alpha \geq 0$ and the solution of $\mathcal{F}(\alpha)$ with respect to the values of P and Q. Consequently, we also deduce the monotonic relationship between α and $T(\alpha)$.

Lemma 1. *Given* $0 \leq \alpha' < \alpha''$ *and let* (P', Q'), $(P'', Q'') \in \mathcal{S}$ *be the solutions of* $\mathcal{F}(\alpha')$ *and* $\mathcal{F}(\alpha'')$, *respectively. Then* $P' \geq P''$ *and* $Q' \leq Q''$. *Moreover,* $T(\alpha') > T(\alpha'')$.

Proof. The optimality of (P', Q') and (P'', Q'') gives

$$P' + \alpha'Q' \geq P'' + \alpha'Q'', \text{ and} \tag{1a}$$

$$P'' + \alpha''Q'' \geq P' + \alpha''Q' \tag{1b}$$

Adding (1a) and (1b) gives $(\alpha' - \alpha'')(Q' - Q'') \geq 0$. Since $\alpha' < \alpha''$, $Q' \leq Q''$. On the other hand, the inequality (1a) implies $P' - P'' \geq \alpha'(Q'' - Q') \geq 0$. Since $P' \geq P''$, $Q' \leq Q''$ and $\alpha' < \alpha''$, we obtain $T(\alpha') = P' - \alpha'Q' \geq P'' - \alpha'Q'' > P'' - \alpha''Q'' = T(\alpha'')$. □

As a result of Lemma 1, if $\alpha' < \alpha''$ and (P', Q') is the solution of both $\mathcal{F}(\alpha')$ and $\mathcal{F}(\alpha'')$ then (P', Q') is the solution of $\mathcal{F}(\alpha)$ for all $\alpha' < \alpha < \alpha''$. Moreover, based on the monotonic relationship between α and $T(\alpha)$, for given $0 \leq \alpha_i < \alpha_j$ and $T(\alpha_i)T(\alpha_j) > 0$, we have $\alpha^{PF} \notin (\alpha_i, \alpha_j)$ because for $\alpha' \in (\alpha_i, \alpha_j)$ and an arbitrary solution (P', Q') of $\mathcal{F}(\alpha')$, $T(\alpha')$ has the same sign as $T(\alpha_i)$ and $T(\alpha_j)$ which implies $T(\alpha') \neq 0$ and then $\alpha' \neq \alpha^{PF}$.

Let α^{sup} be an upper bound of α^{PF} such that $\alpha^{PF} < \alpha^{sup}$ (we will provide a detailed definition for α^{sup} in our algorithm). According to the results of Theorem 1 and Lemma 1, the main idea of our algorithm is based on the binary search algorithm in the interval $[0, \alpha^{sup}]$. More precisely, we use Procedure $SEARCH()$ to identify the PF solution and the PF coefficient α^{PF} in such interval, ensuring that $T(\alpha^{PF}) = 0$. Starting from an interval $[\alpha_i, \alpha_j] \subseteq [0, \alpha^{sup}]$ with $T(\alpha_i) > 0$ and $T(\alpha_j) < 0$, Procedure $SEARCH()$ selects α_s as the midpoint of the interval $[\alpha_i, \alpha_j]$ and solve $\mathcal{F}(\alpha_s)$ to obtain a solution (P_s, Q_s). Then, we use Procedure $Verify\_PF\_sol()$ and Procedure $Verify\_PF\_coeff()$ to verify whether (P_s, Q_s) is the PF solution and whether α_s is the PF coefficient, respectively. If the verification is unsuccessful, the half-interval in which the PF coefficient cannot exist is eliminated, and we retain only one half-interval for further exploration within Procedure $SEARCH()$. The choice is made between $[\alpha_i, \alpha_s]$ and $[\alpha_s, \alpha_j]$, depending on the sign of $T(\alpha_s)$. We continue these steps until we obtain an interval with a length smaller than a positive parameter ϵ defined by the input of the Max-Max BOCO problem. The selection method of ϵ guarantees the absence of the PF coefficient in such an interval. Consequently, our algorithm always converges in a logarithmic number of iterations in terms of ϵ and α^{sup}.

In the following, we present our algorithm's statement and proofs.

3.2   Algorithm Statement and Proofs

In this section, we first introduce Procedure $Verify\_PF\_sol(\alpha_0, P_0, Q_0)$ to verify whether a solution (P_0, Q_0) of $\mathcal{F}(\alpha_0)$ is the PF solution. The correctness of this procedure will be shown in the next lemma.

Procedure 1. Verify whether a solution (P_0, Q_0) of $\mathcal{F}(\alpha_0)$ is the PF solution

---

Input: $\alpha_0 \geq 0$, $(P_0, Q_0) \in \mathcal{S}$ is a solution of $\mathcal{F}(\alpha_0)$.
Output: True if (P_0, Q_0) is the PF solution or False otherwise.
 1: **procedure** $Verify\_PF\_sol(\alpha_0, P_0, Q_0)$
 2:     **if** $P_0 - \alpha_0 Q_0 = 0$ **then** return True
 3:     **else**
 4:         $\alpha' \leftarrow P_0/Q_0$
 5:         solving $\mathcal{F}(\alpha')$ to obtain the solution (P', Q').
 6:         **if** $f_{\alpha'}(P', Q') = f_{\alpha'}(P_0, Q_0)$ **then** return True
 7:         **else** return False
 8:     **end if**
 9:     **end if**
10: **end procedure**

Lemma 2. *Given $\alpha_0 \geq 0$ and $(P_0, Q_0) \in \mathcal{S}$ as a solution of $\mathcal{F}(\alpha_0)$. Let $\alpha' = P_0/Q_0$ and (P', Q') be a solution of $\mathcal{F}(\alpha')$. If $T(\alpha_0) = P_0 - \alpha_0 Q_0 \neq 0$ then (P_0, Q_0) is the PF solution if and only if $f_{\alpha'}(P', Q') = f_{\alpha'}(P_0, Q_0)$.*

Proof. $\Longrightarrow$ If (P_0, Q_0) is the PF solution then (P_0, Q_0) is also a solution of $\mathcal{F}(\alpha')$ due to Theorem 1. Thus, $f_{\alpha'}(P', Q') = f_{\alpha'}(P_0, Q_0)$.

$\Longleftarrow$ If $f_{\alpha'}(P', Q') = f_{\alpha'}(P_0, Q_0)$ then (P_0, Q_0) is also a solution of $\mathcal{F}(\alpha')$. Since $\alpha' = P_0/Q_0$, (P_0, Q_0) is the PF solution due to Theorem 1. $\square$

Then, from a given $\alpha_0 \geq 0$ and a solution (P_0, Q_0) of $\mathcal{F}(\alpha_0)$, we discuss how to construct Procedure *Verify_PF_coeff* (α_0, P_0, Q_0) which aims to verify whether α_0 is the PF coefficient and return the PF solution if the verification is successful. It is important to remind that if $T(\alpha_0) = P_0 - \alpha_0 Q_0 = 0$ then $\alpha_0 = \alpha^{PF}$ and (P_0, Q_0) is necessarily the PF solution due to Theorem 1. However, if $T(\alpha_0) \neq 0$, we may not assert that $\alpha_0 \neq \alpha^{PF}$ as well as (P_0, Q_0) is not the PF solution. In general, although the PF solution is necessary a solution of $\mathcal{F}(\alpha^{PF})$, we might not obtain the PF solution by solving $\mathcal{F}(\alpha^{PF})$. The fact is that the problem $\mathcal{F}(\alpha^{PF})$ may have multiple solutions, and we obtain one solution, which might not be the PF solution. More precisely, we state the following proposition.

Proposition 2. *We might not obtain the PF solution by solving $\mathcal{F}(\alpha)$, $\forall \alpha \geq 0$.*

Proof. To prove this conclusion, we consider an example of the Bi-Objective Spanning Tree Problem (BOSTP) which is also a Max-Max BOCO problem. Let G be an undirected, connected graph, and each edge of G is associated with two positive values: profit and reliability. The BOSTP consists of finding a spanning tree of G, maximizing both the total profit and the minimum edge reliability.

This example of BOSTP with two values on each edge is illustrated in Fig. 1. For example, the profit and reliability associated with edge (14) are 20 and 9.

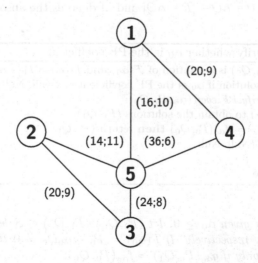

Fig. 1. An example of the BOSTP

Let (P,Q) denote the solution for the total profit and the minimum edge reliability corresponding to a spanning tree solution. We show each distinct spanning tree by listing its edges and the corresponding solution (P, Q) as follows.

- (14) (15) (23) (25) and $(P_1, Q_1) = (70, 9)$
- (14) (15) (23) (35) and $(P_2, Q_2) = (80, 8)$
- (14) (15) (25) (35) and $(P_3, Q_3) = (74, 8)$
- (14) (45) (23) (25) and $(P_4, Q_4) = (90, 6)$
- (14) (45) (23) (35) and $(P_5, Q_5) = (100, 6)$
- (14) (45) (25) (35) and $(P_6, Q_6) = (94, 6)$
- (15) (45) (23) (25) and $(P_7, Q_7) = (86, 6)$
- (15) (45) (23) (35) and $(P_8, Q_8) = (96, 6)$
- (15) (45) (25) (35) and $(P_9, Q_9) = (90, 6)$

Then, we can easily verify that (P_2, Q_2) is the PF solution since

$$\frac{P_i}{P_2} + \frac{Q_i}{Q_2} \leq 2, \forall 1 \leq i \leq 9,$$

Thus, we get $\alpha^{PF} = P_2/Q_2 = 10$. However, if $0 \leq \alpha < 10$ (resp. $\alpha > 10$) then (P_5, Q_5) (resp. (P_1, Q_1)) is the solution of $\mathcal{F}(\alpha)$ and if $\alpha = 10$, solving $\mathcal{F}(\alpha)$ may return (P_1, Q_1) or (P_5, Q_5) instead of the PF solution (P_2, Q_2) because they are simultaneously the solutions of $\mathcal{F}(10)$ due to $P_1 + 10Q_1 = P_2 + 10Q_2 = P_5 + 10Q_5 = 160$. In this case, $T(10) \neq 0$ despite $\alpha^{PF} = 10$.

Generally, if $\mathcal{F}(\alpha^{PF})$ has multiple (distinct) solutions, including the PF solution, we might not obtain the PF solution by solving $\mathcal{F}(\alpha)$.   □

For Procedure *Verify_PF_coeff*(), we present the following optimization problem

$$\mathcal{G}(\alpha) = \max_{(P,Q) \in \mathcal{S}} g_\alpha(P, Q),$$

where $g_\alpha(P, Q) = P + \alpha Q - |P - \alpha Q|$ and $|.|$ denotes the absolute function.

Procedure 2. Verify whether α_0 is the PF coefficient

Input: $\alpha_0 \geq 0$, (P_0, Q_0) is a solution of $\mathcal{F}(\alpha_0)$ and $T(\alpha_0) = P_0 - \alpha_0 Q_0 \neq 0$.
Output: The PF solution if α_0 is the PF coefficient or (Null, Null) otherwise.
1: **procedure** *Verify_PF_coeff*(α_0, P_0, Q_0)
2:     solving $\mathcal{G}(\alpha_0)$ to obtain the solutions (P_1, Q_1)
3:     **if** $g_{\alpha_0}(P_1, Q_1) = f_{\alpha_0}(P_0, Q_0)$ **then** return (P_1, Q_1)
4:     **else** return (Null, Null)
5:     **end if**
6: **end procedure**

Lemma 3. *For a given $\alpha_0 \geq 0$, let $(P_0, Q_0), (P_1, Q_1) \in \mathcal{S}$ be the solutions of $\mathcal{F}(\alpha_0)$ and $\mathcal{G}(\alpha_0)$, respectively. If $T(\alpha_0) = P_0 - \alpha_0 Q_0 \neq 0$ then α_0 is the PF coefficient if and only if $g_{\alpha_0}(P_1, Q_1) = f_{\alpha_0}(P_0, Q_0)$.*

Proof. $\implies$ Suppose that $\alpha_0 = \alpha^{PF}$. According to Theorem 1, there exists the PF solution $(P^{PF}, Q^{PF}) \in S$ such that (P^{PF}, Q^{PF}) is a solution of $\mathcal{F}(\alpha_0)$ and $P^{PF} = \alpha_0 Q^{PF}$. Since both (P_0, Q_0) and (P^{PF}, Q^{PF}) are the solutions of $\mathcal{F}(\alpha_0)$ and $P^{PF} - \alpha_0 Q^{PF} = 0$, we have

$$P_0 + \alpha_0 Q_0 = P^{PF} + \alpha_0 Q^{PF} - |P^{PF} - \alpha_0 Q^{PF}|,$$

The optimality of (P^{PF}, Q^{PF}) gives

$$P^{PF} + \alpha_0 Q^{PF} \geq P_1 + \alpha_0 Q_1,$$

Since $|P_1 - \alpha_0 Q_1| \geq 0$, we deduce $P^{PF} + \alpha_0 Q^{PF} \geq P_1 + \alpha_0 Q_1 - |P_1 - \alpha_0 Q_1|$. Thus,

$$P^{PF} + \alpha_0 Q^{PF} - |P^{PF} - \alpha_0 Q^{PF}| \geq P_1 + \alpha_0 Q_1 - |P_1 - \alpha_0 Q_1|, \qquad (2)$$

Since (P_1, Q_1) is a solution of $\mathcal{G}(\alpha_0)$, we have

$$P_1 + \alpha_0 Q_1 - |P_1 - \alpha_0 Q_1| \geq P^{PF} + \alpha_0 Q^{PF} - |P^{PF} - \alpha_0 Q^{PF}|, \qquad (3)$$

From (2) and (3), we get $P_1 + \alpha_0 Q_1 - |P_1 - \alpha_0 Q_1| = P^{PF} + \alpha_0 Q^{PF} - |P^{PF} - \alpha_0 Q^{PF}| = P_0 + \alpha_0 Q_0$ which implies $g_{\alpha_0}(P_1, Q_1) = f_{\alpha_0}(P_0, Q_0)$.
$\impliedby$ Suppose that $g_{\alpha_0}(P_1, Q_1) = f_{\alpha_0}(P_0, Q_0)$. We obtain $P_1 + \alpha_0 Q_1 - |P_1 - \alpha_0 Q_1| = P_0 + \alpha_0 Q_0$. Since $P_1 + \alpha_0 Q_1 - |P_1 - \alpha_0 Q_1| \leq P_1 + \alpha_0 Q_1 \leq P_0 + \alpha_0 Q_0$, we must have $|P_1 - \alpha_0 Q_1| = 0$ and $P_1 + \alpha_0 Q_1 = P_0 + \alpha_0 Q_0$. Consequently, (P_1, Q_1) is a solution of $\mathcal{F}(\alpha_0)$ and $P_1 = \alpha_0 Q_1$. Thus, (P_1, Q_1) is the PF solution and $\alpha_0 = \alpha^{PF}$ due to Theorem 1. $\qquad \square$

Notice that we can obtain the PF solution by solving $\mathcal{F}(10)$ and $\mathcal{G}(10)$ for the instance of the BOSTP mentioned in Proposition 2. Subsequently, for $0 \leq \alpha_i < \alpha_j$ and $(P_i, Q_i), (P_j, Q_j)$ as the solutions of $\mathcal{F}(\alpha_i)$ and $\mathcal{F}(\alpha_j)$, we present Procedure $SEARCH(\alpha_i, P_i, Q_i, \alpha_j, P_j, Q_j, \epsilon)$ for determining the PF solution where the PF coefficient α^{PF} is in the interval $[\alpha_i, \alpha_j]$. We recall that the parameter ϵ is presented for the stopping condition of Procedure $SEARCH()$ as mentioned in Sect. 3.1.

For Max-Max BOCO, ϵ can be determined as

$$\epsilon = min\ \{|\frac{P' - P''}{Q'' - Q'} - \frac{P'' - P'''}{Q''' - Q''}|\} \qquad (4)$$

where $|.|$ denotes the absolute function, $(P', Q'), (P'', Q''), (P''', Q''') \in S$ are the solutions of $\mathcal{F}(\alpha'), \mathcal{F}(\alpha''), \mathcal{F}(\alpha''')$ for which $0 \leq \alpha' < \alpha'' < \alpha'''$, $P' \geq P'' \geq P''', Q' > Q'' > Q'''$ and $\frac{P' - P''}{Q'' - Q'} \neq \frac{P'' - P'''}{Q''' - Q''}$.

Notice that if the objectives P, Q are positive integers (this hypothesis is natural in combinatorial optimization), from (4) we have

$$|\frac{P' - P''}{Q'' - Q'} - \frac{P'' - P'''}{Q''' - Q''}| = \frac{|(P' - P'')(Q''' - Q'') - (P'' - P''')(Q'' - Q')|}{(Q'' - Q')(Q''' - Q'')} \geq \frac{1}{Q_{max}^2},$$

since $|(P'-P'')(Q'''-Q'')-(P''-P''')(Q''-Q')| \in \mathbb{Z}_+, 0 < Q''-Q', Q'''-Q'' \le Q_{max}$ where Q_{max} is the maximum value of Q. Thus, when the objectives of Max-Max BOCO take positive integer values, we can select ϵ as $1/Q_{max}^2$.

Procedure 3. Determine the PF solution where the PF coefficient is in $[\alpha_i, \alpha_j]$

Input: (α_i, P_i, Q_i) and (α_j, P_j, Q_j) satisfying the following conditions:
- $0 \le \alpha_i < \alpha_j$ such that α_i, α_j are not PF coefficients.
- (P_i, Q_i) and (P_j, Q_j) are solutions of $\mathcal{F}(\alpha_i)$ and $\mathcal{F}(\alpha_j)$, respectively.
- $(P_i, Q_i) \not\equiv (P_j, Q_j)$, (P_i, Q_i) and (P_j, Q_j) are not PF solutions.
- a parameter ϵ as defined in (4).

Output: The PF solution if it exists or Null otherwise.

1: **procedure** $SEARCH(\alpha_i, P_i, Q_i, \alpha_j, P_j, Q_j, \epsilon)$
2:     $\alpha_k \leftarrow \frac{P_i - P_j}{Q_j - Q_i}$
3:     **if** $\alpha_k = \alpha_i$ **or** $\alpha_k = \alpha_j$ **then** return Null
4:     **end if**
5:     solving $\mathcal{F}(\alpha_k)$ to obtain a solution (P_k, Q_k)
6:     **if** $Verify\_PF\_sol(\alpha_k, P_k, Q_k) ==$ True **then** return (P_k, Q_k)
7:     **end if**
8:     $(P', Q') \leftarrow Verify\_PF\_coeff(\alpha_k, P_k, Q_k)$
9:     **if** $(P', Q') \not\equiv$ (Null, Null) **then** return (P', Q')
10:    **else if** $(P_k, Q_k) \equiv (P_i, Q_i)$ **or** $(P_k, Q_k) \equiv (P_j, Q_j)$ **then** return Null
11:    **else**
12:        **if** $\alpha_j - \alpha_i \ge \epsilon$ **then**
13:            $\alpha_s \leftarrow \frac{\alpha_i + \alpha_j}{2}$
14:            solving $\mathcal{F}(\alpha_s)$ to obtain a solution (P_s, Q_s)
15:            **if** $Verify\_PF\_sol(\alpha_s, P_s, Q_s) ==$ True **then** return (P_s, Q_s)
16:            **end if**
17:            $(P'', Q'') \leftarrow Verify\_PF\_coeff(\alpha_s, P_s, Q_s)$
18:            **if** $(P'', Q'') \not\equiv$ (Null, Null) **then** return (P'', Q'')
19:            **else**
20:                $T(\alpha_s) \leftarrow P_s - \alpha_s Q_s$
21:                **if** $T(\alpha_s) > 0$ **then** return $SEARCH(\alpha_s, P_s, Q_s, \alpha_j, P_j, Q_j, \epsilon)$
22:                **else if** $T(\alpha_s) < 0$ **then** return $SEARCH(\alpha_i, P_i, Q_i, \alpha_s, P_s, Q_s, \epsilon)$
23:                **end if**
24:            **end if**
25:        **else** return Null
26:        **end if**
27:    **end if**
28: **end procedure**

It is necessary to select α_k at each iteration before selecting the midpoint α_s. The fact is that we might not obtain the PF solution by only selecting the midpoints of the intervals. For example, in the instance of the BOSTP mentioned in Proposition 2, the PF solution can only be obtained by solving $\mathcal{F}(\alpha)$ and $\mathcal{G}(\alpha)$ with $\alpha = 10$. However, by repeating choosing the midpoints of the intervals, we might not reach $\alpha_s = 10$ (in contrast, for α_k, we can obtain $\alpha_k = 10$).

By the following lemma, we show that our choosing method for α_k at each iteration and the parameter ϵ offer some specific criteria to promptly verify the existence of α^{PF} in the interval $[\alpha_i, \alpha_j]$.

Lemma 4. *Let $[\alpha_i, \alpha_j]$ be an interval such that $0 \leq \alpha_i < \alpha_j$, α_i, α_j are not PF coefficients. Let $(P_i, Q_i) \not\equiv (P_j, Q_j)$ be the solutions of $\mathcal{F}(\alpha_i)$, $\mathcal{F}(\alpha_j)$ and they are not PF solutions. Let $\alpha_k = \frac{P_i - P_j}{Q_j - Q_i}$ and (P_k, Q_k) be a solution of $\mathcal{F}(\alpha_k)$. If one of the following conditions is satisfied, then $\alpha^{PF} \notin [\alpha_i, \alpha_j]$.*

1. Either $\alpha_k = \alpha_i$ or $\alpha_k = \alpha_j$;
2. $\alpha_k \neq \alpha^{PF}$ and either $(P_k, Q_k) \equiv (P_i, Q_i)$ or $(P_k, Q_k) \equiv (P_j, Q_j)$;
3. $\alpha_k \neq \alpha^{PF}$, $(P_k, Q_k) \not\equiv (P_i, Q_i)$, $(P_k, Q_k) \not\equiv (P_j, Q_j)$ and $\alpha_j - \alpha_i < \epsilon$;

Proof. Since $\alpha_i < \alpha_j$, we have $P_i \geq P_j$, $Q_i \leq Q_j$ due to Lemma 1. Suppose that $Q_i = Q_j$. The optimality of (P_j, Q_j) gives

$$P_j + \alpha_j Q_j \geq P_i + \alpha_j Q_i, \tag{5}$$

Since $Q_i = Q_j$, we obtain $P_j \leq P_i$. Thus, $P_i = P_j$ and then $(P_i, Q_i) \equiv (P_j, Q_j)$ which leads to a contradiction.

Hence, $Q_j > Q_i$ and consequently, α_k is well defined.

We then show that $\alpha_k \in [\alpha_i, \alpha_j]$. The optimality of (P_i, Q_i) gives

$$P_i + \alpha_i Q_i \geq P_j + \alpha_i Q_j, \tag{6}$$

From (5) and (6), we obtain $\alpha_i \leq \frac{P_i - P_j}{Q_j - Q_i} \leq \alpha_j$ which leads to $\alpha_i \leq \alpha_k \leq \alpha_j$.

1. If $\alpha_k = \alpha_i$ then $P_i + \alpha_i Q_i = P_j + \alpha_i Q_j$. Thus, (P_i, Q_i) and (P_j, Q_j) are both solutions of $\mathcal{F}(\alpha_i)$. Hence, for all $\alpha \in (\alpha_i, \alpha_j)$, (P_j, Q_j) is the unique solution of $\mathcal{F}(\alpha)$ as a result of Lemma 1. Similarly, if $\alpha_k = \alpha_j$, (P_i, Q_i) is the unique solution of $\mathcal{F}(\alpha)$ for all $\alpha \in (\alpha_i, \alpha_j)$. Since (P_i, Q_i), (P_j, Q_j) are not PF solutions and α_i, α_j are not PF coefficients, we have $\alpha^{PF} \notin [\alpha_i, \alpha_j]$.

2. Let (P_k, Q_k) be a solution of $\mathcal{F}(\alpha_k)$. Without loss of generality, we suppose that $\alpha_k \neq \alpha^{PF}$ and $(P_k, Q_k) \equiv (P_i, Q_i)$. Consequently, (P_i, Q_i) is a solution of $\mathcal{F}(\alpha_k)$. Since $\alpha_k = \frac{P_i - P_j}{Q_j - Q_i}$, $P_i + \alpha_k Q_i = P_j + \alpha_k Q_j$. Thus, (P_j, Q_j) is also a solution of $\mathcal{F}(\alpha_k)$.

As a result of Lemma 1, when $\alpha \in (\alpha_i, \alpha_k)$ (resp. $\alpha \in (\alpha_k, \alpha_j)$), (P_i, Q_i) (resp. (P_j, Q_j)) is the unique solution of $\mathcal{F}(\alpha)$. Consequently, $\alpha^{PF} \notin [\alpha_i, \alpha_j]$.

3. Similar to the proof above, we also have $Q_i < Q_k < Q_j$ and

$$\alpha_i \leq \frac{P_i - P_k}{Q_k - Q_i} \leq \alpha_k \leq \frac{P_k - P_j}{Q_j - Q_k} \leq \alpha_j,$$

According to the definition of ϵ, if $\frac{P_i - P_k}{Q_k - Q_i} > \frac{P_k - P_j}{Q_j - Q_k}$ we obtain $\alpha_j - \alpha_i \geq \frac{P_i - P_k}{Q_k - Q_i} - \frac{P_k - P_j}{Q_j - Q_k} \geq \epsilon$ which leads to a contradiction.

Thus, we have $\frac{P_i - P_k}{Q_k - Q_i} = \frac{P_k - P_j}{Q_j - Q_k}$. Consequently, $\frac{P_i - P_k}{Q_k - Q_i} = \alpha_k = \frac{P_k - P_j}{Q_j - Q_k}$ which implies $P_k + \alpha_k Q_k = P_i + \alpha_k Q_i = P_j + \alpha_k Q_j$. In other words, (P_i, Q_i) and (P_j, Q_j) are also the solutions of $\mathcal{F}(\alpha_k)$. Similar to the case 2, $\alpha^{PF} \notin [\alpha_i, \alpha_j]$. $\square$

Combining these three procedures, our algorithm to determine the PF solution for Max-Max BOCO can be stated as follows.

Algorithm 4. Determine the PF solution for Max-Max BOCO

Input: An instance of Max-Max BOCO, ϵ defined as (4).
Output: PF solution if it exists or Null otherwise.
1: solving $\mathcal{F}(0)$ to obtain a solution (P_0, Q_0)
2: **if** $Verify\_PF\_sol(0, P_0, Q_0) ==$ True **then** return (P_0, Q_0)
3: **end if**
4: $\alpha^{sup} \leftarrow P_0/Q_0 + 1$
5: solving $\mathcal{F}(\alpha^{sup})$ to obtain a solution (P^{sup}, Q^{sup})
6: **if** $Verify\_PF\_sol(\alpha^{sup}, P^{sup}, Q^{sup}) ==$ True **then** return (P^{sup}, Q^{sup})
7: **else** return $SEARCH(0, P_0, Q_0, \alpha^{sup}, P^{sup}, Q^{sup}, \epsilon)$
8: **end if**

Notice that since $0 < \alpha^{PF}$ and $0 < \alpha^{sup}$, $P_0 \geq P^{PF}, Q_0 \leq Q^{PF}$ and $P_0 \geq P^{sup}, Q_0 \leq Q^{sup}$ due to Lemma 1. Thus, $\alpha^{PF} = P^{PF}/Q^{PF} \leq P_0/Q_0 < \alpha^{sup}$. Moreover, $T(0) = P_0 > 0$ and $T(\alpha^{sup}) = P^{sup} - \alpha^{sup}Q^{sup} < P^{sup} - \frac{P_0}{Q_0}Q^{sup} \leq 0$.

Theorem 2. *Algorithm 4 can determine the PF solution in a logarithmic number of iterations in terms of ϵ and α^{sup}.*

Proof. The execution of Algorithm 4 is based on the binary search algorithm for the interval $[0, \alpha^{sup}]$ with a length equals α^{sup}. At each iteration, we divided an interval into two half-intervals with equal length. Then, the half in which the PF coefficient cannot exist is eliminated, and the search continues on the remaining half. Since Algorithm 4 terminated in the worst case when it found an interval with a length smaller than ϵ, the number of iterations for Algorithm 4 is $O(\log_2 \frac{\alpha^{sup}}{\epsilon})$. Consequently, Algorithm 4 can determine the PF solution in a logarithmic number of iterations in terms of ϵ and α^{sup}. $\square$

Due to Theorem 2, notice that if solving $\mathcal{F}(\alpha)$ and $\mathcal{G}(\alpha)$ can be done in polynomial time, then the PF solution can be determined in polynomial time.

4   Experimental Study on the BOSTP

4.1   Definition and Modeling

In this section, we first restate the BOSTP used in Sect. 3.2. The BOSTP is a variant of the spanning tree problem that merges the Maximum STP, which involves maximizing the total profit, and the Max-Min STP, which aims to maximize the minimum edge reliability. Notice that the Maximum STP is algorithmically equivalent to the *Minimum STP*, a fundamental optimization problem that can be solved efficiently in polynomial time [13]. The Max-Min STP - which is also algorithmically equivalent to a variant called *Min-Max STP* mentioned in the prior literature [14] - aims at constructing solutions having a good performance in the worst case. For instance, in network design and optimization, the Max-Min STP can help ensure that the weakest link (edge with minimum reliability) in a communication or transportation network is as strong as possible, minimizing the risk of failure or congestion.

For the BOSTP, we find a spanning tree achieving proportional fairness between two objectives: the total profit and the minimum edge reliability. Notice that profit and reliability are two important criteria in the various applications of the spanning tree problem [15]. Furthermore, for the simplicity of calculation, we suppose that the values of profit and reliability are positive integers. Thus, the objective values of the BOSTP are also positive integers, and then the parameter ϵ can be selected as mentioned in Sect. 3.2.

We consider a finite, connected, undirected graph $G = (V, E)$ where $V = [n] := \{1, ..., n\}$ with $n \geq 2$, $|E| = m$ and $p_e, r_e \in \mathbb{Z}_+$ are two weights associated with edge $e \in E$ representing profit and reliability on this edge, respectively. Let $\mathcal{T}(G)$ denote the set of all spanning trees in G. Let $P, Q > 0$ denote the total profit and the minimum edge reliability in a spanning tree of G, respectively. The BOSTP can be formally formulated as $P = \sum\limits_{e \in T, T \in \mathcal{T}(G)} p_e$ and $Q = \min\limits_{e \in T, T \in \mathcal{T}(G)} r_e$.

As shown in Sect. 3, for determining the PF solution, we aim to solve $\mathcal{F}(\alpha)$ and $\mathcal{G}(\alpha)$ for some $\alpha \in [0, \alpha^{sup}]$. According to Sect. 2, we present the formulation for $\mathcal{F}(\alpha)$.

$$\mathcal{F}(\alpha): \max \ P + \alpha Q \tag{7a}$$

$$\text{s.t.} \quad P = \sum_{e \in E} p_e x_e \tag{7b}$$

$$Q \leq r_e x_e + (1 - x_e) M \qquad \forall e \in E \tag{7c}$$

$$\sum_{e \in E} x_e = n - 1 \tag{7d}$$

$$\sum_{e \in \delta(V')} x_e \geq 1 \qquad \forall V' \subseteq V, \emptyset \neq V' \neq V \tag{7e}$$

$$x_e \in \{0, 1\} \qquad \forall e \in E \tag{7f}$$

where x_e is the binary variables representing the occurrence of edge e in the spanning tree solution. Constraint (7d) is the degree constraint that assures exactly $n-1$ edges in the spanning tree solution. Constraints (7e) are the subtour elimination constraints: $\delta(V')$ is the set of edges crossing the cut (one endpoint in V' and one in $V - V'$).

Constraints (7c) allow bounding Q by the minimum edge reliability in the spanning tree solution. Indeed, in case $x_e = 1$, Constraints (7c) guarantee that Q is smaller than all the edge reliabilities in the tour. Otherwise, when x_e equals 0, the largest edge reliability M assures the validity of Constraints (7c). As $P + \alpha Q$ is maximized, Q will take the minimum edge reliability values.

We present the following formulation, which contains all the constraints from (7b) to (7f) for $\mathcal{G}(\alpha)$. However, to prevent redundancy, these constraints have been omitted.

$$\mathcal{G}(\alpha) : \max \ P + \alpha Q - t \tag{8a}$$
$$\text{s.t. } t \geq P - \alpha Q \tag{8b}$$
$$t \geq \alpha Q - P \tag{8c}$$

Using two constraints (8b) and (8c), the parameter t represents the absolute value of $P - \alpha Q$.

It is important to note that a special-purpose algorithm can be used for solving $\mathcal{F}(\alpha)$ as well as $\mathcal{G}(\alpha)$ in polynomial time, which is similar to the one for solving the Min-Max STP [14]. It is based on the fact that there are at most $O(n^2)$ different values of Q, and the Maximum STP can be solved in polynomial time. However, in this section, we show the computational results by solving directly the MIP formulations of $\mathcal{F}(\alpha)$ and $\mathcal{G}(\alpha)$ due to its simple setting and better running time.

4.2   Computational Results on the Instances of the BOSTP

We investigate the performance of the presented algorithm for the BOSTP on random NetworkX graph. It returns a $G_{n,pro}$ random graph, also known as an Erdos-Renyi graph or a binomial graph [16] where n is the number of nodes and pro is the probability for edge creation. For this paper, we selected the number of nodes from the interval $[15, 40]$ with probability $pro = 0.5$. Moreover, the edge profit and the edge reliability are generated uniformly randomly in the intervals $[100, 900]$ and $[10, 90]$, respectively.

The solutions concerning the values of P, Q for the Maximum STP, Max-Min STP, and the PF solutions for the BOSTP are shown in Table 1. Notice that the Maximum STP and the Max-Min STP solutions are also feasible solutions for the BOSTP. We provided the time calculation and the number of iterations in the columns "Time" and "Iters". For each number of nodes n, we have generated two distinct graphs "GNn_1" and "GNn_2". The values of P, Q in case the PF solution does not exist are denoted as "Null". We use CPLEX 12.10 on a PC Intel Core i5-9500 3.00GHz to solve these MIP formulations with 6 cores and 6 threads.

According to Table 1, we obtained the PF solutions for most instances and they are different from the solutions of the Maximum STP and the Max-Min STP. For the instance "GN20_1", we see that the PF solution has the same value of Q compared to the solution of the Max-Min STP but the value of P is much better. Generally, the PF solutions offer a more favorable compromise

Table 1. Computational results of Maximum STP, Max-Min STP, and BOSTP

Instance	Maximum STP			Max-Min STP			PF solution for BOSTP			
	P	Q	Time	P	Q	Time	P	Q	Time	Iters
GN15_1	10809	18	0.01	7727	59	0.01	9837	57	0.26	2
GN15_2	10812	10	0.02	7186	61	0.01	8587	57	0.48	3
GN20_1	15554	12	0.01	9390	54	0.03	11860	54	0.24	1
GN20_2	15152	14	0.02	9058	62	0.18	13179	61	0.30	2
GN25_1	20300	13	0.05	11046	66	0.12	Null	Null	1.68	3
GN25_2	20334	10	0.04	12052	73	0.23	15743	68	3.29	2
GN30_1	24259	12	0.10	14062	74	0.30	21633	67	3.28	2
GN30_2	24272	16	0.07	13359	74	0.08	18651	71	1.96	2
GN35_1	28329	11	0.08	19314	77	0.24	Null	Null	5.24	4
GN35_2	28554	17	0.05	17944	69	0.25	25138	68	4.42	2
GN40_1	33531	10	0.14	20432	73	0.24	28358	72	6.45	3
GN40_2	33681	12	0.14	17789	79	0.65	29171	68	5.07	2

between two objectives than the solutions of the Maximum STP (resp. Max-Min STP): the significant increase in the values of Q (resp. P) compared to the slight drop in the values of P (resp. Q) in percentage. Table 1 also indicates that our algorithm seems to converge quickly regarding time calculation and number of iterations. It is worth noting that the upper bound on the number of iterations, as specified in Theorem 2, may theoretically be higher due to the determinations of α^{sup} and ϵ. However, in practice, adding the selecting method of α_k helps us quickly verify the existence of the PF solution rather than only using the binary search algorithm, especially when there is no PF solution. Another important remark is that the existence of the PF solution seems to be much related to the edge weights and the structure of the graph rather than to the size of the graph. Although we randomly selected the values of profit and reliability, the PF solutions appeared with a high frequency, approximately 85% over the total tested instances.

5   Conclusion

In this paper, we have applied *proportional fairness* in the context of *Max-Max Bi-Objective Combinatorial Optimization* (Max-Max BOCO) where the two objectives to be maximized take only positive values and the feasible set is discrete, finite and non-convex. We considered a general Max-Max BOCO problem where we looked for a solution achieving proportional fairness between two objectives - which is referred to as *proportional fair solution* (PF solution). We first presented the characterization of the PF solution for Max-Max BOCO. Then, we designed an exact algorithm that converges within a logarithmic number of iterations to determine the PF solution. Finally, computational experiments on some instances of the Bi-Objective Spanning Tree Problem have shown the effectiveness of our algorithm, indicating its rapid convergence.

For future works, in cases the PF solution does not exist, we are interested in modifying our algorithm to provide a near-PF solution that maximizes the product of the objectives, resembling a generalized Nash bargaining solution. Furthermore, the results of this paper could be extended to multi-objective combinatorial optimization involving more than two objectives.

Appendix

Proposition 1. *If* $(P^{PF}, Q^{PF}) \in \mathcal{S}$ *is a PF solution for Max-Max BOCO, then it is the unique solution that maximizes the product PQ.*

Proof. Suppose that $(P^{PF}, Q^{PF}) \in \mathcal{S}$ is a PF solution for Max-Max BOCO. We have

$$\frac{P}{P^{PF}} + \frac{Q}{Q^{PF}} \leq 2, \ \forall (P, Q) \in \mathcal{S},$$

Using Cauchy-Schwarz inequality, we obtain

$$2 \geq \frac{P}{P^{PF}} + \frac{Q}{Q^{PF}} \geq 2\sqrt{\frac{PQ}{P^{PF}Q^{PF}}},$$

Thus, $P^{PF}Q^{PF} \geq PQ, \forall (P, Q) \in \mathcal{S}$.

Now suppose that there exists another PF solution $(P^*, Q^*) \in \mathcal{S}$ such that $P^*Q^* = P^{PF}Q^{PF}$. We also have

$$2 \geq \frac{P^*}{P^{PF}} + \frac{Q^*}{Q^{PF}} \geq 2\sqrt{\frac{P^*Q^*}{P^{PF}Q^{PF}}} = 2,$$

Thus, the equality in the Cauchy-Schwarz inequality above must hold, which implies $P^* = P^{PF}$ and $Q^* = Q^{PF}$. □

Theorem 1. $(P^{PF}, Q^{PF}) \in \mathcal{S}$ *is the PF solution if and only if* (P^{PF}, Q^{PF}) *is a solution of* $\mathcal{F}(\alpha^{PF})$ *with* $\alpha^{PF} = P^{PF}/Q^{PF}$.

Proof. $\implies$ Let (P^{PF}, Q^{PF}) be the PF solution and $\alpha^{PF} = P^{PF}/Q^{PF}$. We have

$$\frac{P}{P^{PF}} + \frac{Q}{Q^{PF}} \leq 2, \ \forall (P, Q) \in \mathcal{S}, \tag{9}$$

Multiplying (9) by $P^{PF} > 0$ and replacing P^{PF}/Q^{PF} by α^{PF}, we obtain

$$P^{PF} + \alpha^{PF}Q^{PF} \geq P + \alpha^{PF}Q, \ \forall (P, Q) \in \mathcal{S},$$

Hence, (P^{PF}, Q^{PF}) is a solution of $\mathcal{F}(\alpha^{PF})$.

$\Longleftarrow$ Let (P^{PF}, Q^{PF}) be a solution of $\mathcal{F}(\alpha^{PF})$ with $\alpha^{PF} = P^{PF}/Q^{PF}$. We have

$$P + \alpha^{PF}Q \leq P^{PF} + \alpha^{PF}Q^{PF}, \ \forall (P,Q) \in \mathcal{S},$$

Replacing α^{PF} by P^{PF}/Q^{PF}, we obtain (2) which implies (P^{PF}, Q^{PF}) is the PF solution. $\qquad\square$

References

1. Kelly, F.P., Maulloo, A.K., Tan, D.K.H.: Rate control for communication networks: shadow prices, proportional fairness and stability. J. Oper. Res. Soc. **49**(3), 237 (1998)
2. Kushner, H.J., Whiting, P.A.: Convergence of proportional-fair sharing algorithms under general conditions. IEEE Trans. Wirel. Commun. **3**(4), 1250–1259 (2004)
3. Nicosia, G., Pacifici, A., Pferschy, U.: Price of Fairness for allocating a bounded resource. Eur. J. Oper. Res. **257**(3), 933–943 (2017)
4. Bertsimas, D., Farias, V.F., Trichakis, N.: The price of fairness. Oper. Res. **59**(1), 17–31 (2011)
5. Nash, John F..: The Bargaining Problem. Econometrica **18**(2), 155 (1950)
6. Sub, C., Park, S., Cho, Y.: Efficient algorithm for proportional fairness scheduling in multicast OFDM systems. In: IEEE 61st Vehicular Technology Conference, Stockholm (2005)
7. Boche, H., Schubert, M.: Nash bargaining and proportional fairness for wireless systems. IEEE/ACM Trans. Netw. **17**(5), 1453–1466 (2009)
8. Brehmer, J., Utschick, W.: On proportional fairness in non-convex wireless systems. In: International ITG Workshop on Smart Antennas - WSA, Berlin (2009)
9. Nguyen, M.H., Baiou, M., Nguyen, V.H., Vo, T.Q.T.: Nash fairness solutions for balanced TSP. In: International Network Optimization Conference (2022). https://doi.org/10.48786/inoc.2022.17
10. Nguyen, M.H., Baiou, M., Nguyen, V.H.: Nash balanced assignment problem. In: Ljubić, I., Barahona, F., Dey, S.S., Mahjoub, A.R. (eds.) ISCO 2022. LNCS, vol. 13526, pp. 172–186. Springer, Cham (2022). https://doi.org/10.1007/978-3-031-18530-4_13
11. Nguyen, M.H., Baiou, M., Nguyen, V.H., Vo, T.Q.T.: Generalized nash fairness solutions for bi-objective minimization problems. Networks **83**(1), 83–99 (2023). https://doi.org/10.1002/net.22182
12. Nguyen, M.H., Baiou, M., Nguyen, V.H.: Determining the generalized nash fairness solution set for bi-objective discrete optimization. In: Submitted to Discrete Applied Mathematics (2023). https://hal.science/hal-04010827v1
13. Kruskal, J.B.: On the shortest spanning subtree of a graph and the traveling salesman problem. Am. Math. Soc. **7**(1), 48–50 (1956)
14. Camerini, P.M.: The min-max spanning tree problem. Inf. Process. Lett. **7**(1) (1978)
15. Ebrahimi, S.B., Bagheri, E.: Optimizing profit and reliability using a bi-objective mathematical model for oil and gas supply chain under disruption risks. Comput. Indust. Eng. **163**, 107849 (2022)
16. Erdos, P., Renyi, A.: On random graphs. Publ. Math. **6**, 290–297 (1959)

Core Stability in Altruistic Coalition Formation Games

Matthias Hoffjan, Anna Maria Kerkmann, and Jörg Rothe[✉][iD]

Heinrich-Heine-Universität Düsseldorf, MNF, Institut für Informatik, Düsseldorf, Germany
{matthias.hoffjan,anna.kerkmann,rothe}@hhu.de

Abstract. Coalition formation games model settings where sets of agents have to partition into groups. We study the notion of core stability in the context of altruistic coalition formation games (ACFGs). While in most commonly studied classes of coalition formation games agents seek to maximize their individual valuations, agents are not completely selfish in ACFGs. Given some underlying network of friendship, the agents also take into account their friends' valuations when comparing different coalitions or coalition structures. The notion of the core has been extensively studied for several classes of (hedonic) coalition formation games. Kerkmann and Rothe [7] initiated the study of core stability in ACFGs. They showed that verifying core stability is coNP-complete for selfish-first ACFGs—their least altruistic case of ACFGs. The complexity of the other two (more altruistic) degrees of altruism, however, remained an open problem. We show that the core stability verification problem is coNP-complete for all cases of ACFGs, i.e., for all three degrees of altruism and for both sum-based and minimum-based aggregation of the friends' preferences.

Keywords: Coalition formation · Hedonic games · Core stability · Cooperative game theory · Altruism

1 Introduction

Coalition formation is a central topic in cooperative game theory. While the agents in such games are mostly modeled to aim for the maximization of their own utilities, there are some approaches that integrate altruistic aspects into the agents' preferences: Nguyen *et al.* [9] (see also Kerkmann *et al.* [6]) introduced *altruistic hedonic (coalition formation) games* where agents incorporate into their utility functions not only their own valuation of their coalition but also the valuations of their friends in the same coalition. Depending on the order in which the agents' own utilities and their friends' utilities are taken into account, three degrees of altruism are modeled, resulting in *selfish-first*, *equal-treatment*, and *altruistic-treatment* preferences. Kerkmann *et al.* [6] study these games both for sum-based and minimum-based aggregation of the friends' valuations.

ⓒ The Author(s), under exclusive license to Springer Nature Switzerland AG 2024
J. A. Soto and A. Wiese (Eds.): LATIN 2024, LNCS 14579, pp. 320–333, 2024.
https://doi.org/10.1007/978-3-031-55601-2_21

Kerkmann *et al.* [5,7] extended this model to a more general framework. In their *altruistic coalition formation games* (ACFGs), agents always integrate the valuations of *all* their friends into their utility functions, even if these friends belong to other coalitions.

The most important question in the context of coalition formation games is which partitions of players into coalitions can be expected to form, and various stability notions have been introduced to study this question. Among all these notions, the perhaps most central one is *core stability*, which is defined via the absence of blocking coalitions of players, where a coalition C blocks a partition Γ if all players in C prefer C to their old coalition in Γ. In general, core stable partitions are not guaranteed to exist and even verifying core stability is a hard problem. Core stability in coalition formation games (most notably in hedonic games) have been intensively studied, see, e.g., Banerjee et al. [1], Dimitrov et al. [3], Sung and Dimitrov [13], Woeginger [14], Peters [10], and Rey et al. [11], to name just a few.

For selfish-first ACFGs, core stability verification was shown to be coNP-complete by Kerkmann *et al.* [5,7]. However, pinpointing the complexity of this problem was still open for equal-treatment and altruistic-treatment ACFGs, both for sum-based and minimum-based aggregation of the friends' valuations. These cases are not only more interesting than selfish-first ACFGs, as they model higher degrees of altruism, they are also—as we will see—technically quite involved and more difficult to handle. In this paper, we solve these four open problems.

These results are significant for various reasons:

First, the complexity of verifying core stability (as pointed out above, one of the most central and well-studied stability notions for cooperative games) is now settled for an appealing new model that introduces certain degrees of altruism into coalition formation games.

Second, the complexity picture is now complete: Verifying core stability in ACFGs is coNP-complete under all three degrees of altruism and both for sum-based and minimum-based aggregation.

Third, modeling altruism in games is a relatively novel branch of game theory and usually requires—depending on the specific model used—much technical effort; a number of such models in noncooperative and cooperative games have been surveyed by Rothe [12] and by Kerkmann *et al.* [6]. In particular, the higher degrees of altruism that we study here are quite natural and robust; interestingly, one of them can be obtained also by a different approach (when restricted to the context of hedonic games where players care only about friends in their own coalition): Minimum-based altruistic hedonic games with equal-treatment preferences coincide with the loyal variant of symmetric friend-oriented hedonic games due to Bullinger and Kober [2].

While our paper settles the complexity of the verification problem for core stability in all variants of ACFGs, it remains open how hard it is to solve the existence problem and the problem of computing a core stable coalition structure in such games.

2 Preliminaries

In a coalition formation game (CFG), a set of players $N = [1, n]$ has to divide into coalitions, i.e., subsets of N. (For any integers i and j with $i \leq j$, we let $[i, j]$ denote the set $\{i, \ldots, j\}$ of integers.) Every partition of N is called a *coalition structure* and $\mathscr{C}_N$ denotes the set of all possible coalition structures. For a coalition structure $\Gamma \in \mathscr{C}_N$, we refer to the coalition containing player i as $\Gamma(i)$. A CFG is a pair $(N, \succeq)$, where $\succeq = (\succeq_1, \ldots, \succeq_n)$ is a profile of preferences of the players in N: $\succeq_i$ is player i's preference relation (i.e., weak order) over $\mathscr{C}_N$. We say that a player i *weakly prefers* a coalition structure Γ to another coalition structure Δ if $\Gamma \succeq_i \Delta$; i *prefers* Γ to Δ if $\Gamma \succeq_i \Delta$ and not $\Delta \succeq_i \Gamma$; and i is *indifferent* between Γ and Δ if $\Gamma \succeq_i \Delta$ and $\Delta \succeq_i \Gamma$.

Kerkmann et al. [5] introduced *altruistic coalition formation games* (ACFGs) as a generalization of the *altruistic hedonic games* due to Kerkmann et al. [6]. In these games, each player splits the other players into friends and enemies and then evaluates any coalition individually depending on the number of friends and enemies in it, while also taking into account the valuations of her friends when comparing two coalition structures.

More concretely, based on the friend-oriented preference extension of Dimitrov et al. [3], an ACFG with player set N can be represented by a *network of friends*, an undirected graph (N, F) that represents the mutual friendship relations among the agents. The (friend-oriented) valuation of player $i \in N$ for a coalition structure $\Gamma \in \mathscr{C}_N$ is then defined as

$$v_i(\Gamma) = n \cdot |F_i \cap \Gamma(i)| - |E_i \cap \Gamma(i)|,$$

where $F_i = \{j \in N \mid \{i, j\} \in F\}$ is the set of i's friends and $E_i = N \setminus (F_i \cup \{i\})$ is the set of i's enemies. Hence, a player's valuation is first determined by the number of friends in her coalition and only in case of a tie by the number of her enemies in it.

Kerkmann et al. [6] distinguish different degrees of altruism. In their *selfish-first* model (SF), the agents' preferences mainly depend on their own valuations. They only consult their friends' valuations if their own valuation is the same for two coalition structures. Under the *equal-treatment* model (EQ), the agents put equal weights on their own and their friends' valuations. And under the *altruistic-treatment* model (AL), a player first considers the valuation of her friends and just in the case of a tie, her own valuation is considered. For ACFGs, the described preferences $\succeq^{SF}$, $\succeq^{EQ}$, and $\succeq^{AL}$ can be expressed by the utilities

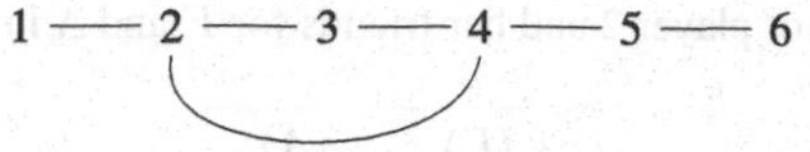

$$1 \text{---} 2 \text{---} 3 \text{---} 4 \text{---} 5 \text{---} 6$$

Fig. 1. Network of friends for Example 1

$$u_i^{SF}(\Gamma) = M \cdot v_i(\Gamma) + \sum_{j \in F_i} v_j(\Gamma),$$

$$u_i^{EQ}(\Gamma) = \sum_{j \in F_i \cup \{i\}} v_j(\Gamma), \text{ and}$$

$$u_i^{AL}(\Gamma) = v_i(\Gamma) + M \cdot \sum_{j \in F_i} v_j(\Gamma),$$

where $M \geq n^3$ is a constant weight on some of the valuations [5].

In addition to the sum-based aggregation of the friends' valuations, Kerkmann *et al.* [6] also define a minimum-based aggregation for these three notions of altruism in hedonic games, and Kerkmann *et al.* [5] do so for coalition formation games in general. These minimum-based preferences $\succeq^{minSF}$, $\succeq^{minEQ}$, and $\succeq^{minAL}$ are defined by

$$u_i^{minSF}(\Gamma) = M \cdot v_i(\Gamma) + \min_{j \in F_i} v_j(\Gamma),$$

$$u_i^{minEQ}(\Gamma) = \min_{j \in F_i \cup \{i\}} v_j(\Gamma), \text{ and}$$

$$u_i^{minAL}(\Gamma) = v_i(\Gamma) + M \cdot \min_{j \in F_i} v_j(\Gamma).$$

Depending on which aggregation method and which degree of altruism is used, we denote our game as sum-based or min-based SF, EQ, or AL ACFG.

Example 1. To get familiar with the different types of ACFGs, consider the coalition formation game that is represented by the network of friends in Fig. 1 and the coalition structures $\Gamma = \{\{1,2,3\},\{4,5,6\}\}$ and $\Delta = \{\{1,2\},\{3,4,5,6\}\}$. Player 2 has two friends and zero enemies in her coalition in Γ and she has one friend and zero enemies in her coalition in Δ. Hence, in a non-altruistic setting or in a SF ACFG, player 2 prefers Γ to Δ. In an altruistic hedonic game, player 2's decision also depends on her friends within the same coalition which are players 1 and 3 in Γ and just player 1 in Δ. Now, in an ACFG, player 2 considers the valuation of all of her friends including players 3 and 4 both in Γ and Δ. Table 1 shows the valuations of player 2 and her friends resulting from the number of friends and enemies in their coalition in Γ and Δ.

Under min-based EQ and AL preferences, player 2 prefers Γ to Δ, as 5 is the minimum valuation in Γ and 4 in Δ. Under sum-based EQ preferences, player 2 is indifferent between Γ and Δ, as both valuations sum up to 27, and under sum-based AL preferences player 2 prefers Δ to Γ.

Table 1. The valuations of player 2 and her friends for Γ and Δ in the CFG of Example 1

i	$v_i(\Gamma)$	$v_i(\Delta)$
1	$1 \cdot 6 - 1$	$1 \cdot 6$
2	$2 \cdot 6$	$1 \cdot 6$
3	$1 \cdot 6 - 1$	$1 \cdot 6 - 2$
4	$1 \cdot 6 - 1$	$2 \cdot 6 - 1$

3 Core Stability in ACFGs

There are multiple common stability notions which indicate whether a coalition structure is accepted by the players. One of the most famous of these notions is *core stability*. For a CFG $(N, \succeq)$, let $\Gamma_{C \to \emptyset}$ denote the coalition structure that arises if all players in a coalition $C \subseteq N$ leave their coalition in coalition structure $\Gamma \in \mathscr{C}_N$ and form a new coalition while all players in $N \setminus C$ remain in their coalition from Γ. We say C *blocks* Γ if all players in C prefer $\Gamma_{C \to \emptyset}$ to Γ, and Γ is *core stable* if no nonempty coalition blocks Γ.

3.1 The Core Stability Verification Problem

We are interested in the complexity of the *core stability verification problem*: Given an ACFG $(N, \succeq)$ and a coalition structure Γ, is Γ core stable? Kerkmann *et al.* [5] have shown that this problem is coNP-complete for sum-based and min-based SF ACFGs. For the two more altruistic cases of EQ and AL ACFGs, however, this problem remained open. We will show that core stability verification is also coNP-complete for sum-based and min-based EQ and AL ACFGs. Membership of core stability verification in coNP is known for any ACFG [5], so it remains to prove coNP-hardness for our four problems. For each of them, we will provide a polynomial-time many-one reduction from the NP-complete problem RESTRICTED EXACT COVER BY 3-SETS (RX3C)—as shown by Gonzalez [4]—to the complement of one of our core stability verification problems. An instance $(B, \mathscr{S})$ of RX3C consists of a set $B = [1, 3k]$ and a collection $\mathscr{S} = \{S_1, \ldots, S_{3k}\}$ where every set S_i for $i \in [1, 3k]$ contains three elements of B and every element in B occurs in exactly three sets in $\mathscr{S}$. The question is whether $\mathscr{S}$ contains an exact cover of B, i.e., if there exists a subset $\mathscr{S}' \subseteq \mathscr{S}$ with $|\mathscr{S}'| = k$ and $\bigcup_{S \in \mathscr{S}'} S = B$.

3.2 Min-Based EQ and AL ACFGs

We start with min-based ACFGs, where the valuations of the friends are aggregated by taking the minimum.

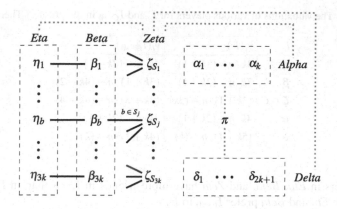

Fig. 2. Network of friends in the proof of Theorem 1

Theorem 1. *Core stability verification in min-based EQ and AL ACFGs is* coNP-*hard.*

Proof. Let $(B, \mathscr{S})$ be an instance of RX3C with $B = \{1, \dots, 3k\}$ and $\mathscr{S} = \{S_1, \dots, S_{3k}\}$. We assume that $k > 2$. From $(B, \mathscr{S})$ we now construct the following ACFG. The set of players is $N = Beta \cup Eta \cup Zeta \cup Alpha \cup Delta \cup \{\pi\}$ with

$$Beta = \{\beta_b \mid b \in B\},$$
$$Eta = \{\eta_b \mid b \in B\},$$
$$Zeta = \{\zeta_S \mid S \in \mathscr{S}\},$$
$$Alpha = \{\alpha_1, \dots, \alpha_k\}, \text{ and}$$
$$Delta = \{\delta_1, \dots, \delta_{2k+1}\}.$$

Fig. 2 shows the network of friends, where a dashed rectangle around a group of players means that all these players are friends of each other and where a dotted gray line between two groups of players indicates that every player from one group is friends with every player from the other group:

- All players in *Beta* are friends with each other.
- All players in *Eta* are friends with each other.
- All players in *Alpha* are friends with each other.
- All players in *Delta* are friends with each other.
- For each $b \in [1, 3k]$, η_b is friends with β_b.
- For each $S \in \mathscr{S}$, ζ_S is friends with every β_b, $b \in S$, with all players in *Alpha*, with all players in *Delta*, and with all players in *Eta*.
- π is friends with all players in *Alpha* and with all players in *Delta*.

Consider the coalition structure $\Gamma = \{R, Q\}$ with $R = Beta \cup Eta$ and $Q = N \setminus R$. We will now show that $\mathscr{S}$ contains an exact cover for B if and only if Γ is not core stable under the min-based EQ or AL model.

Only if: Assume that there is an exact cover $\mathscr{S}' \subseteq \mathscr{S}$ for B. Then $|\mathscr{S}'| = k$. Consider the coalition $C = Beta \cup Eta \cup \{\zeta_S \mid S \in \mathscr{S}'\}$. Table 2 shows the valuations of various players for Γ and $\Gamma_{C \to \emptyset}$. We can see that all players in C prefer $\Gamma_{C \to \emptyset}$ to Γ:

Table 2. The valuations of various players for Γ and $\Gamma_{C\to\emptyset}$ in the proof of Theorem 1

i	$v_i(\Gamma)$	$v_i(\Gamma_{C\to\emptyset})$
η	$3k\cdot n-(3k-1)$	$4k\cdot n-(3k-1)$
β	$3k\cdot n-(3k-1)$	$(3k+1)\cdot n-(4k-2)$
$\zeta\in C$	$(3k+1)\cdot n-(3k)$	$(3k+3)\cdot n-(4k-4)$
α	$4k\cdot n-(2k+1)$	$3k\cdot n-(2k+1)$
δ	$(5k+1)\cdot n-(k)$	$(4k+1)\cdot n-(k)$

1. All players in *Eta*, *Beta*, and *Zeta* have more friends in $\Gamma_{C\to\emptyset}$ than in Γ. Thus, all players in *Eta* and *Beta* prefer $\Gamma_{C\to\emptyset}$ to Γ.
2. The friend with minimum valuation of a $\zeta\in C$ is a β in Γ and an α in $\Gamma_{C\to\emptyset}$. Since this minimum valuation is greater in the case of $\Gamma_{C\to\emptyset}$, all $\zeta\in C$ prefer $\Gamma_{C\to\emptyset}$ to Γ. This holds under min-based EQ and AL preferences.

Hence, C blocks Γ, so Γ is not core stable.

If: Assume that Γ is not core stable and let $C\subseteq N$ be a coalition that blocks Γ. First of all, in the following two claims, we will show that neither π nor any of the players in *Alpha* or *Delta* can be in C.

Claim 1. $\pi\notin C$.

Proof. For a contradiction, assume that $\pi\in C$. Then all players in *Alpha* and in *Delta* have to be in C. Under min-based EQ, this is obvious because in Γ π is the player with the lowest valuation of her friends and herself and therefore π needs at least as many friends in C as in Q in order to prefer $\Gamma_{C\to\emptyset}$ to Γ. Under min-based AL, all players in *Alpha* have to be in C because otherwise they would have at least one friend fewer in $\Gamma_{C\to\emptyset}$ than in Γ, and π would not prefer $\Gamma_{C\to\emptyset}$ to Γ in this case. Then all the players in *Alpha* need to prefer $\Gamma_{C\to\emptyset}$ to Γ, too. Since π is their friend with lowest valuation in Γ, π needs at least as many friends in $\Gamma_{C\to\emptyset}$ as in Γ. Thus, all players in *Delta* have to be in C under min-based AL as well. Now, π is in a coalition with all of her friends in C and Q and therefore needs fewer enemies in C than in Q, so that the players from *Alpha* and *Delta* prefer $\Gamma_{C\to\emptyset}$ to Γ. So there can be between zero and $3k-1$ players from *Zeta* in C. However, the players from *Zeta* that are not in C have zero friends in $\Gamma_{C\to\emptyset}$, and that is why the players in *Alpha* and *Delta* would not prefer such a coalition. So π cannot be in C. □

Claim 2. $\forall i\in[1,k]\ \forall j\in[1,2k+1]:\alpha_i\notin C\wedge\delta_j\notin C$.

Proof. For a contradiction, assume that some player p from *Alpha* or *Delta* is in C. Then π has to be in C because otherwise π would have fewer friends and therefore p would not prefer $\Gamma_{C\to\emptyset}$ to Γ. But this is a contradiction to Claim 1. □

After we have shown which players definitely cannot be in C, we will now focus on the players that actually can be in C. To this end, we will firstly show that if any player from *Beta* or *Eta* is in C, then every player from *Beta* and *Eta* has to be in C, as well as at least k *Zeta* players.

Claim 3. $\exists b \in [1, 3k] : \eta_b \in C \vee \beta_b \in C \Rightarrow (\forall b \in [1, 3k] : \eta_b \in C \wedge \beta_b \in C) \wedge |Zeta \cap C| \geq k$.

Proof. Assume that an arbitrary player m from Eta or $Beta$ is in C. Without loss of generality, we may assume that m is in $Beta$. However, m cannot be in C without the other players in $Beta$, because each player from $Beta$ who is not in C loses at least one friend. Furthermore, every player in $Beta$ is friends with exactly one player in Eta and so every player in Eta has to be in C too, because otherwise there would be one player in $Beta$ who would not prefer $\Gamma_{C \to \emptyset}$ to Γ because of her Eta friend. So we have $R \subset C$. Now we need players who increase the number of friends of every player in R, so that every player in R prefers $\Gamma_{C \to \emptyset}$ to Γ. For that we have the players in $Zeta$. Every player in $Zeta$ is friends with every player in Eta, but only with three players in $Beta$. We need at least so many players from $Zeta$ that every player in $Beta$ has at least one $Zeta$ friend in C. That is why we need at least k $Zeta$ players in C. ☐

In the following claim, we will show that if any player from $Zeta$ is in C, then again all players from $Beta$ and Eta and at least k players from $Zeta$ have to be in C as well.

Claim 4. $\exists S \in \mathscr{S} : \zeta_S \in C \Rightarrow (\forall b \in [1, 3k] : \eta_b \in C \wedge \beta_b \in C) \wedge |Zeta \cap C| \geq k$.

Proof. Assume that an arbitrary player ζ_S from $Zeta$ is in C. Note that ζ_S prefers $\Gamma_{C \to \emptyset}$ to Γ if two conditions are met:

(a) all the players in Eta and all of her $Beta$ friends increase their number of friends in $\Gamma_{C \to \emptyset}$ compared to Γ,[1] and
(b) all the players in $Alpha$ and $Delta$ have more friends or as many friends and fewer enemies in $\Gamma_{C \to \emptyset}$ than all the players in $Beta$ and Eta in Γ.

In particular, to achieve (a), all players from $Beta$ and Eta have to be in C, which implies that at least k players in $Zeta$ have to be in C as shown in the proof of Claim 3. ☐

Next, we show that at most k players from $Zeta$ are in C.

Claim 5. $|Zeta \cap C| \leq k$.

Proof. In the proof of Claim 4, we have stated two sufficient conditions, called (a) and (b), for ζ_S to prefer $\Gamma_{C \to \emptyset}$ to Γ. In particular, to achieve (b), at least $2k$ players from $Zeta$ have to stay in the coalition Q. Thus, every player in $Alpha$ has at least $3k$ friends and just $2k - 1$ enemies in the new coalition which is an improvement compared to the $3k$ friends and $3k - 1$ enemies of all players in $Beta$ and Eta in Γ. Every player in $Delta$ has even more friends. As a consequence of the fact that at least $2k$ players in $Zeta$ have to stay in the coalition Q, at most k $Zeta$ players can be in C. ☐

It follows from the previous claims that we need exactly k players from $Zeta$ in C and we have to ensure that every $Beta$ player gets one new friend in $\Gamma_{C \to \emptyset}$. This is only possible if there exists an exact cover for B in $\mathscr{S}$, which completes the proof of Theorem 1. ☐

---

[1] The number of enemies of these friends cannot be reduced in $\Gamma_{C \to \emptyset}$, because we have shown in Claim 3 that if any player from Eta or $Beta$ is in C, then all of them are, or otherwise none of them is in C, but then all of them are in R as in Γ.

3.3 Sum-Based EQ and AL ACFGs

We now turn to sum-based EQ and AL ACFGs, starting with the former.

Theorem 2. *Core stability verification in sum-based EQ ACFGs is coNP-hard.*

Proof. Let $(B, \mathscr{S})$ be an instance of RX3C with $B = \{1, \ldots, 3k\}$ and $\mathscr{S} = \{S_1, \ldots, S_{3k}\}$. We assume that $k > 3$. From $(B, \mathscr{S})$ we now construct the following ACFG. The set of players is $N = Beta \cup Eta \cup Zeta \cup Alpha \cup Delta \cup Epsilon$ with

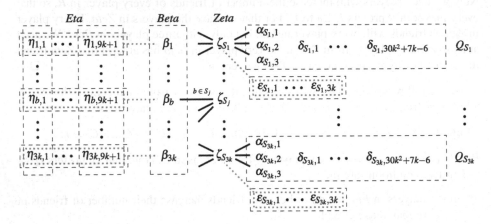

Fig. 3. Network of friends in the proof of Theorem 2 and in the proof of Theorem 3

$$Beta = \{\beta_b \mid b \in B\},$$
$$Eta = \{\eta_{b,1}, \ldots, \eta_{b,9k+1} \mid b \in B\},$$
$$Zeta = \{\zeta_S \mid S \in \mathscr{S}\},$$
$$Alpha = \{\alpha_{S,1}, \alpha_{S,2}, \alpha_{S,3} \mid S \in \mathscr{S}\},$$
$$Delta = \{\delta_{S,1}, \ldots, \delta_{S,30k^2+7k-6} \mid S \in \mathscr{S}\}, \text{ and}$$
$$Epsilon = \{\varepsilon_{S,1}, \ldots, \varepsilon_{S,3k} \mid S \in \mathscr{S}\}.$$

So, we have $n = 90k^3 + 57k^2$ players in total.

Figure 3 shows the network of friends, where a dashed rectangle around a group of players means that all these players are friends of each other and where a dotted gray rectangle around a group of players indicates that every player of this group is friends with the one player who is connected with the group via the dotted gray line:

- All players in *Beta* are friends with each other.
- For every $b \in [1, 3k]$ and $i \in [1, 9k + 1]$, $\eta_{b,i}$ is friends with every $\eta_{b,j}$, where $j \in [1, 9k + 1]$ and $j \neq i$, with every $\eta_{a,i}$, where $a \in [1, 3k]$ and $a \neq b$, and with β_b.

- For every $S \in \mathscr{S}$, all players in $\{\alpha_{S,1}, \alpha_{S,2}, \alpha_{S,3}, \delta_{S,1}, \ldots, \delta_{S,30k^2+7k-6}\}$ are friends of each other.
- For every $S \in \mathscr{S}$, $\varepsilon_{S,1}, \ldots, \varepsilon_{S,3k}$ are friends of each other.
- For every $S \in \mathscr{S}$, ζ_S is friends with every β_b, $b \in S$, with $\varepsilon_{S,1}, \ldots, \varepsilon_{S,3k}$, and with $\alpha_{S,1}$, $\alpha_{S,2}$, and $\alpha_{S,3}$.

Consider the coalition structure $\Gamma = \{R, Q_{S_1}, \ldots, Q_{S_{3k}}\}$ with $R = Beta \cup Eta$ and $Q_S = \{\zeta_S, \alpha_{S,1}, \alpha_{S,2}, \alpha_{S,3}, \varepsilon_{S,1}, \ldots, \varepsilon_{S,3k}, \delta_{S,1}, \ldots, \delta_{S,30k^2+7k-6}\}$ for $S \in \mathscr{S}$. We will now show that $\mathscr{S}$ contains an exact cover for B if and only if Γ is not core stable under the sum-based EQ model.

Only if: Assume that there is an exact cover $\mathscr{S}' \subseteq \mathscr{S}$ for B. Then $|\mathscr{S}'| = k$. Consider the coalition $C = Beta \cup Eta \cup \{\zeta_S \mid S \in \mathscr{S}'\} \cup \{\varepsilon_{S,1}, \ldots, \varepsilon_{S,3k} \mid S \in \mathscr{S}'\}$. We will now show that C blocks Γ by showing that all players in C prefer $\Gamma_{C \to \emptyset}$ to Γ. Hence, Γ is not core stable.

We start with the players in *Beta*: Every player in *Beta* has one more friend in $\Gamma_{C \to \emptyset}$ than in Γ and every player in *Eta* and *Zeta* has as many friends in $\Gamma_{C \to \emptyset}$ as in Γ, so the total number of friends of every *Beta* player and her friends is increased by $3k$. Furthermore, for every player in *Beta*, the increase of enemies of her friends and herself in $\Gamma_{C \to \emptyset}$ compared to Γ is less than n, as their $9k + 1$ *Eta* friends get $k \cdot (3k + 1)$ new enemies each, the players in *Beta* get $k \cdot (3k + 1) - 1$ new enemies each, and the one *Zeta* friend in C even reduces her number of enemies. So the new enemies cannot even make up for one new friend. Thus, all players in *Beta* prefer $\Gamma_{C \to \emptyset}$ to Γ.

Next, we turn to the players in *Eta*: Every player in *Beta* has one more friend in $\Gamma_{C \to \emptyset}$ than in Γ and every player in *Eta* has as many friends in $\Gamma_{C \to \emptyset}$ as in Γ, so the total number of friends of every *Eta* player and her friends is increased by one. Also, for every player in *Eta*, the total number of new enemies of the friends and oneself in $\Gamma_{C \to \emptyset}$ adds up to $36k^3 + 15k^2 + k - 1$, as the one *Beta* friend gets $k \cdot (3k + 1) - 1$ new enemies and the players in *Eta* each get $k \cdot (3k + 1)$ new enemies, and every *Eta* is friends with $12k - 1$ other *Eta* players. The number of new enemies is less than n and therefore the new enemies cannot make up for the one new friend. Thus, all *Eta* players prefer $\Gamma_{C \to \emptyset}$ to Γ.

Now, we turn to the players in $\{\zeta_S \mid S \in \mathscr{S}'\}$: For every ζ_S with $S \in \mathscr{S}'$, the three *Beta* friends of ζ_S gain one friend, the three *Alpha* friends lose one friend, and ζ_S and her *Epsilon* friends keep their number of friends. Hence, the total number of friends of ζ_S and her friends remains the same in $\Gamma_{C \to \emptyset}$ as in Γ. However, the total number of enemies of ζ_S and her friends is reduced. ζ_S and her friends get $90k^3 + 51k^2 + 4k - 7$ new enemies in $\Gamma_{C \to \emptyset}$ (see Table 3), but lose $90k^3 + 51k^2 + 7k - 6$ enemies compared to Γ (see Table 4). Thus, all players in $\{\zeta_S \mid S \in \mathscr{S}'\}$ prefer $\Gamma_{C \to \emptyset}$ to Γ.

Next, we turn to the players in $\{\varepsilon_{S,1}, \ldots, \varepsilon_{S,3k} \mid S \in \mathscr{S}'\}$: For every $\varepsilon_{S,i}$ with $S \in \mathscr{S}'$ and $i \in [1, 3k]$, $\varepsilon_{S,i}$ and all of her friends keep their numbers of friends. Hence, the total number of friends of $\varepsilon_{S,i}$ and her friends remains the same in $\Gamma_{C \to \emptyset}$ as in Γ. However, the total number of enemies of $\varepsilon_{S,i}$ and her friends is reduced. $\varepsilon_{S,i}$ and her friends get $90k^3 + 42k^2 + k - 4$ new enemies in $\Gamma_{C \to \emptyset}$ (see Table 5), but lose $90k^3 + 51k^2 - 2k - 6$ enemies compared to Γ (see Table 6). Thus, also all players in $\{\varepsilon_{S,1}, \ldots, \varepsilon_{S,3k} \mid S \in \mathscr{S}'\}$ prefer $\Gamma_{C \to \emptyset}$ to Γ.

Table 3. Enemies of ζ_S and her friends only in $\Gamma_{C \to \emptyset}$ in the proof of Theorem 2

Group of friends of ζ_S and ζ_S herself	Enemies of these players only in $\Gamma_{C \to \emptyset}$	Total number of enemies
$\varepsilon_{S,1}, \ldots, \varepsilon_{S,3k}$	$\beta_1, \ldots, \beta_{3k}$	$9k^2$
$\varepsilon_{S,1}, \ldots, \varepsilon_{S,3k}$	$\eta_{b,1}, \ldots, \eta_{b,9k+1}$ for $b \in [1,3k]$	$81k^3 + 9k^2$
$\varepsilon_{S,1}, \ldots, \varepsilon_{S,3k}$	ζ_T for $T \in \mathscr{S}' \setminus S$	$3k^2 - 3k$
$\varepsilon_{S,1}, \ldots, \varepsilon_{S,3k}$	$\varepsilon_{T,1}, \ldots, \varepsilon_{T,3k}$ for $T \in \mathscr{S}' \setminus S$	$9k^3 - 9k^2$
β_b for $b \in S$	ζ_T for $T \in \mathscr{S}' \setminus S$	$3k - 3$
β_b for $b \in S$	$\varepsilon_{T,1}, \ldots, \varepsilon_{T,3k}$ for $T \in \mathscr{S}'$	$9k^2$
ζ_S	β_b for $b \notin S$	$3k - 3$
ζ_S	$\eta_{b,1}, \ldots, \eta_{b,9k+1}$ for $b \in [1,3k]$	$27k^2 + 3k$
ζ_S	ζ_T for $T \in \mathscr{S}' \setminus S$	$k - 1$
ζ_S	$\varepsilon_{T,1}, \ldots, \varepsilon_{T,3k}$ for $T \in \mathscr{S}' \setminus S$	$3k^2 - 3k$
Result		$90k^3 + 51k^2 + 4k - 7$

Table 4. Enemies of ζ_S and her friends only in Γ in the proof of Theorem 2

Group of friends of ζ_S and ζ_S herself	Enemies of these players only in Γ	Total number of enemies
$\varepsilon_{S,1}, \ldots, \varepsilon_{S,3k}$	$\delta_{S,1}, \ldots, \delta_{S,30k^2+7k-6}$	$90k^3 + 21k^2 - 18k$
$\varepsilon_{S,1}, \ldots, \varepsilon_{S,3k}$	$\alpha_{S,1}, \alpha_{S,2}, \alpha_{S,3}$	$9k$
$\alpha_{S,1}, \alpha_{S,2}, \alpha_{S,3}$	$\varepsilon_{S,1}, \ldots, \varepsilon_{S,3k}$	$9k$
ζ_S	$\delta_{S,1}, \ldots, \delta_{S,30k^2+7k-6}$	$30k^2 + 7k - 6$
Result		$90k^3 + 51k^2 + 7k - 6$

If: Assume that Γ is not core stable and let $C \subseteq N$ be a coalition that blocks Γ. We now show that then there exists an exact cover $\mathscr{S}' \subseteq \mathscr{S}$ for B, which will follow from a series of claims whose proofs are omitted due to space constraints (but will be contained in the full version of this paper).

First of all, we show in Claims 6 and 7 that if one player of $\zeta_S, \varepsilon_{S,1}, \ldots, \varepsilon_{S,3k}$ with $S \in \mathscr{S}$ is in C, then each of those player is in C.

Claim 6. $\exists S \in \mathscr{S} \; \exists i \in [1,3k] : \varepsilon_{S,i} \in C \Rightarrow \forall j \in [1,3k] : \varepsilon_{S,j} \in C \wedge \zeta_S \in C$.

Claim 7. $\exists S \in \mathscr{S} : \zeta_S \in C \Rightarrow \forall i \in [1,3k] : \varepsilon_{S,i} \in C$.

Now we know that if one of the players $\zeta_S, \varepsilon_{S,1}, \ldots, \varepsilon_{S,3k}$ is in C, then they all are in C. In the following we will use this knowledge to prove the claim that no player from *Delta* can be in C from which we can further conclude that also no player from *Alpha* can be in C.

Claim 8. $\forall S \in \mathscr{S} \; \forall i \in [1, 30k^2 + 7k - 6] : \delta_{S,i} \notin C$.

Claim 9. $\forall S \in \mathscr{S} \; \forall i \in \{1,2,3\} : \alpha_{S,i} \notin C$.

After we have shown which players definitely cannot be in C, we will now focus on the players that actually can be in C. To this end, we will firstly show that if one player

Table 5. Enemies of $\varepsilon_{S,i}$ and her friends only in $\Gamma_{C\to\emptyset}$ in the proof of Theorem 2

Group of friends of $\varepsilon_{S,i}$ and $\varepsilon_{S,i}$ herself	Enemies of these players only in $\Gamma_{C\to\emptyset}$	Total number of enemies
$\varepsilon_{S,1},\ldots,\varepsilon_{S,3k},\zeta_S$	$\eta_{b,1},\ldots,\eta_{b,9k+1}$ for $b\in[1,3k]$	$81k^3+36k^2+3k$
$\varepsilon_{S,1},\ldots,\varepsilon_{S,3k},\zeta_S$	ζ_T for $T\in\mathscr{S}'\setminus S$	$3k^2-2k-1$
$\varepsilon_{S,1},\ldots,\varepsilon_{S,3k},\zeta_S$	$\varepsilon_{T,1},\ldots,\varepsilon_{T,3k}$ for $T\in\mathscr{S}'\setminus S$	$9k^3-6k^2-3k$
$\varepsilon_{S,1},\ldots,\varepsilon_{S,3k}$	$\beta_1,\ldots,\beta_{3k}$	$9k^2$
ζ_S	β_b for $b\notin S$	$3k-3$
Result		$90k^3+42k^2+k-4$

Table 6. Enemies of $\varepsilon_{S,i}$ and her friends only in Γ in the proof of Theorem 2

Group of friends of $\varepsilon_{S,i}$ and $\varepsilon_{S,i}$ herself	Enemies of these players only in Γ	Total number of enemies
$\varepsilon_{S,1},\ldots,\varepsilon_{S,3k},\zeta_S$	$\delta_{S,1},\ldots,\delta_{S,30k^2+7k-6}$	$90k^3+51k^2-11k-6$
$\varepsilon_{S,1},\ldots,\varepsilon_{S,3k}$	$\alpha_{S,1},\alpha_{S,2},\alpha_{S,3}$	$9k$
Result		$90k^3+51k^2-2k-6$

from *Beta* or *Eta* is in C, then every player from *Beta* and *Eta* has to be in C as well. We can show this by proving two claims. One claim is that if one player from *Eta* is in C, then every player from *Eta* and *Beta* is in C, and the other claim is that if one player from *Beta* is in C, then again every player from *Eta* and *Beta* is in C.

Claim 10. $\exists i\in[1,9k+1]\ \exists b\in[1,3k]:\eta_{b,i}\in C\Rightarrow\forall i\in[1,9k+1]\ \forall b\in[1,3k]:\eta_{b,i}\in C\wedge\beta_b\in C$.

Claim 11. $\exists b\in[1,3k]:\beta_b\in C\Rightarrow\forall i\in[1,9k+1]\ \forall b\in[1,3k]:\eta_{b,i}\in C\wedge\beta_b\in C$.

So far we know that if one player from *Beta* or *Eta* is in C, then every other player from *Beta* and *Eta* has to be in C as well. We will now show that in this case also at least k *Zeta* players have to be in C

Claim 12. $\exists i\in[1,9k+1]\ \exists b\in[1,3k]:\eta_{b,i}\in C\vee\beta_b\in C\Rightarrow|Zeta\cap C|\geq k$.

We will now take a look at what happens if a player from *Zeta* is in C. We have already shown that in this case all of her *Epsilon* friends have to be in C. We will now show that also all players in *Beta* and *Eta* have to be in C in this case.

Claim 13. $\exists S\in\mathscr{S}:\zeta_S\in C\Rightarrow\forall i\in[1,9k+1]\ \forall b\in[1,3k]:\eta_{b,i}\in C\wedge\beta_b\in C$.

At this point we know that no player from *Delta* and *Alpha* can be in C, and we further know that if one player from *Beta* or *Eta* or *Zeta* or *Epsilon* is in C, then all players from *Beta* and *Eta* plus at least k players from *Zeta* and all of their corresponding *Epsilon* friends have to be in C. The *Epsilon* players in C will only prefer $\Gamma_{C\to\emptyset}$ to Γ if the total number of enemies of their friends is reduced, as none of their friends can get more friends. In the following, we will demonstrate that this will not happen if there are more than k *Zeta* players in C.

Claim 14. $|Zeta\cap C|\leq k$.

So we need exactly k players from *Zeta* in C and we have to ensure that every *Beta* player gets one new friend in $\Gamma_{C \to \emptyset}$. But this is only possible if there exists an exact cover for B in $\mathscr{S}$, completing the proof of Theorem 2.   ☐

The proof of our last result, which is omitted due to space constraints (but will be contained in the full version of this paper), is based on the same construction as that given in the proof of Theorem 2; however, the proof of correctness of the construction differs and requires some technical work.

Theorem 3. *Core stability verification in sum-based AL ACFGs is* coNP-*hard.*

4   Conclusion

Solving four open problems, we have shown that for sum-based and min-based EQ and AL ACFGs, core stability verification is coNP-complete. Thus, we now know that core stability verification is coNP-complete for *any* ACFG, completing the picture. This is particularly useful since the EQ and AL models are more altruistic than the SF model, and also technically more challenging.

While for altruistic *hedonic* games, core stability verification is known to be coNP-complete in the SF model [6], the complexity for EQ and AL still remains open for them. Yet, we suspect that these cases are coNP-hard as well and it would be interesting to verify this conjecture in future work. It would also be desirable to study the core stability existence problem. Kerkmann *et al.* [5] have shown that this problem is trivial under SF preferences, but again, under EQ and AL preferences the complexity is still open.

Another interesting direction for future research is to study the approximability (or inapproximability) of core stability in ACFGs. For example, in the different context of participatory budgeting, Munagala *et al.* [8] provided approximation results for the problem of how close a given committee—that is computed by some preference aggregation method—is to the core. Similarly, one could define a parameter indicating how close a given coalition structure is to the core of an ACFG and then seek to either compute this parameter approximately, or show that it is hard to approximate.

Acknowledgments. Supported in part by DFG grant RO-1202/21-1.

References

1. Banerjee, S., Konishi, H., Sönmez, T.: Core in a simple coalition formation game. Soc. Choice Welfare **18**(1), 135–153 (2001)
2. Bullinger, M., Kober, S.: Loyalty in cardinal hedonic games. In: Proceedings of the 30th International Joint Conference on Artificial Intelligence, pp. 66–72. ijcai.org (2021)
3. Dimitrov, D., Borm, P., Hendrickx, R., Sung, S.: Simple priorities and core stability in hedonic games. Soc. Choice Welfare **26**(2), 421–433 (2006)
4. Gonzalez, T.: Clustering to minimize the maximum intercluster distance. Theor. Comput. Sci. **38**, 293–306 (1985)
5. Kerkmann, A., Cramer, S., Rothe, J.: Altruism in coalition formation games. Annals Math. Artifi. Intell. (2024). https://doi.org/10.1007/s10472-023-09881-y

6. Kerkmann, A., et al.: Altruistic hedonic games. J. Artif. Intell. Res. **75**, 129–169 (2022)
7. Kerkmann, A., Rothe, J.: Altruism in coalition formation games. In: Proceedings of the 29th International Joint Conference on Artificial Intelligence, pp. 347–353. ijcai.org (2020)
8. Munagala, K., Shen, Y., Wang, K.: Auditing for core stability in participatory budgeting. In: Hansen, K., Liu, T.X., Malekian, A. (eds.) WINE 2022. LNCS, vol. 13778, pp. 292–310. Springer, Cham (2022). https://doi.org/10.1007/978-3-031-22832-2_17
9. Nguyen, N., Rey, A., Rey, L., Rothe, J., Schend, L.: Altruistic hedonic games. In: Proceedings of the 15th International Conference on Autonomous Agents and Multiagent Systems, pp. 251–259. IFAAMAS (2016)
10. Peters, D.: Precise complexity of the core in dichotomous and additive hedonic games. In: Rothe, J. (ed.) Algorithmic Decision Theory. ADT 2017. LNCS, vol. 10576, pp. 214–227. Springer, Cham (2017). https://doi.org/10.1007/978-3-319-67504-6_15
11. Rey, A., Rothe, J., Schadrack, H., Schend, L.: Toward the complexity of the existence of wonderfully stable partitions and strictly core stable coalition structures in enemy-oriented hedonic games. Ann. Math. Artif. Intell. **77**(3), 317–333 (2016)
12. Rothe, J.: Thou shalt love thy neighbor as thyself when thou playest: Altruism in game theory. In: Proceedings of the 35th AAAI Conference on Artificial Intelligence, pp. 15070–15077. AAAI Press (2021)
13. Sung, S., Dimitrov, D.: On core membership testing for hedonic coalition formation games. Oper. Res. Lett. **35**(2), 155–158 (2007)
14. Woeginger, G.: A hardness result for core stability in additive hedonic games. Math. Soc. Sci. **65**(2), 101–104 (2013)

Newton-Type Algorithms for Inverse Optimization: Weighted Span Objective

Kristóf Bérczi[1,2], Lydia Mirabel Mendoza-Cadena[1(✉)],
and Kitti Varga[1,2,3]

[1] MTA-ELTE Matroid Optimization Research Group, Department of Operations Research, ELTE Eötvös Loránd University, Budapest, Hungary
kristof.berczi@ttk.elte.hu, lyd21@student.elte.hu, vkitti@math.bme.hu
[2] HUN-REN-ELTE Egerváry Research Group, Budapest, Hungary
[3] Department of Computer Science and Information Theory, Budapest University of Technology and Economics, Budapest, Hungary

Abstract. In inverse optimization problems, the goal is to modify the costs in an underlying optimization problem in such a way that a given solution becomes optimal, while the difference between the new and the original cost functions, called the deviation vector, is minimized with respect to some objective function. The ℓ_1- and ℓ_∞-norms are standard objectives used to measure the size of the deviation. Minimizing the ℓ_1-norm is a natural way of keeping the total change of the cost function low, while the ℓ_∞-norm achieves the same goal coordinate-wise. Nevertheless, none of these objectives is suitable to provide a *balanced* or *fair* change of the costs.

In this paper, we initiate the study of a new objective that measures the difference between the largest and the smallest weighted coordinates of the deviation vector, called the *weighted span*. We provide a Newton-type algorithm for finding one that runs in strongly polynomial time in the case of unit weights.

Keywords: Algorithm · Inverse optimization · Span

1 Introduction

Informally, in an inverse optimization problem, we are given a feasible solution to an underlying optimization problem together with a linear objective function, and the goal is to modify the objective so that the input solution becomes optimal. Such problems were first considered by Burton and Toint [6] in the context of inverse shortest paths, and found countless applications in various areas ever since. We refer the interested reader to [9,12] for surveys, and to [1] for a quick introduction.

There are several ways of measuring the difference between the original and the modified cost functions, the ℓ_1- and ℓ_∞-norms being probably the most standard ones. Minimizing the ℓ_1-norm of the deviation vector means that the

J. A. Soto and A. Wiese (Eds.): LATIN 2024, LNCS 14579, pp. 334–347, 2024.
https://doi.org/10.1007/978-3-031-55601-2_22

overall change in the costs is small, while minimizing the ℓ_∞-norm results in a new cost function that is close to the original one coordinate-wise. However, these objectives do not provide any information about the relative magnitude of the changes on the elements compared to each other. Indeed, a deviation vector of ℓ_∞-norm δ might increase the cost on an element by δ while decrease it on another by δ, thus resulting in a large relative difference between the two. Such a solution may not be satisfactory in situations when the goal is to modify the costs in a fair manner, and the magnitude of the individual changes is not relevant.

To overcome these difficulties, we consider a new type of objective function that measures the difference between the largest and the smallest weighted coordinates of the deviation vector, called the *weighted span*[1]. Although being rather similar at first glance, the ℓ_∞-norm and the span behave quite differently as the infinite norm measures how far the coordinates of the deviation vector are from zero, while the span measures how far the coordinates of the deviation vector are from each other. In particular, it might happen that one has to change the cost on each element by the same large number, resulting in a deviation vector with large ℓ_∞-norm but with span equal to zero. To the best of our knowledge, this objective was not considered before in the context of inverse optimization.

As a motivation, let us describe an application of the span as an objective. Consider a town consisting of several districts and having a single hospital. The city management would like to improve the public transport system in such a way that the position of the hospital becomes optimal in the sense that it minimizes the maximum distance from the districts. However, the changes in travel times affect different districts, and in order to avoid conflicts, the modifications should be kept as balanced as possible. In this context, the span offers a smooth solution to achieve this goal. This example is a typical appearance of the inverse centre location problem, where we are given as input a graph, a special vertex called the location vertex, and a cost function on the edges. The goal is to modify the costs so that the location vertex becomes optimal, meaning that it minimizes the maximum distance from other vertices. The inverse centre location problem is an fundamental problem with a wide range of applications, such as determining locations for establishing medical clinics, internet and telecommunication towers, police and fire stations. The problem was previously studied with different objective functions, see for example Qian et al. [14], Burkard and Alizadeh [3] and Cai et al. [7].

In this paper, we work out the details of an algorithm for the weighted span objective that is based on the following Newton-type scheme: in each iteration, we check if the input solution is optimal, and if it is not, then we "eliminate" the current optimal solution by balancing the cost difference between them. As it turns out, finding an optimal deviation vector under this objective is highly non-trivial. Intuitively, the complexity of the problem is caused by the fact that the underlying optimization problem may have feasible solutions of different sizes,

---

[1] The notion of span appears under several names in various branches of mathematics, such as *range* in statistics, *amplitude* in calculus, or *deviation* in engineering.

and one has to balance very carefully between increasing the costs on certain elements while decreasing on others to obtain a feasible deviation vector.

Previous Work. In *balanced optimization problems*, the goal is to find a feasible solution such that the difference between the maximum and the minimum weighted variable defining the solution is minimized. Martello, Pulleyblank, Toth, and de Werra [13], Camerini, Maffioli, Martello, and Toth [8], and Ficker, Spieksma and Woeginger [11] considered the balanced assignment and the balanced spanning tree problems. Duin and Volgenant [10] introduced a general solution scheme for minimum deviation problems that is also suited for balanced optimization, and analyzed the approach for spanning trees, paths and Steiner trees in graphs. Ahuja [2] proposed a parametric simplex method for the general balanced linear programming problem as well as a specialized version for the balanced network flow problem. Scutellà [16] studied the balanced network flow problem in the case of unit weights, and showed that it can be solved by using an extension of Radzik's [15] analysis of Newton's method for linear fractional combinatorial optimization problems.

An analogous approach was proposed in the context of ℓ_∞-norm objective by Zhang and Liu [17], who described a model that generalizes numerous inverse combinatorial optimization problems when no bounds are given on the changes. They exhibited a Newton-type algorithm that determines an optimal deviation vector when the inverse optimization problem can be reformulated as a certain maximization problem using dominant sets.

Problem Formulation. We denote the sets of *real* and *positive real* numbers by $\mathbb{R}$ and $\mathbb{R}_+$, respectively. For a positive integer k, we use $[k] := \{1, \ldots, k\}$. Let S be a ground set of size n. Given subsets $X, Y \subseteq S$, the *symmetric difference* of X and Y is denoted by $X \triangle Y := (X \setminus Y) \cup (Y \setminus X)$. For a weight function $w \in \mathbb{R}_+^S$, the total sum of its values over X is denoted by $w(X) := \sum\{w(s) \mid s \in X\}$, where the sum over the empty set is always considered to be 0. Furthermore, we define $\frac{1}{w}(X) := \sum\{\frac{1}{w(s)} \mid s \in X\}$, and set $\|w\|_{-1} := \frac{1}{w}(S)$. When the weights are rational numbers, then the values can be re-scaled as to satisfy $1/w(s)$ being an integer for each $s \in S$. Throughout the paper, we assume that w is given in such a form without explicitly mentioning it, implying that $\frac{1}{w}(X)$ is a non-negative integer for every $X \subseteq S$.

Let S be a finite ground set, $\mathcal{F} \subseteq 2^S$ be a collection of *feasible solutions* for an underlying optimization problem, $F^* \in \mathcal{F}$ be an *input solution*, $c \in \mathbb{R}^S$ be a *cost function*, and $w \in \mathbb{R}_+^S$ be a positive *weight function*. We assume that an oracle $\mathcal{O}$ is also available that determines an optimal solution of the underlying optimization problem $(S, \mathcal{F}, c')$ for any cost function $c' \in \mathbb{R}^S$. Note that $\mathcal{F}$ is usually specified in an implicit fashion. For example, $(S, \mathcal{F}, c)$ can be the problem of finding a minimum-cost perfect matching in a graph $G = (V, E)$ with a cost function $c \in \mathbb{R}^E$ on its edge set, $S = E$ and $\mathcal{F}$ denoting the set of all perfect matchings.

In the *minimum-cost inverse optimization problem under the weighted span objective* $(S, \mathcal{F}, F^*, c, \text{span}_w(\cdot))$, we seek a *deviation vector* $p \in \mathbb{R}^S$ such that

(a) F^* is a minimum-cost member of $\mathcal{F}$ with respect to $c - p$,
(b) $\mathrm{span}_w(p) := \max\{w(s) \cdot p(s) \mid s \in S\} - \min\{w(s) \cdot p(s) \mid s \in S\}$ is minimized.

A deviation vector is called *feasible* if it satisfies condition (a) and *optimal* if, in addition, it attains the minimum in (b).

Our Results. Our main result is a Newton-type algorithm that determines an optimal deviation vector for inverse optimization problems under the weighted span objective. Our algorithm makes $O(\|w\|_{-1}^6)$ calls to the oracle $\mathcal{O}$. In particular, the algorithm runs in strongly polynomial time for unit weights if the oracle $\mathcal{O}$ can be realized by a strongly polynomial algorithm.

The rest of the paper is organized as follows. In Sect. 2, we show that it is enough to look for an optimal deviation vector having a special form. The algorithm is presented in Sect. 3. Due to length constraints, detailed proofs of technical statements are deferred to the full version of the paper [5].

2 Optimal Deviation Vectors

We first verify the existence of an optimal deviation vector that has a special structure, which serves as the basic idea of our algorithm. Assume for a moment that w is the unit weight function. If the members of $\mathcal{F}$ have the same size, then one can always decrease the costs on the elements of F^* by some δ in such a way that F^* becomes a minimum-cost member of $\mathcal{F}$. As another extreme case, if F^* is the unique minimum- or maximum-sized member of $\mathcal{F}$, then one can always shift the costs by the same number Δ in such a way that F^* becomes a minimum-cost member of $\mathcal{F}$. The idea of our approach is to combine these two types of changes in the general case by tuning the parameters δ, Δ, while also taking the weights into account.

Lemma 1. *Every minimum-cost inverse optimization problem* $(S, \mathcal{F}, F^*, c, \mathrm{span}_w(\cdot))$ *is feasible.*

Proof. Clearly, $p \equiv c$ is always a feasible deviation vector since $(c - p)(F) = 0$ holds for any $F \in \mathcal{F}$. $\qquad\square$

Consider an instance $(S, \mathcal{F}, F^*, c, \mathrm{span}_w(\cdot))$ of the minimum-cost inverse optimization problem under the weighted span objective, where $w \in \mathbb{R}_+^S$ is a positive weight function. For any $\delta, \Delta \in \mathbb{R}$, let $p_{[\delta,\Delta]} : S \to \mathbb{R}$ be defined as

$$
p_{[\delta,\Delta]}(s) := \begin{cases} (\delta + \Delta)/w(s) & \text{if } s \in F^*, \\ \Delta/w(s) & \text{if } s \in S \setminus F^*, \end{cases}
$$

The following observation shows that there exists an optimal deviation vector of special form.

Lemma 2. *Let* $(S, \mathcal{F}, F^*, c, \mathrm{span}_w(\cdot))$ *be an inverse optimization problem and let* p *be an optimal deviation vector. Then* $p_{[\delta, \Delta]}$ *is also an optimal deviation vector, where* $\Delta := \min\{w(s) \cdot p(s) \mid s \in S\}$ *and* $\delta := \max\{w(s) \cdot p(s) \mid s \in S\} - \Delta$.

Proof. We show that (a) holds. The definitions of Δ and δ imply that $\Delta/w(s) \leq p(s) \leq (\delta + \Delta)/w(s)$ holds for every $s \in S$. Then for an arbitrary solution $F \in \mathcal{F}$, we get

$$
(c - p_{[\delta, \Delta]})(F^*) - (c - p_{[\delta, \Delta]})(F)
$$

$$
= \left(c(F^*) - \sum_{s \in F^*} p_{[\delta, \Delta]}(s) \right) - \left(c(F) - \sum_{s \in F} p_{[\delta, \Delta]}(s) \right)
$$

$$
= \left(c(F^*) - \sum_{s \in F^*} \tfrac{\delta + \Delta}{w(s)} \right) - \left(c(F) - \sum_{s \in F \cap F^*} \tfrac{\delta + \Delta}{w(s)} - \sum_{s \in F \setminus F^*} \tfrac{\Delta}{w(s)} \right)
$$

$$
= c(F^*) - c(F) - \sum_{s \in F^* \setminus F} \tfrac{\delta + \Delta}{w(s)} + \sum_{s \in F \setminus F^*} \tfrac{\Delta}{w(s)}
$$

$$
\leq c(F^*) - c(F) - \sum_{s \in F^* \setminus F} p(s) + \sum_{s \in F \setminus F^*} p(s)
$$

$$
= \left(c(F^*) - p(F^*) \right) - \left(c(F) - p(F) \right)
$$

$$
\leq 0,
$$

where the last inequality holds by the feasibility of p.

To see that (b) holds, observe that $\mathrm{span}_w(p) = \max\{w(s) \cdot p(s) \mid s \in S\} - \min\{w(s) \cdot p(s) \mid s \in S\} = \delta$ and $\mathrm{span}_w(p_{[\delta, \Delta]}) \leq (\delta + \Delta) - \Delta = \delta$. That is, $p_{[\delta, \Delta]}$ is also optimal, concluding the proof of the lemma. □

3   Algorithm

Consider an instance $(S, \mathcal{F}, F^*, c, \mathrm{span}_w(\cdot))$ of the minimum-cost inverse optimization problem under the weighted span objective. By Lemma 1, the problem is always feasible, and by Lemma 2, there exists an optimal deviation vector of the form $p_{[\delta, \Delta]}$ for some choice of $\delta, \Delta \in \mathbb{R}$. Hence the problem reduces to identifying the values of Δ and $\delta + \Delta$.

The idea of the algorithm is to eliminate bad sets in $\mathcal{F}$ iteratively. Assume for a moment that w is the unit weight function. Let us call a set $F \in \mathcal{F}$ *bad* if it has smaller cost than the input solution F^* with respect to the current cost function, *small* if $|F| < |F^*|$ and *large* if $|F| > |F^*|$. A small or large bad set can be eliminated by decreasing or increasing the cost on every element by the same value, respectively. Note that such a step does not change the span of the deviation vector. Therefore, it is not enough to concentrate on single bad sets, as otherwise it could happen that we jump back and forth between the same pair of small and large bad sets by alternately decreasing and increasing the costs. To avoid the algorithm changing the costs in a cyclic way, we keep track of the small and large bad sets that were found the latest. If in the next iteration we find a bad set F having the same size as F^*, then we drop the latest small and large bad sets, if they exist, and eliminate F on its own. However, if we find a small or large bad set F, then we eliminate it together with the latest large or small bad set, if exists, respectively.

We refer to this algorithm as "Newton-type" since it mimics the idea of Newton's method for finding the root x^* of a function f. In each step, we consider the current optimal solution F (in Newton's method, this is the current vector x). If the cost of F equals that of F^* (in Newton's method, we check if x is a root of f), then we stop. Otherwise, we modify the costs in such a way that F is eliminated (in Newton's method, we move from x to a root of the first-order approximation of f). Note that a similar naming convention was used already by Zhang and Liu [17].

For arbitrary weights, the only difference is that the size of a set is measured in terms of $\frac{1}{w}$, i.e., a set $F \in \mathcal{F}$ small if $\frac{1}{w}(F) < \frac{1}{w}(F^*)$ and large if $\frac{1}{w}(F) > \frac{1}{w}(F^*)$. At each iteration i, we distinguish five main cases depending on the bad sets eliminated in that iteration. In Case 1, we eliminate a bad set F_i having the same size as F^*. In Case 2, we eliminate a small bad set F_i alone. In Case 3, we eliminate a small bad set F_i together with a large bad set Z_i that was found before. In Case 4, we eliminate a large bad set F_i alone. Finally, in Case 5, we eliminate a large bad set F_i together with a small bad set X_i that was found before.

The algorithm is presented as Algorithm 1. By convention, undefined objects are denoted by $*$. The algorithm and its analysis is rather technical, and requires the discussion of the five cases. Nevertheless, it is worth emphasizing that in each step we apply the natural modification to the deviation vector that is needed to eliminate the current bad set or sets. At this point, the reader may rightly ask whether it is indeed necessary to consider all these cases. The answer is unfortunately yes, as the proposed Newton-type scheme may run into any of them; for examples, see Fig. 1.

For ease of discussion, we introduce several notation before stating the algorithm. Let

$$\mathcal{F}' := \left\{ F' \in \mathcal{F} \,\middle|\, \tfrac{1}{w}(F') < \tfrac{1}{w}(F^*) \right\},$$
$$\mathcal{F}'' := \left\{ F'' \in \mathcal{F} \,\middle|\, \tfrac{1}{w}(F'') = \tfrac{1}{w}(F^*),\ F'' \neq F^* \right\},$$
$$\mathcal{F}''' := \left\{ F''' \in \mathcal{F} \,\middle|\, \tfrac{1}{w}(F''') > \tfrac{1}{w}(F^*) \right\}.$$

In each iteration, the algorithm computes an optimal solution of the underlying optimization problem, and if the cost of the input solution F^* is strictly larger than the current optimum, then it updates the costs using a value determined by one of the following functions:

(f1) $f_1(c, F'') := \dfrac{c(F^*) - c(F'')}{\frac{1}{w}(F^* \setminus F'')},$

(f2) $\quad f_2(c, F) := \dfrac{c(F^*) - c(F)}{\frac{1}{w}(F^*) - \frac{1}{w}(F)},$

(f3) $\quad f_3(c, F', F''') := \dfrac{\dfrac{c(F^*) - c(F')}{\frac{1}{w}(F^*) - \frac{1}{w}(F')} - \dfrac{c(F^*) - c(F''')}{\frac{1}{w}(F^*) - \frac{1}{w}(F''')}}{\dfrac{\frac{1}{w}(F^* \setminus F')}{\frac{1}{w}(F^*) - \frac{1}{w}(F')} - \dfrac{\frac{1}{w}(F^* \setminus F''')}{\frac{1}{w}(F^*) - \frac{1}{w}(F''')}},$

(f4) $\quad f_4(c, F', F''') := \dfrac{c(F^*) - c(F') - f_3(c, F', F''') \cdot \frac{1}{w}(F^* \setminus F')}{\frac{1}{w}(F^*) - \frac{1}{w}(F')},$

(f5) $\quad f_5(c, F', F''') := \dfrac{c(F^*) - c(F''') - f_3(c, F', F''') \cdot \frac{1}{w}(F^* \setminus F''')}{\frac{1}{w}(F^*) - \frac{1}{w}(F''')}.$

In the definitions above, we have $F \in \mathcal{F}' \cup \mathcal{F}'''$, $F' \in \mathcal{F}'$, $F'' \in \mathcal{F}''$, $F''' \in \mathcal{F}'''$, and $c \in \mathbb{R}^S$.

Table 1 gives a summary of the cases occurring in Algorithm 1 which might be helpful when reading the proof. Figure 1 provides toy examples for the different cases, while Example 1 discusses a more complex instance where all five cases appear.

For proving the correctness and the running time of the algorithm, we need the following lemmas. The proofs of these statements follow from the definitions in a fairly straightforward way, and are deferred to the full version of the paper [5].

Our first lemma shows that if F^* is not optimal with respect to the current cost function, then in the next step it has the same cost as the current optimal solution with respect to the modified cost function.

Lemma 3. *If F^* is not a minimum c_i-cost member of $\mathcal{F}$, then $c_{i+1}(F^*) = c_{i+1}(F_i)$.*

The following lemmas together imply an upper bound on the total number of iterations.

Lemma 4. *Let $i_1 < i_2$ be indices such that both steps i_1 and i_2 correspond to Case 1. Then $\frac{1}{w}(Y_{i_1+1} \cap F^*) < \frac{1}{w}(Y_{i_2+1} \cap F^*)$.*

Lemma 5. *Let (a_1, a_2, a_3, a_4) be a 4-tuple of integers satisfying $0 \le a_k \le \|w\|_{-1}$ for any $k \in [4]$.*

(i) *There is at most one index i such that F^* is not a minimum c_i-cost member of $\mathcal{F}$, $\frac{1}{w}(F_i) < \frac{1}{w}(F^*)$, $Z_i \ne *$ and $\left(\frac{1}{w}(F_i), \frac{1}{w}(Z_i), \frac{1}{w}(F_i \cap F^*), \frac{1}{w}(Z_i \cap F^*)\right) = (a_1, a_2, a_3, a_4)$.*

(ii) *There is at most one index i such that F^* is not a minimum c_i-cost member of $\mathcal{F}$, $\frac{1}{w}(F_i) > \frac{1}{w}(F^*)$, $X_i \ne *$ and $\left(\frac{1}{w}(X_i), \frac{1}{w}(F_i), \frac{1}{w}(X_i \cap F^*), \frac{1}{w}(F_i \cap F^*)\right) = (a_1, a_2, a_3, a_4)$.*

Algorithm 1: Algorithm for the minimum-cost inverse optimization problem under the weighted span objective.

Input: An instance of a minimum-cost inverse optimization problem $(S, \mathcal{F}, F^*, c, \mathrm{span}_w(\cdot))$ and an oracle $\mathcal{O}$ for the optimization problem $(S, \mathcal{F}, c')$ with any cost function c'.
Output: An optimal deviation vector.

1 $d_0 \leftarrow 0, \quad D_0 \leftarrow 0$
2 $X_0 \leftarrow *, \quad Y_0 \leftarrow *, \quad Z_0 \leftarrow *$
3 $c_0 \leftarrow c$
4 $F_0 \leftarrow$ minimum c_0-cost member of $\mathcal{F}$ determined by $\mathcal{O}$
5 $i \leftarrow 0$
6 **while** $c_i(F^*) > c_i(F_i)$ **do**
7   **if** $\frac{1}{w}(F_i) = \frac{1}{w}(F^*)$ **then**
8      $X_{i+1} \leftarrow *, \quad Y_{i+1} \leftarrow F_i, \quad Z_{i+1} \leftarrow *$
9      $\delta_{i+1} \leftarrow f_1(c_i, F_i), \quad \Delta_{i+1} \leftarrow 0$ // Case 1
10   **if** $\frac{1}{w}(F_i) < \frac{1}{w}(F^*)$ **then**
11      $X_{i+1} \leftarrow F_i, \quad Y_{i+1} \leftarrow Y_i, \quad Z_{i+1} \leftarrow Z_i$
12   **if** $Z_i = *$ **then**
13        $\delta_{i+1} \leftarrow 0, \quad \Delta_{i+1} \leftarrow f_2(c_i, F_i)$ // Case 2
14   **else**
15        $\delta_{i+1} \leftarrow f_3(c_i, F_i, Z_i), \quad \Delta_{i+1} \leftarrow f_4(c_i, F_i, Z_i)$ // Case 3
16   **if** $\frac{1}{w}(F_i) > \frac{1}{w}(F^*)$ **then**
17      $X_{i+1} \leftarrow X_i, \quad Y_{i+1} \leftarrow Y_i, \quad Z_{i+1} \leftarrow F_i$
18   **if** $X_i = *$ **then**
19        $\delta_{i+1} \leftarrow 0, \quad \Delta_{i+1} \leftarrow f_2(c_i, F_i)$ // Case 4
20   **else**
21        $\delta_{i+1} \leftarrow f_3(c_i, X_i, F_i), \quad \Delta_{i+1} \leftarrow f_5(c_i, X_i, F_i)$ // Case 5
22    $d_{i+1} \leftarrow d_i + \delta_{i+1}, \quad D_{i+1} \leftarrow D_i + \Delta_{i+1}$
23    $c_{i+1} \leftarrow c - p_{[d_{i+1}, D_{i+1}]}$
24    $F_{i+1} \leftarrow$ minimum c_{i+1}-cost member of $\mathcal{F}$ determined by $\mathcal{O}$
25    $i \leftarrow i + 1$
26 **end**
27 **return** $p_{[d_i, D_i]}$

Lemma 6. *(i) If $\frac{1}{w}(F_i), \frac{1}{w}(F_{i+1}) < \frac{1}{w}(F^*)$ and $Z_i, Z_{i+1} = *$, then at least one of $\frac{1}{w}(F_i) \neq \frac{1}{w}(F_{i+1})$ and $\frac{1}{w}(F_i \cap F^*) < \frac{1}{w}(F_{i+1} \cap F^*)$ holds.*
(ii) If $\frac{1}{w}(F_i), \frac{1}{w}(F_{i+1}) > \frac{1}{w}(F^*)$ and $X_i, X_{i+1} = *$, then at least one of $\frac{1}{w}(F_i) \neq \frac{1}{w}(F_{i+1})$ and $\frac{1}{w}(F_i \cap F^*) < \frac{1}{w}(F_{i+1} \cap F^*)$ holds.

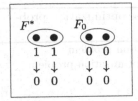

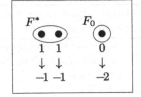

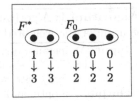

(a) Case 1: $\delta_0 = 0$, $\Delta_0 = 0$,   (b) Case 2: $\delta_0 = 0$, $\Delta_0 = 0$,   (c) Case 4: $\delta_0 = 0$, $\Delta_0 = 0$,
$\delta_1 = 1$, $\Delta_1 = 0$. $\delta_1 = 0$, $\Delta_1 = 2$. $\delta_1 = 0$, $\Delta_1 = -2$.

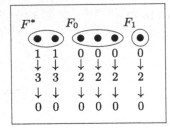

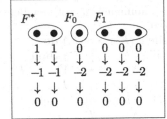

(d) Case 3: $\delta_0 = 0$, $\Delta_0 = 0$, $\delta_1 = 0$,   (e) Case 5: $\delta_0 = 0$, $\Delta_0 = 0$, $\delta_1 = 0$,
$\Delta_1 = -2$, $\delta_2 = 1$, $\Delta_1 = 2$ (first step is $\Delta_1 = 2$, $\delta_2 = 1$, $\Delta_1 = -2$ (first step is
in Case 4).   in Case 2).

Fig. 1. Toy examples illustrating the different cases occurring in Algorithm 1, where the values of the original cost function c are presented in the top row.

Table 1. Summary of the cases occurring in Algorithm 1.

Case	Conditions	δ_{i+1}	Δ_{i+1}
1	$\frac{1}{w}(F_i) = \frac{1}{w}(F^*)$	$f_1(c_i, F_i)$	0
2	$\frac{1}{w}(F_i) < \frac{1}{w}(F^*)$ $Z_i = *$	0	$f_2(c_i, F_i)$
3	$\frac{1}{w}(F_i) < \frac{1}{w}(F^*)$ $Z_i \neq *$	$f_3(c_i, F_i, Z_i)$	$f_4(c_i, F_i, Z_i)$
4	$\frac{1}{w}(F_i) > \frac{1}{w}(F^*)$ $X_i = *$	0	$f_2(c_i, F_i)$
5	$\frac{1}{w}(F_i) > \frac{1}{w}(F^*)$ $X_i \neq *$	$f_3(c_i, X_i, F_i)$	$f_5(c_i, X_i, F_i)$

Now we are ready to prove the correctness and discuss the running time of the algorithm.

Theorem 1. *Algorithm 1 determines an optimal deviation vector for the minimum-cost inverse optimization problem* $(S, \mathcal{F}, F^*, c, \mathrm{span}_w(\cdot))$ *using* $O(\|w\|_{-1}^6)$ *calls to* $\mathcal{O}$.

Proof. We discuss the time complexity and the correctness of the algorithm separately.

Time Complexity. Recall that $w \in \mathbb{R}_+^S$ is scaled so that $\frac{1}{w}(X)$ is an integer for each $X \subseteq S$. We show that the algorithm terminates after at most $O(\|w\|_{-1}^6)$ iterations of the while loop. By Lemma 4, there are at most $O(\|w\|_{-1})$ iterations corresponding to Case 1. By Lemma 5, there are at most $O(\|w\|_{-1}^4)$ iterations corresponding to Cases 3 and 5. Between two such iterations, there are at most $O(\|w\|_{-1}^2)$ iterations corresponding to Cases 2 and 4 by Lemma 6. Hence the total number of iterations is at most $O(\|w\|_{-1}^6)$.

Optimality. The feasibility of the output of Algorithm 1 follows from the fact that the while loop ended. If F^* is a minimum c_0-cost member of $\mathcal{F}$, then we are clearly done. Otherwise, there exists an index q such that F^* is a minimum c_{q+1}-cost member of $\mathcal{F}$. Suppose to the contrary that $p_{[d_{q+1}, D_{q+1}]}$ is not optimal. By Lemma 2, there exists $\delta, \Delta \in \mathbb{R}$ such that $\delta < d_{q+1}$, the deviation vector $p_{[\delta, \Delta]}$ is optimal. If all steps correspond to Cases 2 and 4, then $\mathrm{span}_w(p_{[d_{q+1}, D_{q+1}]}) = d_{q+1} = 0 \leq \delta$, a contradiction. Otherwise, let q' be the largest index for which $\delta_{q'+1} \neq 0$. Note that $d_{q+1} = d_{q'+1}$. We arrive to a contradiction using different arguments, depending on which case step q' belongs to.

Case 1. By Lemma 3, we have

$$0 \geq (c - p_{[\delta, \Delta]})(F^*) - (c - p_{[\delta, \Delta]})(F_{q'})$$

$$= \left(c(F^*) - \sum_{s \in F^*} \frac{\delta}{w(s)} - \sum_{s \in F^*} \frac{\Delta}{w(s)} \right) - \left(c(F_q) - \sum_{s \in F_q \cap F^*} \frac{\delta}{w(s)} - \sum_{s \in F_q} \frac{\Delta}{w(s)} \right)$$

$$= c(F^*) - c(F_{q'}) - \delta \cdot \tfrac{1}{w}(F^* \setminus F_{q'}) - \Delta \cdot \left(\tfrac{1}{w}(F^*) - \tfrac{1}{w}(F_{q'}) \right)$$

$$= c(F^*) - c(F_{q'}) - \delta \cdot \tfrac{1}{w}(F^* \setminus F_{q'}) - \Delta \cdot 0$$

$$> c(F^*) - c(F_{q'}) - d_{q'+1} \cdot \tfrac{1}{w}(F^* \setminus F_{q'}) - D_{q'} \cdot 0$$

$$= c(F^*) - c(F_{q'}) - d_{q'+1} \cdot \tfrac{1}{w}(F^* \setminus F_{q'}) - D_{q'} \cdot \left(\tfrac{1}{w}(F^*) - \tfrac{1}{w}(F_{q'}) \right)$$

$$= c_{q'+1}(F^*) - c_{q'+1}(F_{q'})$$

$$= 0,$$

a contradiction.

Case 3. We have

$$0 \geq (c - p_{[\delta, \Delta]})(F^*) - (c - p_{[\delta, \Delta]})(F_{q'})$$

$$= c(F^*) - c(F_{q'}) - \delta \cdot \tfrac{1}{w}(F^* \setminus F_{q'}) - \Delta \cdot \left(\tfrac{1}{w}(F^*) - \tfrac{1}{w}(F_{q'}) \right)$$

$$> c(F^*) - c(F_{q'}) - d_{q'+1} \cdot \tfrac{1}{w}(F^* \setminus F_{q'}) - \Delta \cdot \left(\tfrac{1}{w}(F^*) - \tfrac{1}{w}(F_{q'}) \right)$$

and

$$0 \geq (c - p_{[\delta,\Delta]})(F^*) - (c - p_{[\delta,\Delta]})(Z_{q'})$$
$$= c(F^*) - c(Z_{q'}) - \delta \cdot \tfrac{1}{w}(F^* \setminus Z_{q'}) - \Delta \cdot \left(\tfrac{1}{w}(F^*) - \tfrac{1}{w}(Z_{q'})\right)$$
$$> c(F^*) - c(Z_{q'}) - d_{q'+1} \cdot \tfrac{1}{w}(F^* \setminus Z_{q'}) - \Delta \cdot \left(\tfrac{1}{w}(F^*) - \tfrac{1}{w}(Z_{q'})\right).$$

These together imply

$$\frac{c(F^*) - c(F_{q'}) - d_{q'+1} \cdot \tfrac{1}{w}(F^* \setminus F_{q'})}{\tfrac{1}{w}(F^*) - \tfrac{1}{w}(F_{q'})} < \Delta, \text{ and}$$

$$\frac{c(F^*) - w(Z_{q'}) - d_{q'+1} \cdot \tfrac{1}{w}(F^* \setminus Z_{q'})}{\tfrac{1}{w}(F^*) - \tfrac{1}{w}(Z_{q'})} > \Delta.$$

Therefore, we have

$$d_{q'+1} > \frac{\dfrac{c(F^*) - c(F_{q'})}{\tfrac{1}{w}(F^*) - \tfrac{1}{w}(F_{q'})} - \dfrac{c(F^*) - c(Z_{q'})}{\tfrac{1}{w}(F^*) - \tfrac{1}{w}(Z_{q'})}}{\dfrac{\tfrac{1}{w}(F^* \setminus F_{q'})}{\tfrac{1}{w}(F^*) - \tfrac{1}{w}(F_{q'})} - \dfrac{\tfrac{1}{w}(F^* \setminus Z_{q'})}{\tfrac{1}{w}(F^*) - \tfrac{1}{w}(Z_{q'})}}$$

$$= \frac{\dfrac{c_{q'}(F^*) - c_{q'}(F_{q'}) + d_{q'} \cdot \tfrac{1}{w}(F^* \setminus F_{q'}) + D_{q'} \cdot \left(\tfrac{1}{w}(F^*) - \tfrac{1}{w}(F_{q'})\right)}{\tfrac{1}{w}(F^*) - \tfrac{1}{w}(F_{q'})}}{\dfrac{\tfrac{1}{w}(F^* \setminus F_{q'})}{\tfrac{1}{w}(F^*) - \tfrac{1}{w}(F_{q'})} - \dfrac{\tfrac{1}{w}(F^* \setminus Z_{q'})}{\tfrac{1}{w}(F^*) - \tfrac{1}{w}(Z_{q'})}}$$
$$- \frac{\dfrac{c_{q'}(F^*) - c_{q'}(Z_{q'}) + d_{q'} \cdot \tfrac{1}{w}(F^* \setminus Z_{q'}) + D_{q'} \cdot \left(\tfrac{1}{w}(F^*) - \tfrac{1}{w}(Z_{q'})\right)}{\tfrac{1}{w}(F^*) - \tfrac{1}{w}(Z_{q'})}}{\dfrac{\tfrac{1}{w}(F^* \setminus F_{q'})}{\tfrac{1}{w}(F^*) - \tfrac{1}{w}(F_{q'})} - \dfrac{\tfrac{1}{w}(F^* \setminus Z_{q'})}{\tfrac{1}{w}(F^*) - \tfrac{1}{w}(Z_{q'})}}$$

$$= \frac{\dfrac{c_{q'}(F^*) - c_{q'}(F_{q'})}{\tfrac{1}{w}(F^*) - \tfrac{1}{w}(F_{q'})} - \dfrac{c_{q'}(F^*) - c_{q'}(Z_{q'})}{\tfrac{1}{w}(F^*) - \tfrac{1}{w}(Z_{q'})}}{\dfrac{\tfrac{1}{w}(F^* \setminus F_{q'})}{\tfrac{1}{w}(F^*) - \tfrac{1}{w}(F_{q'})} - \dfrac{\tfrac{1}{w}(F^* \setminus Z_{q'})}{\tfrac{1}{w}(F^*) - \tfrac{1}{w}(Z_{q'})}} + d_{q'}.$$

By the above, we obtain

$$\delta_{q'+1} > \frac{\dfrac{c_{q'}(F^*) - c_{q'}(F_{q'})}{\tfrac{1}{w}(F^*) - \tfrac{1}{w}(F_{q'})} - \dfrac{c_{q'}(F^*) - c_{q'}(Z_{q'})}{\tfrac{1}{w}(F^*) - \tfrac{1}{w}(Z_{q'})}}{\dfrac{\tfrac{1}{w}(F^* \setminus F_{q'})}{\tfrac{1}{w}(F^*) - \tfrac{1}{w}(F_{q'})} - \dfrac{\tfrac{1}{w}(F^* \setminus Z_{q'})}{\tfrac{1}{w}(F^*) - \tfrac{1}{w}(Z_{q'})}} = f_3(c_{q'}, F_{q'}, Z_{q'}) = \delta_{q'+1},$$

a contradiction.

Case 5. Similarly as in Case 3, we obtain

$$\frac{c(F^*) - c(X_{q'}) - d_{q'+1} \cdot \frac{1}{w}(F^* \setminus X_{q'})}{\frac{1}{w}(F^*) - \frac{1}{w}(X_{q'})} < \Delta, \text{ and}$$

$$\frac{c(F^*) - c(F_{q'}) - d_{q'+1} \cdot \frac{1}{w}(F^* \setminus F_{q'})}{\frac{1}{w}(F^*) - \frac{1}{w}(F_{q'})} > \Delta.$$

Therefore, we get

$$\delta_{q'+1} > \frac{\frac{c_{q'}(F^*) - c_{q'}(X_{q'})}{\frac{1}{w}(F^*) - \frac{1}{w}(X_{q'})} - \frac{c_{q'}(F^*) - c_{q'}(F_{q'})}{\frac{1}{w}(F^*) - \frac{1}{w}(F_{q'})}}{\frac{\frac{1}{w}(F^* \setminus X_{q'})}{\frac{1}{w}(F^*) - \frac{1}{w}(X_{q'})} - \frac{\frac{1}{w}(F^* \setminus F_{q'})}{\frac{1}{w}(F^*) - \frac{1}{w}(F_{q'})}} = f_3(c_{q'}, X_{q'}, F_{q'}) = \delta_{q'+1},$$

a contradiction. This finishes the proof of the theorem. □

To provide a better understanding of the algorithm, we present its running step by step on a more complex instance where all five cases appear; see Fig. 2.

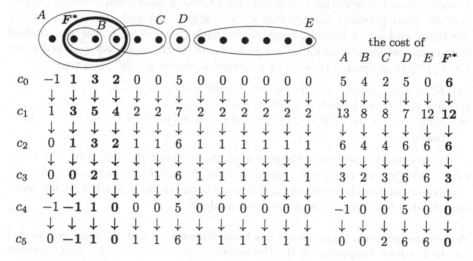

Fig. 2. An illustration of Algorithm 1, where $\mathcal{F} = \{F^*, A, B, C, D, E\}$, $w \equiv 1$, and the costs $c = c_0$ are in the first row. Recall $X_0 = *$, $Y_0 = *$, and $Z_0 = *$.

Example 1. Consider the instance of the problem on Fig. 2. The set E has minimum cost with respect to c_0 and $|E| > |F^*|$. Thus, the first step belongs to Case 4, $X_1 = *$, $Y_1 = *$, $Z_1 = E$, and $\delta_1 = 0$, $\Delta_1 = -2$. The set D has minimum cost with respect to c_1 and $|D| < |F^*|$. Thus, the second step belongs to Case 3, $X_2 = D$, $Y_2 = *$, $Z_2 = E$, and $\delta_2 = 1$, $\Delta_2 = 1$. The set C has minimum cost with respect to c_2 and $|C| = |F^*|$. Thus, the third step belongs to Case 1, $X_3 = *$,

$Y_3 = D$, $Z_3 = *$, and $\delta_3 = 1$, $\Delta_3 = 0$. The set B has minimum cost with respect to c_3 and $|B| < |F^*|$. Thus, the fourth step belongs to Case 2, $X_4 = B$, $Y_4 = D$, $Z_4 = *$, and $\delta_4 = 0$, $\Delta_4 = 1$. The set A has minimum cost with respect to c_4 and $|A| > |F^*|$. Thus, the fifth step belongs to Case 5, $X_5 = B$, $Y_5 = D$, $Z_5 = A$, and $\delta_5 = 1$, $\Delta_5 = -1$. The set F^* has minimum cost with respect to c_5, so the algorithm terminates with $d_5 = 3$, $D_5 = -1$, i.e., the span of an optimal deviation vector is 3.

4   Conclusions

In this paper, we introduced an objective for minimum-cost inverse optimization problems that measures the difference between the largest and the smallest weighted coordinates of the deviation vector, thus leading to a fair or balanced solution. We presented a purely combinatorial algorithm that efficiently determines an optimal deviation vector, assuming that an oracle for solving the underlying optimization problem is available. The running time of the algorithm is pseudo-polynomial due to the presence of weights, and finding a strongly polynomial algorithm remains an intriguing open problem.

The proposed algorithm in fact extends to the more general setting when the coordinates of the deviation vector are ought to fall within given lower and upper bounds. More precisely, assume that $\ell \colon S \to \mathbb{R} \cup \{-\infty\}$ and $u \colon S \to \mathbb{R} \cup \{+\infty\}$ are lower and upper bounds, respectively, such that $\ell \leq u$. In the *constrained* minimum-cost inverse optimization problem under the weighted span objective $(S, \mathcal{F}, F^*, c, \ell, u, \mathrm{span}_w(\cdot))$, we seek a deviation vector $p \in \mathbb{R}^S$ such that

(a) F^* is a minimum-cost member of $\mathcal{F}$ with respect to $c - p$,
(b) p is within the bounds $\ell \leq p \leq u$, and
(c) $\mathrm{span}_w(p) := \max \{w(s) \cdot p(s) \mid s \in S\} - \min \{w(s) \cdot p(s) \mid s \in S\}$ is minimized.

In an extended version of the paper [4], we work out the details of a Newton-type algorithm for this more general setting. However, the presence of bound constraints makes the problem, and hence the resulting algorithm, much more complicated that requires a careful case-analysis.

Acknowledgement. This research has been implemented with the support provided by the Lendület Programme of the Hungarian Academy of Sciences – grant number LP2021-1/2021, by the Ministry of Innovation and Technology of Hungary – grant number ELTE TKP 2021-NKTA-62, and by Dynasnet European Research Council Synergy project – grant number ERC-2018-SYG 810115. The authors have no competing interests to declare that are relevant to the content of this article.

References

1. Ahmadian, S., Bhaskar, U., Sanità, L., Swamy, C.: Algorithms for inverse optimization problems. In: 26th Annual European Symposium on Algorithms (ESA 2018). Leibniz International Proceedings in Informatics, LIPIcs, vol. 112. Schloss Dagstuhl-Leibniz-Zentrum für Informatik (2018)

2. Ahuja, R.K.: The balanced linear programming problem. Eur. J. Oper. Res. **101**(1), 29–38 (1997)

3. Alizadeh, B., Burkard, R.E., Pferschy, U.: Inverse 1-center location problems with edge length augmentation on trees. Computing **86**(4), 331–343 (2009)

4. Bérczi, K., Mendoza-Cadena, L.M., Varga, K.: Newton-type algorithms for inverse optimization II: weighted span objective. arXiv preprint arXiv:2302.13414 (2023)

5. Bérczi, K., Mendoza-Cadena, L.M., Varga, K.: Newton-type algorithms for inverse optimization: weighted span objective. Technical report, TR-2023-16, Egerváry Research Group, Budapest (2023). egres.elte.hu

6. Burton, D., Toint, P.L.: On an instance of the inverse shortest paths problem. Math. Program. **53**, 45–61 (1992)

7. Cai, M.C., Yang, X.G., Zhang, J.Z.: The complexity analysis of the inverse center location problem. J. Global Optim. **15**, 213–218 (1999)

8. Camerini, P.M., Maffioli, F., Martello, S., Toth, P.: Most and least uniform spanning trees. Discret. Appl. Math. **15**(2–3), 181–197 (1986)

9. Demange, M., Monnot, J.: An introduction to inverse combinatorial problems. In: Paradigms of Combinatorial Optimization: Problems and New Approaches, 2nd edn., pp. 547–586. Wiley (2014)

10. Duin, C., Volgenant, A.: Minimum deviation and balanced optimization: a unified approach. Oper. Res. Lett. **10**(1), 43–48 (1991)

11. Ficker, A.M.C., Spieksma, F.C.R., Woeginger, G.J.: Robust balanced optimization. EURO J. Comput. Optim. **6**(3), 239–266 (2018)

12. Heuberger, C.: Inverse combinatorial optimization: a survey on problems, methods, and results. J. Comb. Optim. **8**(3), 329–361 (2004)

13. Martello, S., Pulleyblank, W.R., Toth, P., de Werra, D.: Balanced optimization problems. Oper. Res. Lett. **3**(5), 275–278 (1984)

14. Qian, X., Guan, X., Jia, J., Zhang, Q., Pardalos, P.M.: Vertex quickest 1-center location problem on trees and its inverse problem under weighted ℓ_∞ norm. J. Global Optim. **85**(2), 461–485 (2023)

15. Radzik, T.: Parametric flows, weighted means of cuts, and fractional combinatorial optimization. In: Complexity in Numerical Optimization, pp. 351–386. World Scientific (1993)

16. Scutellà, M.G.: A strongly polynomial algorithm for the uniform balanced network flow problem. Discret. Appl. Math. **81**(1–3), 123–131 (1998)

17. Zhang, J., Liu, Z.: A general model of some inverse combinatorial optimization problems and its solution method under ℓ_∞ norm. J. Comb. Optim. **6**(2), 207–227 (2002)

Author Index

Printed in the United States
by Baker & Taylor Publisher Services